Musimathics

Musimathics

The Mathematical Foundations of Music

Volume 1

Gareth Loy

The MIT Press
Cambridge, Massachusetts
London, England

MIT Press books may be purchased at special quantity discounts for business or sales promotional use. For information, please e-mail <special_sales@mitpress.mit.edu> or write to Special Sales Department, The MIT Press, 5 Cambridge Center, Cambridge, MA 02142.

This book was set in Times Roman by Interactive Composition Corporation.
Printed and bound in the United States of America.

Library of Congress Cataloging-in-Publication Data

Loy, D. Gareth.
Musimathics : the mathematical foundations of music / Gareth Loy.
p. cm.
Includes bibliographical references and indexes.
Contents: v. 1. Musical elements
ISBN 0-262-12282-0—ISBN 978-0-262-12282-5 (v. 1 : alk. paper), 978-0-262-51655-6 (pb.)
1. Music in mathematics education. 2. Mathematics—Study and teaching. 3. Music theory—Mathematics.
4. Music—Acoustics and physics. I. Title.

QA19.M87L69 2006
781.2—dc22

2005051090

This book is dedicated to the memory of John R. Pierce

Contents

Foreword

Musimathics by Gareth Loy is a guided tour-de-force of the mathematics and physics of music. It pulls no punches in presenting the scientific fundamentals needed to really understand music, but at the same time it is so clearly written that readers willing to spend time can learn all they need to know to do basic research in modern technical music. Advanced placement courses in math and science in any good high school are plenty of background—from there on Loy leads readers to wherever they want to go.

Loy has always been a brilliantly clear writer. In *Musimathics* he is also an encyclopedic writer. He covers everything needed to understand existing music and musical instruments or to create new music or new instruments.

Loy's book and John R. Pierce's famous *The Science of Musical Sound* belong on everyone's bookshelf, and the rest of the shelf can be empty.

Max Mathews

Preface

To start a great enterprise requires at the beginning only the first step.[1]

Mathematics can be as effortless as humming a tune, if you know the tune. But our culture does not prepare us for appreciation of mathematics as it does for appreciation of music. Though we start hearing music very early in life, the same cannot be said of mathematics, even though the two subjects are twins. This is a shame; to know music without knowing its mathematics is like hearing a melody without its accompaniment.

If you are drawn to mathematics because of your love of music, then this book is for you. It provides a commonsense, self-contained, self-consistent, self-referential introduction to these subjects for nonspecialist readers. It is designed for musicians who find their art increasingly mediated by technology, and it is written for anyone who desires to understand the intersection between art and science.

It has been my experience that there are many who want a deeper understanding of the mathematics of music if the subject could be presented in a manner accessible to them. This book aims to meet that need. My goal is always to sustain readers' motivation while competence is gradually built up in mathematical fundamentals.

Readers will need only average experience with mathematics and music—advanced high school math or college freshman algebra and some basic music theory. No knowledge of the calculus, apart from a small amount supplied in volume 2, is required. Some physics background is helpful, but the text supplies almost everything necessary for understanding.

Virtually all of this book is focused on the mathematics of music:

- The topics are all subjects that contemporary composers, musicians, and music engineers have found to be important.
- The examples are all practical problems in music and audio.
- Even the fundamentals are cast in terms of the goal: I try to make it clear up front why a foundation is relevant and what readers will be able to do with it once it is mastered.

This is not a book for the mathematically inexperienced, nor is it for experts. My aim is balance. I travel at a somewhat leisurely pace through this very remarkable material, examining not just its mathematical content but its aesthetic and philosophical qualities as well.

Musimathics presents the story of music engineering by examining its mathematics. Since engineering is basically about applying human values to nature, readers will discover a lot about themselves, about the world of sound and music, and about what human cultures have valued. However, because I approach these values from an abstract perspective, they can be seen objectively, giving a better vantage point from which readers can make their own choices.

There are three main directions of inquiry in volume 1:

- The materials of music: notes, intervals, scales
- The physical properties of music: frequency, amplitude, duration, timbre
- The perception of music and sound: how we hear
- Music composition

Volume 2 presents a deeper cut into the underlying mathematics of music and sound, including

- Digital audio, sampling, binary numbers
- Complex numbers and how they simplify representation of musical signals
- Fourier transform, convolution, and filtering
- Resonance, the wave equation, and the behavior of acoustical systems
- Sound synthesis
- The short-time Fourier transform, phase vocoder, and the wavelet transform

The Web site, http://www.musimathics.com/, contains additional source material, animations, figures, and sources for other program examples in this book. Also, try saying "Musimathics" to your favorite Web browser and see what happens.

About the Author

This section is here to give readers a sense of comfort that they are in good hands. I received my Doctor of Musical Arts (DMA) degree from Stanford University in 1980 in composition of computer music. I did my graduate work at the Stanford Center for Computer Research in Music and Acoustics (CCRMA), one of the premier institutions for the study of this subject, then housed in the Stanford Artificial Intelligence Laboratory. I have been a performing musician all my life (violin, guitar, lute, sitar, and voice) and am an award-winning composer (Bourges prize) and a National Endowment for the Arts grant recipient. I spent over a decade conducting research and teaching computer music, electronic music, and musical acoustics at the University of California, San Diego, as Director of Research at the Center for Music Experiment. More recently I've been a computer programmer, software architect, and digital audio systems engineer in various companies in Silicon Valley. I am president of a (very) small corporation, http://www.GarethInc.com/, which provides engineering consulting services internationally.

But there's more about me that you should know. Mathematics has never been an easy subject for me; I am a composer by training, not a mathematician. My academic career suffered badly in proportion to the amount of mathematics included in the syllabus. The aim of confessing this is paradoxically to give readers confidence. I know what it's like not to comprehend mathematics easily, and I also know what it's like not to give up.

Notwithstanding my inability to add a column of figures and come up with the same answer twice, I found that mathematics was the lion in my path, the invariant obstacle to the realization of my artistic visions. So it was more out of necessity than facility that I came to study mathematics. The composer Harry Partch constructed an entire orchestra of novel instruments to realize his artistic vision and once called himself "a composer seduced into carpentry." By analogy, I suppose I'm a composer seduced into mathematics.

I considered subtitling this book, "Everything I wanted to know about music when I was eleven." At that age I prowled the stacks of a nearby university library in search of answers to my burning questions, only to discover that they were out of reach because I didn't understand the jargon in which the answers were written. At that age we are still intellectually fearless. In my experience as a child and as a father and teacher, I've come to believe that there is nothing an eleven-year-old can't understand given the right explanation. But by the time most of us have reached adulthood, this inquisitive quality is in eclipse, in large part because the right explanations are very hard to come by. This book is my gift to myself all those years ago, of all the best explanations I've been able to find or invent for many of the questions I had. And this book is my gift to you; may it help throw open the doors to the mathematics of music, one of the crown jewels of our civilization.

C. G. Jung (1962) wrote, "The decisive question for man is: Is he related to something infinite or not? In the final analysis, we count for something only because of the essential we embody, and if we do not embody that, life is wasted."

In the storm called life, mathematics and music are two sure guides to that essential that we all embody.

Acknowledgments

This work was supported in part by a generous grant of love and encouragement from my wife, Lisa, and my children, Morgan, Greta, and Tutti.

Thanks to all those whose passion for the subject has helped inflame my own, including my teachers Herbert Bielawa, John Chowning, Andy Moorer, John Grey, Loren Rush, Leonard Ratner, and Leland Smith. Thanks to those who have helped keep the dream in focus: Connie Strohbehn, Shari Carlson, Linda Grahm.

I am grateful to all whose scholarship and research have fed into the rich stream of knowledge that this book can at best sample and summarize. The enormous list of these individuals begins with the bibliography of this book and extends recursively through all the influences they cite. If there is anything to praise in this work, it is because it reflects the wisdom of these antecedents; if there is fault, it is mine alone.

Thanks to those courageous individuals who reviewed chapters of this book prior to publication: Charles Seagrave, Stan Green, Dana Massie, Mark Kahrs, Richard Kavinoky, Malcolm Slaney, John Strawn, Dan Freed, Herbert Bielawa, Stephen Pope, Roy Harvey, Julius Smith, Ted Marsh, Mark Dolson, Andy Moorer, Robert Owen.

Thanks also to the mockingbird outside my window whose song at this late hour reminds me of the universality of music.

Gareth Loy
Corte Madera, California

1 Music and Sound

"How did you know how to do that?" he asks.
"You just have to figure it out."
"I wouldn't know where to start," he says.
I think to myself, That's the problem, all right, where to start. To reach him you have to back up and back up, and the further back you go, the further back you see you have to go, until what looked like a small problem of communication turns into a major philosophic enquiry."
—Robert M. Pirsig, *Zen and the Art of Motorcycle Maintenance*

The problem of finding the right place to begin an explanation is rather like finding the right fulcrum point to move a stone with a lever. Putting the fulcrum point too close to the stone provides great leverage but little range of movement (figure 1.1a). Putting it too far from the stone provides great range of movement but no leverage (figure 1.1b). The fulcrum point of an explanation is the knowledge and assumptions the reader must already have in order to make sense of the explanation. The assumptions are like the axioms in geometry: a short list of simple, self-evident facts from which the entire subject can ultimately be derived.

This chapter is such a fulcrum for the rest of this book, and it therefore runs the greatest risk of overwhelm or underwhelm. Given the choice, I've decided to err on the side of underwhelm. The rest of this chapter introduces some basic properties of sound that will become immediately useful in chapter 2. If it looks like there are no surprises here, skip this chapter.

And if this subject is new to you, I have a suggestion: if any of the material seems beyond you at times, just read it like a mystery novel. Seriously! I recommend this approach based on years of personal experience reading things I didn't at first understand. You don't have to speak fluent French in order to enjoy Paris, but you'll certainly get more out of it if you pick some up along the way.

1.1 Basic Properties of Sound

If you were to strike a tuning fork and hold it next to your ear, you would hear one of nature's purest, simplest sounds. What you hear is a result of the periodic changes in air pressure at your ear drum caused by the vibration of the air set in motion by the tines of the fork (figure 1.2a). Figure 1.2b is a representation of the air molecules in the vicinity of the fork, showing areas of greater and lesser

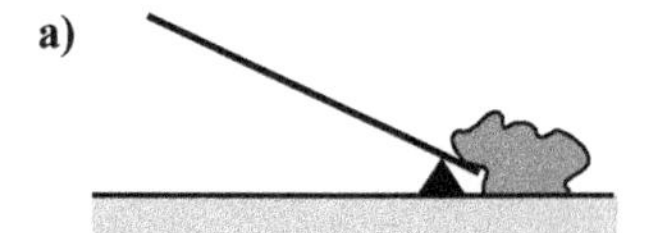

Easy to lift but the rock hardly moves

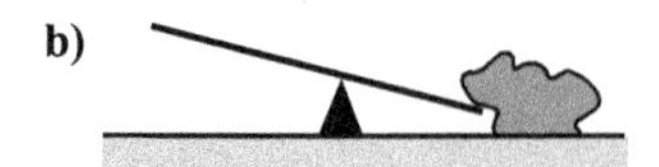

Hard to lift but the rock moves farther

Figure 1.1
Fulcrum.

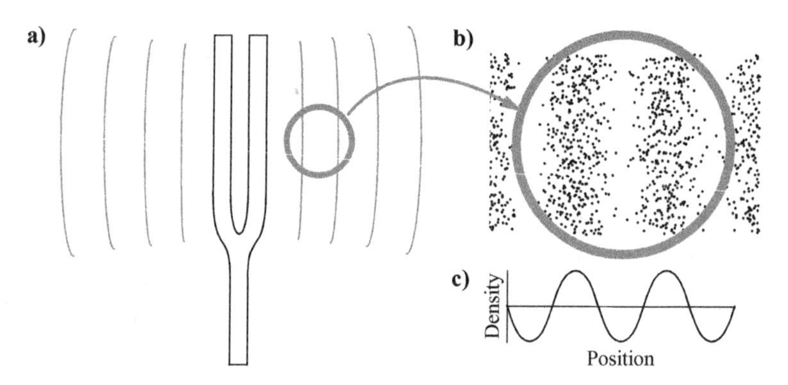

Figure 1.2
Sound wave from a vibrating tuning fork.

air pressure radiating away from the fork as it vibrates, similar in some respects to the way water waves radiate away from a stone thrown into a pond.

1.1.1 Physical Properties

The rate of periodic pressure change is frequency, and the strength of pressure fluctuations is intensity. The onset is the time when the sound begins, and its duration is the length of time we can hear it. The characteristic way in which the intensity of a sound changes through time is its envelope.

One final attribute, wave shape, completes the basic list of the physical properties of sound. Our hearing uses the shape of sound waves to characterize sound quality. We use words like "pure," "shrill," and "muffled" to describe wave shapes. We also use wave shape to identify the type of sound source, for instance, a trumpet or an oboe.

There are many other important properties of sound, such as the direction it comes from and what it means to us. But frequency, intensity, onset, duration, envelope and wave shape are enough to start with.

Frequency is measured as cycles per second. The unit of one cycle per second is hertz (Hz) (see section 4.3.1). Humans can hear sound over the range of about 17 Hz to about 17,000 Hz. Sound intensity is measured in decibels (dB) (see section 4.24.1). From soft to loud, intensity of sound ranges from the threshold of hearing at about 40 dB in very quiet rooms up to the limit of hearing at about 120 dB, also called the threshold of pain. Duration is measured in seconds.

1.1.2 Perception of Sound

Even though our senses are connected directly to the world, our inner experience of phenomena is not identical to the stimuli we receive. Our perception depends upon a multitude of interacting factors, including the sensitivity of our sense organs and the various ways our brains can be wired; even the culture of our birth and our location in time and space affect our experience of the world. So our language has developed terms that relate our inner experience to outer phenomena.

For simple sounds such as a tuning fork, the principal physical properties of sound are pretty closely related to what we hear. When the high- and low-pressure waves from the tuning fork have propagated through the air to the ears, they push and pull on the ear drum at the same rate that the tuning fork created them (just as the reeds at the edge of a pond rock back and forth from the waves created by a stone thrown into the water). The ears report the frequency of these air pressure changes to the brain as pitch. The intensity of the pressure changes is reported to the brain as loudness. If there are no changes in air pressure around the ears (that is, if the atmospheric pressure remains unchanged), we hear silence. In a musical context, onset and duration of sounds are perceived as elements of rhythm.

Loudness, pitch, onset, and duration seem to be relatively straightforward one-dimensional measures of our experience. A sound gets louder or softer; higher or lower; faster or slower, much the same way as a thermometer rises and falls with temperature. Measuring timbre, on the other hand, is not so simple.

Later I explain that the physical and psychological aspects of sound cannot be compartmentalized quite as neatly as I've suggested here, and that timbre is not as hard to study as it at first seems.

1.2 Waves

A *wave* is an organized traveling disturbance in a medium, such as air. The medium itself does not flow because of the wave; rather, a disturbance in the medium travels through the medium. *Waves transmit energy without transmitting matter.* For instance, part of the energy from the vibrating tuning fork is transferred to the ear.

1.2.1 Wave Shape

When I describe a wave as organized, I mean that it has a characteristic shape. Our ears are very sensitive to the shape of pressure changes in sound waves as they strike our ears. Throughout our lives we learn to associate particular wave shapes with particular sound sources. We also use this

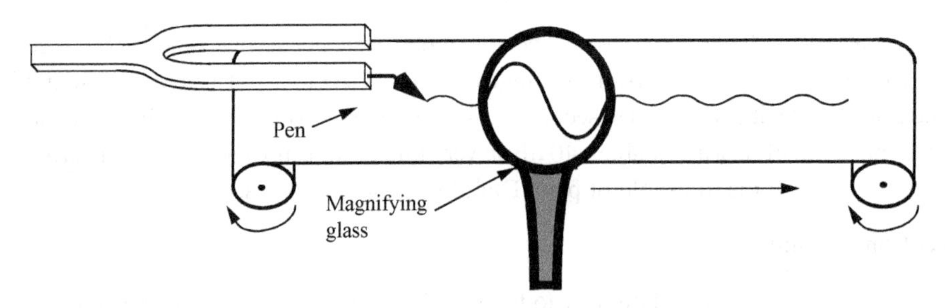

Figure 1.3
The wave shape of a vibrating tuning fork.

information to identify a sound's relative location and important characteristics about our environment. The wave shape of a tuning fork is very simple in comparison to most other sounds. If we graph the average particle density of the tuning fork sound shown in figure 1.2b, we see a shape similar to figure 1.2c.

The vibration of the tines of a tuning fork is very small and too rapid for the eye to see. But suppose we could view this motion, for example, by attaching a miniscule pen to one of its tines and then quickly passing a roll of paper underneath while it vibrates. Under magnification the vibration might be seen to leave a wavy mark on the paper (figure 1.3). The wave shape would be similar to the one in figure 1.2c.

1.2.2 Simple Harmonic Motion

The back and forth motion of the tuning fork tine shown in figures 1.2 and 1.3 is known as *simple harmonic motion*. Understanding this motion is fundamental to understanding all kinds of vibration, including music, the quantum mechanical motion of an atom, and the celestial music of the spheres. This motion is easiest to visualize when it is made up of the interplay of inertia of a mass and the elastic force of a spring. For the tuning fork, the mass and the spring are just different aspects of the same metallic substance: the metal has both inertia and elastic force. But we can better visualize simple harmonic motion by suspending a large mass from the end of a spring (figure 1.4a). This allows us to neglect the mass of the spring and the elasticity of the mass.

If left undisturbed, the mass will eventually come to rest at its *point of equilibrium,* where the downward force of gravity equals the upward-lifting spring force. But if it is disturbed from its equilibrium position, the mass will vibrate up and down in simple harmonic motion (figure 1.4b).

1.2.3 Guided Tour of Simple Harmonic Motion

If I pull down on the mass and release it, the force of the stretched spring lifts the mass upward against gravity and against the inertia of the weight, attempting to restore it to its equilibrium position. As its velocity increases, momentum tends to keep the mass traveling upward. The spring begins to go

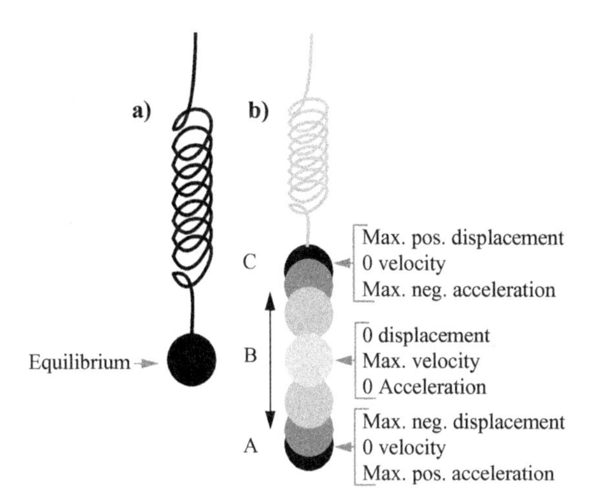

Figure 1.4
Simple harmonic motion.

slack as the mass rises, and when the mass reaches the equilibrium point, the spring no longer lifts the mass upward. But the mass continues to rise above the equilibrium point in spite of the slack spring, though its velocity slows. When its momentum is exhausted, the mass stops at a point of maximum positive displacement from equilibrium, and its velocity momentarily goes to zero.

The slack spring cannot hold the mass above its equilibrium point, so with its upward momentum spent, the force of gravity takes over and begins to pull the mass downward. Its velocity increases until it reaches its equilibrium point again. The mass continues to fall below the equilibrium point, though it slows because it is increasingly opposed by the tightening spring. The mass stops at a point of maximum negative displacement from equilibrium, and its velocity momentarily goes to zero. Then the cycle repeats.

Now go back to the initial moment, while I was still holding the mass below its equilibrium point. At that moment, the mass had zero velocity and zero acceleration. The moment I released it, it had zero velocity, but maximum acceleration. As the mass rose to approach its equilibrium point, its acceleration diminished, but its velocity continued to grow. At the equilibrium point, acceleration was zero, but velocity was maximum. Above equilibrium, the mass decelerated and velocity diminished, until at maximum positive displacement, velocity was zero.

Then the same process took place in reverse. At the moment it began its downward movement, acceleration was maximum, velocity was zero. As the mass approached its equilibrium point, its acceleration diminished, but velocity continued to grow. At the equilibrium point, acceleration was zero, but velocity was maximum. Below equilibrium, the mass decelerated and velocity diminished, until at maximum negative displacement, velocity was again zero.

Figure 1.5 shows the motion of the spring/mass system through time. Points marked A, B, and C in figure 1.4 are shown as lines in figure 1.5 for reference. The mass achieves its maximum velocity

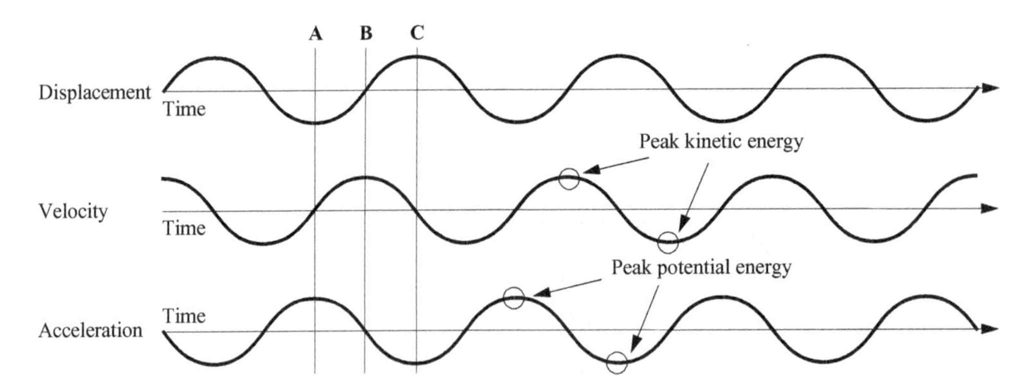

Figure 1.5
Displacement, velocity, and acceleration of simple harmonic motion.

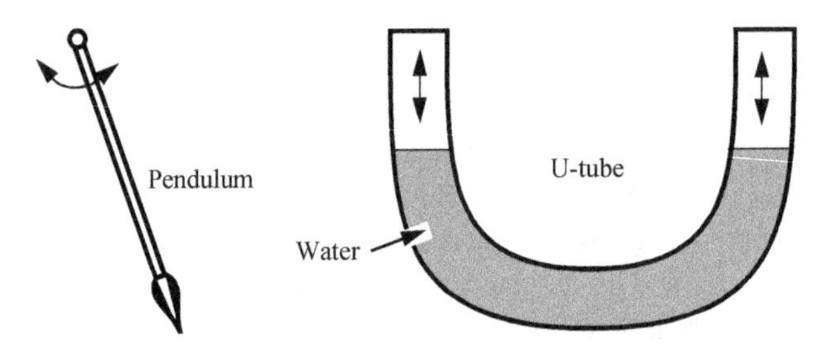

Figure 1.6
Other sources of simple harmonic motion.

in the instant it crosses its equilibrium point (B), and at this point it has zero acceleration. The mass achieves its maximum acceleration in the instant it reaches its point of maximum displacement from equilibrium (A and C), and at this point it has zero velocity. When the mass has maximum velocity (and zero acceleration) we say it has peak kinetic energy, and when the mass has maximum acceleration (and zero velocity) we say it has peak potential energy.

If we changed the inertia of the mass or the elasticity of the spring, we'd change its characteristic speed of vibration. If we used a heavier weight, the frequency would go down; if we used a stiffer spring, the frequency would go up. But the characteristic shape of the motion would remain. If we stretched the spring farther before letting it go, we'd increase the total potential and kinetic energy of the vibration, giving it a larger amplitude. But again, the characteristic harmonic motion would remain.

There are many examples of simple harmonic motion in the universe. The tuning fork and the spring/mass example and the examples in figure 1.6 are all simple mechanical vibrating systems. Even the basilar membrane, which is the organ within our hearing system that converts acoustic energy into nerve impulses, vibrates using the same principle of simple harmonic motion. Simple

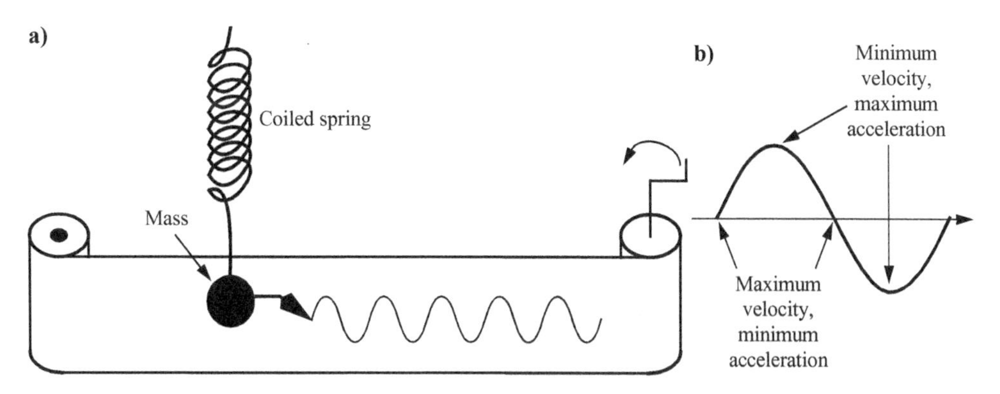

Figure 1.7
Sinusoid—simple harmonic motion through time.

harmonic motion can also be studied in electrical, optical, chemical, thermal, atomic, and other natural systems.

1.2.4 Sine and Sinusoid

Look again at figure 1.3. Tracing the shape made by a body moving in simple harmonic motion through time, we observe it makes a characteristic curve. Such a curve is a *sinusoid*. Simple harmonic motion is sinusoidal motion.

Figure 1.7b shows one period of the sinusoid generated by the spring and weight apparatus shown in figure 1.7a. Notice that the spring and weight make the pen move fastest when the wave crosses the centerline. This point is also where its acceleration reverses (going from acceleration to deceleration). Thus, sinusoidal motion captures all the salient features of simple harmonic motion through time.

The term *sinusoidal* means having the shape of a sine wave. Sine motion is a mathematical abstraction of simple harmonic motion, just as a point is a geometrical abstraction of a location in space. We can make an ink dot on a piece of paper and say it represents a geometrical point; similarly, a particular sinusoidal motion can be said to represent sine motion. But both sine motion and geometrical point really exist only in our minds, and the sinusoid and ink dot are their real-world counterparts.

Here's the difference: as we will see in chapter 5, sine motion has a precise mathematical definition in terms of circular motion. Because it is based on the circle, sine motion is a timeless description of motion having no beginning or end. Thus, sine motion is a mathematical ideal, an infinite, perfect motion that cannot exist outside of our imaginations. On the other hand, any reasonable approximation of sine motion (such as the one shown in figure 1.7) can be called sinusoidal. Because no physical motion can more than approximate ideal sine motion, all such real-world approximations are by definition sinusoidal.

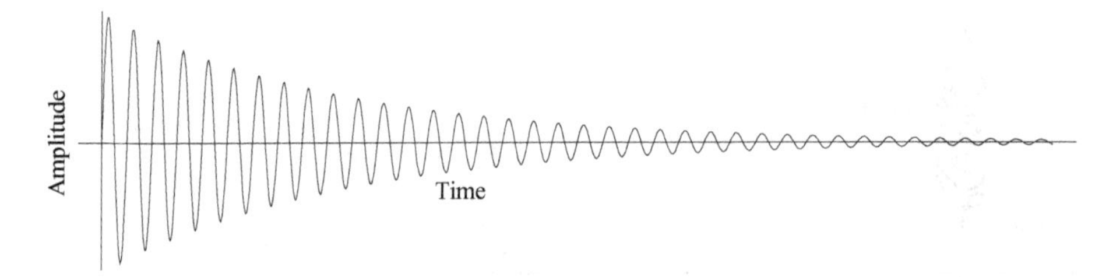

Figure 1.8
Damped waveform of a plucked musical instrument.

1.2.5 Conservative and Nonconservative Forces

Unless we continually supply energy to an object vibrating in simple harmonic motion, it will eventually come to rest at its equilibrium position because its energy is constantly being dissipated, radiated away as heat and/or sound. The effect of energy dissipation on a vibrating system is *damping*. Figure 1.8 shows how a sinusoid generated by the system in figure 1.7 might look through time because of the interplay of vibratory forces and dissipative forces.

If all the energy drains away at once, there can be no vibration, because then there's no energy left with which to vibrate. But even if the energy drains away slowly, all the energy will eventually dissipate completely. This suggests that there are conservative and nonconservative forces at work simultaneously in vibrating systems. The conservative forces operate within the system to perpetuate vibration, while the nonconservative forces operate between the system and its surroundings to dissipate energy through friction, and radiate energy through heat and sound. The balance between these two kinds of forces determines how the system vibrates.

- A spring's elastic force is a conservative force that is constantly transforming the spring's up and down movement from potential to kinetic energy and back again as the system vibrates.
- The external frictional force of air resistance and the internal friction of the spring itself are nonconservative forces that dissipate the system's energy into its surroundings, until total energy in the system has returned to its equilibrium.

Note in figure 1.8 that only the amplitude of the damped waveform changes through time, while the frequency (here represented as the distance covered by each repeated waveform) remains the same throughout.

In common usage, the terms "oscillate" and "vibrate" are often interchanged. But they are not the same: a system *vibrates* when it moves or swings from side to side regularly; a system *oscillates* if it moves or swings from side to side continuously and regularly. Hence, a sinusoid oscillates, whereas a plucked string vibrates.

1.3 Summary

The physical properties of sound include frequency, intensity, onset, duration, and wave shape. Frequency, onset, and duration are time-based aspects of sound, and intensity is a measure of the energy in a sound. These physical properties of frequency and intensity correspond to the perceptual cues of pitch and loudness. Onset and duration largely determine musical rhythm.

A wave is an organized traveling disturbance in a medium that transmits vibrating energy without transmitting matter. The simplest wave shape is the sinusoid, generated by simple harmonic motion. This motion is created by the interplay of elastic forces and inertia. The velocity of an object moving in simple harmonic motion is greatest near its equilibrium point; acceleration is greatest near the extremities of its excursion. If we graph simple harmonic motion in time, it makes a sinusoidal shape.

The forces that sustain vibration are conservative forces; the forces that cause damping are nonconservative forces.

2 Representing Music

Both mathematical notation and musical notation point to universes quite different from the one in which ordinary language functions so well. But, in each too, there is genius in the very notation that has developed for giving representation to ideas that seem to lie beyond ordinary language. There are times in mathematics when the similarities in notation is the first clue to a deeper relationship. Similarly musical notation not only created a structure within which Western music could develop but also shows something other than just the sounds being made. It indicates how the various elements stand in relation to one another, how sound creates a space, it shows how different musical voices move against and through each other. The notation in both subjects can make visible the hidden connections within each subject that reveal hidden connections among outside phenomena.
—Edward Rothstein, *Emblems of the Mind*

Just as music comes alive in the performance of it, the same is true of mathematics. The symbols on the page have no more to do with mathematics than the notes on a page of music. They simply represent the experience.
—Keith Devlin, *Mathematics: The Science of Patterns*

Our ears are continuously bombarded with a stream of pressure fluctuations from the surrounding air, not unlike the way ocean waves ceaselessly beat upon the shore. Nonetheless, our ears discern discrete events in this continuous flow of sound and assign them meaning, such as footsteps, a baby's cry, or a musical tone.

Just as the geometrical point is a mental construct that helps us navigate the underlying continuity of space, so the musical tone is a free creation of the human mind that we apply to the unbroken ocean of sound to help us organize and make sense of what we hear. Though its definition has been stretched to the breaking point by recent musical trends, tone is still the fundamental unit of musical experience. This chapter lays out the basics of music representation from a mathematical perspective, laying the groundwork for subsequent chapters.

2.1 Notation

The realm of personal musical experience lies entirely within each one of us, and we cannot share our inner experiences directly with anyone. However, many world cultures have developed systems for communicating musical experience by representing it in symbolic written and verbal forms. As members of a particular culture, we learn from childhood to map our inner experiences of music onto particular symbols which carry meaning that all members share. This allows us to speak and write about music, learn and perform the works of others, transcribe and analyze musical performances, and teach music, among other things. All this is possible because of the innate human capacity to abstract musical tones from the continuous stream of sound and to represent these tones symbolically.

This chapter characterizes one such system: the *Western common music notation system* (CMN). Its prevalence today makes it a good entry point to a broader discussion of the mathematical basis of tuning systems (see chapter 3). Understanding CMN will help us to fully appreciate its relationship to other musical traditions as well as to understand the history of tuning systems and current musical research.

2.2 Tones, Notes, and Scores

In CMN a *tone* is characterized by three sonic qualities: pitch, musical loudness, and timbre. When a tone is combined with two additional temporal qualities, onset and duration, the result is a *note*. A note is a tone placed in a particular temporal context.

Notes are combined in temporal order to create a musical *score,* which provides the necessary context to correctly interpret the performance of the notes. Roughly speaking, when notes are performed in sequence, the result is *melody,* and when notes are performed simultaneously, the result is *harmony.* The context provided by a score includes the sequence order of the notes and their timing as well as other details of how the notes are to be played on particular musical instruments.

Figure 2.1 shows a complete score written in CMN consisting of a single note. The score is written out on a *staff* of five horizontal lines that serves as a grid indicating pitch range. The relative pitch and duration of a note are indicated by placing *note symbols* such as ♩ (quarter note), ♪ (eighth note), 𝅗𝅥 (half note), and 𝅝 (whole note) on the staff lines. The mapping of pitches to staff lines is determined by the type and placement of the *clef* sign, placed at the left of each staff. The clef mark in figure 2.1 is the G clef, 𝄞. The spiral in this symbol encircles the second-to-bottom line, indicating that this staff line corresponds—by ancient convention—to the pitch G. This pitch is one whole step below A440, the reference pitch used to tune all modern Western instruments. Another common clef is the F clef, 𝄢. When placed on a staff, its two vertical dots bracket the second-to-top line, indicating that this staff line corresponds—by the same ancient convention—to the pitch F, a fifth below middle C.

Figure 2.1
A score of a single note in Western common music notation.

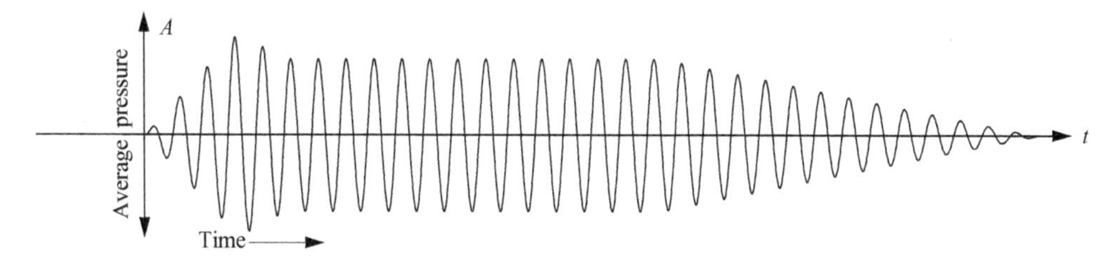

Figure 2.2
Amplitude function of the score in figure 2.1. The waveform has been shortened to make it fit the page.

If we were to record a musical instrument performing this score, the waveform might look like the one in figure 2.2, which shows fluctuating air pressure *A* as a function of time *t*. Figures 2.1 and 2.2 are just different views of the same information, the former describing the sound symbolically, the latter describing it physically.

Each view has advantages and disadvantages. The functional view provides a great deal of information about how a particular performer realized (performed) the note, allowing us to analyze the physical vibration of the instrument. But it is generally not useful to give such a representation to another player to describe how to play the same note. For this, the symbolic approach is superior.

There are many useful representations of tones, each of which has advantages and disadvantages in different contexts. For instance, although we can easily derive pitch, loudness, and duration information from either a musical score or from a functional representation like figure 2.2, neither gives much direct insight into timbre (see chapter 6).

2.3 Pitch

Frequency is a physical measure of vibrations per second. *Pitch* is the corresponding perceptual experience of frequency.

Pitch has been defined as "that auditory attribute of sound according to which sounds can be ordered on a scale from low to high" (ANSI 1999). Unfortunately, stipulating precisely what "that auditory attribute" is turns out to be a complex scientific affair that has spanned across centuries of research. While our sense of pitch is proportional to frequency, it is also influenced by frequency range, loudness, and the presence of other higher or lower frequencies. Pitch is limited to sounds within the range of human hearing, but frequency is not.

There are at least two motivations for developing measurements of pitch: scientific curiosity and the requirements of music engineering. I take up the scientific interests in chapter 6. Meanwhile, there is the more pragmatic problem of engineering the pitch range of human hearing for musical purposes so that we may communicate musically about pitch.

2.3.1 Frequency and Pitch

If we restrict ourselves to simple tones such as might come from a flute or tuning fork, then for some tone with frequency f we hear some corresponding pitch p. For instance, if the frequency of a tuning fork is $f = 440$ Hz, then the pitch p that we hear is conventionally called *A440*, the pitch commonly used by modern Western orchestras to tune all instruments together. The reference pitch used by orchestras has not always been set at 440 Hz but has varied through the ages. It became standardized at 440 vibrations per second in the early part of the twentieth century (see section 3.2.3).

2.3.2 Intervals and Frequency

An *interval* is the difference in pitch between two tones. The sensitivity of our ears to intervals is the basis of melody and harmony.

If a reference tone has frequency f_R, then a tone with frequency $f_R \cdot 2^1$ is said to be one *octave* higher. If the frequency is $f_R \cdot 2^2$, then it is two octaves higher. Generalizing, the frequency f_x of any octave x of the reference frequency f_R is

$$f_x = f_R \cdot 2^x, \qquad x \in \mathbf{I}. \qquad \textit{Octaves} \ (2.1)$$

This equation says, "The frequency x octaves above reference frequency f_R is equal to the reference frequency times 2 raised to the power of x." The expression $x \in \mathbf{I}$ means that x is an element of the set of all *integers*—all possible positive and negative whole numbers. Here it suffices to say that $x \in \mathbf{I}$ means that x can be any integer. The significance of requiring x to be an integer is that frequency f_x will only be an octave of f_R if x is an integer value.

If $x = 0$, the frequency of f_x is in *unison* with f_R because $f_x = f_R \cdot 2^0 = f_R$. If $x = -1$, the frequency of f_x is an octave below f_R because then $f_x = f_R \cdot 2^{-1} = f_R/2$. If we allow x to be any integer, all octaves of f_R can be realized.

2.3.3 Character of Intervals

Our ears are extremely sensitive to the intervals of unison and octave, and virtually all cultures organize their music primarily around these intervals. The unison has the musical quality of *identity*. For example, if two flutes intone A440, we say their pitch is identical.

Octaves have a musical quality of *equivalence*. If identity means that two pitches sound the same, equivalence means that we can tell them apart but each can serve the same musical purpose equivalently. In virtually every musical culture, pitches in any octave can perform the same musical function, a principle known as *octave equivalence*.

If the range of x in equation (2.1) is expanded to include all real numbers, then we can obtain the frequency f_x of any arbitrary interval x of reference frequency f_R:

$$f_x = f_R \cdot 2^x, \qquad x \in \mathbf{R}. \qquad \textit{Interval} \ (2.2)$$

The expression $x \in \mathbf{R}$ means x is an element of the set of real numbers (in other words, x can be any real number). *Real numbers* include all integers and all possible fractional and irrational values. Real values in the range $0 \leq x < 1$ select frequencies within the first octave above f_R. Values $x < 0$ select frequencies below f_R, values $x \geq 1$ select frequencies beyond the first octave above f_R, and so forth.

An exponent appears in equations (2.1) and (2.2) as the independent variable; it seems that our neural anatomy is wired to perceive an exponential relation between pitch and frequency. Frequency f goes up exponentially as pitch p goes up linearly: to double pitch, we must quadruple frequency.[1]

2.3.4 Interval Ratios

The frequencies of tones that make up an interval can be compared by making a ratio of their frequencies. For instance,

The interval of a unison is 1/1.

The interval corresponding to one octave up is 2/1.

The interval corresponding to one octave down is 1/2.

Consider the interval formed by the frequencies 880 Hz and 440 Hz. This ratio can be reduced to the lowest common denominator:

$$\frac{880}{440} : \frac{2}{1}.$$

The same is true of 132/66, 34/17, and so on. The advantage of expressing intervals as ratios in the lowest common denominator is that the kind of interval can be seen directly without the complication of the actual frequencies involved.

2.3.5 Categorizing Intervals

If the unison expresses identity and the octave expresses equivalence, the rest of the intervals signify *individuality.* Each of the intervals has a unique character to its sound—like a unique personality—that the ear can readily detect regardless of wide variations in frequency, amplitude, duration, or timbre. Our hearing seems to organize intervals by a subjective sense of distance that can be characterized as height or width: the interval of a fifth ($3/2 = 1\frac{1}{2}$) is experienced as "higher" or "wider" than a fourth ($4/3 = 1\frac{1}{3}$). In chapter 6 this quality is called chroma. Intervals figure prominently in music because they are so readily distinguished by our hearing.

2.3.6 Organizing Pitch Space

Equation (2.1) shows that there are an infinite number of pitches because we can assign any values to reference frequency f_R or octave x. But to engineer a practical scale system requires that we take into account the realistic limits imposed by our hearing.

Determining the Range of Pitch Space First, we can only hear frequencies between about 17 Hz and about 17,000 Hz (higher generally for youths and women, lower for rock concert aficionados, people who listen to music over headphones at elevated levels, people who drove Volkswagens in the 1960s, and the aged—especially the aged who drove Volkswagens to rock concerts while wearing headphones).

Even within this frequency range, pitches above about 4000 Hz are difficult to tell apart. Recognizing this, the musical engineers of the world's musical traditions have historically set realistic limits on the frequency range used by musical instruments to represent distinct musical pitches. The piano has one of the widest pitch ranges of traditional instruments. Its lowest pitch is about 27 Hz, and its highest is a little less than 4000 Hz.

Determining the Density of Pitch Space If pitches are crowded too closely together in frequency, we have a hard time telling them apart. Because of this, the world's musical engineers have limited the total number of pitches that cover the range of pitch space so that each can be easily identified. In some traditions there are as few as a dozen pitches altogether. The Western orchestra provides only about 90 total pitches to work with. So even though there are thousands of potentially identifiable pitches in the range of human hearing, relatively few are actually selected for use in musical scales.

Assigning Pitches To communicate about music, we must be able to name the pitches and associate them with frequencies. This is not an engineering problem so much as a design question, and each culture has answered it in a manner that speaks to what is important to that culture. In the West the choices have been profoundly influenced by the ideas of Pythagoras (see chapter 3).

2.4 Scales

A musical scale is an ordered set of pitches, together with a formula for specifying their frequencies. Each individual pitch of a scale is called a *degree*. The degrees are an ordered set of names and positions for the scale pitches.

Most musical traditions have acknowledged the importance of the unison and octave intervals by organizing their scales around them like anchor points. Most scales associate names of the degrees with their frequencies in one octave only, with the understanding that because of octave equivalence, degrees of the scale can be played in any other octave yet still perform the same musical function in the scale.

In an unfortunate twist of terminology, the degrees of the scale are also sometimes called pitch classes. (I'd rather they'd been called something like degree classes.) In any event, each degree is a member of a class that it shares with the same degree in all other octaves because of octave equivalence.

2.4.1 Gamut

A term related to scale is *gamut,* the entire range of notes reachable by an instrument or voice. Whereas a scale theoretically has no limits in frequency, a gamut does, as it is always tied to a particular instrument that can play only so high or so low.

"Gamut" is actually a compound of two other terms: the Greek letter gamma, Γ, used as a symbol for the lowest tone of the medieval musical scale, and "ut," the first syllable of a then-well-known hymn to St. John, the melody of which has the peculiarity of beginning one degree higher with each successive phrase. "Gamut" thus represents "all the tones from gamma onward" (Apel 1944).

2.4.2 Diatonic Scale

The prototype of all scale systems in the West is the *diatonic scale*. It has seven pitches per octave, named with the seven letters C, D, E, F, G, A, and B corresponding to the seven degrees of this scale.[2] The degrees of the diatonic scale are named tonic (1), supertonic (2), mediant (3), subdominant (4), dominant (5), submediant or superdominant (6), and subtonic (7). They are represented in CMN as shown in figure 2.3. This scale may also be familiar as the scale that goes with the *solmization syllables* do, re, mi, fa, sol, la, ti.[3]

The diatonic scale contains two interval sizes, the *half step* and the *whole step*. A whole step contains exactly two half steps. The whole step and the half step are also called *whole tone* and *semitone*. Chapter 3 details the frequencies that go with each diatonic scale degree and the frequency size of the half and whole steps. Here I focus only on the order of the interval sizes. The *interval order* of the diatonic scale is the sequence of whole and half steps in the scale.

The interval order and the starting degree are the two primary identifying characteristics of the diatonic scale that hold regardless of the pitch the scale starts on.

Figure 2.4 shows the interval order of the diatonic scale. Note the characteristic order of interval sizes: {2, 2, 1, 2, 2, 2, 1}, and observe that the scale starts on the first degree. For our purposes, these two characteristics completely define the diatonic scale. Note the asymmetrical structure of the interval order: there's a group {2, 2, 1} followed by {2, 2, 2, 1}. The unique order of whole and half steps provides a crucial asymmetry that our hearing exploits in order to orient ourselves

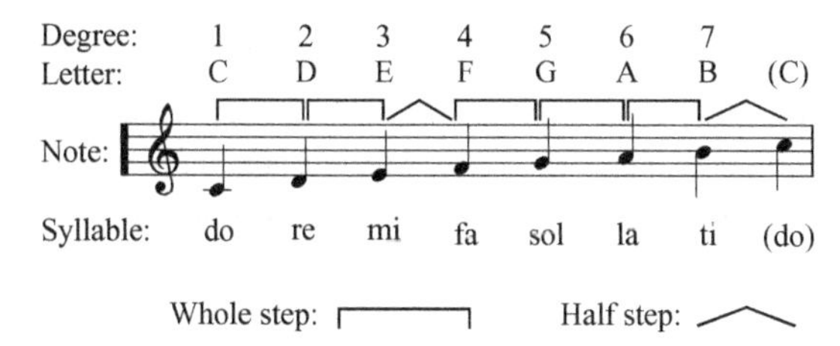

Figure 2.3
Diatonic scale.

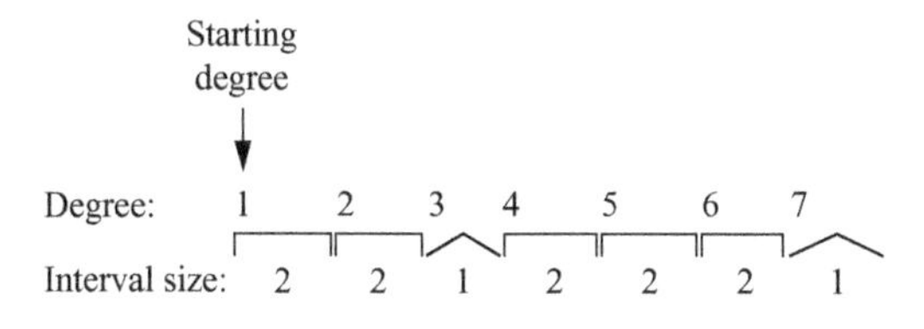

Figure 2.4
Interval order of the diatonic scale.

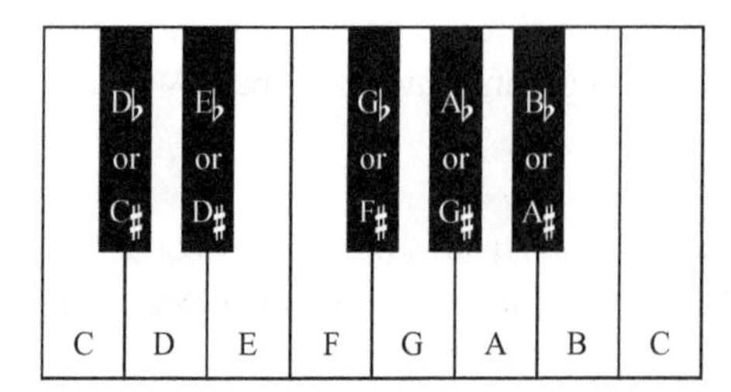

Figure 2.5
Piano keyboard.

to the music we're hearing. If the interval pattern were not asymmetrical, it would be impossible for us to orient ourselves in the scale.

2.4.3 Staff Lines and the Piano Keyboard

Look at figure 2.3 again and notice that the staff lines hide the asymmetry of the diatonic interval order visually. Each successive degree of the scale moves vertically up the staff by the same distance regardless of whether the interval between the successive degrees is a semitone or a whole tone. However, the asymmetry can't be hidden in the layout of the piano keyboard (figure 2.5). When starting from C, the interval pattern of the keyboard is the same as the diatonic interval order.

2.5 Interval Sonorities

Groups of intervals share *sonorities,* common traits that allow us to group them together (table 2.1). The sonorities correspond to the sonic character of the intervals. *Perfect intervals* have a quality that has been described as clear, pristine, structural, or astringent. *Major intervals* and *minor intervals* supply a warmth or feelingful character. *Augmented intervals* and *diminished intervals* provide a piquancy or strangeness that can be disturbing. Table 2.1 shows the classification of the intervals. Intervals can also be classified as consonant or dissonant (see section 3.10).

2.5.1 Major and Minor Scales

Another name for the diatonic scale is the *major scale*. The *minor scale* uses the standard diatonic interval order but starts on degree 6. Table 2.2 shows three octaves of the diatonic scale from left to right. The diatonic interval order is highlighted in the middle row, and the minor interval order is shown below it.

If we project one octave of the diatonic scale clockwise on a circle, as in figure 2.6, we see that the minor scale is the same as the major scale started two diatonic degrees counterclockwise around the circle. So the major and minor scales are related by the underlying diatonic order and are distinguished only by their starting degrees.

Table 2.1
Interval Classification by Sonority

Class	Name	Semitones	Description
Perfect	Unison	0	Provides harmonic anchoring and framework.
	Octave	12	
	Fourth	5	
	Fifth	7	
Major	Third	4	Provides expansive emotional color.
	Sixth	9	
	Seventh	11	
	Second	2	
Minor	Third	3	Upper pitch is one semitone smaller than major intervals. Minor intervals provide a contractive emotional color.
	Sixth	8	
	Seventh	10	
	Second	1	
Diminished		6	Upper pitch is one half step less than a minor or a perfect interval. A diminished fifth is called a tritone.
Augmented		6	Upper pitch is one half step greater than a major or a perfect interval. An augmented fourth is also called a tritone.

Table 2.2
Diatonic and Minor Scale Interval Order

Diatonic Degree	. . .	1	2	3	4	5	6	7	1	2	3	4	5	6	7	1	2	3	4	5	6	7	. . .
Diatonic interval order	. . .	2	2	1	2	2	2	1	2	2	1	2	2	2	1	2	2	1	2	2	2	1	. . .
Minor interval order	. . .	2	2	1	2	2	2	1	2	2	1	2	2	2	1	2	2	1	2	2	2	1	. . .

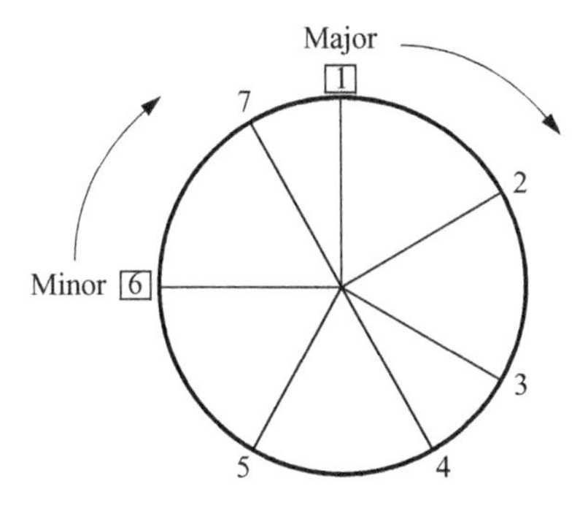

Figure 2.6
Major and minor scales.

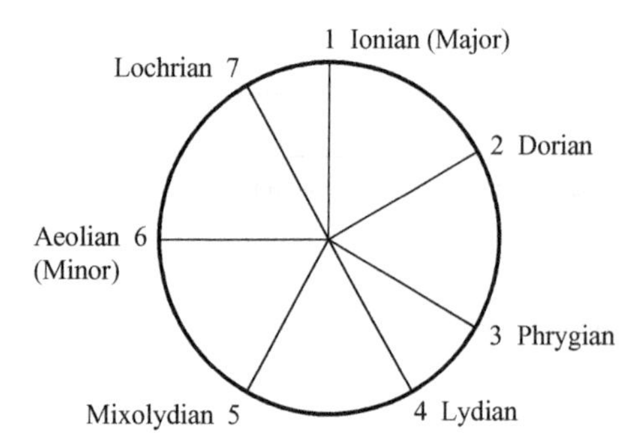

Figure 2.7
Starting the diatonic scale on other degrees to create modes.

2.5.2 Modes

Starting a scale from other than degree 1 or 6 produces scales that are other than major or minor but that share the diatonic interval order. Called *modes,* these variations of the diatonic scale order are shown in figure 2.7. The initial degree of a mode is its *final* because typically music in a mode would end on that note. So the final of Ionian mode is 1 and the final of Aeolian is 6.

The names derive, evidently, from seventeenth-century French music theorists, who named the modes arbitrarily after regions of Greece (Apel 1944). (The music theory of the ancient Greeks bears no resemblance to these modes.) The diatonic modes are the tonal basis of Gregorian chant and of early Western music (until about 1600 C.E.).

Notice that the major and minor scales are synonyms for Ionian and Aeolian modes, respectively. The various modes can be played on the white keys of a piano simply by starting the mode on the degree indicated in the figure. For example, starting on degree 4 produces the Lydian mode. The Lochrian mode is purely a theoretical mode, considered unusable by conventional music theory because of the tritone that exists between its final (7) and its fourth degree.

The listener may notice that some of the modes, especially Phrygian and Mixolydian, have a kind of antique quality to their sound. Before the advent of tempered tunings (see chapter 3), composers exploited the modes as an important source of tonal contrast. Shifting between modes was a way to add structure and shape to a composition. However, with the arrival of transposable instruments in the Baroque period, interest in modes declined, as key transposition took over the role of the modes to structure music. This left only the major and minor scales in common use. Hence, music built upon modal scales can sometimes suggest an ancient quality to the Western ear.

2.5.3 Chromatic Scale

The chromatic scale extends the diatonic scale by breaking up the whole steps into half steps and adding these new half steps to the scale. It uses the standard diatonic letter names A–G but adds symbols that raise or lower each diatonic degree by a semitone to indicate these in-between half

Figure 2.8
Chromatic scale in common Western notation.

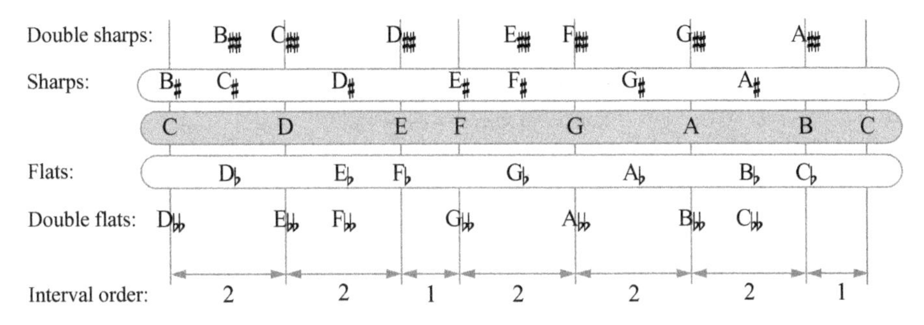

Figure 2.9
Diatonic scale names with chromatic and enharmonic inflections.

steps. The symbol ♯ (*sharp*) raises a diatonic degree by a semitone, and the symbol ♭ (*flat*) lowers it a semitone. The symbol ♮ (*natural*) restores a previously sharped or flatted pitch to its diatonic degree. Sharp, flat, and natural are *accidentals*. Given the order of half and whole steps in the diatonic scale from which it is constructed, there are thus 12 semitones in the chromatic scale:

{A, (A♯ | B♭), B, C, (C♯ | D♭), D, (D♯ | E♭), E, F, (F♯ | G♭), G, (G♯ | A♭)},

where the symbol | means *or*. Thus, one may write either A♯ or B♭, since they are *enharmonic equivalents*—they sound the same pitch. On the piano, for example, A♯ and B♭ are the same physical key (see figure 2.5).

The musical representation of all 12 pitches of the chromatic scale in CMN is given in figure 2.8. This scale can equivalently be written using flats instead of sharps (or any mixture). The fact that the degrees of the chromatic scale are named by their position with respect to the degrees of the diatonic scale shows again that the chromatic scale was derived from it.

In addition to the standard chromatic enharmonic spelling using sharps and flats, degrees can also be represented using double sharps (×) and double flats (♭♭), which raise or lower their respective degrees by two semitones (figure 2.9). The degree names in each column are enharmonic equivalents, thus $C_{\times}$ = D = $E_{♭♭}$.

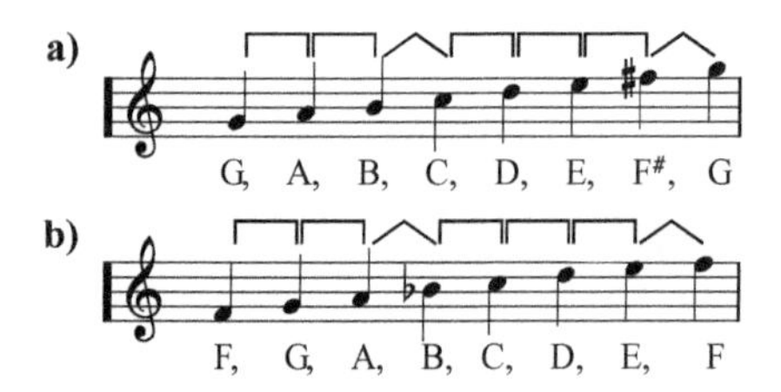

Figure 2.10
Diatonic scale in keys of G and F.

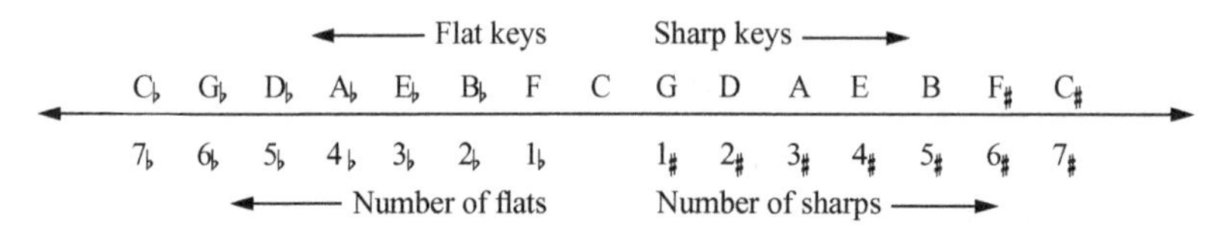

Figure 2.11
Transposition versus accidentals required for the diatonic scale.

2.5.4 Transposing

If a scale is started on any chromatic degree but C, it is said to be *transposed*. The diatonic scale can be transposed to any chromatic degree so long as the diatonic interval order of whole and half steps is preserved. For instance, if we begin the diatonic scale on G, then F must be sharped to preserve diatonic interval order; similarly, if we start it on F, then B must be flatted. Figure 2.10a shows the diatonic scale transposed to G, and figure 2.10b shows it transposed to F.[4] The degree to which the diatonic scale is transposed is called the *key*. For example, the diatonic scale transposed to G by the introduction of F♯ is the *key of G*. The untransposed diatonic scale is the *key of C*.

2.5.5 Key Signature

Notice that F is a fifth below C, while G is a fifth above C. Transposing the diatonic scale to begin on F requires one flat: B♭. Transposing to G requires one sharp: F♯. As we go down by fifths from C, the scale built on each subsequent transposed degree requires the introduction of one more flat in order to preserve the interval order of the diatonic scale. Correspondingly, as we go up by fifths from C, the scale built on each subsequent pitch requires the introduction of another sharp. This result is shown pictorially in figure 2.11.

A major or minor scale can be erected on any of the chromatic degrees by appropriate application of accidentals to establish the correct major or minor interval order. The accidentals required to start a major or minor scale on each chromatic degree are shown in figure 2.12. These are called *key signatures* because they stipulate the association between the key (the chromatic degree that the scale starts on) and the accidentals required for the corresponding diatonic scale.

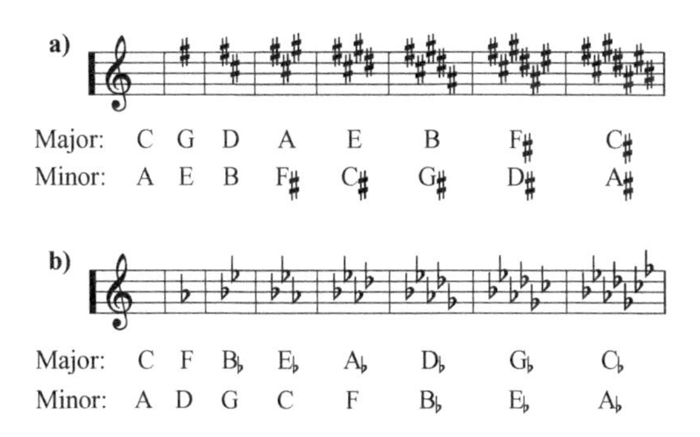

Figure 2.12
Key signatures.

Figure 2.12 allows us to infer from a score what the key should be. For example, if we observe three sharps in a score, we can infer that its corresponding major scale must start on A and its corresponding minor scale must start on F♯.[5] Since the major and minor keys that share a key signature are related by the underlying diatonic interval order, they are called the *relative major* and *relative minor.* For example, the relative major of B♭ minor is D♭ major, while the relative minor of A major is F♯ minor.

2.5.6 Circle of Fifths

As we move farther away from the key of C in figure 2.11, enharmonically equivalent keys start to crop up. In particular, the key of D♭ is enharmonically identical to the key of C♯, the key of G♭ is the same as the key of F♯, and the key of C♭ is the same as the key of B. This suggests that there is a circularity involved in the key structure, which becomes apparent if we twist the key sequence shown in figure 2.11 into a spiral, as shown in figure 2.13.

This is the *circle of fifths,* although it is easier to represent as a spiral, since it could continue into the double sharps and double flats, and so on. There are only 15 useful mappings of the diatonic interval order onto the chromatic scale, namely the ones shown in figure 2.11.

2.5.7 Nondiatonic Scale Orders

Of course, the diatonic scale specifies but one of many possible orderings of intervals. While diatonic ordering has had immense influence on music of cultures around the world, we're free to choose any ordering that serves our needs. The following is a select sampling of some nondiatonic scales. More are considered in chapter 3.

Pentatonic Scale If the diatonic scale is the father of scales, the pentatonic scale must be the grandfather, for it appears in virtually every culture worldwide. Its interval order is {2, 3, 2, 2, 3}. The black keys on a piano are an instance of the pentatonic scale. Like the diatonic scale, one can create pentatonic modes by choosing a different starting degree (figure 2.14).

Harmonic Minor Scale This scale (figure 2.15) uses the interval order of the minor scale but raises the seventh degree by one semitone. Its interval order is {2, 1, 2, 2, 1, 3, 1}. The minor scale

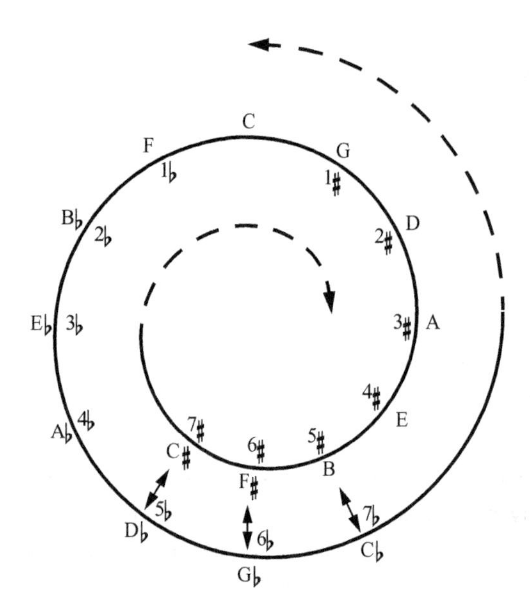

Figure 2.13
Spiral (circle) of fifths.

Figure 2.14
Pentatonic scale.

Figure 2.15
Harmonic minor scale.

(see section 2.5.1) is sometimes called the *natural minor scale* to differentiate it from the harmonic minor. The seventh degree of the diatonic scale is sometimes called the *leading tone* because it seems to lead the ear to the tonic. Raising the seventh degree of the natural minor lends this important harmonic function to the minor scale.

Melodic Minor Scale This scale (figure 2.16) varies its order depending upon the melodic function of the music—hence its name. It has an ascending order, which is used when the music rises up the scale, and a descending order, which is used when the music goes down the scale. The

Figure 2.16
Melodic minor scale.

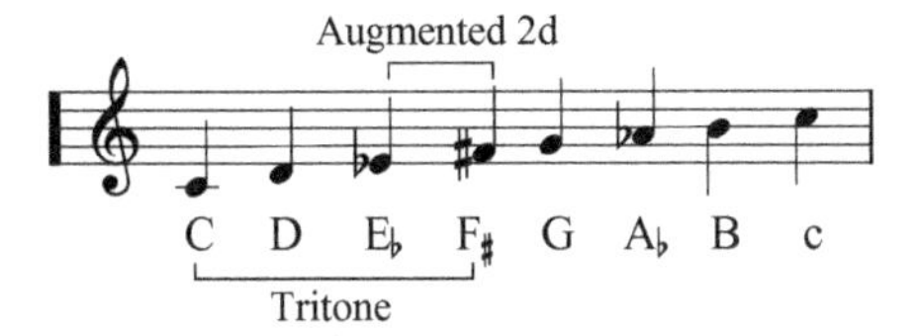

Figure 2.17
Hungarian minor.

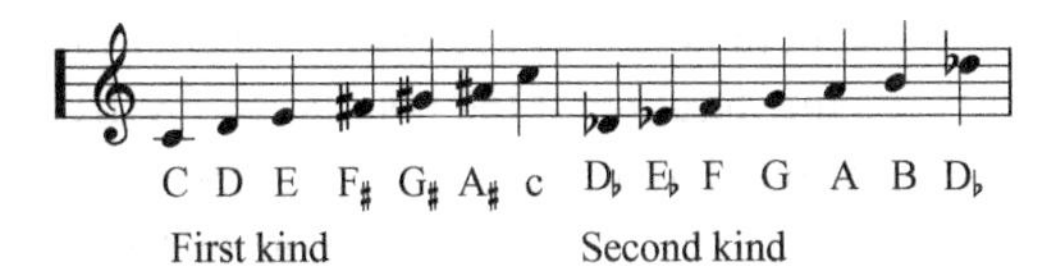

Figure 2.18
Whole-tone scales.

ascending order of this scale is like the harmonic minor but with the sixth degree also raised by a semitone. The descending form is identical to the natural minor.

Hungarian Minor This minor scale (figure 2.17) has an augmented second between the third and fourth degrees, and an augmented fourth (tritone) from first to fourth, lending it a spicy, rakish quality.

Whole-Tone Scale As there are 12 chromatic degrees per octave, picking every other semitone yields a scale containing only six degrees (excluding the octave), all of them whole tones. Its interval order is symmetrical: {2, 2, 2, 2, 2, 2}. Since we pick every other degree, there are necessarily two kinds of whole-tone scale (figure 2.18). The chromatic degrees of the first kind are $2n$, $n = 0, 1, \ldots, 5$, and the degrees of the second kind are $2n + 1$ over the same range (counting the first degree of the chromatic scale as 0).

Because the whole-tone scale interval order is symmetrical, it does not provide the ear with the anchoring asymmetry supplied by, for example, the diatonic interval order, leaving listeners harmonically "at sea." An obvious compositional device is to alternate between the two whole-tone scales for contrast. A falling whole-tone scale gives a particularly vulnerable and "slippery" feeling to the fall. Composers as various as Claude Debussy[6] and Thelonious Monk[7] have featured this scale in their compositions.

2.6 Onset and Duration

The *duration* of a note is the number of beats it lasts. The *beat* is the fundamental unit of time measurement and corresponds to the pulse of the music—in other words, what you tap your foot to. Beats are grouped into *measures,* set off from other measures in a score by *bar lines*.

The *onset* of a note is the moment stipulated by the score for it to begin, counted in beats from the beginning of the score. The *onset time* of a note is the same moment counted in seconds from the beginning of the score. Onset time can be calculated by multiplying the number of beats from the beginning times the duration of a beat.

2.6.1 Relative Duration

Musical symbols for relative note duration are given in the upper row of table 2.3. The symbols in the lower row indicate the duration of *rests,* the silences between notes. In table 2.3, each symbol indicates a duration one half as long as the symbol to its left. Shorter durations, such as one thirty-second can be created by adding more flags (𝅮) to the stem of the note.

Additional relative durations can be derived from those in table 2.3 as needed by the addition of dots to the right of notes or rests. A single dot extends the duration of the note or rest by 1/2. For example, ♩. = ♩ + ♪, and 𝄽. = 𝄽 + 𝄾. A second dot increases the duration of the note or rest by an additional 1/4. For example, ♩.. = ♩ + ♪ + 𝅘𝅥𝅯, and 𝄽.. = 𝄽 + 𝄾 + 𝄿. In general, n dots after a note or rest of duration D indicate that the total duration T is

$$T = D \cdot \left(\frac{1}{2^0} + \frac{1}{2^1} + \cdots + \frac{1}{2^n} \right).$$

2.6.2 Absolute Duration

The absolute duration of any note is determined by a metronome mark on the score in conjunction with the duration symbols in table 2.3. The *metronome mark* indicates which duration symbol gets the beat and how many beats there are per minute. For example, the metronome mark ♩ = 60MM indicates that the quarter note gets the beat and that there are 60 beats per minute. Thus, each quarter note lasts for one second.

The *tempo* is the number of beats per minute. *Rubato,* small perturbations in the tempo, may be employed by performers informally to emphasize a phrase or delineate a symmetry in the music.

Table 2.3
CMN Symbols for Relative Duration

	Whole	Half	Quarter	Eighth	Sixteenth
Note	𝅝	𝅗𝅥	♩	♪	𝅘𝅥𝅯
Rest	𝄻	𝄼	𝄽	𝄾	𝄿

Table 2.4
Time Signatures

Two quarters per measure	Three quarters per measure	Four quarters per measure
$\frac{2}{4}$	$\frac{3}{4}$	$\frac{4}{4}$
Common time[a]	Six eighths per measure	Nine 32ds per measure
c	$\frac{6}{8}$	$\frac{9}{32}$

Note: a. Same as 4/4 time.

The suffix MM on the metronome mark has an interesting history. It stands for "Mälzel Metronome." Johann Nepomuk Mälzel was not the inventor of the metronome, which honor is in fact due to Diedrich Nikolaus Winkel (1773–1826) of Germany. But Mälzel was a shrewd businessman who patented Winkel's invention in England and France before Winkel could do so. So successful was his marketing effort that only Mälzel's name remains commonly associated with the metronome (Tiggelen 1987).

2.6.3 Time Signatures

The rhythm of a score is determined by the time signature in much the same way that the scale is determined by the key signature. The *time signature* stipulates how many beats there are per measure and what beats are stressed to establish the rhythm (table 2.4). *Common time* groups four quarter notes per measure. It is notated with a capital letter C.

Not all beats have an equal stress when performed. Often the first beat is stressed, while other beats in a measure receive less stress. A few conventional stress patterns are associated with the most common time signatures. For example, common time and 4/4 time stress beat 1 the strongest and beat 3 somewhat less; the other beats are unstressed. For 3/4 time, typically beat 1 is the strongest, beat 3 is stressed less, and beat 2 is unstressed. Like the asymmetrical structure of the diatonic scale, the asymmetry in stress patterns helps orient the listener in the measure.

2.7 Musical Loudness

The sound intensity of many musical instruments can be adjusted over a certain range, depending upon their construction. The range from the softest to loudest sound for an instrument is its *dynamic range*. Some instruments, such as the harpsichord, are fixed at one loudness level. The oboe has a small dynamic range, and the pipe organ has quite a wide dynamic range. Loudness depends upon a number of perceptual and acoustical factors, and is not easy to characterize in general terms (see section 6.5).

Nonetheless, CMN provides a very simple notation for dynamic levels. Part of every musician's training is to learn how to translate the CMN symbols for dynamic level to the appropriate loudness

Table 2.5
CMN Indications for Dynamic Range

Pianississimo	*ppp*	As soft as possible	Mezzo forte	*mf*	Moderately loud
Pianissimo	*pp*	Very soft	Forte	*f*	Loud
Piano	*p*	Soft	Fortissimo	*ff*	Very loud
Mezzo piano	*mp*	Moderately soft	Fortississimo	*fff*	As loud as possible

level for his or her instrument, depending upon musical context. The nuances of this context are quite subtle and extensive, usually requiring years to master.

The CMN indications for dynamic range are shown in table 2.5. The Italian names are universally used, I suppose because they invented the usages, which were subsequently adopted by other European countries. The dynamic range indications in table 2.5 are entirely subjective. I describe how to relate them to objective measurements in section 4.24.

For instruments that can change dynamic level over the course of time, the "hairpin" symbol $<$ indicates a gradual increase in loudness, while $>$ indicates a gradual decrease. Bowed and blown instruments can usually effect a change in dynamic level during the course of a single note. Struck instruments including pianos generally can't change the dynamic level of a note after it is sounded but can change dynamic levels over the course of several notes. The proper interpretation of these cues is part of every musician's training.

2.8 Timbre

In musical scores, *timbre* means the type of instrument to be played, such as violin, trumpet, or bassoon. But timbre also is used in a general sense to describe an instrument's sound quality as sharp, dull, shrill, and so forth.

How quickly an instrument speaks after the performer starts a note, whether it can be played with vibrato, and many other instrumental qualities are also lumped together as timbre. Timbre also gets mixed up with loudness because some instruments, like the trombone, get more shrill as they get louder. As a consequence, it's easier to say what timbre isn't than what it is: timbre is everything about a tone that is *not* its pitch, *not* its duration, and *not* its loudness. However, negative definitions are slippery and provide no new information.

There are other ways of representing tones that shed positive light on timbre. Just as colors can be shown to consist of mixtures of light at various frequencies and strengths, sounds can be shown to consist of mixtures of sinusoids at various frequencies and strengths (see volume 2, chapter 3). For instance, when we hear a note played on a trumpet, even though our ears tell us we are hearing a single tone, in fact we are hearing simpler tones mixed together in a characteristic way that our minds—perhaps through long experience, perhaps through some intrinsic capability—fuse into the perception of a trumpet sound.

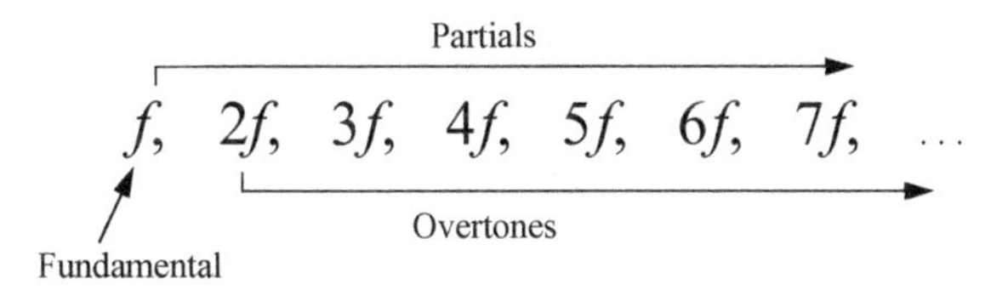

Figure 2.19
Harmonic overtone series.

The individual sinusoids that collectively make up an instrumental tone are called its *partials* because each carries a partial characterization of the whole sound. Partials are also known as *components,* and I will use these terms interchangeably. The principal properties of the partials are their frequencies and amplitudes. The way the partials manifest in frequency, amplitude, and time is what our ears use to determine what kind of instrument made a particular sound.

2.8.1 The Fundamental, Harmonics, Overtones, and Partials

The lowest pitched partial in a tone is called the *fundamental.* It is generally what our ears pick out as the pitch of the tone. Since, by definition, the remaining partials in the tone are pitched higher, they are called *overtones*.[8] Our ears use the pattern of overtone frequencies as an important cue to recognize timbres. The overtone frequencies of wind and string instruments are positive integer multiples of the fundamental, where the positive integers are 1, 2, 3, and so on. For instance, if a flute or violin has fundamental frequency f, then the frequencies of its overtones will be positive integer multiples of f (figure 2.19). The partials of such instruments are called *harmonics*. Note that because the positive numbers start at 1, and because $1 \times f = f$, therefore the first harmonic is the same as the fundamental.

Instruments with harmonic partials are usually chosen to carry the melody and harmony of music because frequencies of the harmonics tend to agree in frequency with the pitches of the diatonic scale. Instruments with *inharmonic partials* such as drums and bells are usually not used to carry melody and harmony because for the most part the frequencies of their partials do not agree with the diatonic scale.

The amplitudes and frequencies of the partials of musical instruments tend to vary in a characteristic way over the duration of a tone, depending upon the instrument and performance style of the performer. Though the variation may be slight, the precise amplitude and frequency ballistics of the partials help our ears to fuse a single percept of an instrument out of its individual partials, and help identify the type of instrument.

2.8.2 Vibration Modes

Each partial is created by a specific part of the vibrating system of the instrument. Consider a vibrating string, for example. Its fundamental frequency is created by that portion of the total

Figure 2.20
Vibration modes.

energy in the string that vibrates coherently along its entire length (mode 1 in figure 2.20). Vibration along the entire length of a string is called mode 1 vibration.

Not all the energy in a string vibrates in mode 1; some energy pushes one part of the string down while the other end counters it by rising (mode 2 in figure 2.20). The frequency of this vibration is twice the frequency of the fundamental, corresponding to the second harmonic. Some of the string's energy causes it to vibrate in three balanced regions (mode 3 in figure 2.20) corresponding to the third harmonic. For many vibrating systems (but not all), the higher the mode, the less energy it has. Stringed instruments can have dozens of vibration modes with significant energy.

Not all vibrating systems contain all possible modes. The clarinet has energy only at the fundamental and odd-numbered harmonics. Some vibrating systems do not divide the vibrating medium into integer ratios as the string does. The inharmonic partials of instruments such as bells and drums are not integer multiples of a fundamental.

2.8.3 Spectra

When we project sunlight through a prism, the resulting rainbow of colors, its *spectrum,* reveals the individual colors of sunlight. The prism distributes the colors into a linear sequence from low to high frequencies. The intensity of each color in the rainbow indicates the contribution of that color to the quality of sunlight.

So, too, the spectrum of a sound shows the intensities and frequencies of the sinusoids that make up the sound. A spectrum shows the *energy distribution* of a waveform in frequency.

The spectrum comprises the set of all possible frequencies from $-\infty$ to ∞ Hz at all possible intensities from 0 to ∞ dB (measuring up from silence). The spectrum of a particular sound will be a subset of this infinite two-dimensional space.

For example, figure 2.21 shows four waveforms and their corresponding spectra. The top waveform is a single sinusoid. Its spectrum shows a single vertical line. The line's horizontal position gives the sinusoid's frequency, and its height gives the sinusoid's intensity. The spectrum of the second waveform shows it contains two sinusoids, the fundamental at frequency f and the third

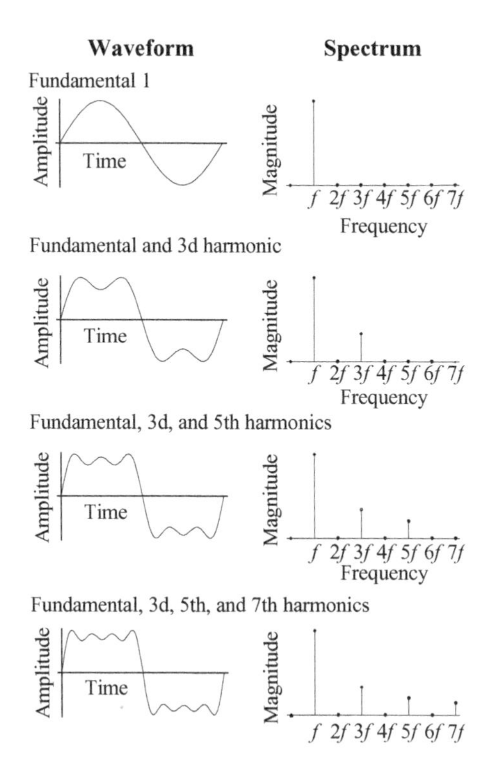

Figure 2.21
Harmonic waveforms and spectra.

harmonic at frequency $3f$. The harmonic has less energy because its line is shorter than the fundamental. The last two waveforms show additional odd-numbered harmonics being added, each with higher frequency and less energy than their predecessors. If we could hear the last waveform, it would sound somewhat like a clarinet. Since all frequencies are integer multiples of the fundamental, these are *harmonic spectra*. Because the components in figure 2.22 are noninteger multiples of the fundamental, this spectrum is an *inharmonic spectrum*. Percussion instruments such as bells, gongs, and drums produce inharmonic spectra.

Static and Dynamic Spectra In the foregoing discussion, I have conveniently neglected time as a required element. In order to compute the spectrum of a sound, we must have some length of it to analyze. If we wish to capture all the spectral information available in a waveform, the mathematics of spectral analysis requires us to observe the sound not just over its full duration but actually over all of time, from minus infinity to positive infinity. This is clearly a physical impossibility. Fortunately, there are mathematical techniques that allow us to analyze sounds with limited length. However, the shorter the waveform, the less precisely we can characterize its spectrum. So there is some inherent uncertainty between the temporal and spectral views

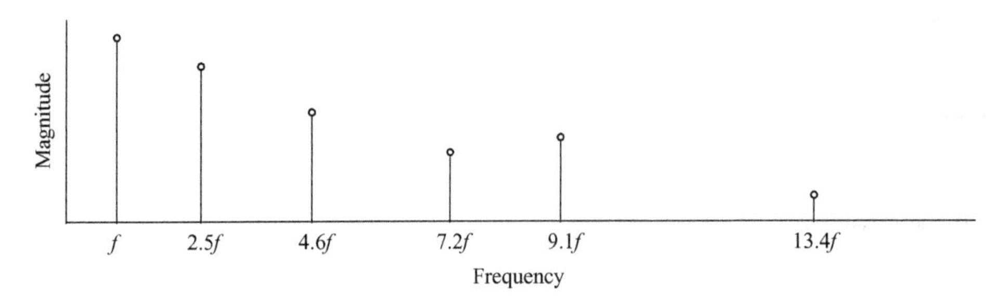

Figure 2.22
An inharmonic spectrum.

of waveforms of finite length. This subject is related to Heisenberg's uncertainty principle (see volume 2, chapter 3).

The length of sound available for spectral analysis determines the kind of spectrum we can create. A *static spectrum* shows the energy distribution of partials averaged over a fairly long period of time, such as the duration of an entire note. Figures 2.21 and 2.22 represent static spectra because they show the average intensities of the partials over the duration of an entire note. Because static spectra show averages, they cannot show how the energy distribution of a sound changes dynamically over the duration of the note. Static spectra can be useful, for instance, to confirm whether a sound is harmonic or inharmonic.

Dynamic Spectra Our ears are highly attuned to the way the spectra of sounds change through time, and we rely on this information to help us identify the type of instrument making a sound. The vibrational energy radiated by musical instruments evolves through time in a characteristic way based on the physical properties of the instrument and how the musician performs it. The dynamic elements in an instrument's spectrum that are contributed by the performance include vibrato, tremolo, glissando, crescendo, and decrescendo. There are also dynamic properties of the instrument's vibration that are largely determined by the interaction of the physics of the instrument and the physics of the performer's touch. Clearly, it would be very useful if we could capture the way spectra evolve through time.

Suppose we have a musical note lasting a few seconds. We can observe how its energy distribution evolves through time as follows:

1. Break the note down into a sequence of short sound segments each lasting a small fraction of a second.
2. Take the static spectrum of each sound segment separately.
3. Assemble the spectra in time order.

Imagine printing each static spectrum on a pane of glass, then assembling the panes in time order. Looking through the panes, we can observe how the spectrum of the sound changes through

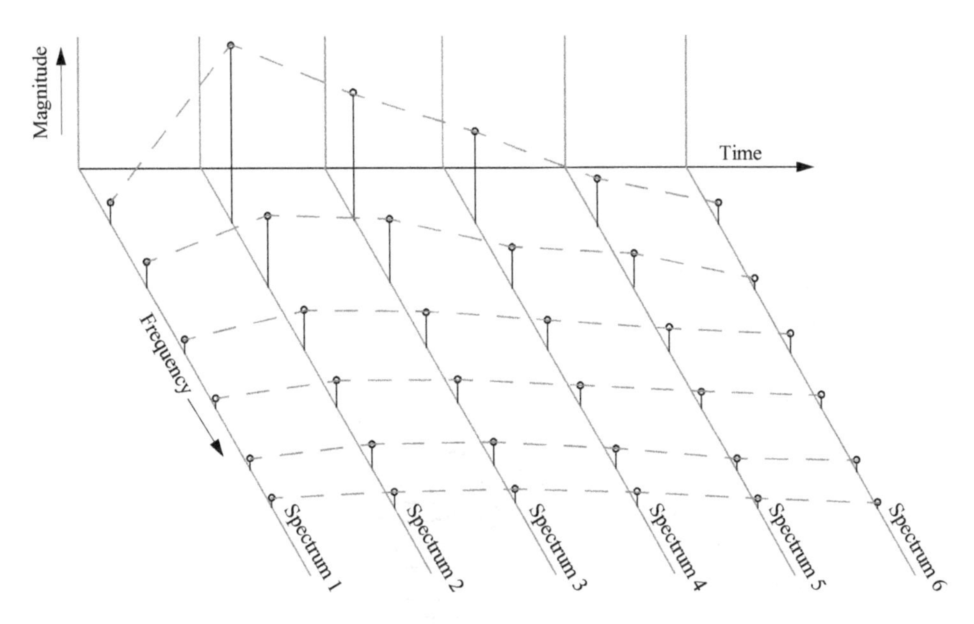

Figure 2.23
Dynamic spectrum.

time. This three-dimensional result is a *dynamic spectrum* because it shows spectral evolution through time. Figure 2.23 shows an idealized dynamic spectrum as a set of static spectra in time order. The x-axis shows time, the y-axis intensity, and the z-axis frequency. Dashed lines connect partials at the same frequency in adjacent spectral slices, showing how each partial's amplitude changes through time.

Figure 2.24 shows the spectral evolution of a string tone. We can tell a great deal about a sound by looking at its spectrum through time. For instance, the even spacing of the partials along the frequency axis suggests a harmonic spectrum. There are relatively few partials with significant energy. Most energy is concentrated in the lowest partials, and energy drops quickly with increasing partial number. The lower harmonics start sounding rather more quickly than the higher harmonics, as indicated by the broad grey line across the components at the beginning, and higher harmonics drop out more quickly, as indicated by the broad grey line across the components at the end.

Much of the aliveness we hear in a musical tone is communicated to us by the way the instrument's timbre changes instant by instant. The scrape of the bow on a violin string before the note sounds, or the puff of air that precedes an alto saxophone tone, or the characteristic way the overtones of a trumpet tone change strength during the course of a note provide important clues about what we are hearing.

The *sonogram* is another way to graph dynamic spectra (figure 2.25). Time is shown on the x-axis and frequency on the y-axis, and the thickness of the line shows the intensity of the spectral

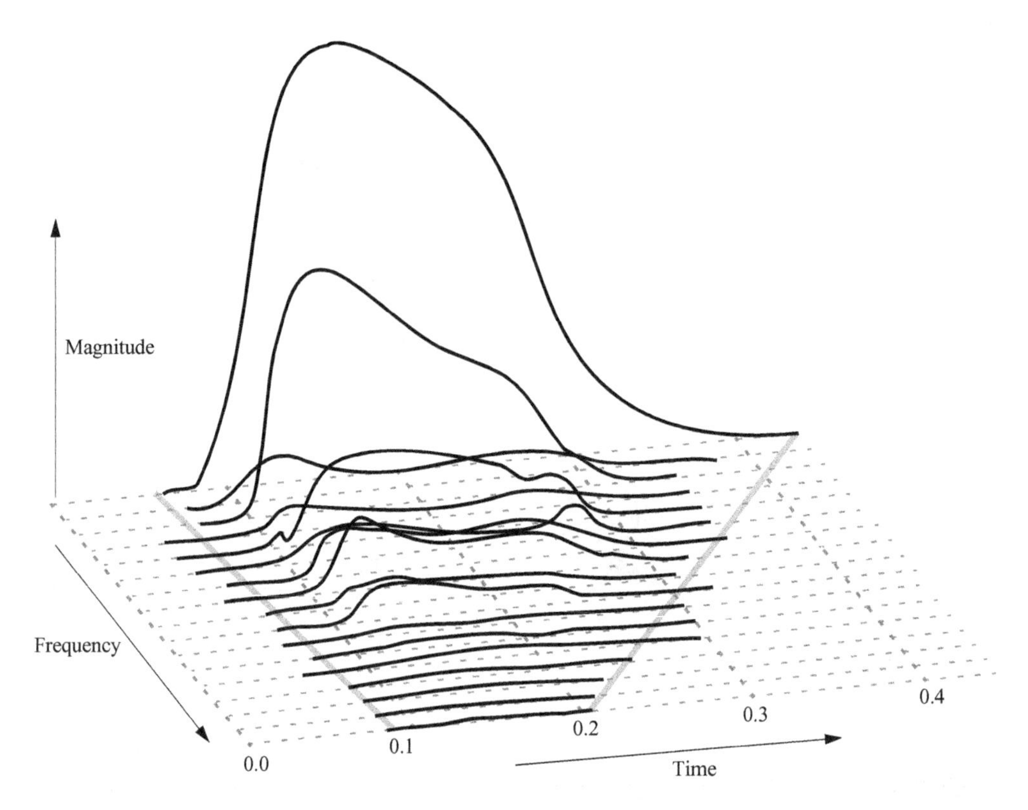

Figure 2.24
Amplitude, frequency, and time plot of a stringed instrument tone. (Adapted from a drawing in Grey 1975.)

components. The result is a two-dimensional image of the sound that has three-dimensional information. This sonogram represents four distinct bird chirps.

2.8.4 Amplitude Envelope

A tone's partials can be represented using just amplitude, frequency, and time.

- If we look at these three attributes together, we see the tone's spectral envelope in three dimensions (figure 2.24). But we can reduce this information to two dimensions by averaging.
- Averaging the amplitude of *each* partial *separately* through time, we get the tone's static spectrum in two dimensions: amplitude vs. frequency (figures 2.21 and 2.22).
- Averaging the amplitude of *all* partials *together* through time, we get the tone's *amplitude envelope* in two dimensions: amplitude vs. time (figure 2.26). Figure 2.26 follows the amplitude contour of the waveform in figure 2.2.

Clearly, these are just three different views of the same information.

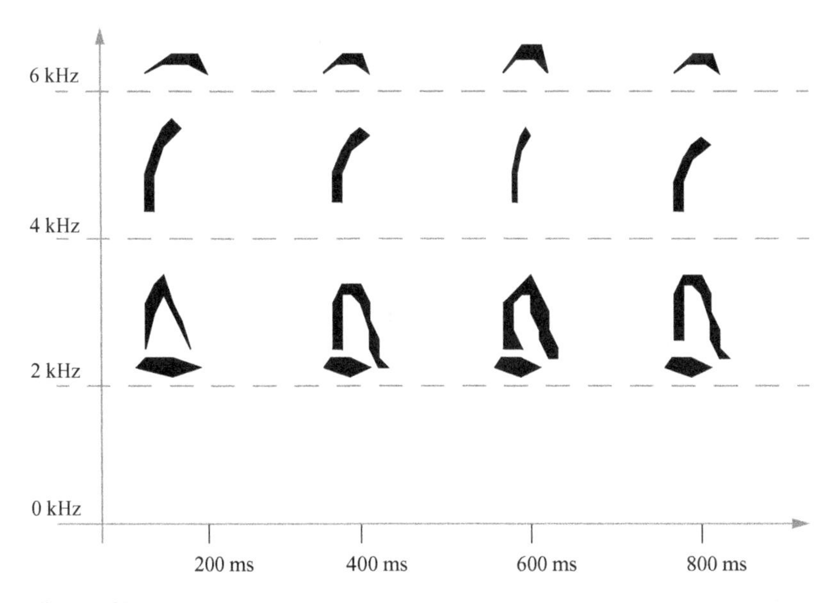

Figure 2.25
Sonogram of four bird calls.

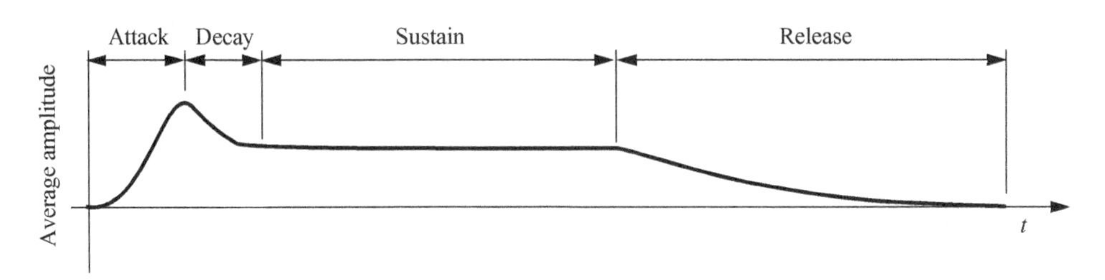

Figure 2.26
Amplitude envelope of waveform shown in figure 2.2.

The amplitude envelope of a note reveals in general how an instrument dissipates the energy it receives from the player through time as sound. Amplitude envelopes are conventionally divided into four segments:

- *Attack,* the period of time from silence, when exciting energy is first applied to the instrument, until the instrument is maximally dissipating its energy. Typical attack times are about 10 ms to 50 ms for most instruments. Energy may flood unevenly through the instrument at first, resulting in vibrational instabilities that produce, for instance, a scratching sound in violins or a warbling in brass tones. The ear is highly attuned to these instabilities and uses information about how the sound starts, grows, and stabilizes to identify the source of the tone.

- *Decay,* which follows the attack. Some instruments (brasses in particular) decline back to a sustainable level based on the amount of exciting energy being applied continuously.
- *Sustain,* the period following the decay. The instrument stabilizes so that the amount of energy being dissipated matches the exciting force.
- *Release,* the final portion of the sound from the moment no more energy is injected into the instrument until all energy is dissipated and it becomes silent.

Together, this classification is denoted *ADSR,* named for the initial letters of each segment of the amplitude envelope. These categories are completely arbitrary and by no means fit the amplitude envelopes of most real instruments playing real music. For example, struck instruments such as the piano have no sustain segments because they receive no sustaining force after the hammer strikes the string. *Legato* performance effects, where an instrument plays overlapping notes, are not well modeled by this system either. Nonetheless, it is sometimes a convenient shorthand and quite commonly found in sound synthesizers.

2.8.5 Bands and Bandwidth

A *band* is a range of frequencies within a spectrum. The *bandwidth* of a sound is the distance between upper and lower frequency limits of a sound. The *band center* of a band is its mean frequency. The bandwidth of human hearing is approximately 17 Hz to 17 kHz.

Sounds vary enormously in bandwidth. The bandwidth of a jet engine or a waterfall exceeds the audible spectrum. These are called *broadband* sounds. The tuning fork has a very narrow bandwidth and is called *narrowband.* Most musical instrument tones lie somewhere between.

2.8.6 Resonance

How is it that a musical instrument or a voice can strengthen one partial and attenuate another? The answer is that musical instruments are not as efficient at producing some frequencies as others. Where an instrument has a *resonance,* it is efficient at producing that frequency, but where it has an *antiresonance,* it may be inefficient or unable to vibrate at all. When we make different vowel sounds with our mouths, we are amplifying certain partials of the broadband waveform generated in the larynx and attenuating others. By some innate capacity or long experience (or both), our minds associate a certain profile of strong and weak partials with a particular vowel.

A *formant* is a group of frequencies of some particular bandwidth that is emphasized by a resonant system. Vowels are vocal formants. Formants may be fixed or variable. For example, good violins often have a fixed formant, sometimes called the singing formant, with a band center of approximately 1000 Hz. Diphthongs in speech are actually formant ranges that shift up and down in frequency, emphasizing higher or lower partials of the sound made by the glottis.

Resonance is involved in the production of sound for virtually every musical instrument. A flute is driven by the breathy broadband noise coming from the player's mouth through its fipple. The

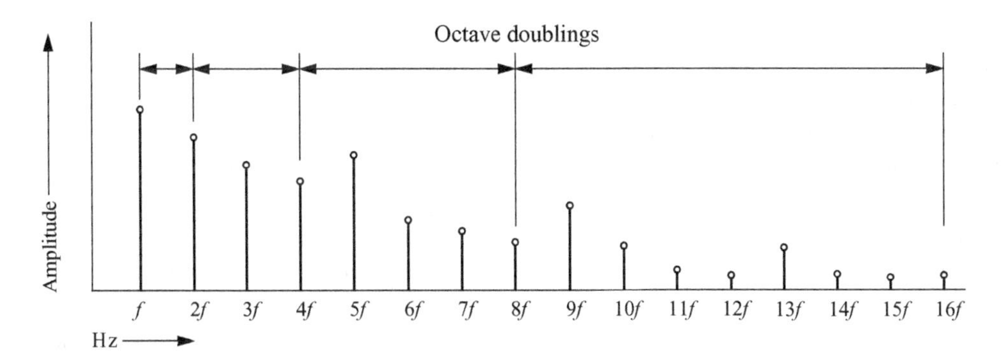

Figure 2.27
Harmonic spectrum and octaves.

air trapped in the body of the flute tends to resonate only at particular frequencies and captures the energy from the broadband noise only at these frequencies.

To take a nonmusical example, consider a car driving down a corrugated dirt road. There is a certain speed of travel that makes the car shudder the most violently from the corrugations: this is the car's resonant frequency, that is, the frequency at which the most up-and-down energy from the wheels passing over the corrugations can be transmitted to the rest of the car and its occupants.

2.8.7 Overtones and Octaves

As shown in figure 2.19, the harmonic series is a linear factor n times the fundamental frequency, producing a series of harmonics such as $f, 2f, 3f, 4f, 5f, 6f, 7f, \ldots$. The octave series is an exponential factor 2^n times the fundamental: $f, 2f, 4f, 8f, 16f, 32f, \ldots$. Figure 2.27 shows the relation between partials and octaves. Notice that there are many more harmonics within the compass of the higher octaves.

2.9 Summary

Amazingly, we are able to parse discrete notes out of the ocean of sound surrounding us. And in spite of the fact that we can't directly share our private experiences, we've developed symbolic systems to communicate about many things, including music. Common music notation represents notes as pitch, loudness, timbre, onset, and duration. A score is a collection of notes in time order. Notes are written on a staff, which also provides clef, key signature, time signature, and metronome mark.

Pitch is how our ears register frequency. Loudness is how our ears register intensity. Timbre describes either the kind of instrument making a sound or the sound's quality.

Intervals are characterized by the frequency ratio of two pitches. Intervals include the unison, octave, perfect, and imperfect intervals, and the dissonances. Scales are made up of collections of intervals in particular patterns. The diatonic scale is the prototype of modern Western music and also the foundation for many other musical systems in the world. The modes are simply the diatonic scale started on a different degree of the scale. The chromatic scale has 12 semitones per octave. Scales can be played on any starting pitch by using sharps or flats to preserve the interval order. There are many other nondiatonic scales besides the chromatic scale, including pentatonic, harmonic minor, melodic minor, Hungarian minor, and the whole-tone scale.

Rhythms are written in terms of how many beats they occupy. Tempo is the beat rate.

Timbre is the spectrum of frequencies in a tone. Harmonic spectra have an integer multiple spacing between components. Partials are generated by the vibration modes of the instrument. Static spectra average the strengths of the partials through time; dynamic spectra show each partial at each moment in time. An amplitude envelope shows the average intensity of all partials through time. The voice and most instruments have resonances that amplify or attenuate certain vibration modes.

3 Musical Scales, Tuning, and Intonation

Alterations of pitch in melodies take place by intervals and not by continuous transitions. We consequently find the most complete agreement among all nations that use music at all, from the earliest to the latest times, as to the separation of certain determinate degrees of tone from the possible mass of continuous gradations of sound, all of which are audible, and these degrees form the scale in which the melody moves. But in selecting the particular degrees of pitch, deviations of national taste become immediately apparent. The number of scales used by different nations and at different times is by no means small.
—Hermann Helmholtz, *On the Sensations of Tone*[1]

Why are musical scales organized the way they are? Why is most Western music based on scales made up of seven tones when there are twelve tones per octave? What does "equal-tempered" mean, and why after all these centuries is it still controversial? What choices have other cultures made about intonation, and why? What can we learn about ourselves, our music, and our culture by taking a careful look at the underlying mathematics? This chapter examines one of the most basic issues of music technology: musical scales, tuning, and intonation.

Certainly, tones and intervals are the primary materials of music. Virtually all music depends upon playing tones in certain intervals to convey musical ideas. A flexible and convenient way of describing tones and intervals is therefore fundamental, and this constitutes the main focus of this chapter. However, what starts out like a walk in the park becomes a surprisingly twisty trail with some deep insights into the choices our culture has made about the music we want to hear.

3.1 Equal-Tempered Intervals

The modern equal-tempered scale is a good place to begin because it is so ubiquitous and so simple. We can use it to develop some basic tools and terminology that will lead the way into a wider discussion of intonation.

As described in chapter 2, modern Western instruments divide the octave into 12 equal-sized semitones. This system of tuning is called *equal temperament* because the frequencies of all intervals are based on one uniform semitone interval.

We can use equation (2.2), $f_x = f_R \cdot 2^x$, $x \in \mathbf{R}$, to compute the frequencies of the equal-tempered scale. For some reference frequency f_R, we obtain the frequency f_k of any equal-tempered interval k ($k = 0, 1, \ldots, 11$) within the first octave by computing

$$f_k = f_R \cdot 2^{k/12}. \qquad \textit{Equal-Tempered Intervals} \quad (3.1)$$

For example, the pitch one semitone above $f_R = 440$ Hz is $f_1 = f_R \cdot 2^{(1/12)} \cong 466.16$ Hz. The size of the tempered semitone itself can be expressed as the ratio

$$\frac{2^{1/12}}{1} = \frac{\sqrt[12]{2}}{1} \cong 1.05946. \qquad \textit{Semitone Interval} \quad (3.2)$$

The nomenclature $\sqrt[x]{z}$ means the xth root of z, so $\sqrt[12]{2}$ is the twelfth root of 2.

3.2 Equal-Tempered Scale

Table 3.1 shows the conventional assignment of alphabetic letters to the frequencies of the equal-tempered scale. The table was generated by setting $f_R = 440$ Hz in equation (3.1) and calculating the frequencies of all 12 values of k.

A slight modification of equation (3.1) enables us to create equal-tempered intervals outside of an octave. In this version,

$$f_{k,v} = f_R \cdot 2^{v+k/12}, \qquad (3.3)$$

$f_{k,v}$ is the frequency of equal-tempered interval k in octave v. The values of k are the integers between 0 and 11, and the value of v is any integer. Note that the octaves that v selects are relative to the reference pitch, f_R. That is, $v = 0$ selects the same octave as f_R, while $v > 0$ selects octaves above f_R and $v < 0$ selects octaves below f_R.

This is unfortunately at odds with the common Western practice of naming octaves after the order of their appearance on a standard 88-key piano keyboard. In this practice, A440 is in the fourth piano octave and hence can also be called A4. C4 is called middle C in this system. The

Table 3.1
Frequencies of the Equal-Tempered Scale

k	Name	Frequency (Hz)	k	Name	Frequency (Hz)
0	A	440.000	7	E	659.255
1	A♯, B♭	466.163	8	F	698.456
2	B	493.883	9	F♯, G♭	739.988
3	C	523.251	10	G	783.990
4	C♯, D♭	554.365	11	G♯, A♭	830.609
5	D	587.329	12	A	880.000
6	D♯, E♭	622.253			

88-key keyboard ranges from A0 to C8. All we have to do to adopt this practice is to subtract 4 from the exponent of equation (3.3):

$$f_{k,v} = f_R \cdot 2^{(v-4)+k/12}. \tag{3.4}$$

For example, given $f_R = 440$ Hz, the frequency of the pitch A4 is $440 \cdot 2^{(4-4)+0/12}$, and the pitch an octave and a semitone above is B♭5, and its frequency is $440 \cdot 2^{(5-4)+1/12}$.

3.2.1 Constructing an Equal-Tempered Scale

To construct an equal-tempered scale, we must

1. Tie it to a reference frequency like A440
2. Name the intervals of the scale
3. Calculate the frequencies of the intervals from the reference

Choosing the Reference Frequency Piano keys are named by combining their pitch class and their octave. The octaves start at 0 at the bottom of the keyboard, and the lowest pitch is called A0. Counting octaves up from A0, middle C corresponds to C4. By convention, we use A440 as the reference and assign it to the piano key A4.

The Reference Octave Now we must establish a reference octave. Here there is a small difficulty. If the first pitch class in an octave were named A, the first letter in the alphabet, we could use the A440 reference as both the pitch A4 and the pitch of the start of each octave. But historically, new octaves begin with the pitch class C. Why the pitch class A was not chosen for this honor is a mystery shrouded in an enigma, but we're stuck with it.

The solution is to use equation (3.3) to compute the frequency of C4 based on the pitch of A4. Then we can use C4 as the reference frequency to deduce all the rest of the frequencies of the equal-tempered scale.

We can figure out the frequency of middle C this way: if A4 is 440 Hz, then by equation (2.2), A3 will be 220 Hz. Middle C is three semitones above A3 on the piano. So by (3.3), the frequency of middle C is

$$C4 = \frac{440 \cdot 2^{3/12}}{2}, \qquad \textit{Middle C} \tag{3.5}$$

which is about 261.626 Hz. To make the following equations a little simpler, let's define $R = C4 = 261.6$ Hz. The purpose of introducing R is to let it stand for the reference frequency no matter what actual frequency it is. For the following examples, we set the reference R to C4, but it could just as easily be any other frequency, and we'll choose different values for R when we study other scales.

Defining Scale Intervals Using reference frequency R, we can construct all other equal-tempered pitches in any octave. To make this slightly more convenient, let's define the function

$$f(k,v) = f_{k,v}, \tag{3.6}$$

where $f_{k,v}$ is as defined in (3.4). This function takes two arguments:

- k is an integer signifying one of the 12 pitch classes from C to B numbered 0 to 11.
- v is the desired octave; octave number 4 corresponds to the fourth piano octave.

We can define a set of symbols for all equal-tempered pitches in all octaves using equation (3.6) to specify their proper frequencies. For example, we can define the chromatic pitches playable on a piano as follows:

A0 = $f(0, 9)$, As0 = $f(0, 10)$, B0 = $f(0, 11)$, C1 = $f(1, 0)$, Cs1 = $f(1, 1)$, . . .

C4 = $f(4, 0)$, Cs4 = $f(4, 1)$, D4 = $f(4, 2)$, E4 = $f(4, 3)$, F4 = $f(4, 4)$, . . . B7 = $f(7, 11)$, C8 = $f(8, 0)$.

3.2.2 Equal-Tempered Semitone as a Ratio

In discussing equation (3.1), we saw that in the equal-tempered scale the number 1.05946 . . ., which corresponds to $\sqrt[12]{2}$, is the factor by which the frequency of a tone must be raised in order to obtain a frequency one semitone higher. Another way to say this is that the *interval of a semitone* is the ratio 1.05946 : 1. The advantage of this representation is that it is independent of any particular frequency. When any frequency is multiplied by the factor 1.05946 . . ., the next semitone in sequence is automatically produced. For example, if $A = 440$ Hz, then A♯ = 440 · 1.0595 . . . and so on.

3.2.3 Nonstandard Reference Frequencies

Using the equal-tempered semitone as a ratio allows for construction of scales on nonstandard reference frequencies as well. For example, we can find a semitone above 450 Hz by multiplying 450 · 1.0595. This can be used to construct equal-tempered scales for antique and nonstandard instruments that used this reference frequency.

The use of A440 as a standard pitch is a comparatively recent development. Agreement is still so fragile among musicians that in 1986 the Piano Technicians Guild, an international nonprofit organization of more than 3500 piano technicians, felt compelled to adopt a resolution calling for continued worldwide acceptance of A440 as the standard pitch. The Guild summarized the situation as follows:

> The history of musical pitch over the last three centuries has been one of confusion and misunderstanding. The pitch of A has ranged from 312 hertz used in a seventeenth-century church organ to a high of 464 used by some British military bands at the end of the nineteenth century.
>
> As early as 1834, a congress in Stuttgart, Germany, unsuccessfully attempted to standardize pitch at A-440. In the early years of this century, a number of groups in the United States formally adopted A-440 as a standard pitch.
>
> The United States Bureau of Weights and Measures adopted A-440 in 1920, and it was adopted as the worldwide standard in a treaty signed during an International Standards Association meeting in London in 1939.

Nonetheless, instrumentalists and orchestras continue to demand alternative pitch references, either to perform antique music or to satisfy the vanity of a particular virtuoso.

3.3 Just Intervals and Scales

Just intervals are intervals made from the ratio of small whole numbers. The only interval that is just in the equal-tempered scale is the octave, 2/1. But the *just scales* are based entirely on such small whole-number ratios. While the creation of scales from small integer ratios is a very ancient practice,[2] the equal-tempered scale emerged from the just scales only in recent centuries.

3.3.1 Origins of the Just Intervals

Ordinarily, when we hear a musical instrument, our ears fuse its many harmonics into a single percept that we identify with the source of the sound. However, if we treat the harmonics not as elements of a composite tone but as simple individual tones, we can view the harmonic series as a set of intervals. Figure 3.1 shows a harmonic spectrum containing a fundamental at frequency f and five overtones at integer multiples. The intervals between adjacent harmonics are simply the ratios of their frequencies, as shown in the figure.

I think it's amazing that the most important musical intervals are embodied in just the first six components of the harmonic series. The octave, fifth, and fourth are *perfect intervals,* and the major and minor thirds are *imperfect intervals* (see section 3.8.2).

3.3.2 Adding and Subtracting Intervals

We can use equation (3.1) to add and subtract intervals. If $k = 2$ in that equation, then frequency f_k will be two octaves above frequency f_R. By the distributive law, we can rewrite this as

$$
\begin{aligned}
f_k &= f_R \cdot 2^1 \cdot 2^1 \\
&= f_R \cdot \frac{2}{1} \cdot \frac{2}{1}.
\end{aligned}
$$

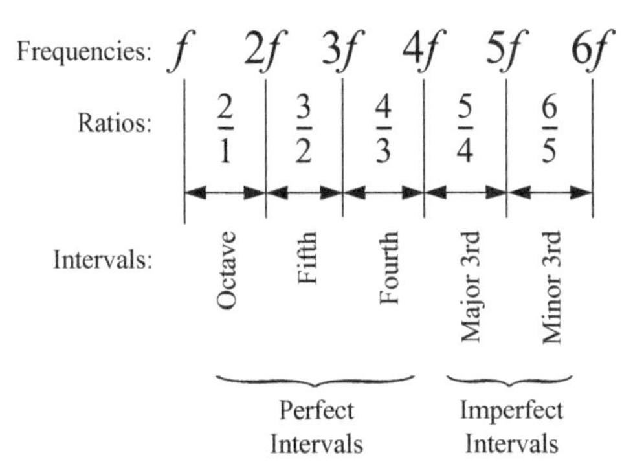

Figure 3.1
Intervals of the harmonic series.

So to add two octaves to f_R, we multiply it by the octave ratio 2/1 twice. This suggests that

Intervals are added by multiplying their ratios.

Let's test it. The sum of a fifth plus a fourth should be an octave. If we multiply the ratios of the fifth and fourth:

$$\frac{3}{2} \cdot \frac{4}{3} = \frac{2}{1},$$

the result is indeed an octave.

If multiplying ratios corresponds to adding intervals, then dividing ratios should correspond to subtracting them. From the example we'd expect that subtracting a fifth from an octave should yield a fourth, and indeed

$$\frac{2}{1} \div \frac{3}{2} = \frac{4}{3}.$$

So it follows that

Intervals are subtracted by dividing their ratios.

These rules are a consequence of the exponential relationship between pitch and frequency.

Subtracting an interval from an octave produces its *inversion*. Thus, in the previous example, the fifth and the fourth intervals are each other's inversions.

We can add or subtract an interval to or from itself n times simply by raising its ratio to the power of n (where n is an integer). For example, $(2/1)^2 = 4$ ascends two octaves, and $(1/2)^2 = 1/4$ descends two octaves. Similarly, $(3/2)^n$ ascends by n fifths, and $(2/3)^n$ descends by the same amount.

If we add or subtract an interval to every pitch in a score, we transpose that score. For example, to raise a melody by a fifth, multiply the ratios of all its pitches by 3/2. To lower it by a fifth, multiply the ratios of all its pitches by $2/3 = (3/2)^{-1}$.

3.3.3 Just Pentatonic Scale

The simplest just scale—one that seems to exist in every human culture—is the *just pentatonic scale*. It is very consonant because it has no minor second. We can get a reasonably good idea of what this scale sounds like by playing only the black keys of a piano. However, the original just pentatonic scales were based on ratios of small integers, not on the homogenized divisions of the octave given by the equal-tempered scale as used in pianos.

The just pentatonic scale can be constructed entirely from the interval of the fifth (3/2). However, there is a more intuitive way of constructing this scale, involving the fifth and its inversion the fourth (4/3):

1. Start with some pitch, such as C.
2. Multiply the frequency of C by 4/3 to find the frequency for F.
3. Mutiply C by 3/2 to find the frequency for G.

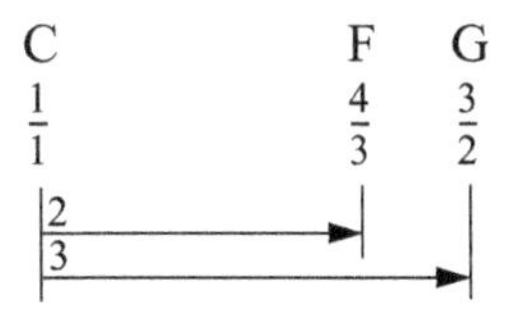

Figure 3.2
Pentatonic scale, first step.

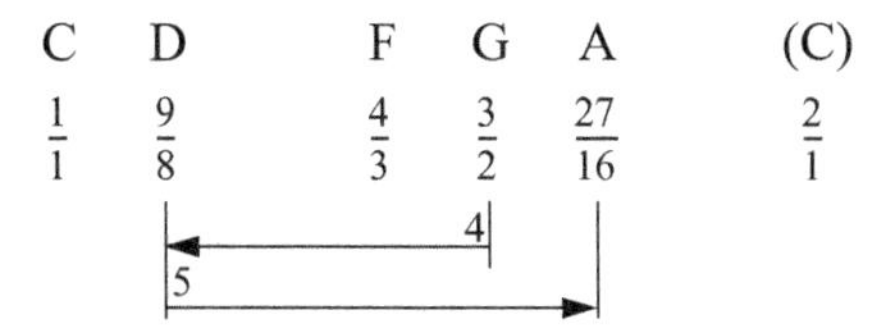

Figure 3.3
Just pentatonic scale.

So far we have three pitches, C, F, and G (figure 3.2). We create the remaining two pitches of the scale, D and A, from the ones we have so far.

4. To get D, go down a fourth from G. If the upward-going fourth is 4/3, the downward-going fourth is 3/4. Expressed in ratios, $D = (3/2) \cdot (3/4) \cdot C$, which simplifies to $D = 9/8 \cdot C$.

5. To get A, go up a fifth from D: $A = (9/8) \cdot (3/2) \cdot C$, which equals $(27/16) \cdot C$. Notice that the interval $(27/16) \cdot C$ is a major sixth up from C.

The full pentatonic scale is shown in figure 3.3 with the octave added.

3.4 The Cent Scale

The cent scale is a simple means for comparing the size of intervals.[3] Where the equal-tempered chromatic scale divides the octave into 12 degrees, the cent scale divides the octave into 1200 degrees, supplying 100 times the pitch resolution of the equal-tempered chromatic scale. Recalling the definition of the semitone given in equation (3.2), we can define the interval of 1 cent as

$$2^{1/1200} = 1.0005778\,. \qquad \textit{Cent} \ (3.7)$$

As a consequence, one semitone is exactly 100 cents. The pitch distance between adjacent cent intervals is not noticeable to the ear (see sections 6.4.3 and 6.4.5). So, the cent scale serves as a pragmatic way to compare any musical intervals regardless of how the intervals are derived.

If r is an interval, then the cent size c of that interval is

$$c = 1200 \cdot \frac{\log_{10} r}{\log_{10} 2}, \qquad \textit{Cent Interval} \ (3.8)$$

where $\log_{10} x$ is the logarithm base 10 of x (see appendix A). For example, consider $r = 2/1$, the octave. Then we have

$$c = 1200 \cdot \frac{\log_{10} 2}{\log_{10} 2} = 1200.$$

Let's use this to compare the ratios of the just fifth, 3/2, and the tempered fifth, $2^{7/12}$. The tempered fifth is exactly 700 cents. By (3.8) the just fifth is 701.955 cents. So the tempered fifth is almost 2 cents flat of a perfect fifth.

To go the other direction from an interval in cent to a ratio,

$$r = 10^{\left(\frac{c}{1200/\log_{10} 2}\right)} \qquad \textit{Inverse Cent} \ (3.9)$$

Trivially, if $c = 1200$, $r = 2/1$, the octave.

3.5 A Taxonomy of Scales

In order to talk sensibly about all kinds of scales, let's define the *dodecaphonic scale* as any scale with 12 degrees. *Dodeca* is Greek for "twelve." Then the equal-tempered scale, also known as the chromatic scale, is just a kind of dodecaphonic scale.

Similarly, let the *heptatonic scale* be any scale with seven degrees. By this definition, the scale made by the white keys of the piano is the equal-tempered heptatonic scale. The diatonic scale (see section 2.4.2) is a heptatonic scale with a particular order of scale degrees.[4] Similarly, the *pentatonic scale* is any scale with five degrees, and the black notes on the piano are the equal-tempered pentatonic scale. Any pentatonic scale built on just ratios is an instance of the just pentatonic scale.

With these definitions in place, a simple taxonomy of scales can be based on the number of degrees and whether the scale system is tempered or just (table 3.2).

Table 3.2
Simple Taxonomy of Scales

	Intonation	
No. of Degrees	Just	Equal-Tempered
Pentatonic	Just pentatonic	Equal-tempered pentatonic
Heptatonic	Just heptatonic	Equal-tempered heptatonic
Dodecaphonic	Just dodecaphonic	Equal-tempered dodecaphonic
⋮	⋮	⋮

3.6 Do Scales Come from Timbre or Proportion?

In section 3.3.1, we saw how the perfect intervals and the major and minor thirds are all present in the first six partials. This is such a striking coincidence that it has led some to wonder if perhaps the goal of the early music engineers might have been to fashion scales from these ratios. I call this the deductive scale conjecture—that scales were deduced from the nature of the harmonics. This conjecture is disputed by some. In his book *Genesis of a Music* (1947, 87), Harry Partch states, "Long experience . . . convinces me that it is preferable to ignore partials as a source of musical materials. The ear is not impressed by partials as such. The faculty—the prime faculty—of the ear is the perception of small-number intervals, 2/1, 3/2, 4/3, etc. and the ear cares not a whit whether these intervals are in or out of the overtone series."

The earliest known research in the West on musical scales was conducted by Pythagoras (ca. 580–500 B.C.E.) and his followers. We know that the Pythagoreans viewed music as a branch of science and believed that the construction of musical scales should proceed out of an analogical process that related, for example, the periodic movements of a string to the periodic movements of the planets. They weighed the distances between planets the same way they weighed the divisions of a musical string, namely by the study of ratio and proportion. Figure 3.4 shows an interpretation by Robert Fludd (a contemporary of Johannes Kepler) of the relation between the harmony of the spheres and the proportional divisions of a string.[5]

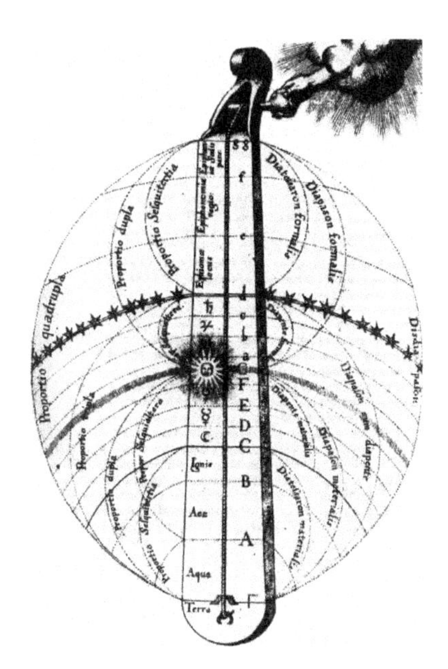

Figure 3.4
The cosmic monochord of Robert Fludd.

From the Pythagorean perspective (shared by Partch), the important thing about a musical scale is its proportionality—how it divides up the unity of a string—not the relationship between that proportionality and any physical artifact such as the overtone series.

From this evidence, one might argue that scales developed out of the mathematics of proportion. I call this the inductive scale conjecture—that scales are a free creation of the human mind, based on ratio and proportion. According to this conjecture, the scales are Platonic archetypes, and physical musical instruments are imperfect instances of these archetypes that are manifested in the world by way of human creativity.

Of course, these are only conjectures. The truth of how the scales actually developed is lost in the mists of time. Are the scales derivative of the overtone series or derivative of mental constructions of proportionality? Is the prime faculty of the ear the perception of small-number intervals or the perception of harmonics? I argue it both ways in this chapter because there is plenty of evidence for both perspectives.

It is evident that musical scales are free creations of the human mind because they do not occur in nature. It is at least a striking coincidence that they align in their principal dimensions with the harmonic sequence. Perhaps it was the very numinosity of this coincidence that compelled the Pythagoreans to study this subject in the first place.

3.7 Harmonic Proportion

Pythagoras is credited by ancient Greek writers with having discovered the intervals of the octave, fifth, fourth, and double octave (4/1). Pythagoras and his followers attached great numerological significance to the fact that these most harmonious intervals were constructed strictly from ratios of the consecutive integers 1, 2, 3, and 4. They were also impressed by the fact that these intervals formed a sequence of *superparticular ratios,* that is, ratios of the form $(n + 1)/n$: 2/1 (octave), 3/2 (fifth), and 4/3 (fourth). They found mystical significance in the fact that by their nature superparticular ratios pair an even and an odd number. They also noted that small integer superparticular ratios seemed to be the most harmonious. These observations became permanent fixtures in the minds of music theorists for the next two thousand years.

The means Pythagoras used to construct his scale can be stated as follows. He started with a division of the string into 12 equal parts.

1. The octave is the ratio 12:6.

2. The fifth is found by taking the *arithmetic mean* of the octave, defined as $x = (a + b)/2$. Thus, $(12 + 6)/2 = 9$, and the ratio $9:6 = 3:2$ is the fifth.

3. The fourth is found via the *harmonic mean,* defined as $x = 2ab/(a + b)$. Thus, $(2 \cdot 12 \cdot 6)/(12 + 6) = 8$, and the ratio $8:6 = 4:3$ is the fourth.

Pythagoras combined these results into what he called the *harmonic proportion,*

$$12:9::8:6, \tag{3.10}$$

which he took to be the foundation of all music.

3.8 Pythagorean Diatonic Scale

The scale that eventually came to be associated with Pythagoras adds two more pitches, E and B, to the just pentatonic scale to produce the *Pythagorean diatonic scale*. Although it can be built entirely from fifths, using its inversion the fourth helps keep its construction simple.

1. Construct a pentatonic scale.
2. Add pitch E by going down a fourth from A:

$$E = \frac{27}{16} \cdot \frac{3}{4} = \frac{81}{64}.$$

3. Add pitch B by going up a fifth from E:

$$B = \frac{81}{64} \cdot \frac{3}{2} = \frac{243}{128}.$$

The Pythagorean scale is shown in figure 3.5.

We can create a set of functions to produce the frequencies of the Pythagorean diatonic scale just as we did for the equal-tempered scale (see section 3.2.1). As before, we need a reference frequency, a reference octave, and the intervals.

1. Start from A440. The reference frequency $R = 440$ Hz.
2. Build the scale so that when $v = 4$ frequencies are in the fourth piano octave. We want to create a function that takes the octave v as its argument and gives Pythagorean C in any octave. How do we go from A440 to Pythagorean C? The answer is in figure 3.5. We subtract the interval of a major sixth, the distance from A down to C, by multiplying A by 16/27:

$$C_\pi(v) = R \cdot \frac{16}{27} \cdot 2^{v-4}. \quad (3.11)$$

Because we are using integer ratios, we end up with a different frequency for middle C than the equal-tempered scale (260.741 Hz). I introduce the notation C_π to distinguish the "πthagorean" C from the equal-tempered C. *Pythagorean middle C* is $C_\pi 4$.

C	D	E	F	G	A	B	(C)
$\frac{1}{1}$	$\frac{9}{8}$	$\frac{81}{64}$	$\frac{4}{3}$	$\frac{3}{2}$	$\frac{27}{16}$	$\frac{243}{128}$	$\frac{2}{1}$

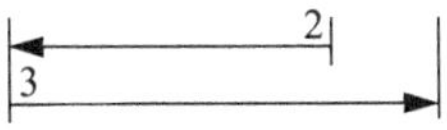

Figure 3.5
Pythagorean scale.

3. Finally, create interval frequency functions:

$$F_\pi(v) = C_\pi(v) \cdot \frac{4}{3},$$

$$G_\pi(v) = C_\pi(v) \cdot \frac{3}{2},$$

$$D_\pi(v) = C_\pi(v) \cdot \frac{9}{8}, \ldots,$$

where v is the desired octave.

Hearing early music played with just intervals can sound transcendentally beautiful, especially if the intervals are played accurately. Music in the Middle Ages was mostly written using the Pythagorean scale, and the just ratios seem to lend this music a refreshing, crisp air.

But there are two significant problems with the Pythagorean scale that musicians have historically disliked: some of its intervals are not musically pleasing because they do not align with the harmonic series, and it is awkward to transpose.

3.8.1 Intervals of the Pythagorean Diatonic Scale

Figure 3.6 shows the Pythagorean scale with intervals between the pitches. The top row shows the intervals built up from C_π. The bottom row shows the sizes of the intervals, that is, the difference between adjacent intervals. Recall that intervals are subtracted by dividing their ratios. For example, the interval of the whole step C:D is

$$\frac{9/8}{1/1}.$$

The whole step D:E is

$$\frac{81/64}{9/8} = \frac{9}{8}.$$

The half step E:F is

$$\frac{4/3}{81/64} = \frac{256}{243}.$$

The rest of the intervals follow this pattern.

3.8.2 The Syntonic Comma

The interval of the third in the Pythagorean scale was considered a dissonance in the Middle Ages, and as a result compositions would typically omit the third in the final chord of a composition so as to end only with perfect intervals—fourths, fifths, and octaves—an effect that sounds hollow to modern ears.

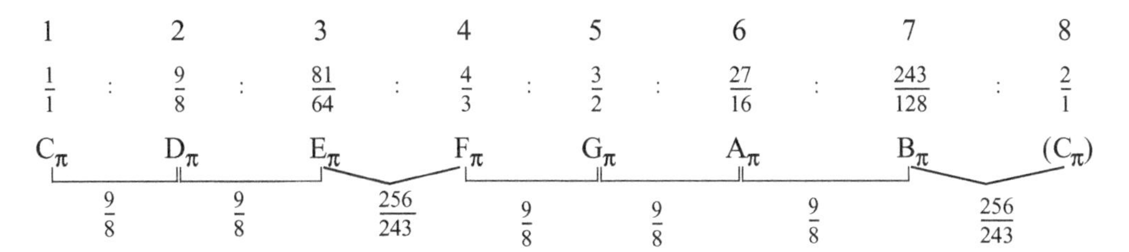

Figure 3.6
Pythagorean scale with intervals.

The reason the third was considered dissonant is that all the Pythagorean major thirds (C:E, F:A, and G:B) use the 81/64 ratio, which is not the same as the 5/4 major third that occurs naturally in the overtone series. The three Pythagorean major thirds are a little sharp of the 5/4 major third; hence they don't line up perfectly with the overtones of harmonic instruments, causing a roughness in the sound because of beats (see section 6.7). This imperfection in the otherwise beautifully symmetrical edifice of the Pythagorean scale was irritating enough to be given a name. The ratio of

$$\frac{81}{64} \div \frac{5}{4} = \frac{81}{80}$$

is the *Syntonic comma,* also known as the comma of Didymus. It is the amount by which the Pythagorean major thirds are out of tune with the 5/4 major third of the overtone series. The Pythagorean major third is about 21.5 cents sharp, about a fifth of a semitone, which is easily noticed. The same problem afflicts the Pythagorean minor third, the major and minor sixths, and the major seventh and minor second. Only the perfect intervals are exactly aligned with the overtone series. Perhaps this is where the nomenclature of "perfect/imperfect" originated.

3.9 The Problem of Transposing Just Scales

Suppose we have a song that was arranged for a high female voice, but we only have a low female voice available. Unless, trivially, we could just drop the pitch of the song an entire octave to solve the problem, it is necessary to transpose the music by some interval so that it lies within the available vocalist's range. If all we have is the diatonic Pythagorean scale, we have only two less-than-ideal work-arounds:

- Retune all accompanying instruments to a new reference frequency *R*.
- Transpose to a different key within the Pythagorean scale.

Retuning instruments is at least nontrivial, and for some instruments impossible, and is to be avoided. So the only realistic alternative is transposition.

To achieve a transposable tuning system, one might naively think that all we must do is extend the Pythagorean scale to 12 degrees using the method of adding and subtracting intervals. Then one could transpose music to any chromatic degree as we do with the modern equal-tempered scale. Let us test this idea by constructing the dodecaphonic Pythagorean scale.

3.9.1 Pythagorean Dodecaphonic Scale

All the intervals in the Pythagorean dodecaphonic scale can be generated from the interval of the fifth (3/2) raised to integer powers.

1. Beginning with $(3/2)^0 = 1$, labeled C, ascend and descend by six fifths in both directions. The spelling of the scale degrees (whether they are sharp, flat, or natural) is determined by the direction of interval movement. Since we start at C, we move up a fifth to G, and so forth. Eventually the interval of a fifth above B is F♯. Similarly, going down by fifths from C, the fifths below F are B♭, A♭, and so forth. Note that at the extremes we have a low G♭ at 64/729 and a high F♯ at 729/64.

Powers:	$\left(\frac{3}{2}\right)^{-6}$	$\left(\frac{3}{2}\right)^{-5}$	$\left(\frac{3}{2}\right)^{-4}$	$\left(\frac{3}{2}\right)^{-3}$	$\left(\frac{3}{2}\right)^{-2}$	$\left(\frac{3}{2}\right)^{-1}$	$\left(\frac{3}{2}\right)^{0}$	$\left(\frac{3}{2}\right)^{1}$	$\left(\frac{3}{2}\right)^{2}$	$\left(\frac{3}{2}\right)^{3}$	$\left(\frac{3}{2}\right)^{4}$	$\left(\frac{3}{2}\right)^{5}$	$\left(\frac{3}{2}\right)^{6}$
Ratios:	$\frac{64}{729}$	$\frac{32}{243}$	$\frac{16}{81}$	$\frac{8}{27}$	$\frac{4}{9}$	$\frac{2}{3}$	$\frac{1}{1}$	$\frac{3}{2}$	$\frac{9}{4}$	$\frac{27}{8}$	$\frac{81}{16}$	$\frac{243}{32}$	$\frac{729}{64}$
Degrees:	G♭	D♭	A♭	E♭	B♭	F	C	G	D	A	E	B	F♯

2. Add or subtract octaves from these intervals until they lie within the compass of one octave (remembering that adding intervals is multiplying their ratios).

$$\frac{64}{729}\frac{2^4}{1} \quad \frac{32}{243}\frac{2^3}{1} \quad \frac{16}{81}\frac{2^3}{1} \quad \frac{8}{27}\frac{2^2}{1} \quad \frac{4}{9}\frac{2^2}{1} \quad \frac{2}{3}\frac{2}{1} \quad \frac{1}{1} \quad \frac{3}{2} \quad \frac{9}{4}\frac{1}{2} \quad \frac{27}{8}\frac{1}{2} \quad \frac{81}{16}\frac{1}{2^2} \quad \frac{243}{32}\frac{1}{2^2} \quad \frac{729}{64}\frac{1}{2^3}$$

$$\frac{1024}{729} \quad \frac{256}{243} \quad \frac{128}{81} \quad \frac{32}{27} \quad \frac{16}{9} \quad \frac{4}{3} \quad \frac{1}{1} \quad \frac{3}{2} \quad \frac{9}{8} \quad \frac{27}{16} \quad \frac{81}{64} \quad \frac{243}{128} \quad \frac{729}{512}$$

3. Arrange the intervals in ascending order of magnitude, and add the unison and octave.

Observe in figure 3.7 that the dodecaphonic Pythagorean scale contains within it all the intervals of the just pentatonic scale and the Pythagorean diatonic scale. This shows that the interval of the fifth underlies all of these scales. This method of generating fifth-based just scales can be extended to any number of degrees. Interestingly, the magnitude of the ratio for F♯ (1.42) makes it sharper than G♭ (1.40).

Note that there are actually 13 degrees in this scale as constructed, because we have two kinds of tritone intervals that are slightly different (F♯ and G♭). In the equal-tempered scale, the augmented fourth F♯ and diminished fifth G♭ are equal (see section 2.5), but in the Pythagorean dodecaphonic scale they are not, and it is ambiguous which should serve as the tritone. On some historical keyboard instruments, the black key between F and G was actually split in two, with F♯ on one side and G♭ on the other, rather than throwing one of them out. More often, one or the other was simply

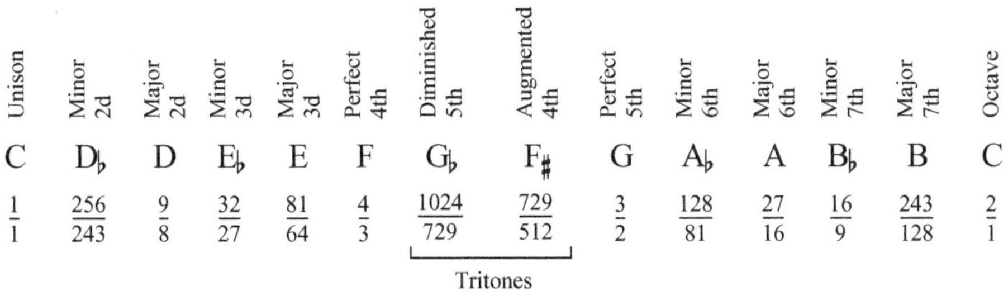

Unison
Minor 2d
Major 2d
Minor 3d
Major 3d
Perfect 4th
Diminished 5th
Augmented 4th
Perfect 5th
Minor 6th
Major 6th
Minor 7th
Major 7th
Octave
C D♭ D E♭ E F G♭ F♯ G A♭ A B♭ B C
1/1 256/243 9/8 32/27 81/64 4/3 1024/729 729/512 3/2 128/81 27/16 16/9 243/128 2/1
Tritones

Figure 3.7
Pythagorean chromatic scale.

Unison
Minor 2d
Major 2d
Minor 3d
Major 3d
Perfect 4th
Diminished 5th
Augmented 4th
Perfect 5th
Minor 6th
Major 6th
Minor 7th
Major 7th
Octave
C C♯ D E♭ E F G♭ F♯ G A♭ A B♭ B C
1/1 256/243 9/8 32/27 81/64 4/3 1024/729 729/512 3/2 128/81 27/16 16/9 243/128 2/1
256/243 2187/2048 256/243 2187/2048 256/243 256/243 531,441/524,288 256/243 256/243 2187/2048 256/243 2187/2048 256/243
2187/2048 2187/2048

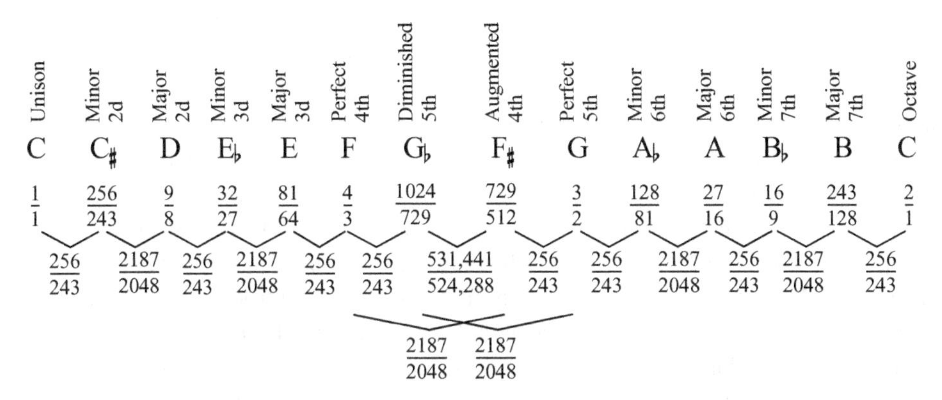

Figure 3.8
Chromatic Pythagorean scale with intervals.

left out. But if F♯ is left out, the fifth between G♭ and B is not a just 3/2 fifth. It is called a *wolf fifth* because the beating between the interval and the overtones makes it sound unpleasantly like wolves howling. And if G♭ is left out, some of the thirds and sixths are not harmonious either.

The tritone was called by medieval music theorists the *diabolus en musica,* "the devil in music," not just because of its dissonant sound but because of the ambiguity of its ratios and the enormous numeric sizes of those ratios.

4. For the final step, determine the interval sizes by subtracting the lower interval from its upper neighbor (remembering that subtracting intervals is dividing their ratios).

Notice in figure 3.8 that there are two semitone intervals, a smaller interval with ratio 256/243, called the *Pythagorean diatonic semitone,* or *limma,* and a larger interval with ratio 2187/2048, called the *Pythagorean chromatic semitone,* or *apotome*. The ratio of these two semitones is

$$\frac{2187}{2048} \div \frac{256}{243} = \frac{531{,}441}{524{,}288},$$

the *Pythagorean comma*. The difference between these two semitones is 23.46 cents, about a fifth of a tempered semitone. Notice that this is also the ratio between G♭ and F♯, the two tritones. Coincidently, this is also the amount by which the interval of 12 fifths differs from seven octaves as we will subsequently see.

The intervals from F to F♯ and from G♭ to G both use the 2187/2048 semitone.

Studying figure 3.8, we see that if we could only get rid of that pesky Pythagorean comma and somehow make G♭ = F♯, we would have a self-consistent circular scale system built out of just ratios. Then it would be possible to transpose to any key and remain in tune. This possibility underlies the entire motivation for the development of tempered tunings.

3.9.2 Impact of Polyphony on Just Scales

Besides bringing music into a more playable range, transposition has become a powerful organizing principle in music over time. Throughout the last eight centuries, Western composers have become increasingly enamored of *polyphony,* the art of sounding more than one melody line at the same time. In the process, they have sorted out which combinations of pitches sound good together and which don't, and figured out how to harmonize multiple musical lines and chords. Out of this arose *harmony theory,* which is the art of arranging multiple concurrent musical lines to reinforce a feeling of harmonic movement and arrival, suspension and resolution. Most classical music, and virtually all popular music, still follows rules of harmony first set down centuries ago.

The effective key signature of a musical work can change through the introduction of accidentals not in the original key signature. This is musical *modulation*. For example, a melody started in the key of C major might modulate to the key of G major by introducing F♯, and then eventually modulate back to C major by reintroducing F♮ (see section 2.5.5). Modulation became an important organizing principle for music in the Baroque and later eras. Over time, composers sought to modulate to remote keys with more sharps and flats. But the irregular interval sizes of the dodecaphonic Pythagorean scale limited music from being freely transposable to arbitrary keys because playing music in some keys sounded better than in others. As modulation became increasingly important to composers, the need for freely transposable tuning systems became urgent. Theorists began searching for solutions to the problems of the Pythagorean scale.

3.9.3 Natural Chromatic Scale

It has been well known to music theorists from antiquity that if left to their own devices, singers (and other performers, if their instruments would allow) eschew the Pythagorean thirds and sixths where possible and prefer intervals that align with the harmonic series to improve the sonority of the performance. As early as the second century, the Greek scientist, mathematician, and geographer Claudius Ptolemy proposed a just intonation system that would reflect what musicians actually played.[6]

Following Ptolemy's lead, let's find out just how far from the 5/4 major third the Pythagorean major third actually is. The answer is

$$\frac{81}{64} \div \frac{5}{4} = \frac{81}{80},$$

the Syntonic comma.[7] Then how far is the Pythagorean minor third from the 6/5 minor third? The answer is

$$\frac{6}{5} \div \frac{32}{27} = \frac{81}{80},$$

again the Syntonic comma. In fact, the out-of-tune Pythagorean major and minor sixths as well as the too-sharp major seventh and too-flat minor second are all *exactly* one Syntonic comma away from ratios of much smaller integers that are in the overtone series and have a more agreeable sound.

What if we subtracted a Syntonic comma from all the Pythagorean intervals that are too sharp and added it to the ones that are too flat? This would rectify all the intonational difficulties of the Pythagorean scale in one fell swoop. Mathematically, we'd substitute ratios of much smaller integers, and musically we'd align the scale degrees with the harmonic series. Ptolemy called this the *Syntonic diatonic scale* (table 3.3). The Pythagorean diatonic scale and the interval differences between the two scales are shown in the table.

Ptolemy's practical concern in designing this scale was to make the intervals agree with musical practice. But he also noted approvingly that the ratios of the scale are all superparticular ratios (see section 3.7). Ptolemy combined the best of both worlds: a practical scale that also contains more superparticular ratios than does the Pythagorean scale (Berkert 1972).

The chromatic version of this scale is shown in table 3.4, together with the dodecaphonic Pythagorean scale. The third row shows the interval differences between them. I call this the

Table 3.3
Ptolemy's Syntonic Diatonic Scale

	C	D	E	F	G	A	B	(C)
Syntonic diatonic	$\frac{1}{1}$	$\frac{9}{8}$	$\frac{5}{4}$	$\frac{4}{3}$	$\frac{3}{2}$	$\frac{5}{3}$	$\frac{15}{8}$	$\frac{2}{1}$
Pythagorean diatonic	$\frac{1}{1}$	$\frac{9}{8}$	$\frac{81}{64}$	$\frac{4}{3}$	$\frac{3}{2}$	$\frac{27}{16}$	$\frac{243}{128}$	$\frac{2}{1}$
Difference	$\frac{1}{1}$	$\frac{1}{1}$	$\frac{80}{81}$	$\frac{1}{1}$	$\frac{1}{1}$	$\frac{80}{81}$	$\frac{80}{81}$	$\frac{1}{1}$

Table 3.4
The Natural Chromatic Scale

	1	2	3	4	5	6	7	8	9	10	11	12	(13)
Semitone	C	C♯	D	E♭	E	F	F♯	G	A♭	A	B♭	B	(C)
Natural chromatic	$\frac{1}{1}$	$\frac{16}{15}$	$\frac{9}{8}$	$\frac{6}{5}$	$\frac{5}{4}$	$\frac{4}{3}$	$\frac{64}{45}$	$\frac{3}{2}$	$\frac{8}{5}$	$\frac{5}{3}$	$\frac{16}{9}$	$\frac{15}{8}$	$\frac{2}{1}$
Pythagorean dodecaphonic	$\frac{1}{1}$	$\frac{256}{243}$	$\frac{9}{8}$	$\frac{32}{27}$	$\frac{81}{64}$	$\frac{4}{3}$	$\frac{729}{512}$	$\frac{3}{2}$	$\frac{128}{81}$	$\frac{27}{16}$	$\frac{16}{9}$	$\frac{243}{128}$	$\frac{2}{1}$
Difference	$\frac{1}{1}$	$\frac{81}{80}$	$\frac{1}{1}$	$\frac{81}{80}$	$\frac{80}{81}$	$\frac{1}{1}$	$\frac{32{,}768}{32{,}805}$	$\frac{1}{1}$	$\frac{81}{80}$	$\frac{80}{81}$	$\frac{1}{1}$	$\frac{80}{81}$	$\frac{1}{1}$

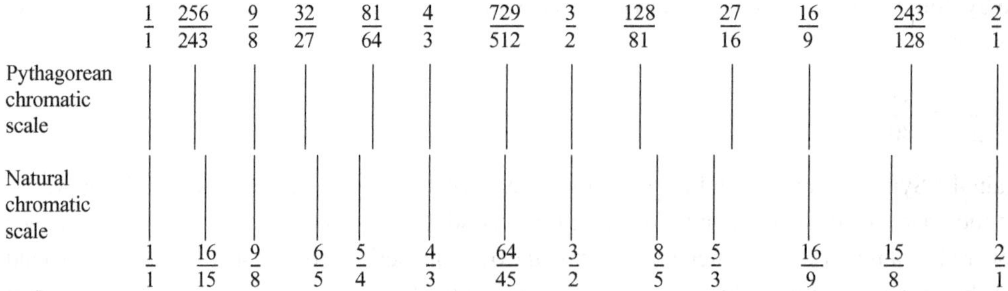

Pythagorean chromatic scale	$\frac{1}{1}$	$\frac{256}{243}$	$\frac{9}{8}$	$\frac{32}{27}$	$\frac{81}{64}$	$\frac{4}{3}$	$\frac{729}{512}$	$\frac{3}{2}$	$\frac{128}{81}$	$\frac{27}{16}$	$\frac{16}{9}$	$\frac{243}{128}$	$\frac{2}{1}$
Natural chromatic scale	$\frac{1}{1}$	$\frac{16}{15}$	$\frac{9}{8}$	$\frac{6}{5}$	$\frac{5}{4}$	$\frac{4}{3}$	$\frac{64}{45}$	$\frac{3}{2}$	$\frac{8}{5}$	$\frac{5}{3}$	$\frac{16}{9}$	$\frac{15}{8}$	$\frac{2}{1}$

Figure 3.9
Pythagorean chromatic and natural chromatic scales compared.

natural chromatic scale. It was championed by Bartolomé Ramos (1482). Figure 3.9 provides a visualization of the differences.

For various religious and political reasons, Ptolemy's proposal was ignored and even suppressed during the next dozen centuries or so. Pope John XXII even issued a papal bull in 1324 that banished from the church music using such lascivious intervals (see appendix A).

The natural chromatic scale sounds very consonant. But ultimately it fares no better than the Pythagorean scale for modulation and transposition. Consider the fifth from D to A, which is

$$\frac{5}{3} \div \frac{9}{8} = \frac{40}{27},$$

about 21.5 cents flat of the 3/2 perfect fifth. A triad built on D certainly sets the wolf tones howling.

3.10 Consonance of Intervals

I've said that the intervals signify such qualities as identity, equality, and individuality (see section 2.3.3). Another important way we characterize the intervals is by how pleasing or disagreeable their sound is to us. While some intervals are harmonious, others, such as the wolf fifth, set our teeth on edge. Table 3.5 shows the just intervals ordered from most to least pleasant, based on the conventions of Western music theory. The musical term for "pleasant" is *consonant,* which comes from Latin *consonare,* "sounding well together." The intervals toward the top of table 3.5 are consonant; the intervals toward the bottom are *dissonant.*

3.10.1 Foundations of Consonance

What is the basis for the effect of consonance or dissonance? Is it something inherent in the intervals, or is it in our perception? If we believe consonance is in the intervals, we should examine their mathematical properties. If we believe that consonance is in our perception, we should examine how we hear the intervals. I take up the latter approach in chapter 6. Here let's pursue two questions: Is there a mathematical basis for the ordering of intervals from consonant to dissonant? Is

Table 3.5
Just Intervals Ordered by Decreasing Consonance

	Name	Ratio	Sum	Prime Factor	Limit
Perfect Intervals					
1	Unison	1/1	$1 + 1 = 2$	1	
2	Octave	2/1	$2 + 1 = 3$	2	3-limit
3	Fifth	3/2	$3 + 2 = 5$	$3/2$	
4	Fourth	4/3	$4 + 3 = 7$	$2^2/3$	
Imperfect Intervals					
5	Major sixth	5/3	$5 + 3 = 8$	$5/3$	
6	Major third	5/4	$5 + 4 = 9$	$5/2^2$	
7	Minor third	6/5	$6 + 5 = 11$	$(2\cdot3)/5$	
8	Minor sixth	8/5	$8 + 5 = 13$	$2^3/5$	
Dissonant Intervals					
9	Major second	9/8	$9 + 8 = 17$	$3^2/2^3$	5-limit
10	Major seventh	15/8	$15 + 8 = 23$	$(3\cdot5)/2^3$	
11	Minor seventh	16/9	$16 + 9 = 25$	$2^4/3^2$	
12	Minor second	16/15	$16 + 15 = 31$	$2^4/(3\cdot5)$	
13	Tritone	64/45	$64 + 45 = 109$	$2^6/(3^2\cdot5)$	

there a mathematical basis for the categorization of the intervals into perfect, imperfect, and dissonant?

A successful metric of consonance must

- Decrease monotonically in proportion to increasing dissonance[8]
- Self-evidently partition intervals into the relevant categories, such as perfect, imperfect, and dissonant

Can we discover or invent an analysis of the traditional interval order (table 3.5) that explains the order and classification numerically?

Concurrence Giovanni Battista Benedetti (1530–1590) is perhaps the first to relate pitch and consonance to frequencies of vibration. In two letters he wrote around 1563 to composer Cipriano de Rore, he related interval consonance to the frequency of wave coincidence between two tones. He observed that an interval consists of a shorter wavelength (higher pitch) and a longer wavelength (lower pitch), and argued that the wavelengths of more consonant intervals coincide more often than do those of more dissonant intervals.

Let's call the time required for the waveforms of an interval to coincide its *precession time*. For example, if one bicycle wheel requires two seconds to turn once around and another requires three seconds, their frequencies form the interval of a fifth, 3/2, and the wheels *precess* against each other (that is, the faster one overtakes the slower one) every $2 \cdot 3 = 6$ seconds (figure 3.10). Benedetti's hypothesis is that consonance decreases as precession time increases. When the

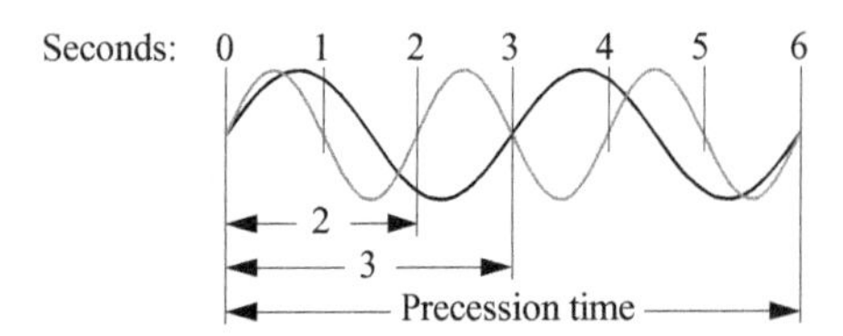

Figure 3.10
Precession of 2 against 3.

intervals are ordered by this criterion, their sequence from consonant to dissonant is 2:1, 3:2, 4:3, 5:3, 5:4, 6:5, 7:5, 8:5, and so on (figure 3.11). Note that these ratios are not strictly superparticular and that by Benedetti's metric, the unused interval 7:5 is more consonant than the major sixth.

Benedetti's theory challenged two ancient dogmas. First, his theory suggested that consonance and dissonance are relative, not categorical, terms. Second, his theory implied that superparticular ratios were not somehow tonally superior to other ratios.

Benedetti's ideas were later developed by Isaac Beeckman (1588–1637) and by Marin Mersenne (1588–1648) in *Harmonie Universelle* (1635). Benedetti's approach shows an orderly progression from consonance to dissonance, so it passes our first criterion for consonance. But it does not suggest a way to partition the intervals into perfect, imperfect, and dissonant; indeed, it predicts that there is no such criterion.

Additive Dissonance Metric The Sum column in table 3.5 shows the sums of the numerator and denominator of the ratio of each interval appearing in the Ratio column. For instance, the ratio of the fifth is 3/2, and $3 + 2 = 5$.

This additive dissonance metric is monotonically related to dissonance. Figure 3.12 plots the interval number ordered by dissonance (in the order given in the first column in table 3.5) from unison to minor second on the x-axis against the sum of each numerator and denominator on the y-axis. The curve takes a significant jump upward from the minor second (31) to the tritone (109), so I indicated the tritone to the side rather than plotting it. The fitted curve in the background is just an aid to help join the points.[9]

Because this additive dissonance metric increases monotonically with increasing dissonance, it meets the first criterion for a dissonance metric. However, because the curve is gradual (until it gets to the tritone), it does not suggest how to partition the intervals into perfect, imperfect, and dissonant, so it fails the second criterion.

Partitioning Dissonance Metric Any whole number greater than 1 can be factored into a product of primes raised to powers, for example, $8 = 2^3$, $47 = 47^1$, $48 = 2^4 \cdot 3^1$, $49 = 7^2$.

Prime numbers are whole numbers greater than 1 that are not divisible by any other number besides themselves and 1. (By convention, 1 itself is not considered to be prime.) For example, 2, 3, 5, and 7 are primes, but 4, 6, 8, and 9 are not because at least one prime divides them evenly. Similarly, 47 is prime, but 48 and 49 are not.

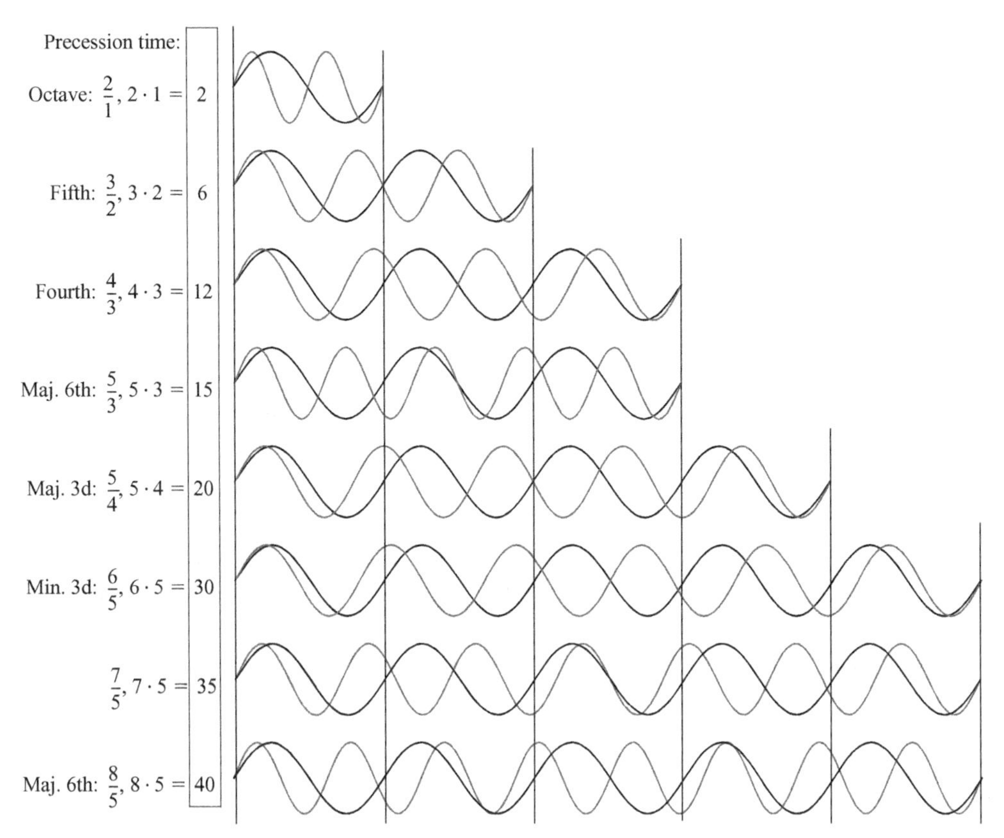

Figure 3.11
Precession time for various intervals.

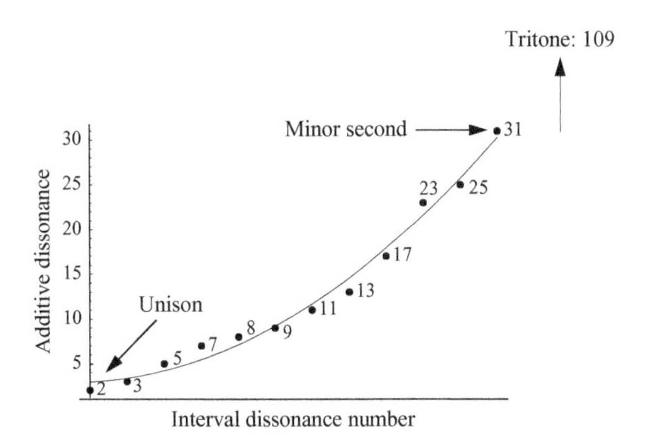

Figure 3.12
Additive dissonance metric.

The Prime Factor column in table 3.5 shows each interval as a ratio of the products of prime numbers raised to powers. For example, the major second ratio is 9/8, therefore the prime factor of the major second is $3^2/2^3$. Notice that the more dissonant intervals tend to involve larger primes and higher powers. The perfect intervals involve only the small prime numbers 2 and 3. The more dissonant intervals involve 5 as well (with the exception of the major second and minor seventh). None of the just intervals shown in table 3.5 use 7 or the higher primes.

In spite of its limitations, there seems to be some historical justification for this metric. The perfect intervals—those built from primes 2 and 3 only—were the first ones favored by early scale builders. Ratios of prime factor 5 began appearing around 400 B.C.E. The exclusion of primes higher than 5 to build musical ratios is called the *five-limit* by the composer Harry Partch in his book *Genesis of a Music* (1947). The five-limit has only been transcended in recent centuries. Partch used an eleven-limit system of ratios in the construction of his scales. These days, if a scale is said to be *n*-limit, this means that the highest prime factor of any interval in the scale is *n*.

Attempts to order and classify consonance using strictly numeric rules are fine as far as they go. But while we generally agree as to the consonance of the perfect intervals, opinions vary widely as to the relative consonance or dissonance of the others, and no one metric seems to sum it all up.

Consonance appears to be influenced, but not determined, by underlying psychophysical principles we all share. It seems as well to be a matter of taste decided differently by each musical culture and each age. The harmonies in the chorales of J. S. Bach, for example, do not strike the modern ear as particularly dissonant; however, listeners of his age sometimes found them shocking. A similar progression has occurred with the music of Mozart, Beethoven, Wagner, Mahler, Debussy, Stravinsky, Schoenberg, among others. So where intervals are concerned, it seems that familiarity breeds consonance.

Its highly contextual nature suggests that attempts to classify consonance without regard to the fundamentals of auditory perception are doomed. So let's defer further judgment until chapter 6.

3.10.2 Natural Major Scale

Ptolemy's idea of a natural musical scale, first revived by Ramos, were rediscovered again in the early Renaissance and championed by medieval theorists, including Lodovico Fogliano in *Musica Theoretica* (1529). Around that time, the famous Renaissance music theoretician Gioseffo Zarlino (1517–1590), in *Institutioni Armoniche* (1558), used the same basic ideas to create a scale based on the ratios 4:5:6, which form a just major triad. If we take 4/4 as the root of the triad, the major third above is 5/4, and the fifth above is 6/4. This triad incorporates the major third (5/4), minor third (6/5), and perfect fifth (6/4 = 3/2). While the Pythagorean scale was built from the integers 1 to 4, this scale uses integers 1 to 6. Zarlino called this set the *numero senario* and, like the Pythagoreans, found a mystical significance in it and sought to establish it as the proper foundation of harmony.

There are three major triads in the just diatonic scale: C:E:G, F:A:C, and G:B:D (figure 3.13). In Zarlino's scale, the frequencies of these three triads are perfectly in agreement with the harmonic overtone series. Notice the presence of the prime number 5 in the 4:5:6 ratio, making this a five-limit scale.

$$\begin{bmatrix} C_1 & D_1 & E & F & G & A & B & C_2 & D_2 \\ 4 & : & 5 & : & 6 & & & & \\ & & & 4 & : & 5 & : & 6 & \\ & & & & 4 & : & 5 & : & 6 \end{bmatrix}$$

Figure 3.13
Natural major scale.

To construct the frequencies for the natural major scale, we create new pitches out of ones we've established already:

1. Find E from C:

$$\frac{E}{C} = \frac{5}{4}, \quad \text{or} \quad E = \frac{5}{4}C.$$

2. Find G from C:

$$\frac{G}{C} = \frac{6}{4}, \quad \text{or} \quad G = \frac{6}{4}C = \frac{3}{2}C.$$

3. Find F from C_2:

$$\frac{C_2}{F} = \frac{6}{4}, \quad \text{or} \quad F = \frac{4}{6}C_2 = \frac{4}{6} \cdot \frac{2}{1}C = \frac{4}{3}C.$$

4. Find A from F:

$$\frac{A}{F} = \frac{5}{4}, \quad \text{or} \quad A = \frac{5}{4}F = \frac{5}{4} \cdot \frac{4}{3} \cdot C = \frac{5}{3}C.$$

5. Find B from G:

$$\frac{B}{G} = \frac{5}{4}, \quad \text{or} \quad B = \frac{5}{4}G = \frac{5}{4} \cdot \frac{3}{2} \cdot C = \frac{15}{8}C.$$

6. Find D_2 from B:

$$\frac{D_2}{B} = \frac{6}{5}, \quad \text{or} \quad D_2 = \frac{6}{5}B = \frac{6}{5} \cdot \frac{15}{8} \cdot C = \frac{90}{40}C = \frac{9}{4}C.$$

7. Find D:

$$\frac{D_2}{D} = \frac{2}{1}, \quad \text{or} \quad D = \frac{9}{8}C.$$

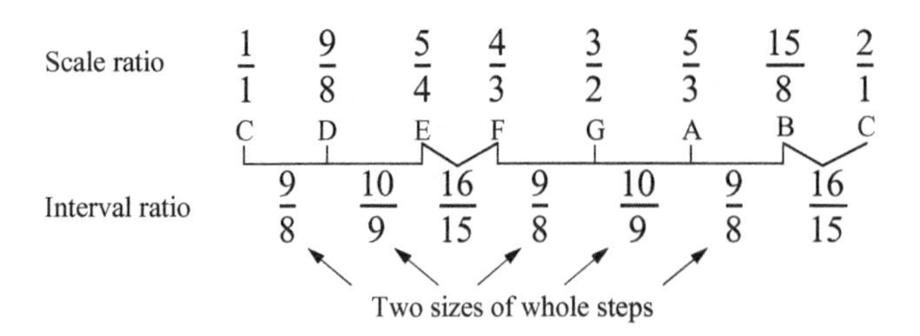

Figure 3.14
Natural major scale with interval sizes.

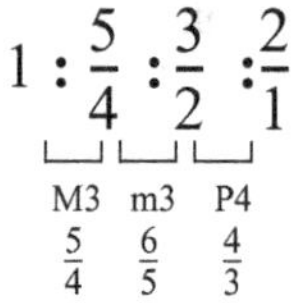

$$1 : \frac{5}{4} : \frac{3}{2} : \frac{2}{1}$$

M3 m3 P4

$\frac{5}{4}$ $\frac{6}{5}$ $\frac{4}{3}$

Figure 3.15
Major triad.

Figure 3.14 shows the natural major scale with intervals between the pitches in the bottom row.

Although the natural major scale succeeds at making the thirds consonant with the harmonic series, it does so at the expense of the whole steps, which now are uneven in size. Some whole steps are 9/8, but others are 10/9. Whereas in the Pythagorean scale the major thirds were "too big," here some of the whole steps are "too small."

3.10.3 Natural Minor Scale

As we saw with the natural major scale, the ratios of the major triad are the ratios 4 : 5 : 6. The major triad consists of a reference frequency R plus a major third up, $R \cdot (5/4)$, plus another minor third up,

$$R \cdot \frac{5}{4} \cdot \frac{6}{5} = \frac{30}{20} R = \frac{3}{2} R \,.$$

Figure 3.15 shows the pitch ratios of a major triad plus the octave. Notice that the order of the intervals is

$$\frac{5}{4} : \frac{6}{5} : \frac{4}{3} ,$$

that is, a major third, a minor third, and a perfect fourth.

$$\begin{bmatrix} C_1 & D_1 & E\flat & F & G & A\flat & B\flat & C_2 & D_2 \\ 10 & : & 12 & : & 15 & & & & \\ & & & 10 & : & 12 & : & 15 & \\ & & & & 10 & : & 12 & : & 15 \end{bmatrix}$$

Figure 3.16
Just minor scale.

We could create a just minor scale if we could reverse the order of the 5/4 and the 6/5 intervals, creating a triad in the order minor third, major third, perfect fourth. Then we'd have something like this:

1 : ? : ? : 2

m3 M3 P4.

$\frac{6}{5}$ $\frac{5}{4}$ $\frac{4}{3}$

But what are the ratios of the pitches in this case? We're looking for something like the integer ratio 4:5:6 but that produces a minor triad. Suppose we just stack up what we want the order to be, like this:

$$1 : \frac{6}{5} : \left(\frac{6}{5} \cdot \frac{5}{4} = \frac{3}{2}\right) : \left(\frac{3}{2} \cdot \frac{4}{3} = \frac{2}{1}\right).$$

This produces the right sequence of minor third, major third, and perfect fourth, but the ratios don't come out as whole numbers:

$$1 : \frac{6}{5} : \frac{3}{2} : \frac{2}{1},$$

expressed as decimal fractions is 1:1.2:1.5:2.

Since this is not a ratio of integers, it can't be the basis of a proper just scale. But we could salvage this and make it into a ratio of integers just by multiplying all ratios by 10, like this: 10:12:15:20. With this ratio, we can properly form the just minor scale (figure 3.16).

3.10.4 Mean-Tone Tempered Scale

Another transitional attempt to create a transposable scale based on simple integer ratios was the *mean-tone tempered scale*. It is a fascinating exercise in music engineering.

Temperament represented a radical departure from the just scales of the past. I've already used the term to refer to the equal-tempered scale. In this context, tempering means the practice of adjusting some of the degrees of the scale to "irrational" values so as to fit within an overarching

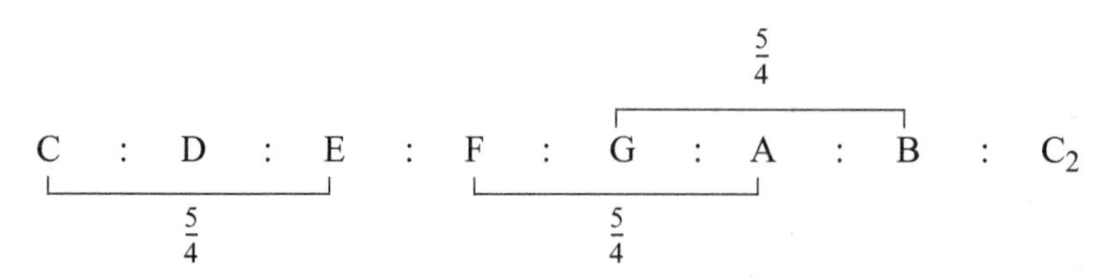

Figure 3.17
Constructing the mean-tone scale, step 1.

order that is still based on simple integer ratios. The meaning of temperament is the same for the equal-tempered scale, but there the application is to all pitches of the scale uniformly.

But what does "irrational" mean? A rational number is a number that can be represented as a *ratio* of two integers. The value of π is irrational because there is no ratio of integers that can precisely represent it. Another example of an irrational number is $\sqrt{2}$.

Constructing a Mean-Tone Tempered Scale The mean-tone tempered scale starts with the same three natural major thirds that were used for the natural major scale. Five whole tones and two semitones are derived from the thirds. The goal is to use only perfect 5/4 major thirds so as to preserve consonance across transposition and modulation. The intended improvement over the natural major scale is to do something about those pesky uneven whole steps by bending, or tempering, them to fit.

We can develop the mean-tone tempered scale in the following way:

1. As with the natural major scale, we want to have three pure 5/4 major thirds between C:E, F:A, and G:B (figure 3.17). We still need to nail down the relation between D and its neighbors C and E, and we must do the same for G and its neighbors F and A.

2. We tackle the major seconds between C:D:E, F:G:A, and G:A:B. Here's where the tempering comes in. What if we simply cut the interval of the pure 5/4 major third in half to create two whole steps, that is, if we took the *mean value* of a pure major third? (This is where the scale gets its name.) What is its mean value? It wouldn't be 5/8, the arithmetic mean, because pitch is exponential in frequency. To add intervals we must multiply their ratios, and we are looking for one ratio that when multiplied by itself (that's the clue) adds up to a 5/4 major third. Such a ratio would be a uniform division of the major third. What we are looking for is $\sqrt{5/4}$, the *geometric mean*. This allows us to fill in the major seconds (figure 3.18).

3. We must figure out the interval size of the two minor seconds, E:F and B:C. Until we define them, we have two disconnected islands of tonality, C:D:E and F:G:A:B. We must create two equal-sized half steps that fill in the difference between the sum of the whole steps and the octave. Fortunately, the minor seconds yield to the same logic that created the major seconds.

There are two gaps in our scale that we want to fill with minor seconds. Let s be the (as yet undefined) size of a minor second. We need two such minor seconds, or s^2, because when we add intervals

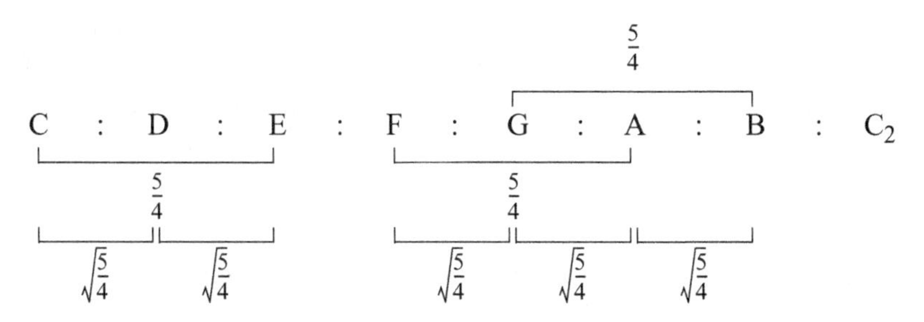

Figure 3.18
Constructing the mean-tone scale, step 2.

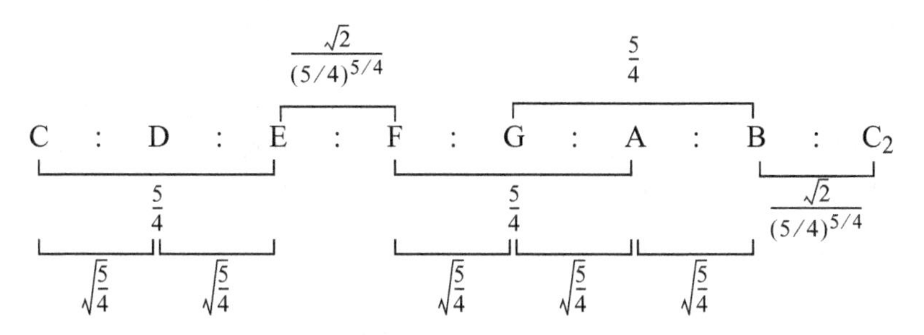

Figure 3.19
Mean-tone tempered scale.

we multiply their ratios. We observe that there are five whole steps of size $\sqrt{5/4}$. We want two semitones of size s plus five whole steps of size $\sqrt{5/4}$ to add up to an octave of size 2/1. An informal equation for this might read, 2 semitones + 5 whole steps = octave. That translates into the equation

$$s^2 \cdot \left(\sqrt{\frac{5}{4}}\right)^5 = \frac{2}{1}.$$

Now we solve this for s, as follows.

Take the square root of both sides:

$$s \cdot \left(\sqrt{\frac{5}{4}}\right)^{5/2} = \sqrt{2}.$$

Isolate s:

$$s = \sqrt{\frac{2}{(5/4)^{5/2}}} = \frac{\sqrt{2}}{(5/4)^{5/4}}. \tag{3.12}$$

The entire scale can now be constructed (figure 3.19).

So, after all this mathematical heavy lifting, what does this scale sound like? Was it worth the effort? Well, the improved uniformity does allow for greater transposition, but in the end (of course) we still have problems: the fifth is no longer a simple 3/2. The half steps and whole steps are not simple either, so we're really no closer to having a scale that can transpose and that also lines up with musical instrument harmonic overtones.

3.11 The Powers of the Fifth and the Octave Do Not Form a Closed System

If we step back to look at all these efforts over the centuries to build the perfect scale, it's as though we were trying to build a bridge but couldn't ever find a design that was sufficiently proportional. There's always a piece that doesn't fit. My impression of the mean-tone scale is that it's like a carpentry project gone awry: the main boards are cut right, but the carpenter had to bend the rest into place and forcefully nail them down or they would spring loose again.

The problem is, simple integer ratios don't line up the way we'd like. For instance, as we transpose around the circle of fifths, we logically expect to come back to our starting key. That is, starting on C, if we go up by fifths, we expect to return to C in a higher octave:

C, G, D, A, E, B, F♯, C♯, A♭, E♭, B♭, F, C.

But if we use the simple 3/2 ratio to go up by fifths, and use the 2/1 to go up by octaves, *the two series don't end up on the same frequency for C at the top*. As we go through the 12 keys, we're adding fifths, which means we multiply their ratios. Twelve fifths would be $(3/2)^{12} = 129.746$, which is just a little over seven octaves. But seven octaves exactly would be $(2/1)^7 = 128$. So they don't line up. Stated another way,

$$\frac{(3/2)^{12}}{(2/1)^7} \neq 1.$$

In fact, it can be proven that there are no integers m and n such that

$$\left(\frac{3}{2}\right)^m = \left(\frac{2}{1}\right)^n, \tag{3.13}$$

apart from the trivial solution $m = n = 0$. Contrary to the wishes of scale builders and musicians from antiquity to the present, the powers of the integer ratios 3/2 and 2/1 do not form a closed system.

If there is no exact solution to (3.13), then what about approximate solutions? How close to equal can we get for any possible combination of m and n? The optimal solution appears to be $m = 12$, $n = 7$. The interval corresponding to this choice of m and n is

$$\frac{(3/2)^{12}}{(2/1)^7} = 1.01364 = 23.46 \text{ cents} \tag{3.14}$$

Recall from section 3.9.1 that the ratio in (3.14) is known from antiquity as the Pythagorean comma. While the distance by which the interval of 12 fifths misses seven octaves is a mere 23.46 cents, in this case, a miss is as good as a mile.

It seems that simple integer ratios raised to arbitrary powers don't necessarily form a closed system and that the particular case of interest, $(3/2)^m = (2/1)^n$, has no solution. The significance of this is that making a closed cyclic scale system based on multiples of fifths and octaves *can't be done* with simple integer ratios. A closed scale system is required in order to allow music to be transposed to any key and still sound in tune, so a transposable scale based on small integer ratios is impossible, and a tempered scale must be used if transposing is really that important.

The less the intervals of a scale are tempered the better, because then the tempered intervals will sound less dissonant against the harmonic overtone series. The Pythagorean comma suggests to the tempered scale developer where best to close the cycle of fifths and octaves. If 12 fifths are flatted to equal seven octaves, the overall distortion in the fifths will be only 23.46 cents. This is the rationale for building the equal-tempered scale with 12 semitones.

Is there any other combination of m and n that comes closer to unity than the Pythagorean comma? Suppose we evaluate

$$\frac{(3/2)^m}{(2/1)^n}$$

for values of m and n over some range, say, 0 to 100 each, looking for scale systems that come as close or closer to unity than does the scale system for $m = 12$, $n = 7$. Some candidate entries are

m	n	Cents	
12	7	23.46	Pythagorean comma
41	24	−19.85	All fifths would have to be stretched
53	31	3.62	Very close to unity
94	55	−16.23	All fifths would have to be stretched

A positive cents value indicates that the fifths are sharp by that amount, and a negative value indicates they are flat. Perhaps the most interesting result is that 53 fifths are only 3.62 cents sharp of 31 octaves. Both 31 and 53 have been used to build scales.

3.12 Designing Useful Scales Requires Compromise

Given the limitations of the just tuning systems, we find ourselves at a fork in the road:

- We can move toward our original goal of transposing while retaining the just ratios—but with compromises.
- We can abandon the goal and choose another.

Although we still want to engineer a scale that meets our needs, now we know it's just a design problem, not a quest for a holy grail that we now know doesn't exist.

Some common choices that have been made at this juncture include the following:

- Extend the use of tempering (see section 3.13).
- Add more degrees to the just scales, allowing musicians to use alternative ratios when transposing (see section 3.14).
- Avoid transposing and modulation (see Hindustani Scales in section 3.14.2).

3.13 Tempered Tuning Systems

Tempering is a compromise that abandons some aims in order to achieve others. If we give up the goal of just ratios, we'd still like to have a scale that

- Is transposable to all 24 major and minor keys
- Sounds close enough to the just diatonic scale
- Has intervals reasonably close to their small-integer ratio prototypes
- Has 12 half steps to the octave
- Can be transposed around the circle of fifths
- Has no strange differences between supposedly same-sized intervals

To implement this compromise, we use tempering to close the cycle of fifths and octaves. What if we spread the Pythagorean comma across a number of intervals so that it would become unnoticeable?

3.13.1 Origins of Tempering

The concept of a tempered scale arose in the fourth century B.C.E. with Aristoxenus of Tarentum, one of Aristotle's students. Aristoxenus argued empirically that precise ratios should be less important to music theory than what musicians actually use, and suggested that the octave be divided on a subjective basis into an equal number of intervals. To the same effect, the great mathematician Leonhard Euler (1766) wrote, "The sense of hearing is accustomed to identify with a single ratio, all the ratios which are only slightly different from it, so that the difference between them be almost imperceptible." What Euler is referring to is now called the just noticeable difference (JND) of pitch (see chapter 6). Another perspective on Euler's insight is the power of our minds through conditioning and learning to generalize a rule across similar instances (see section 9.22).

Perhaps the first practical tempering system was proposed by Vincenzo Galilei, father of Galileo and a one-time student of Zarlino. Like many, including the Pythagoreans, Zarlino believed that certain proportions had a mystical significance that revealed the hand of God. Vincenzo Galilei, true to his Renaissance culture, believed that all scales were free creations of the human mind and hence could be anything that pleased their creators (V. Galilei 1581; Strunk 1998). He proposed solving the conundrum of intonation by using the integer ratio 18/17 as an approximation of the semitone. At 98.96 cents, this ratio provides a usable tempered tuning system that has been employed by fretted instrument makers ever since (see section 3.15).

Between the Renaissance and the modern age, Western music theorists tried many ways to hide the Pythagorean comma and yet salvage as many just intervals as possible—usually the fifths and major thirds—while excluding the wolf tones by the use of tempering. There are an unlimited number of possible temperings, but the available solutions tend to cluster around a few common aims, depending upon what one wants to optimize:

- *Mean-tone* Optimize the thirds and fifths in selected keys, and never mind the rest.
- *Well-tempered* Make all keys usable, but make some more purely intoned than others.
- *Equal-tempered* Make all keys sound the same.

3.13.2 Well Tempering

The term *well tempered* covers all tuning systems that temper at least some intervals or that have reasonably equal-sized semitones.

Andreas W. Werckmeister (1645–1706) developed a number of tempered tunings, including Werckmeister temperament III, which he developed in 1691. Roughly speaking, this scale leaves the black notes in Pythagorean just intonation and tempers the white notes, resulting in various-sized major and minor intervals and either true or nearly true fifths and fourths. Such *irregular tempering* essentially scatters bits of the Pythagorean comma widely, though not evenly, across the scale, allowing fairly graceful transposition and modulation to remote keys.

Other irregular temperaments of the time included

- Kirnberger temperament III (1779), by Johann Philip Kirnberger (1721–1783); some fifths are tempered, some are pure.
- Valotti temperament (1728), by Francesco Antonio Vallotti (1697–1780); the "front" six fifths of the circle of fifths (F, C, G, D, A, E, B) are tempered by 1/6 of a Pythagorean comma, whereas the fifths on the "back" side are tuned pure.[10]
- Young temperament II (1800), by Thomas Young (1773–1829); similar to Vallotti's but starting on C rather than F.

3.13.3 Tonal Palette

As a consequence of the uneven distribution of the Pythagorean comma in irregular temperaments, each key was imbued with a unique *tonal palette* or coloration based on the placement of the various-sized intervals in its scale. Far from being a problem, this aspect of irregular temperaments was appreciated by composers and performers of those times as lending character to the different keys. Modulating around the circle of fifths in irregular temperaments alters the tension in the triads and dominant seventh chords in characteristic ways that they found musically useful.

In the literature on tuning systems, the arguments for and against the various tuning systems sound as though they were referring to wine tasting. Werkmeister III is pure in the best keys and excellent for organs because many fourths and fifths are in tune, but it is irregular and quixotic in how it handles modulation, and uneven in key color. Vallotti is smooth and regular, perhaps with too little key contrast. It is clear that the choice of tuning system was a matter of taste.

A common misconception about J. S. Bach's famous *Das Wohltemperierte Klavier*[11] is that it was written as a demonstration piece for equal-tempered tuning. Bach almost certainly did not use equal temperament, which did not come into practical use until after he died. He undoubtedly used a mean-tone or irregular temperament of some sort, possibly one of Werckmeister's or one of his own devising. Which exact tuning he used is unknown, but it is certain that Bach used this composition as a vehicle to systematically explore the tonal palettes of the keys of the temperament he was using (Barbour 1947; Barnes 1979; Kellner 1979).

3.13.4 Equal Tempering

The attempt with irregular temperaments to include some pure ratios only hides the intonational problems in remote keys. But as composers developed and extended functional harmonization and modulation, eventually there were no "remote" keys left in which to hide the wolves. Why then not try tempering every degree of the scale in the same amount? Perhaps that would spread out the Pythagorean comma to the extent that it would become unnoticeable because the "out-of-tune-ness" would be everywhere the same.

What if we shrank the interval of a fifth just a little so that 12 of them would equal seven doublings of the starting pitch? Let's name the tempered fifth T_5. Then we would be looking for a value of T_5 such that $(T_5)^{12} = 2^7$. Solving for T_5 gives $T_5 = 2^{7/12} = 1.498$, which is pretty close to $3/2 = 1.5$ (although the fifths are a little flat). To generate the 12 steps of the scale, all we would have to do is form successive intervals of T_5, and after creating 12 of them, we would be back to where we started, a few octaves higher.

While the equal-tempered scale takes the approach of tempering the fifth according to $2^{7/12}$, another equally valid approach is to shrink the semitone according to $\sqrt[12]{2} = 1.0594631$, which is reasonably close to the minor second, $16/15 = 1.0666667$. The two approaches are equivalent, since the result either way is that the octave is divided into 12 equal intervals.

Curiously, this quintessentially Western scale appears to have been first invented in China. In 1596, Prince Chu Tsai-yu (or Zhu Zai-You) apparently calculated the degrees of the equal-tempered chromatic scale without benefit of logarithms (Barbour 1953; Kuttner 1975; Yasser 1932). However, it evidently did not catch on in China as it did in the West. The idea was apparently put forward first in Europe by Simon Stevin (1548–1620).[12] The theory became widely known through the work of Mersenne (1635). But equal temperament did not become generally established in practice until 1800, first in Germany, later in England and France.

3.13.5 Interval Error of Equal-Tempered Tuning

Astonishingly, the equal-tempered intervals are close enough to the natural major scale that most Western composers and musicians from the 1800s to the present have been satisfied with the

Table 3.6
Comparison of Natural and Equal-Tempered Chromatic Intervals

Degree	Name	Error	Degree	Name	Error
1	Unison	0.0	7	Tritone	–9.7763
2	Minor second	–11.731	8	Perfect fifth	–1.955
3	Major second	–3.910	9	Minor sixth	–13.686
4	Minor third	–15.641	10	Major sixth	15.641
5	Major third	13.686	11	Minor seventh	3.910
6	Perfect fourth	1.955	12	Major seventh	11.730

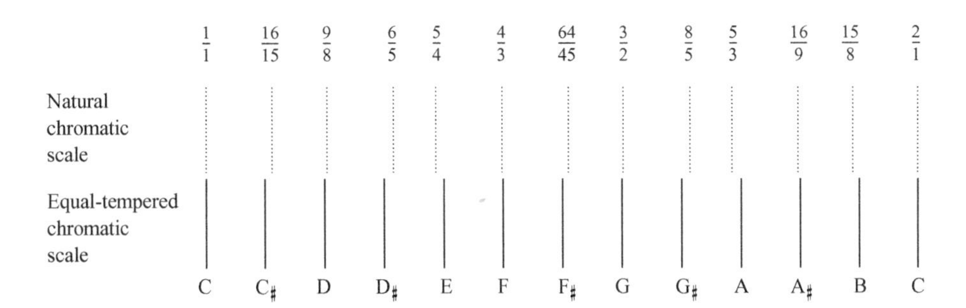

Figure 3.20
Natural and equal-tempered chromatic intervals.

equal-tempered chromatic scale, and a very large body of music has been composed using it. Ironically, however, there is not a single small integer ratio left in the scale (apart from the unison and octave). Thus, one of the principal aims of the early scale builders has been lost. Clearly, the desire for transposability won out over justness of intonation in Western music after the advent of tempered tunings.

But just how badly out of tune is equal temperament? Table 3.6 shows the size of the error in cents between each equal-tempered degree and its natural chromatic scale equivalent. The sign of each value in the Error column shows the cents by which the equal-tempered scale is sharp (positive) or flat (negative) with respect to its just equivalent. Note that the worst errors are for the minor and major thirds and sixths (figure 3.20).

3.13.6 Goodness-of-Fit Metric

We can get a crude quantitative idea of how closely aligned these two scales are by adding the magnitudes of the Error column in table 3.6. Doing so shows that the sum total by which all tempered intervals miss their natural chromatic scale equivalents is 103.624 cents. Is 103.624 cents accumulated error good or bad? Are these differences significant? That analysis is postponed until section 3.14, so that more scales can be evaluated.

Meanwhile, we can adapt this goodness-of-fit measure to other scales discussed in this chapter to show in quantitative terms how closely aligned they are with the intervals of the natural chromatic scale. For scales with many more degrees than the chromatic scale, the method is to first pick the degrees that are closest to their natural chromatic equivalents, then sum the magnitude of the errors.

3.13.7 The Grand Solution

The equal-tempered scale inherits nearly all the important components of the Pythagorean scale and can also transpose. Now every key sounds as in tune (or out of tune), as every other key, just as we wanted, but at the expense of the pure integer ratios, which have been virtually banished. It is somewhat reminiscent of the modern practice where an oak grove is ripped out to build a shopping center and then the shopping center is named Oak Grove. We are left with the impression of the pure intervals but not with their reality. We get the advantage of the modern conveniences (transposition) but at the expense of the reason we wanted it. Isn't it interesting that not even music is immune to the inevitable downside of technological advance? The moral: nothing is free.

Other cultures have made other choices. For instance, classical Hindustani and Arabic music is still firmly rooted in small integer ratio scales, and that music scintillates with a pleasurable harmonicity that has touched a deep longing in the Western ear, as evidenced by their popularity in the West in recent times. The symmetry between the overtones of their instruments and the scales they play upon is deeply satisfying. On the other hand, don't expect an oud or a sitar to transpose.

3.14 Microtonality

As described in the previous section, the compromise of tempered tunings is to give up the use of small integer ratios except for the unison and octave. The compromises of microtonality are not as neatly assessed because of the greater number of directions that can be taken.

One of the main thrusts of early Western microtonal tunings was to increase the number of scale degrees on keyboards. The original aim was to supply alternative choices of intervals when modulating or transposing so as to retain as much as possible the simple integer ratios of the just scales. Such a scale system would then contain *microtones,* which are scale degrees that are smaller than a semitone.

Once again, however, we confront basic design questions. For instance, are the microtones to be organized as a set of tempered intervals or as a collection of small integer ratios? Of course, there are exponents of both approaches, and I consider each in turn.

3.14.1 Tempered Microtonal Scales

What if we simply increased the number of equal divisions of the octave from 12 to a larger number? As the number of equal divisions of the octave goes up, not only will there be more scale degrees to choose from but there is also an increased likelihood that some of them will land closer to the just intervals than their chromatic tempered cousins do. A trivial modification of equation (3.3) allows us to create arbitrary tempered divisions of the octave:

$$f_{k,v} = f_R \cdot 2^{v+k/N}, \tag{3.15}$$

where N is the desired number of degrees per octave, k is the integer degree number, and v is the octave.

A few such tempered scale systems approximate the just intervals better than the chromatic scale.

19-Tone Scale Another relatively close encounter between the series of fifths and octaves occurs at 19 fifths above 11 octaves, where the fifths exceed the octaves by 137.145 cents. When N in (3.15) is set to 19, the size of the equitempered scale division is 63.16 cents.

Why 19? The 19-tone major and minor thirds and major and minor sixths are all closer than the corresponding equal-tempered intervals. The minor third is quite pure. The major third is flat, although closer than the equal-tempered major third (see figure 3.21).

To temper using this scale, the fifths must be flatted by a total of 137.145, which is worse than the tempering required for the chromatic scale. Since there are 19 fifths, each fifth is flat by 7.218 cents, making the fifths far from perfect.

Applying the goodness-of-fit metric to the 19-tone scale results in 109.31 cents accumulated error, not as good as the chromatic scale's 103.624 cents. In spite of the improved thirds and sixths, this scale has not been favored over chromatic equal temperament for good reasons.

Quarter-Tone Scale When N in (3.15) is set to 24, we arrive at the quarter-tone scale, and the size of the equitempered interval is 50 cents, or exactly one-half of a chromatic tempered semitone.

While all microtonal scales can produce exotic-sounding harmonies, the quarter-tone scale is special because it is a superset of the equal-tempered scale. Or, we can think of it as two equal-tempered scales combined, tuned 50 cents apart. A common arrangement for quarter-tone music is to tune two pianos 50 cents apart. Listen, for example, to *Three Quarter-Tone Pieces* by Charles Ives, or the compositions of Alois Hába (1893–1973).

Depending upon how the additional resources are used by a composer, quarter tones can extend the tonal palette of the equal-tempered scale so that it ranges from strictly harmonic (using either of the equal-tempered scale subsets) to mixtures that are reminiscent of the irregular temperaments, to highly dissonant when using all the quarter tones together. The composer is given additional possibilities of harmonic tension.

As one might expect, the goodness-of-fit metric for the quarter-tone scale is the same as for the equal-tempered scale, 103.624 cents.

53-Tone Scale The next close encounter of the fifths and the octaves occurs at 53 fifths and 31 octaves. Here the cycle of fifths ends up merely 3.615 cents above the octave. Each tempered fifth is therefore 3.615/53 = 0.068 cents flat. According to Helmholtz (1863), this scale was first proposed in 1608 by Nicolaus Mercator (1620–1687) as a system for measuring scales.

Even Partch (1947) is impressed with this scale. He says it gives "a degree of falsity that might really be called—and for the first time I use the word without quotation marks—inconsequential."

Table 3.7
Comparison of Natural Chromatic Scale and Cent Scale

Natural Chromatic	Cent	Error	Natural Chromatic	Cent	Error
1	1	0.0	7	611	–0.22
2	113	–0.27	8	703	–0.04
3	205	–0.09	9	815	–0.31
4	317	–0.36	10	885	0.36
5	387	0.31	11	997	0.09
6	499	0.04	12	1089	0.27

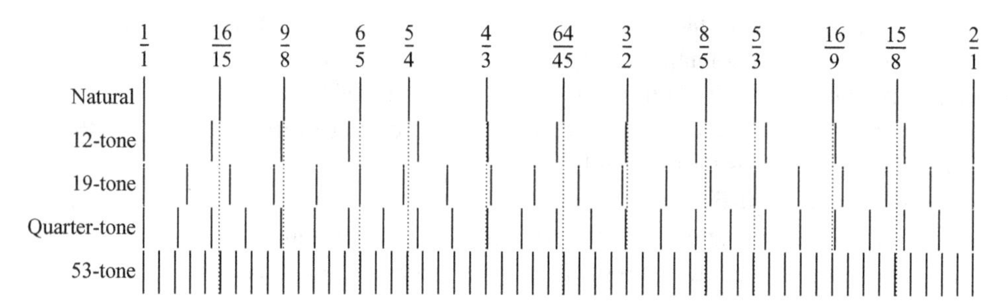

Figure 3.21
Tempered microtonal tunings compared to the natural chromatic scale.

His estimation agrees with the goodness-of-fit metric, which is 10.402 cents accumulated error for the 53-tone scale, which far surpasses the chromatic scale's 103.624 cents.

The Cent Scale as the Ultimate Tempered Microtonal Tuning The cent scale itself is the logical *reductio ad absurdum* of this progression of tempered microtonal scales. With its 1200 degrees per octave, it can be thought of as the ultimate tempered microtonal scale. Why not simply compose directly in cents? Table 3.7 shows which cent degrees correspond most closely to the natural chromatic scale. As might be expected, the goodness-of-fit metric for the cent scale is by far the best of the bunch: 2.38 cents accumulated error.

Comparing the Tempered Microtonal Scales Figure 3.21 compares tempered microtonal tunings to the natural chromatic scale, which is shown as a ruler in the background for comparison with the other scales. The fairly crude resolution of this visual aid still reveals a lot about the accuracy of the approximations these scales make to just ratios. It is evident, for instance, how much better the 19-tone scale's thirds and sixths are than those of the equal-tempered scale. It also shows how much better the 53-tone scale is than all the rest at approximating the just intervals.

Table 3.8 summarizes the goodness-of-fit metric for the tempered scales considered above. As expected, increasing the number of divisions of the octave makes it possible to approximate ever more closely the just diatonic scale by judicious choice of tempered microtonal intervals.

Table 3.8
Goodness of Fit

12-tone	103.624
19-tone	109.310
Quarter-tone	103.624
53-tone	10.402
Cent	2.37599

The fact that even the cent scale has an accumulated error, however small, is noteworthy. Human hearing can't distinguish between adjacent cents (which is one reason it was developed). So does it matter that the cent scale has a nonzero accumulated error, especially if it is much smaller than human hearing can detect? Haven't we provided ourselves with a way to temper a scale that is for all practical purposes *indistinguishable* from the just intervals? Remember that Western musical culture has lived happily with the errors in the equal-tempered scale for centuries. Nonetheless, there are those who criticize the whole approach to tempering intervals on principle.

3.14.2 Just Microtonal Scales

No matter how close they come to the just intervals, tempered microtonal scales do not meet the needs of what I call the intonation rationalists like Partch because they are nothing more than approximations (albeit sometimes pretty good approximations) to the just intervals. From the perspective of the intonation rationalists, the whole idea of tempered intervals is like the difference between 3.14 and π. It's like chopping down a forest and replacing it with telephone poles. They are not the same as the trees, no matter how close they might stand to where trees once stood.

Just tuning systems using microtones are quite widespread, including fifteenth-century European scales, tuning systems from cultures around the world, and systems constructed by contemporary theorists and composers. In Europe microtonal just scales were originally developed to improve transposability. In the classical music of Hindustan and the traditional music of Islamic countries, microtonal just scales are used without transposition. The American theorist Harry Partch also developed an elaborate just microtonal scale. This section explores a small sampling of tuning systems using just microtonal intervals.

Historical European Microtonal Scales According to Murray Barbour, just-intonation microtonal scales manifested in Europe in the late fifteenth century with the introduction of keyboards with split keys, for instance, for E♭ and D♯, to avoid the bad effects of transposing on just keyboards. Barbour (1953) writes,

> The theory was simple enough: provide at least four sets of notes, each set being in Pythagorean tuning and forming just major thirds with the notes in another set; construct a keyboard upon which these notes may be played with the minimum of inconvenience. Only in the design of the keyboards did the inventors show their ingenuity, an ingenuity that might better have been devoted to something more practical. (113)

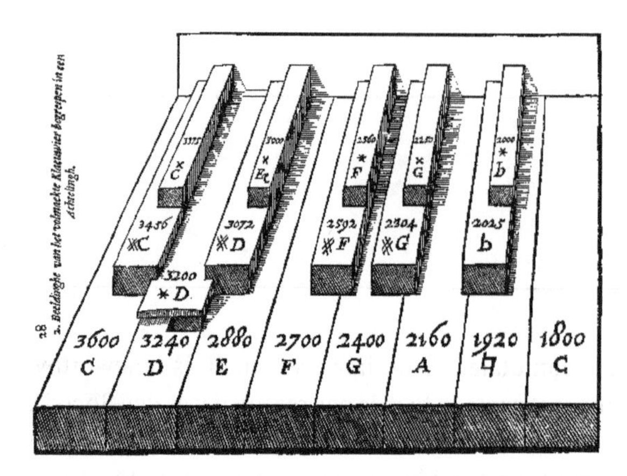

Figure 3.22
Just keyboard by Joan Albert Ban.

Figure 3.22 shows the keyboard developed by Joan Albert Ban (1597–1644) in Haarlem in 1639, based on the theories of Fogliano and Mersenne. Through the addition of floating keys and split keys, each natural key is provided with all possible justly intoned triads, major and minor. The floating D♯ provides the 9/8 above C, while the natural D below it provides the 10/9. The D-major triad D:F♯:A starts on D natural and is spelled 3240 : 2592 : 2160, and the G major triad starts on G natural and is spelled 2400 : 1920 : 3200 (requiring use of the floating D).

But adding microtones to the keyboard proved to be a dead end. They were difficult and temperamental to build and to play, and no common scheme emerged as a rallying point. Electronic keyboards that became available in the twentieth century helped revive interest in microtonal scales. Harry Partch built an entire orchestra of acoustic instruments using various microtonal layouts. However, all have remained idiosyncrasies. With the introduction of the personal computer, it finally became possible to experiment with these scales without having to construct elaborate physical keyboards, and there has been a resurgence of research interest. If a new tolerance for diversity develops, this music may yet get its proper hearing (Keislar 1988).

Partch's 43-Tone Scale Harry Partch is arguably the father of modern microtonality. His fundamental reexamination of the foundations of music theory and his consequent radical departure from musical conventions are described in minute detail in his book *Genesis of a Music* (1947). The direction of his thinking required that he create an entire orchestra of original instruments and compose a body of musical works for it. He said of himself, "I am a composer seduced into carpentry,"[13] but he was also a brilliant theorist. Though he took issue with many accepted musical dogmas of his day, he is principally remembered for his stance on intonation.

He felt that the approximations that the chromatic equal-tempered tuning system made to the pure small integer ratios were a travesty to the ear. For instance, he wrote,

> After hearing an absolutely true triad one feels that the tempered triad throws its weight around in a strangely uneasy fashion, which is not at all remarkable, for what it wants to do more than anything else is to go off

and sit down somewhere—it actually requires resolution! Thus has the composition of music for the tempered scale become one long harried and constipated epic, a veritable and futile pilgrimage in search of that never-never spot—a place to sit! (179)

He recognized that his vitriol could become excessive. "In attempting to correct an illogical situation a man tends to become an extremist" (97). But he was a man with a mission.

He considered the pure untempered ratios to be unique individualities, which the tempered tunings could only approximate. He created orders of tonalities out of small integer ratios of various numerical limits based on scholarship, reasoning, and his own ear. By basing his system on integer ratios, he necessarily discarded closed, transposable, common tempered tuning for an open system populated by a plethora of ratios that were as individualistic as himself.

Table 3.9 shows Partch's 43-tone scale. The table gives the degree number, the ratio, the cents from unison of the ratio, the ratio of the interval to the previous degree (the size of the step), and the interval size in cents. Because of the increased intervalic resources, Partch categorized ranges of his intervals as having various emotional functions roughly analogous to those commonly

Table 3.9
Partch's 43-Tone Scale

No.	Ratio	Cents	Step Size	Step Cents	No.	Ratio	Cents	Step Size	Step Cents
1	1	0			23	10/7	617.49	50/49	34.98
2	81/80	21.50	81/80	21.51	24	16/11	648.68	56/55	31.19
3	33/32	53.27	55/54	31.77	25	40/27	680.45	55/54	31.77
4	21/20	84.47	56/55	31.19	26	3/2	701.96	81/80	21.51
5	16/15	111.73	64/63	27.26	27	32/21	729.20	64/63	27.26
6	12/11	150.64	45/44	38.91	28	14/9	764.90	49/48	35.70
7	11/10	165.00	121/120	14.37	29	11/7	782.49	99/98	17.58
8	10/9	182.40	100/99	17.40	30	8/5	813.69	56/55	31.19
9	9/8	203.91	81/80	21.51	31	18/11	852.59	45/44	38.91
10	8/7	231.17	64/63	27.26	32	5/3	884.36	55/54	31.77
11	7/6	266.87	49/48	35.70	33	27/16	905.87	81/80	21.51
12	32/27	294.14	64/63	27.26	34	12/7	933.13	64/63	27.26
13	6/5	315.64	81/80	21.51	35	7/4	968.83	49/48	35.70
14	11/9	347.40	55/54	31.77	36	16/9	996.09	64/63	27.26
15	5/4	386.31	45/44	38.91	37	9/5	1017.60	81/80	21.51
16	14/11	417.51	56/55	31.19	38	20/11	1035.00	100/99	17.40
17	9/7	435.08	99/98	17.58	39	11/6	1049.36	121/120	14.37
18	21/16	470.78	49/48	35.70	40	15/8	1088.27	45/44	38.91
19	4/3	498.05	64/63	27.26	41	40/21	1115.53	64/63	27.26
20	27/20	519.55	81/80	21.51	42	64/33	1146.73	56/55	31.19
21	11/8	551.32	55/54	31.77	43	160/81	1178.49	55/54	31.77
22	7/5	582.51	56/55	31.20	44	2/1	1200.00	80/81	21.51

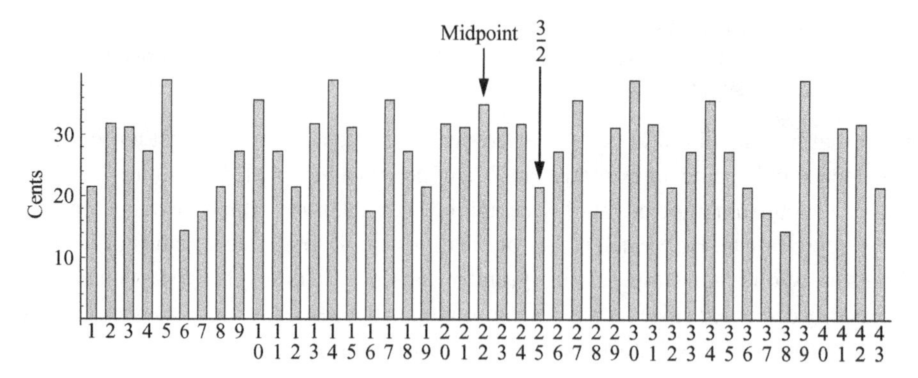

Figure 3.23
Interval sizes of Partch's 43-tone scale.

attributed to the just diatonic scale:

- *Intervals of power,* the perfect intervals—unison (#1), octave (#44), fourth (#19), and fifth (#26), shown with heavy outline
- *Intervals of suspense,* the intervals in the region of the tritone from the fourth (#19) to the fifth (#26), shown with light shading
- *Emotional intervals,* the intervals in the regions of the thirds (#11 to #18) and sixths (#27 to #34), shown with heavy shading
- *Intervals of approach,* the intervals in the regions of the seconds (#2 to #10) and sevenths (#35 to #43), shown with light outline

It is interesting to observe the symmetric regularity of interval size between steps of this scale (figure 3.23). The scale is not symmetrical at the fifth, but at three degrees below the fifth, at number 23—the midpoint of the interval order (see table 3.9). Note the plethora of different step sizes in figure 3.23.

Figure 3.24 compares Partch's scale and the equal-tempered chromatic scale, with the natural chromatic scale shown as a background ruler.

Hindustani Scales Whereas Western music has emphasized harmonic practices requiring transposition and modulation, classical Hindustani music has emphasized melodic practices that are based on just intervals and do not transpose. The degrees of the classical Hindustani scale are called *sruti.* The most common scale has 22 sruti per octave. Continuous-pitch instruments such as the voice or sarod can adapt intonation as needed to play any subset of this scale. Fretted instruments such as the vina, sitar, and esraj are supplied with adjustable frets that can be shifted to adapt to different subsets of sruti intervals. The principal playing strings of these fretted instruments can be pulled sideways across the frets, stretching the string to achieve other sruti as needed, and for ornamentation.

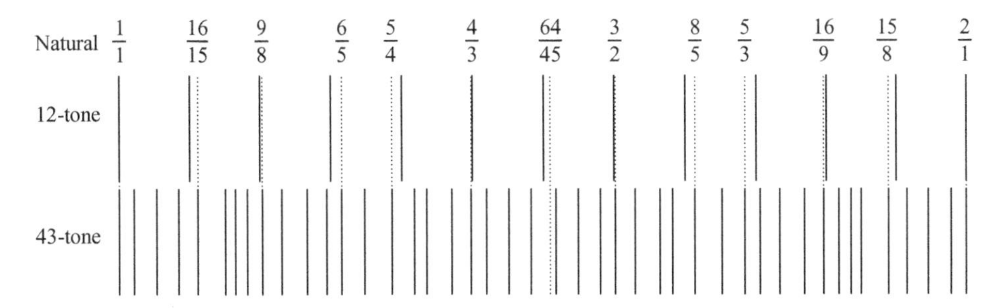

Figure 3.24
Partch's scale and equal-tempered chromatic scale compared.

Table 3.10
Hindustani 22-Sruti Scale

Degree	Ratio	Cents	Interval	Size	Degree	Ratio	Cents	Interval	Size
1	1/1	0	–	–	12	45/32	590.22	25/24	70.67
2	256/243	90.23	256/243	90.23	13	729/512	611.73	81/80	21.51
3	16/15	111.73	81/80	21.51	14	3/2	701.96	256/243	90.23
4	10/9	182.40	25/24	70.67	15	128/81	792.18	256/243	90.23
5	9/8	203.91	81/80	21.51	16	8/5	813.69	81/80	21.51
6	32/27	294.14	256/243	90.23	17	5/3	884.36	25/24	70.67
7	6/5	315.64	81/80	21.51	18	27/16	905.87	81/80	21.51
8	5/4	386.31	25/24	70.67	19	16/9	996.09	256/243	90.23
9	81/64	407.82	81/80	21.51	20	9/5	1017.60	81/80	21.51
10	4/3	498.05	256/243	90.23	21	15/8	1088.27	25/24	70.67
11	27/20	519.55	81/80	21.51	22	243/128	1109.78	81/80	21.51

Barbour (1953, 113) assumes that the Hindustani sruti scale is based on an equal division of the octave into 22 parts, much as one of the common Arabic scales is an equal division into 17 parts. He writes, "If these are considered equal, a new system arises with 'practically perfect' major thirds . . . and very sharp fifths" (116). Judging from their music, it seems very unlikely that Hindustani musicians would settle for sharp fifths, however. Many sources give the sruti scale as an extended just system. This is a more satisfying explanation because it would give a high degree of consonance between the scale and the rich harmonic content of many Hindustani instruments.

Table 3.10 shows the intervals commonly given for the 22-sruti scale. Figure 3.25 compares the 22-sruti scale with the natural chromatic scale and the Pythagorean dodecaphonic scale. The sruti that are in neither of these other scales are shaded in the table and figure. According to table 3.10 and figure 3.25, the 22-sruti scale contains both the natural chromatic and Pythagorean chromatic scales as subsets, and contains four additional intervals that are not in either of the

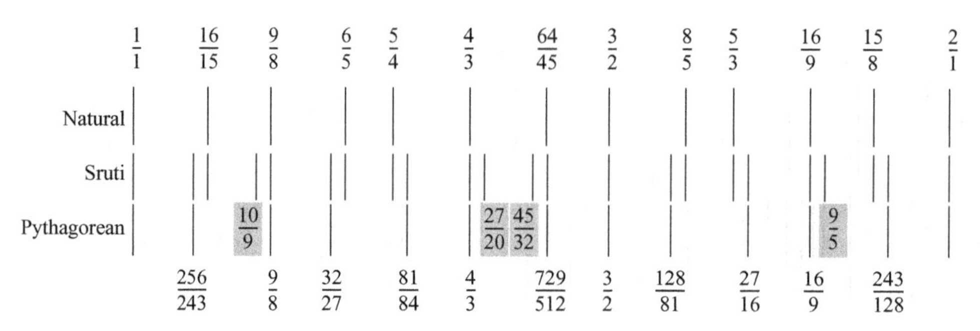

Figure 3.25
Natural chromatic and 22-sruti scales compared.

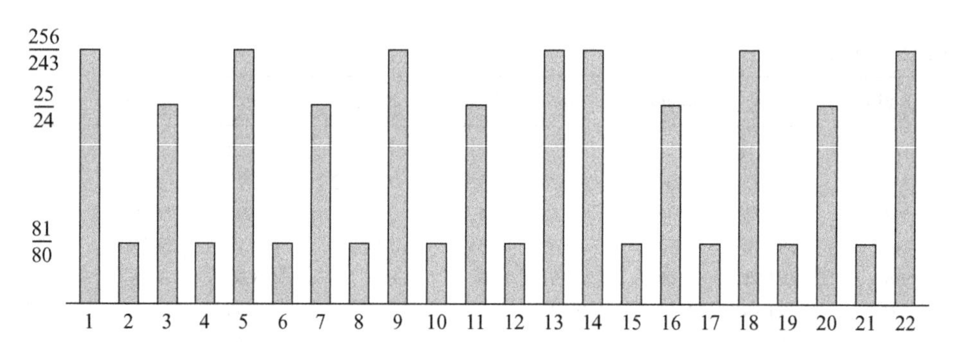

Figure 3.26
Interval structure of the 22-sruti scale.

others. One can select from this any combination of just or Pythagorean scales, plus a variety of other scales. A prime factor analysis of the 22 sruti ratios shows that this is a five-limit scale.

Figure 3.26 shows the interval structure of the 22-sruti scale. There are three interval sizes: 256/243, 25/24, and 81/80. Pingle (1962, 31) calls the smallest intervals murchanas. Interestingly, the size of the murchana interval corresponds to the Pythagorean comma.

Why are there 22 srutis? I was told by my Hindu music teachers that the 22-sruti scale is basically chromatic. It contains both the natural and the Pythagorean chromatic scales.[14] The 22 degrees come from taking all chromatic intervals except the unison and fifth, which are fixed, and splitting them into a lower and an upper microtonal interval. And, indeed, $2 \cdot (12 - 2) + 2 = 22$ degrees altogether. While this is a good description of what we see in figure 3.25, it is not an explanation. Another conjecture I've heard is that 22 was chosen because the ratio of the 22 sruti to the diatonic scale degrees that anchor it is

$$\frac{22}{7} = 3.14286\ldots \cong 3.14159\ldots \cong \pi.$$

Although the ratio of 22/7 was indeed used in ancient times as a rational approximation to π, this is not a particularly compelling *musical* explanation (Beckman 1976).

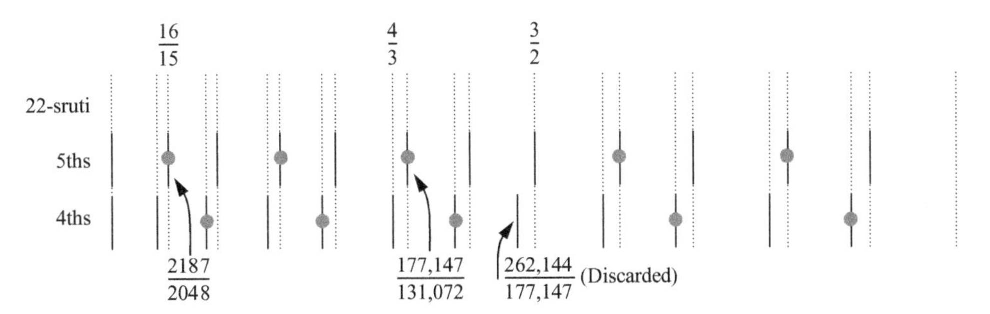

Figure 3.27
22-sruti scale as circle of fifths and circle of fourths.

The most satisfying explanation I've heard so far comes from Lentz (1961), who characterized the scale as a combination of the cycle of fourths and the cycle of fifths. The process, which is much like that described for the Pythagorean dodecaphonic scale, goes like this:

1. Create a set of intervals $(3/2)^m$ for $0 \le m < 12$.
2. Create another set of intervals $(4/3)^n$ for $0 \le n < 12$.
3. Subtract as many octaves as necessary to position each interval within the compass of one octave.

This creates a set of 23 unique intervals (not 24 because the unison is repeated in both series). Figure 3.27 shows the sruti scale of table 3.10 compared to the circle of fifths and circle of fourths. The interval 262,144/177,147 in the circle of fourths (just below the 3/2) must be discarded, leaving 22 sruti.

At first glance, Lentz's combination of fifths and fourths looks very close to the 22-sruti scale. However, small discrepancies are evident even in this crude graphic: some of the powers of fifths and fourths do not line up with the intervals given in the literature but are a little sharp or flat. In fact, the intervals that miss their mark are off by exactly 32,805/32,768, an interval historically called a *schisma*. For instance, while the third degree in table 3.10 is given as 16/15, the third degree by the circle of fifths is 2187/2048, which is a difference of 32,805/32,768.

Lentz's method has the advantage of being a simple and elegant construction, but like the Pythagorean scale, the result may please theorists more than musicians. Who is to say whether an oriental equivalent of Pareja didn't argue for a version of the 22-sruti scale made simpler by adjusting the sruti up and down by schismas to nearby smaller integer ratios, leaving the conventional ratios given in table 3.10?

The 22-sruti scale described here does not by any means exhaust the Hindustani interest in the number 22. An interesting just diatonic scale given by Pingle (1962) consists of the following seven intervals:

22/22, 26/22, 29/22, 31/22, 35/22, 39/22, 42/22, 44/22.

Figure 3.28 compares Pingle's scale with the 22-sruti scale and the natural just diatonic scale. It is an 11-limit scale with a most exotic sound, as all of its intervals are quite sharp in comparision to the just diatonic scale.

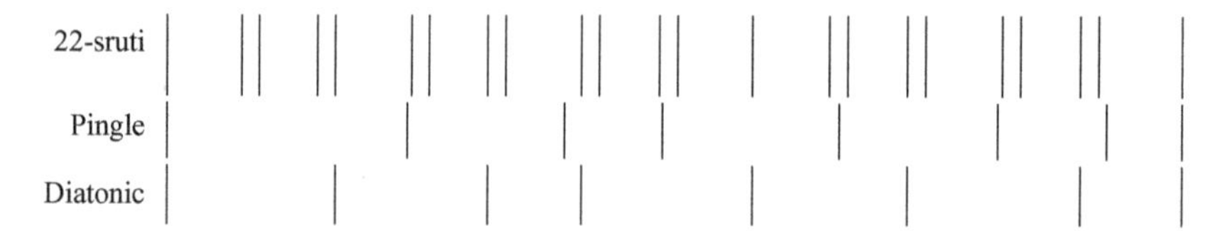

Figure 3.28
B. A. Pingle's diatonic scale.

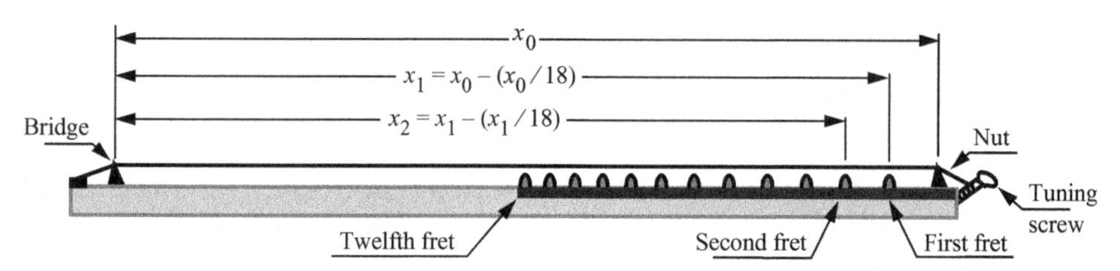

Figure 3.29
Rule of 18 for placing frets.

3.15 Rule of 18

The rule of 18 has been used by Western stringed instrument builders to construct the scales their instruments play since it was first proposed as a tempered scale by Vincenzo Galilei (see section 3.13.1). It highlights a number of interesting mathematical principles.

It so happens that the size of a tempered semitone, the irrational number $\sqrt[12]{2}$, is fairly closely approximated by the rational ratio 18/17, that is,

$$\sqrt[12]{2} \approx \frac{18}{17} \approx 1.0588.$$

It is much easier in practice for builders to work with ratios of integers than irrational ratios when dividing up a linear distance. As shown in figure 3.29, each string of a fretted instrument is suspended between two points, the *bridge* and the *nut*. The frequency of the open string is determined by a peg or screw arrangement near the nut, which tightens or loosens it, varying the tension of the string. The performer varies the frequency by stopping off different lengths of the string against the fingerboard, thereby changing the mass of the part of the string that can vibrate.

3.15.1 Fret Calculations

Fret wires placed along the fingerboard perpendicular to the string help the performer stop off exactly the right length to sound intervals in the scale that the instrument is built to play. Unfretted stringed instruments such as the violin are played similarly but do not have frets to guide the performer's fingers. Frets are foremost an aid to intonation, but they also make it possible to correctly stop multiple strings simultaneously, a useful feature for polyphony. Historically, the frets are placed using the 18/17 tempered scale of Galilei.

To place the frets, the rule of 18 states

Each subsequent fret should be located 1/18 of the remaining distance to the bridge of the instrument.

Let's take for an example a string of length $x_0 = 1$ meter from bridge to nut (figure 3.29). Then the rule of 18 says that the distance x_1 from the bridge to the first fret should be

$$\begin{aligned} x_1 &= x_0 - \frac{x_0}{18} \\ &= 1 - \frac{1}{18} \\ &= \frac{17}{18} \text{ m.} \end{aligned} \qquad \textit{First Fret} \quad (3.16)$$

In order to sound a semitone higher, the rule of 18 says that the length of the string from the bridge to the first fret must be 17/18 of the length of the entire string x_0.

The distance from the bridge to the second fret, x_2, is calculated from the "remaining distance," which is x_1. So we subtract 1/18 of the string from the length of x_1:

$$x_2 = x_1 - \frac{x_1}{18} = \frac{289}{324} \text{ m.} \qquad \textit{Second Fret} \quad (3.17)$$

3.15.2 The Flaw in the Rule of 18

If we continue to apply the rule of 18 twelve times, then the twelfth fret will end up being placed near the midpoint of the string. However, when the string is stopped at the twelfth fret, although ideally it should sound exactly an octave higher than the whole string, it will actually sound slightly flat because $18/17 < \sqrt[12]{2}$. Each fret placed by the rule of 18 will sound slightly flat, and the error will compound for higher-numbered frets because the position of each subsequent fret is derived from the previous one. For example, if the length of the open string is $x_0 = 1$ m, then the position of the twelfth fret is approximately $x_{12} = 0.504$ m instead of the desired 0.5 m, which is where it should be to sound exactly an octave above the open string.

Happily, another artifact of stringed instruments comes to the rescue to a certain extent. Fretting a string bends it, decreasing its elasticity slightly, which raises its pitch slightly. By the nature of their construction, strings must be bent progressively more the higher the fret, which counteracts the progressive flattening of the rule of 18. The precise amount by which the string's pitch is raised by this stretching depends upon the geometry of the instrument and the dimensions and tension of the string. In practice, many additional factors must be taken into account by a stringed instrument maker, a process called (appropriately enough) compensation.

Alternatively, if we shave off a little from the rule of 18 and instead use the "rule of 17.81715," we get fret distances that nearly match the equal-tempered scale, and $x_{12} = 0.500$.

3.15.3 Recursion

The rule of 18 is an example of *recursion,* in which the next value in a sequence depends upon the previous value (or values) in a well-defined way. Suppose we let f_0 be the frequency sounded when the open string in figure 3.29 is played. Then the frequency of the string stopped at the first fret would be $f_1 = f_0 \cdot 18/17$, and the frequency at the second fret would be $f_2 = f_1 \cdot 18/17$. Generalizing, we can find the frequency of any fret:

$$f_n = f_{n-1} \cdot \frac{18}{17}. \tag{3.18}$$

This means that f_3 depends upon the value of f_2, which depends on the value of f_1, which depends upon the value of f_0. In other words,

$$\begin{aligned} f_3 &= f_2 \cdot \frac{18}{17} \\ &= \left(f_1 \cdot \frac{18}{17}\right) \cdot \frac{18}{17} \\ &= \left(\left(f_0 \cdot \frac{18}{17}\right) \cdot \frac{18}{17}\right) \cdot \frac{18}{17}. \end{aligned}$$

This means we can compute f_3 in terms of f_0 just by multiplying f_0 by $(18/17)^3$. Now that we see the pattern, we can compute the frequency at the nth fret in terms only of f_0:

$$f_n = f_0\left(\frac{18}{17}\right)^n. \tag{3.19}$$

In (3.19) the frequency of the nth fret depends only upon the frequency of the open string instead of on the frequency of the fret that came before it, so this equation implements a direct calculation, not a recursive one. If we set $f_0 = 440$ Hz, then by either (3.18) or (3.19) the value of f_3 comes out to be 522.3 Hz.

Where a direct equivalent to a recursive formula can be found, it is generally to be preferred.

- It avoids the problem of compounding errors in calculation.
- It is generally faster because we do not need to calculate all the values between the starting value and the value of interest.

This can be important if, for example, we must calculate values of a function that are far from where we started.

One can often find a way to convert between recursive and direct representations of a formula. For instance, we can write the rule for generating the equal-tempered scale recursively as follows:

$$f_n = f_{n-1} \cdot 2^{1/12},$$

and by similar reasoning, its direct form is

$$f_n = f_0 \cdot 2^{n/12},$$

which is equivalent to equation (3.1).

The rule of 18 also describes an iterative process. If x_n represents the distance of the nth fret from the bridge, then the rule of 18 can be expressed as

$$x_n = x_{n-1} - \frac{x_{n-1}}{k}, \tag{3.20}$$

where k is a constant factor, either 18 or 17.81715, as discussed. Equation (3.20) says, "The distance from the bridge to the next fret (x_n) equals the distance from the bridge to the previous fret (x_{n-1}) minus that distance divided by k." Using (3.20) to compute the distance from the bridge to the third fret, x_3, we proceed as follows:

$$\begin{aligned} x_3 &= x_2 - \frac{x_2}{k} \\ &= x_1 - \frac{x_1}{k} - \frac{x_1 - \frac{x_1}{k}}{k} \\ &= x_0 - \frac{x_0}{k} - \frac{x_0 - \frac{x_0}{k}}{k} - \frac{x_0 - \frac{x_0}{k} - \frac{x_0 - \frac{x_0}{k}}{k}}{k}. \end{aligned} \tag{3.21}$$

Assuming the distance from the bridge to the first fret is $x_0 = 1$ m, and using the modified rule of 18 ($k = 17.81715$), then $x_3 = 0.84$. Notice the interesting way the terms stack up in (3.21). These are called *continued fractions*.

3.16 Deconstructing Tonal Harmony

Back when the Pythagorean scale ruled the day, the degrees each had a unique character and function, like chess pieces. The asymmetry of the scale oriented the ear as the music unfolded. The tonic degree was king, and a hierarchy of tones surrounded it like courtiers. The system was called *tonal harmony.*

Even after the advent of the chromatic equal-tempered scale, composers persisted (as they still do today) in exploring functional harmony based on the expectations of listeners trained to hear the characteristic intervals of the diatonic scale. But the adjustments made over the centuries to facilitate transposition had the eventual effect of disconnecting the pitches from their harmonic function.

By the end of the late Romantic era, functional harmonization had reached its expressive limits because, as its vocabulary expanded, the listener's roots in the old diatonic scheme gradually weakened, until all that was left were the 12 pitches, all of which were now equivalent both in function and in tonal palette.

A century after the equal-tempered tuning system was widely adopted in the West, the composer Arnold Schoenberg and his associates (the so-called Second Viennese School) were inspired to extend the idea of pitch equality further. They believed the old functional harmonic practices lingered on only as a historical artifact of the old just scales and should now be discarded. They devised *atonal* compositional strategies to remove key-centeredness from their music and so to thwart the ear's trained habit of organizing music harmonically. They eventually developed the *12-tone* compositional methodology by giving all pitches equal prominence (see section 9.10). Interestingly, this compositional motivation bears certain resemblances to political experiments in radical democracy, communism, and socialism that occurred in Europe around the same time. Alignments between political economy and musical aesthetics have existed throughout the ages, and transitions in one often presage a transition in the other (Atali 1985). Plato noticed this effect long ago. He said pessimistically, "A change to a new type of music is something to beware of as a hazard of all our fortunes. For the modes of music are never disturbed without unsettling the most fundamental political and social conventions" (*Republic* 424c).

Here, once again, we arrive at the nexus between society, aesthetics, and technology. It seems that the deconstruction of tonal harmony at the end of the Romantic era was the inevitable result of the availability of effective transposable key schemes. This means that advances in musical scale engineering had profound reflexive consequences on musical aesthetics. Circularly, the desire for transposable key schemes was originally motivated by aesthetic requirements, but the consequence of their development was a fundamental transformation in aesthetics.

Thus music takes its place in the pantheon of human pursuits: no activity is immune from our reflexive and self-redefining capacities, which is perhaps our most unique characteristic as a species.

3.17 Deconstructing the Octave

Every true revolution encompasses the paradigm it overthrows, even as it supersedes it. The revolution of the Second Viennese School led to the deconstruction of tonal harmony, but the octave remained sacred. The revolution of the microtonalists led to the deconstruction of the chromatic scale, but the octave likewise remained sacred. The octave has been an invariant feature of virtually all historical scales because of *octave equivalence,* which is our tendency to hear pitches played at octaves as functionally identical. The equivalence is felt so strongly that musical scales around the world are almost invariably organized around the 2:1 ratio of the octave, and pitches related by octaves are virtually always given the same name. Octave equivalence is deeply rooted in our perceptual system (see section 6.4.6).

The invariance of the octave is hard-wired in equation (2.2), $f_x = f_R \cdot 2^x$, $x \in \mathbf{R}$, because of the constant 2 in that equation. If we generalize it,

$$f_x = f_R \cdot \kappa^x, \qquad \kappa \in \mathbf{I}, \qquad x \in \mathbf{R}, \tag{3.22}$$

we can construct scales that are not bound by the octave. (It is customary but not strictly necessary to limit κ in (3.22) to be an integer.) The value of κ defines what I call the *compass interval*. Let the compass interval be $\kappa^{x+1}:\kappa^x$ for any real x. For example, when $\kappa = 2$, the compass interval is 2:1, the octave. When $\kappa = 3$, the compass interval is 3:1, an octave plus a fifth, otherwise known as a twelfth. The value x is typically a rational fraction indicating a division of the compass interval. For the equal-tempered scale, $x = k/12$, where k indexes a particular division of the compass interval.

The inclusion of non-octave-based scales vastly widens the scale possibilities we must consider. However, there are two important characteristics of octave-based scales that we would do well to preserve when evaluating the suitability of non-octave-based scales for musical purposes. Candidate scales should have

- A high degree of consonance for as many of the intervals as possible
- A high degree of internal order, that is, a regular pattern of steps and step sizes

3.17.1 The Bohlen-Pierce Scale

A non-octave-based scale that arguably meets the above criteria and has a number of other interesting features as well was developed by several music researchers in the latter part of the twentieth century. Heinz Bohlen (1978), an electronics and communications engineer without formal musical training (which fact was probably an asset to his accomplishment) was the first to consider building a scale from a triad not based on the familiar 4:5:6 ratios of the natural major scale, but upon the ratios 3:5:7 and the compass of an octave and a fifth. As the compass interval of 2:1 is called the octave, the compass interval of the twelfth was dubbed the *tritave* by John Pierce, who independently discovered this scale system (Mathews, Roberts, and Pierce 1984; Mathews and Pierce 1980).[15]

Because the scale is made from simple integer ratios that are harmonic by definition, it meets the first criterion. But because it does not include an octave and duplicates but two of the octave-based just intervals, it is completely incompatible with any octave-based scale. As for the second criterion, it does have a high degree of internal order.

3.17.2 Constructing the Bohlen-Pierce Just Scale

We can construct this scale using the standard method of adding and subtracting intervals, beginning with the 3:5:7 triad.

1. Take as the first degree of the scale the unison 3/3. Positioning the root of the 3:5:7 triad on the first degree yields scale intervals 3/3 : 5/3 : 7/3. The tritave corresponds to 9/3 = 3/1, giving the degrees shown in figure 3.30.

2. Starting a new root on the 5/3, we can spell another triad with the ratios 5/3 : 7/3 : 9/3. This 5:7:9 triad is shown in figure 3.31.

In the next two steps, we extend the scale to seven degrees.

3. Transpose the 3:5:7 triad in figure 3.30 so that its top pitch equals the 9/3 (figure 3.32). The top of the figure shows the 3:5:7 triad rooted on the first degree, and beneath it is the transposed 3:5:7 triad with its top pitch aligned with the tritave. To find the new root of the transposed triad,

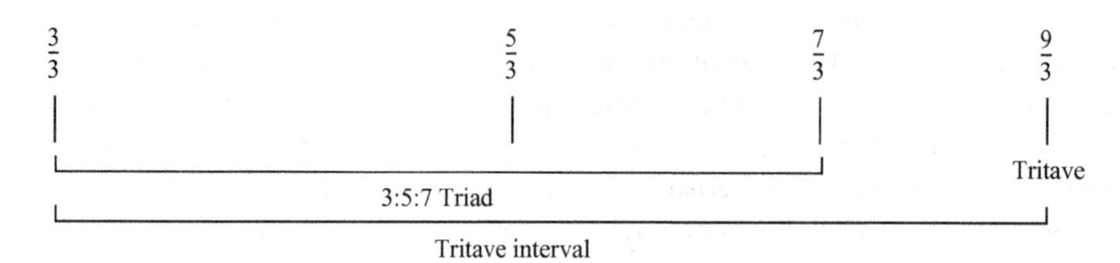

Figure 3.30
Bohlen-Pierce just scale, 3:5:7 triad and tritave.

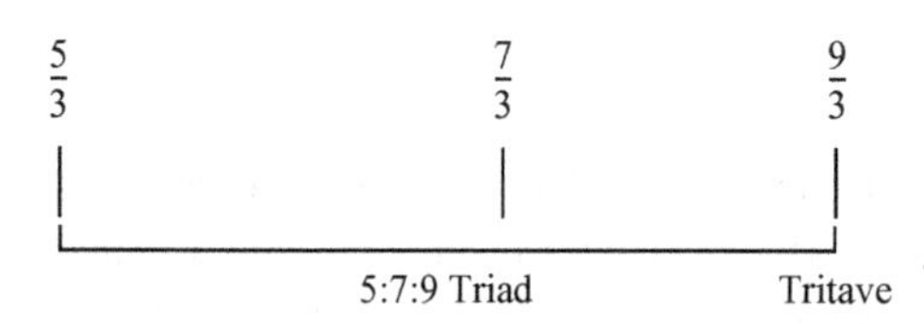

Figure 3.31
Bohlen-Pierce just scale, 5:7:9 triad.

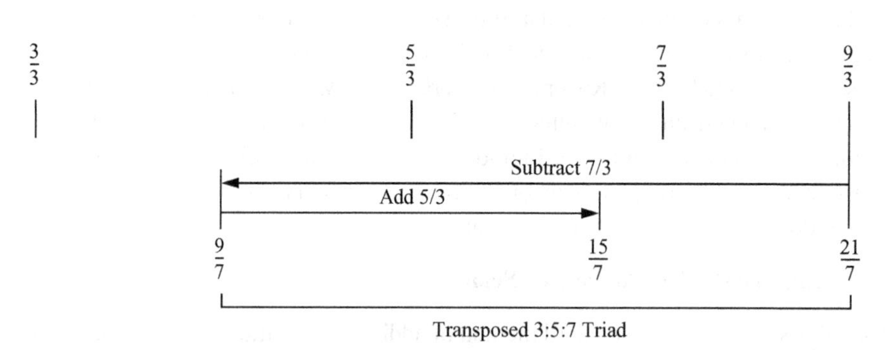

Figure 3.32
Bohlen-Pierce just scale, 9/7 and 15/7 intervals.

we subtract the interval 7/3 from the interval of the tritave:

$$\frac{3}{1} \div \frac{7}{3} = \frac{9}{7}.$$

We find the middle pitch by adding the interval 5/3 to the root:

$$\frac{9}{7} \cdot \frac{5}{3} = \frac{45}{21} = \frac{15}{7}.$$

The root and middle pitches of the transposed 3:5:7 triad thus add two new scale degrees at 9/7 and 15/7.

4. Take the 5:7:9 triad from figure 3.31 and position its root on the first degree of the scale. To do so, subtract the interval 5/3 from each interval:

$$\left(\frac{5}{3} : \frac{7}{3} : \frac{9}{3}\right) \div \frac{5}{3} = \frac{5}{5} : \frac{7}{5} : \frac{9}{5}.$$

We derive two new intervals this way, 7/5 and 9/5.

Figure 3.33 shows the resulting scale. The largest prime is 7, so this is a seven-limit scale.

3.17.3 Constructing the Bohlen-Pierce Chromatic Scale

Figure 3.34 shows the interval sizes of the Bohlen-Pierce just diatonic scale. Observe the symmetrical arrangement around the fourth degree. It is useful to classify the sizes of intervals as small

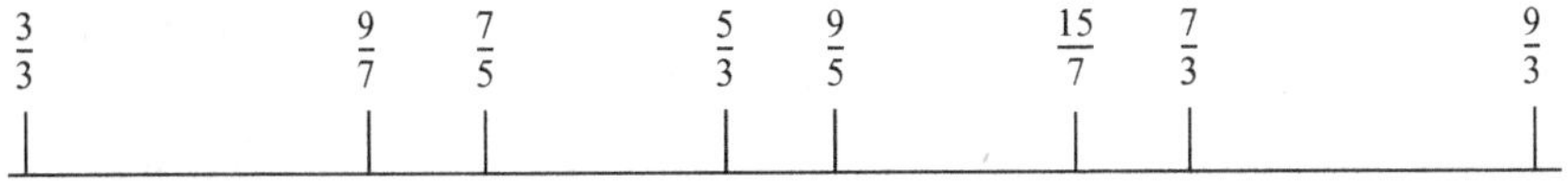

$\frac{3}{3}$ $\frac{9}{7}$ $\frac{7}{5}$ $\frac{5}{3}$ $\frac{9}{5}$ $\frac{15}{7}$ $\frac{7}{3}$ $\frac{9}{3}$

Figure 3.33
Bohlen-Pierce just diatonic scale.

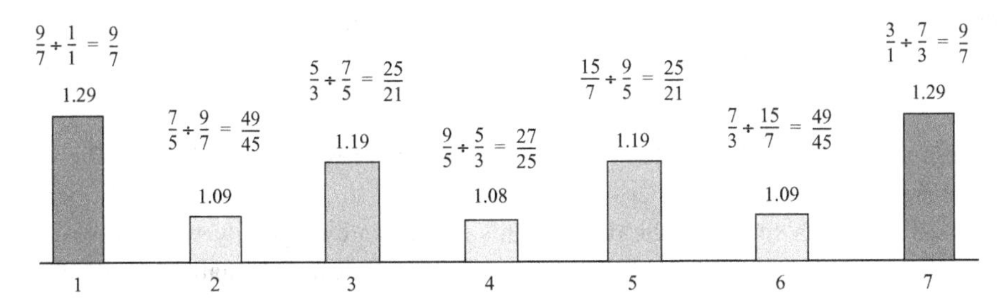

Figure 3.34
Bohlen-Pierce step sizes.

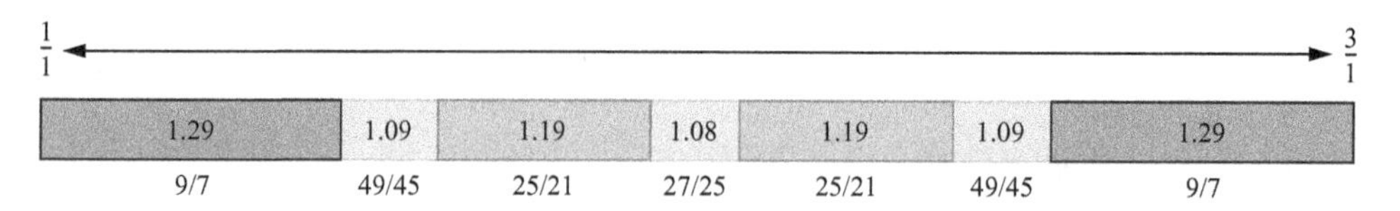

Figure 3.35
Bohlen-Pierce step sizes on their sides.

(1.08 and 1.09), medium (1.19), and large (1.29), shown in the figure from light to dark grey, respectively. Remembering that we are comparing ratios, it seems that the medium interval is about twice as large as the small intervals because $1.09^2 \approx 1.19$. Also, the sum of a small and a medium interval is about equal to the large one because $1.08 \cdot 1.19 \approx 1.29$. So these interval sizes are roughly in the order 1:2:3. We can better visualize their relative sizes if we lay the intervals over on their sides (figure 3.35).

We could devise a chromatic scale from these ratios as follows. First, we replace the large intervals with the combination of a small and a medium interval. This leaves a scale containing only small and medium steps, analogous to the half and whole steps of the equal-tempered scale. Then we replace each medium interval with two small intervals, resulting in a scale containing only small steps, analogous to the equal-tempered semitone scale.

1. Since (27/25)(25/21) = 9/7, we can exactly replace the two large (9/7) steps with the combination of a small (27/25) and a medium (25/21) step.

2. The existing small steps (49/45 and 27/25) needn't change. They constitute semitones in the scale.

3. Unfortunately, neither size small step exactly divides the medium step into two equal parts. For instance, subtracting a 49/45 semitone from a 25/21 whole step leaves 375/343. Similarly, subtracting a 27/25 semitone from a 25/21 whole step leaves 625/567. Altogether then, we end up with four semitones from smallest to largest: 27/25, 49/45, 375/343, and 625/567.

4. We must choose the order in which we substitute the smaller intervals into larger ones. Shall we break them up as {small, large} or {large, small}? Recalling the symmetry in the just Bohlen-Pierce scale in figure 3.34, we can divide the intervals in a correspondingly symmetrical way.

Applying these principles to the Bohlen-Pierce diatonic scale results in the Bohlen-Pierce chromatic just scale with ratios and step sizes as shown in figure 3.36. The result is a nicely symmetrical scale of 13 degrees spanning the tritave. This is the scale originally worked out by Bohlen (1978). Figure 3.37 shows the Bohlen-Pierce chromatic scale and the natural chromatic scale for comparison. The only points of contact between the two scale systems are the 1/1 and the 5/3 (major sixth).

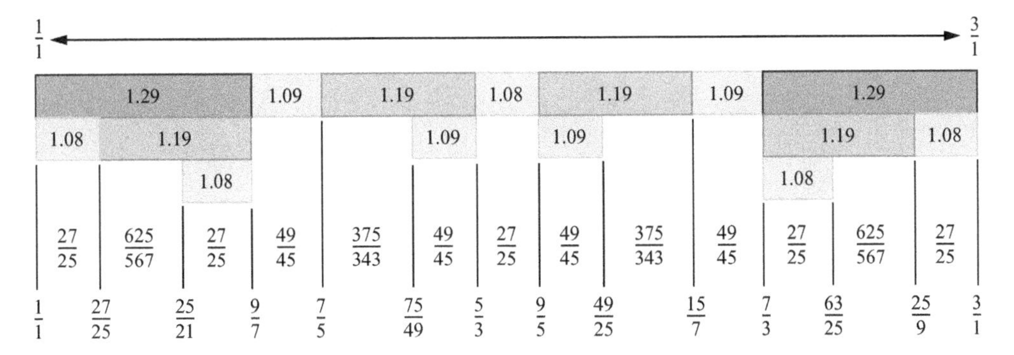

Figure 3.36
Bohlen-Pierce chromatic just scale.

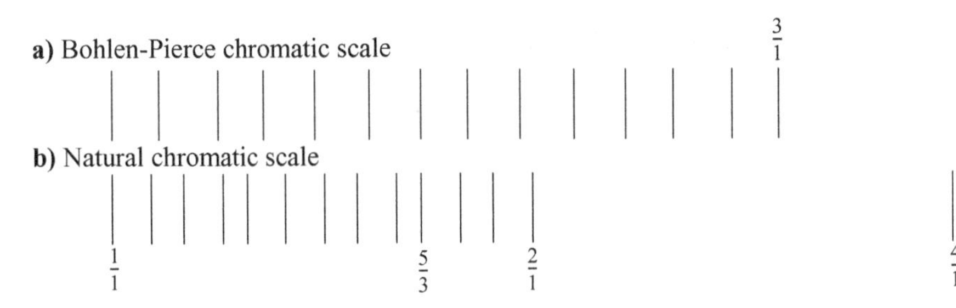

Figure 3.37
Bohlen-Pierce and natural chromatic scales compared.

3.17.4 The Bohlen-Pierce Equal-Tempered Scale

There is an equal-tempered version of the Bohlen-Pierce chromatic scale, just as there is an equal-tempered version of the natural chromatic scale. All we must do to create it is to set $\kappa = 3$ and $x = k/13$ in equation (3.22), yielding

$$f_k = f_R \cdot 3^{k/13} \,. \qquad \textit{Bohlen-Pierce Equal-Tempered Scale} \quad (3.23)$$

As shown in figure 3.38, the degrees of the equal-tempered Bohlen-Pierce scale are much closer to their just counterparts than the octave-based equal-tempered scale degrees are to their just counterparts. The equal-tempered Bohlen-Pierce scale has a goodness-of-fit metric of 81.56 cents to its chromatic just counterpart, compared to the 103.624 cents goodness-of-fit metric for the equal-tempered scale.

3.17.5 Evaluating the Bohlen-Pierce Scale

Given the odd-numbered basis of the Bohlen-Pierce scale, Pierce suggested performing the scale using only timbres with odd harmonics, such as a clarinet, to help emphasize the consonance of

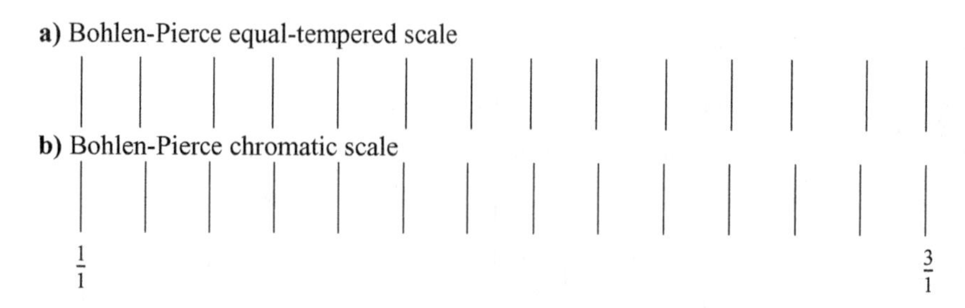

These are much closer to each other than their chromatic counterparts are.

Figure 3.38
Bohlen-Pierce chromatic and equal tempered scales compared.

the primary chords of the scale, and to help the usual expectation of octave equivalence give over to the experience of tritave equivalence. Bohlen created an electronic organ with a clarinetlike square wave timbre to experiment with the scale.

Whether through some combination of neural wiring, or a lifetime of conditioning, or both, it is hard for most listeners to hear past octave equivalence when listening to non-octave-based scales. How, then, can we objectively compare the consonance of the Bohlen-Pierce scale with other scales? Roberts and Mathews (1984) proposed *intonation sensitivity* as a way of evaluating the perceptibility of consonance of a chord. They defined intonation sensitivity as the way in which preference for a chord varies with the tuning or mistuning of the center note of the triad. Their study determined that the 4:5:6 triad had a high degree of intonation sensitivity (as would be expected) and that the intonation sensitivities of the 3:5:7 and 5:7:9 triads were very close to the 4:5:6. Indeed, they are more like diatonic major triads in the way that preference varies with tuning than diatonic minor triads are.

Mathews and Pierce (1989) investigated the consonance of the various triads available in the Bohlen-Pierce scale. Musicians and nonmusicians judged the consonance/dissonance of the 78 triads that can be formed in the span of a tritave. Tones used odd harmonics only. They found that listeners scored the triads over a wide range, indicating that consonance is a salient property of the scale.

Mathews and Pierce also asked trained musicians and nonmusicians to judge the similarity of Bohlen-Pierce chords and octave-based just chords. Here, the respondents diverged in their rankings: whereas musicians and nonmusicians alike judged similarity primarily on pitch height, musicians also ranked inversions of octave-based diatonic chords as similar whereas the nonmusicians did not. Mathews and Pierce concluded from this, "It seems reasonable that training with the [Bohlen-Pierce] scale would make it possible for listeners to recognize and respond to its structure, just as trained musicians recognize and respond to the structure of the diatonic scale and diatonic chords." Richard Boulanger's work *Solemn Song for Evening* is a fine example of the use of this scale system.

3.18 The Prospects for Alternative Tunings

Certainly one liability of non-octave-based scales and of scales with other than 12 degrees per octave is finding instruments to play them. The quarter-tone scale, for instance, requires two pianos tuned a quarter tone apart. Interesting and elaborate keyboard constructions have been proposed or built by various theoreticians over the centuries for different scale systems, both tempered and rational (Keislar 1988). Perhaps Partch had the most imaginative and ambitious approach with his orchestra of various instruments of his own design.

But the problems are not just theoretical; they are also economic. In order to construct a living music one must have instruments, trained players, a body of musical work, and last but not least, an interested and financially involved public. Although Partch did what he could within the span of his lifetime to put his music on a sustainable basis, his instruments are now in danger of becoming museum pieces, rarely played in public.

The advent of electronic and computer musical instruments certainly offers a new opportunity for microtonality and non-octave-based scales (Wilkinson 1988). Music synthesizer manufacturers sometimes include a means for microtonal experimentation in their hardware. Numerous computer music programs are available that allow precise frequencies to be generated. However, this addresses only the instrument need and does not guarantee players, works, or audience.

3.19 Summary

Intervals made from the ratios of small whole numbers are called the just intervals. Some believe that the just intervals arose first from the harmonic series of musical instruments; others, that they arose from the study of proportion by the Pythagoreans.

Intervals are added by multiplying their ratios and subtracted by dividing their ratios.

The cent scale divides the octave into 1200 equal parts; each cent is one hundredth of a tempered semitone.

We can classify scales as to how many degrees they have per octave and whether they are tempered or just.

The just pentatonic scale, diffused throughout the world, is perhaps the oldest scale. The Pythagorean just scale is the prototype for modern Western scales. Though it is highly desirable for the intervals of the scale to be based on small integer ratios, like the harmonic series, some of the Pythagorean intervals are harmonically dissonant.

The Pythagorean just scale can be expanded to 12 degrees to facilitate transposition and modulation, but we end up with two tritones and two sizes of semitone.

Ptolemy suggested modifying the Pythagorean intervals to better suit what musicians actually played. However, his ideas were suppressed until the Middle Ages. In any event, this did not solve the fundamental problem of a transposable scale system with small integer ratios.

Consonance means "to sound well together." Dissonance is its opposite. Though it is tempting to look for consonance metrics in the mathematics of their ratios, the subject also depends upon culture and era as well as psychophysical response.

The natural major scale developed by Zarlino was based on the pure 4:5:6 ratio. It succeeds at making the thirds perfectly consonant, but it does so at the expense of the whole steps, which now are uneven in size.

The mean-tone tempered scale regularized the size of the steps in the natural major scale using tempering. But odd-sized intervals made the scale degrees fail to line up exactly with the harmonic series.

The underlying problem with all just scales is that the powers of the integer ratios 3/2 and 2/1 do not form a closed system. It turns out that 12 fifths above seven octaves is one of the best approximations to a closed system, yielding a system with 12 degrees per octave, but it is not a closed system.

To close the octave so as to allow arbitrary transposition and modulation, we must use tempering. Or we can throw out modulation and transposition and use a just scale. Or we can continue to add scale degrees in an effort to throw additional scale degrees at the problem, increasing the odds that some of them will be less dissonant. Mean-tone temperament optimizes only the thirds and fifths in selected keys. Well-tempered scales make all keys usable but make some more purely intoned than others. Equal-tempered scales make all keys sound the same. The idea that different keys have a unique tonal palette stems from the well-tempered scales, which actually did sound different in different keys.

The original aim of microtonal tuning was to supply alternative choices of intervals when modulating or transposing so as to retain as much as possible the simple integer ratios of the just scales. Examples of tempered microtonal scales include the 19-tone scale, the quarter-tone scale, and the 53-tone scale. Originally developed in the eighteenth century, just microtonal scales didn't catch on because of the difficulties of constructing instruments to play them. Many cultures, such as classical Hindustani music, are satisfied not to transpose but incorporate 22 just microtones called sruti in the scale to provide a rich tonal palette. In the twentieth century, Harry Partch built an entire orchestra to play music using his 43-tone just microtonal scale.

The hierarchical system of diatonic harmonicity began to break down after the equal-tempered scale provided free transposition to any key. All keys were now alike because all intervals were identical. With identical keys, after a while, composers no longer felt the compulsion to obey the older tonal hierarchy. Arnold Schoenberg and his school devised a way to remove any key-centeredness from their music by giving all pitches equal prominence. The result was the deconstruction of tonal harmony in Western music. This was followed by the deconstruction of the octave in the late twentieth century, for example, by the Bohlen-Pierce scale.

We live in an unbelievably rich time when all the musical traditions of the world, both current and historical, are available to us, and we also have the means to construct new scales and build new instruments to play. However, to construct a living music requires more than just a theory: instruments, trained players, a body of musical work, and an interested and financially involved public are also necessary.

3.20 Suggested Reading

Backus, John. 1969. *The Acoustical Foundations of Music*. New York: W. W. Norton.

Barbour, J. Murray. 1947. "Bach and the Art of the Temperament." *Musical Quarterly* 33 (January): 64–89.

———. 1948. "Irregular Systems of Temperament." *Journal of the American Musicological Society* 1: 20–26.

———. 1953. *Tuning and Temperament: A Historical Survey*. East Lansing: Michigan State College Press.

Barnes, John. 1979. "Bach's Keyboard Temperament: Internal Evidence from the Well-Tempered Clavier." *Early Music* 7 (April): 236–249.

Benade, Arthur H. 1976. *Fundamentals of Musical Acoustics*. New York: Oxford University Press.

Blood, William. 1979. "'Well-Tempering' the Clavier: Five Methods." *Early Music* 7 (October): 491–495.

Bobbitt, Richard. 1959. "The Physical Basis of Intervallic Quality and Its Application to the Problem of Dissonance." *Journal of Music Theory* 3 (November): 173–207.

———. 1980. "Das Wohltemperierte Clavier: Tuning and Musical Structure." *English Harpsichord Magazine* 2 (April): 137–140.

Bohlen, Heinz. 1978. "13 Tonstufen in der Duodezime." *Acustica* 39. English translation: "13 Tone Steps in the Twelfth." *Acustica* 87 (2001, no. 5): 617–624.

Bosanquet, Robert. 1876. "Musical Intervals and Temperament." In *Tuning and Temperament Library*, vol. 4. Ed. Rudolf Rasch. Utrecht: Diapason Press, 1987.

Carlos, Wendy. 1987. "Tuning: At the Crossroads." *Computer Music Journal* 11 (1): 29–43.

Carr, Dale C. 1974. "A Practical Introduction to Equal Temperament." *Diapason* 65 (3): 6–8.

Danielou, Alain. 1994. *Music and the Power of Sound*. Rochester, Vt.: Inner Traditions.

Fletcher, N. H., and T. D. Rossing. 1991. *The Physics of Musical Instruments*. New York: Springer-Verlag.

Hamilton, James A. 1844. *Hamilton's Practical Introduction to the Art of Tuning the Pianoforte*. London: R. Cocks.

Hellegouarch, Yves. 2002. "A Mathematical Interpretation of Expressive Intonation." In *Mathematics and Art*. Ed. Claude P. Bruter. New York: Springer-Verlag.

Helmholtz, Hermann. 1863. *On the Sensations of Tone*. Second English edition, 1885. Trans. A. J. Ellis based on the fourth German edition, 1877. New York: Dover, 1954.

Jorgenson, Owen H. 1977. *Tuning the Historical Temperaments by Ear*. Marquette: Northern Michigan University Press.

———. 1978. "In Tune with Old Tunings." *Clavier* 17 (November): 26–28.

———. 1991. *Tuning: Containing the Perfection of Eighteenth-Century Temperament, the Lost Art of Nineteenth Century Temperament, and the Science of Equal Temperament*. East Lansing: Michigan State University Press.

Kellner, Herbert Anton. 1979. "A Mathematical Approach Reconstituting J. S. Bach's Keyboard Temperament." *Bach* 10: 2–9.

Lindley, Mark. 1974. "Early Sixteenth-Century Keyboard Temperaments." *Musica Disciplina* 28: 129–151.

Lloyd, Llewellyn. 1979. *Intervals, Scales, and Temperaments*. Ed. Hugh Boyle. New York: St. Martin's Press.

Mathews, Max V., and John R. Pierce. 1980. "Harmony and Nonharmonic Partials." *Journal of the Acoustical Society of America* 68: 1252–1257.

Mathews, Max V., John R. Pierce, A. Reeves, and L. A. Roberts. 1988. "Theoretical and Experimental Explorations of the Bohlen-Pierce Scale." *Journal of the Acoustical Society of America* 84: 1214–1222.

Mekiel, Joyce. 1960. "The Harmonic Theories of Kirnberger and Marpurg." *Journal of Music Theory* 4 (November): 169–193.

Pierce, John. R. 1966. "Attaining Consonance in Arbitrary Scales." *Journal of the Acoustical Society of America* 40: 249.

Sargent, George. 1969. "Eighteenth-Century Tuning Directions: Precise Intervallic Determinations." *Music Review* 30 (February): 27–34.

Sethares, W. A. 1993. "Local Consonance and the Relationship between Timbre and Scale." *Journal of the Acoustical Society of America* 94 (3): 1218–1228. Also, without the math, in "Relating Tuning and Timbre." *Experimental Music Instruments* 9 (1993, no. 2) and at http://eceserv0.ece.wisc.edu/~sethares/consemi.html.

Slaymaker, F. H. 1968. "Chords from Tones Having Stretched Partials," *Journal of the Acoustical Society of America* 47: 1469–1571.

Werckmeister, Andreas. 1691. "Musicalische Temperatur." In *Tuning and Temperament Library,* vol. 1. Ed. Rudolf Rasch. Utrecht: Diapason Press, 1983.

Wilkinson, S. R. 1988. *Tuning In*. Milwaukee: Hal Leonard Books.

4 Physical Basis of Sound

Music is a science which should have definite rules; these rules should be drawn from an evident principle; and this principle cannot really be known to us without the aid of mathematics. Notwithstanding all the experience I may have acquired in music from being associated with it for so long, I must confess that only with the aid of mathematics did my ideas become clear and did light replace a certain obscurity of which I was unaware before.
—Jean-Philippe Rameau, *Traite de l'Harmonie*

This book uses the international system of standard units defined by the Système International d'Unités, abbreviated SI. It is also known as the MKS system of measurement, which stands for "meter, kilogram, second." This system is used almost universally by the scientific community as well as by most countries of the world except the United States. As an American, I may occasionally slip back into old habits and use the so-called English "foot, pound, second" system. But since even the English have abandoned it, I'm trying to do so as well.

4.1 Distance

The fundamental SI unit of distance is the meter. The SI system multiplies the meter by exponents of 10 to create other named magnitudes (table 4.1). Notice that from the millimeter on down, the exponent decreases by 3 for each succeeding unit. Units larger than the kilometer, such as the megameter and gigameter, will not arise much in the study of music and sound.

4.2 Dimension

Vectors convey both a direction and a magnitude. Vectors are usually drawn as an arrow whose length represents the vector's magnitude and whose orientation indicates its direction.

A *coordinate system* is any method of specifying points. A set of vectors set at right angles to each other defines the *cartesian coordinates*. A single such vector defines one-dimensional space, two vectors at right angles define two-dimensional space, and so on.

Two vectors are *orthogonal* if they maximize the area they delineate. Three vectors are orthogonal if they maximize the volume they delineate.

Table 4.1
SI Units of Distance

Kilometer	km	10^3 m = 1000 m	Thousand
Meter	m	10^0 m = 1 m	
Decimeter	dm	10^{-1} m = 0.1 m	(Little used)
Centimeter	cm	10^{-2} m = 0.01 m	Hundredth
Millimeter	mm	10^{-3} m = 0.001 m	Thousandth
Micrometer	µm	10^{-6} m = 0.000001 m	Millionth
Nanometer	nm	10^{-9} m = 0.000000001 m	Billionth

An *area* is the product of two orthogonal distances; the area of a circle is πr^2. A *volume* is the product of three orthogonal distances. The volume of a sphere is $4\pi r^3/3$.

4.3 Time

Sir Isaac Newton (1643–1727) provided the first published mathematical model of time in 1687 in his *Philosophiæ Naturalis Principia Mathematica,* commonly known as *The Principia.* He modeled time as a line that stretched continuously from the infinite past to the infinite future. Time was thus eternal, having no beginning and no end. This approach to modeling time makes the mathematics of music and sound tractable, but it raises a number of problems. For instance, modern astronomy suggests that time had a beginning and will possibly have an end, depending upon whether the universe will collapse, reach a steady state, or expand forever. But if time is eternal, how can it be limited by the duration of our universe? And if time is not eternal, then what was happening before time began?

Such confusions provide an object lesson on the limitations of mathematical models. They are useful insofar as they accurately characterize the behavior of real systems. But in science reality trumps a model's view of reality. Scientific revolutions come about when the limits of a model are overcome by a more encompassing model. Newton's perspective on time has the advantage of simplicity; it can still be used so long as we remain aware of its limitations.

The fundamental SI unit of time is the second. As with distance, the SI system creates other named magnitudes by multiplying the second by exponents of 10 (table 4.2), but SI time units greater than 1 second are not in decimal organization. Instead, we have years, weeks, days, hours, and minutes.

4.3.1 Period and Frequency

There are two ways to use time as a measurement:

- *Period* The amount of time T elapsed between the start and end of a single event is the *period* of the event. When a train moves past at a constant speed, the time it takes for one car to pass by

Table 4.2
SI Units of Time

Kilosecond	ks	10^3 s = 1000 s	Thousand
Second	s	10^0 s = 1 s	
Decisecond	ds	10^{-1} s = 0.1 s	(Little used)
Centisecond	cs	10^{-2} s = 0.01 s	(Little used)
Millisecond	ms	10^{-3} s = 0.001 s	Thousandth
Microsecond	µs	10^{-6} s = 0.000001 s	Millionth
Nanosecond	ns	10^{-9} s = 0.000000001 s	Billionth

a stationary observer is the car's period. Analogously, the period of vibration, *periodicity,* is the time it takes one cycle of a wave to return to its starting point.

- *Frequency* The number of events f occurring in a single elapsed time interval is the *frequency* of the event. The number of trains passing through a train station per day tells how frequently the trains run.

The two measurement strategies are *reciprocal,* that is, for some frequency f and period T,

$$f = \frac{1}{T}, \qquad \textit{Frequency}\ (4.1)$$

and conversely,

$$T = \frac{1}{f}. \qquad \textit{Period}\ (4.2)$$

Periodicity of sound is typically measured in seconds (s). Frequency is measured in cycles per second. The SI unit for one cycle per second of vibration is *hertz* (Hz). The standard reference pitch for Western orchestras is A440, corresponding to a periodic sound vibration of 440 Hz. The period of one cycle of A440 is 1/440 = 0.00227 s, or 2.27 ms. It is convenient to express frequencies above 1000 Hz in kilohertz; thus 1000 Hz = 1 kHz.

4.4 Mass

Together with time and distance, a basic measurement in the MKS system is *mass,* the quantity of matter contained in an object. *Matter* is anything that occupies space and exhibits inertia. *Inertia* is the tendency of a body to impede acceleration. Your body presses against the seat of your car as you accelerate from a stop because the inertia of your body resists (impedes) the acceleration. We can compare one mass to another using, for example, a beam balance. Or we can measure it by applying a force and measuring the resulting acceleration.

Mass and weight are not the same thing. Mass is a quantitative measure of inertia. As such, mass is an intrinsic quality of matter, unchanged by such things as the location of the object. *Weight,* on the other hand, is the force of gravity acting on the object, and it depends upon the position of the object with respect to other objects around it. For instance, you would weigh less standing on the moon than you do on the earth because your weight depends upon your position. But the mass of your body and the force required to accelerate it at a certain rate are the same regardless of whether you are on the moon or the earth.

The rest of the physical concepts in this section are derived from these three primary measurements.

4.5 Density

Density measures how tightly packed together the material in a body is. Density comes in one-, two-, and three-dimensional versions:

- *Linear density* describes mass per unit of distance, for example, the density of a rope or guitar string. For length l and mass m, the linear density μ is

$$\mu = \frac{m}{l}. \tag{4.3}$$

- *Area density* describes mass per unit of area, for example, the density of a drum head. For mass m and area a, the area density γ is

$$\gamma = \frac{m}{a}. \tag{4.4}$$

- *Cubic density* indicates mass per unit of volume. For mass m and volume v, the cubic density ρ is

$$\rho = \frac{m}{v}. \tag{4.5}$$

Three-dimensional density is measured in kg/m^3 for large bodies or g/cm^3 for small bodies.

4.6 Displacement

In this book I have many occasions to describe the motion of an object, such as a vibrating string, air column, particle of air, or loudspeaker cone, so a careful explanation of motion is appropriate. *Displacement,* the most basic attribute of motion, indicates distance from a starting point, or *origin.* Distance in and of itself has nothing to do with motion, but insofar as displacement relates to a starting position, it is an attribute of motion. When I use the term "displacement," it will always carry this technical sense.

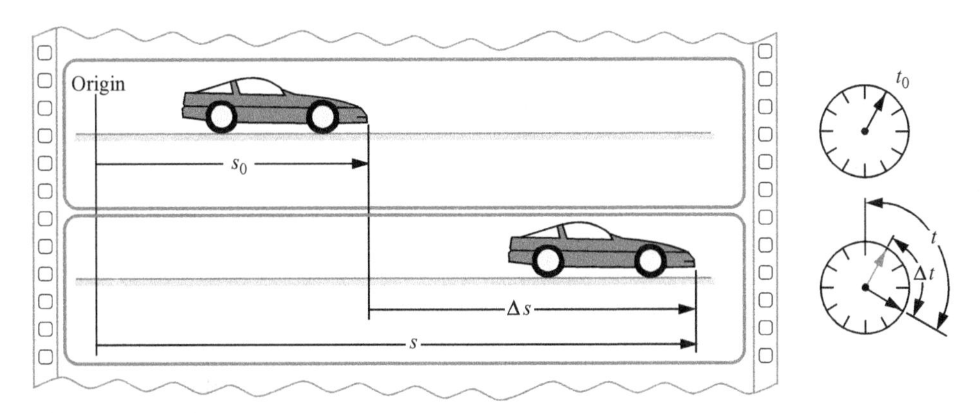

Figure 4.1
Displacement.

Suppose I start taking a movie of a car when it is some distance s_0 from an arbitrary point of origin (figure 4.1). The value of s_0 indicates the distance of the car from the origin in the first frame. Successive frames of the movie show the successive displacement of the car as it moves. The difference between the car's position in the second frame and its position in the first frame is its displacement, Δs. The Greek letter delta (Δ) is used to signify that the variable to which it is attached describes a difference between other values, in this case, the difference between s and s_0. We can also take Δs to mean "the amount of change in s." Because Δ is so commonly used in this way, it is called the *first backward difference operator,* so that in general, if we have a measurement x_n and a previous measurement x_{n-1}, then

$$\Delta x = x_n - x_{n-1}. \qquad \textit{First Backward Difference} \quad (4.6)$$

We can describe displacement as a vector. If we say that the entire distance that the car travels in figure 4.1 is the vector s, we can define the displacement of the car from the origin in the second frame as $s = s_0 + \Delta s$. Rearranging, we get

$$\Delta s = s - s_0, \qquad \textit{Displacement} \quad (4.7)$$

which says that the displacement Δs is the difference between the final position s and the initial position s_0.

The standard SI unit of displacement is the meter (m), but any SI unit of distance can be used so long as the appropriate conversion factors are used.

4.7 Speed

The ratio of distance to time is *speed*. More precisely, we speak of *average speed* as the distance traveled divided by the time elapsed. Here's why we must call it average speed: suppose it takes a jogger five minutes to run three blocks and then walk two more. Although moment-for-moment her speed is uneven, we can still say that her average speed is one block per minute.

We can express average speed $\bar{v}$ in terms of distance s and time t as

$$\bar{v} = \frac{s}{t}. \qquad \textit{Average Speed} \ (4.8)$$

A bar over a variable indicates that it represents an average.

4.8 Velocity

Velocity, like speed, relates distance to time. But velocity also specifies direction; directionless velocity is just speed. For example, "the speed of sound" does not stipulate a direction for the sound to travel. Speed is simply the magnitude component of a velocity vector without respect to its direction. Velocity and speed are measured as the ratio of distance to time, such as meters per second: m/s.

Suppose the position of the car at displacement s_0 (figure 4.1) corresponds to some time t_0. Then if the car reaches displacement s at time t, we say that the elapsed time Δt is the difference between those times:

$$\Delta t = t - t_0. \qquad \textit{Elapsed Time} \ (4.9)$$

The ratio of distance covered to elapsed time is the *average velocity:*

$$\bar{v} = \frac{\Delta s}{\Delta t} = \frac{s - s_0}{t - t_0}. \qquad \textit{Average Velocity} \ (4.10)$$

If the displacement $s - s_0 \geq 0$, the velocity is positive, otherwise it is negative. (A velocity of zero is technically a positive value.) Ordinarily, positive velocity is indicated on the page as going to the right. Note that there is no such thing as negative speed, so speed is always an unsigned value.

4.9 Instantaneous Velocity

Consider again the case of a jogger who runs a few blocks and then walks a few. Her average velocity clearly does not give a good indication about her speed moment-to-moment. It would be nice if we could determine the *instantaneous velocity* of the jogger at any particular moment.

Suppose we have made a movie of the jogger as we did of the car. If we look at a single frame, the motion is arrested and we can't get a sense of her movement, but if we take the difference of her displacement between two adjacent frames, we can. For instance, if she runs past a meter stick, we can estimate her velocity during the moment between the two frames by measuring the displacement and dividing by the elapsed time. Suppose the camera snaps a picture every 1/24 of a second and the distance she covered was 0.1 m between frames; then her instantaneous velocity is 2.4 m/s. Still, there may be some variation in her speed even during this time interval, however slight. We can generally reduce variation and improve accuracy by measuring velocity over smaller and smaller time intervals.[1]

We can continually refine an estimate of velocity by looking at ever smaller intervals of elapsed time Δt by having the camera take successive pictures more rapidly. Since the distance the jogger covers between frames Δs will also be correspondingly smaller, we begin to lose the big picture, but we do get a clearer picture of the jogger's velocity during the time between measurements.

But at some point we'll reach the *limit* of the camera's fastest shutter speed, faster than which the camera can't snap successive images. In the limit when the time elapsed between adjacent movie frames ($\Delta t = t - t_0$) is infinitesimally small, the distance the jogger covers ($\Delta s = s - s_0$) will also be infinitesimally small. But (and this is important) the ratio $\Delta s/\Delta t$ will *not* be infinitesimally small because it is a ratio of two small but nonzero values.[2] As we snap pictures at a faster and faster rate, and as both the time elapsed and the distance covered decrease, their ratio, which is distance divided by time, approaches closer and closer to the value of the instantaneous velocity.

Suppose we have an unbelievably fast camera. When we have increased the rate at which it takes pictures so that the elapsed time is *infinitely close to zero,* we say we have actually reached the instantaneous velocity. We memorialize this by saying that the instantaneous velocity is

$$v = \lim_{\Delta t \to 0} \frac{\Delta s}{\Delta t}, \qquad \textit{Instantaneous Velocity} \quad (4.11)$$

which means "in the limit as Δt approaches infinitely close to zero, the instantaneous velocity v equals the ratio of $\Delta s/\Delta t$."

It's worth thinking for a moment about what happens if we go too far with this shrinking process. Though it is clearly impossible, suppose we had a camera that could take successive snapshots with zero elapsed time, that is, $\Delta t = t - t_0 = 0$. Then the jogger would have covered no distance, and $\Delta s = s - s_0 = 0$. Then successive images of the jogger would be identical; it would be like looking at the same picture. We wouldn't be able to distinguish any motion, thereby defeating the purpose of the measurement. So for (4.11) to yield meaningful results, we can't say $t = 0$; we must say $t \to 0$, that is, t approaches zero (it just never quite gets there).

I always found the idea of limits to be a slippery concept to hang onto because the idea of a number's approaching zero seems indefinite. A number either is zero or it is not zero, right? As I was writing this section I remembered a Zen meditation practice where the novice is instructed to meditate upon the "middle distance," that is, to focus on the space that is not too close nor too far away. I recommend a variation of that perspective here. If we can just let Δt in equation (4.11) become infinitely close to zero without reaching it, many otherwise unexpected truths can emerge—a sort of mathematical satori!

We can define *instantaneous speed* to mean "magnitude of the instantaneous velocity." In this book, when I say "speed," I mean instantaneous speed, and when I say "velocity," I mean instantaneous velocity.

4.10 Acceleration

When the velocity of an object increases, we say it accelerates. When its velocity decreases, we say it has negative acceleration, or decelerates. *Average acceleration* is change in velocity per unit of time:

$$\text{Average acceleration} = \frac{\text{Change in velocity}}{\text{Elapsed time}}. \quad (4.12)$$

Suppose I bow a string on a violin with starting velocity v_0 at time t_0 and end with final velocity v at time t. During the elapsed time interval $\Delta t = t - t_0$, the change in velocity is $\Delta v = v - v_0$. The average acceleration is:

$$\bar{a} = \frac{\Delta v}{\Delta t} = \frac{v - v_0}{t - t_0}. \quad \textit{Average Acceleration } (4.13)$$

The average acceleration $\bar{a}$ is a vector that points in the same direction as Δv.

We can determine the acceleration at a particular moment of time using the approach taken above for instantaneous velocity (4.11). We define instantaneous acceleration as

$$\bar{a} = \lim_{\Delta t \to 0} \frac{\Delta v}{\Delta t}. \quad \textit{Instantaneous Acceleration } (4.14)$$

When the time interval Δt becomes extremely small so that it approaches zero in the limit, the average acceleration and the instantaneous acceleration become equal. Thus, instantaneous acceleration is the limiting case of the average acceleration. The ratio $\Delta v/\Delta t$ will *not* be infinitesimally small because it is a ratio of two small but nonzero values. If acceleration is constant, the acceleration will have the same value at any instant of time. In this book, when I say "acceleration," I mean instantaneous acceleration.

The SI unit of average acceleration is meters per second squared, or m/s^2. To see why, let us take an example. Suppose I accelerate my violin bow from a velocity of 0.1 m/s to a final velocity of 1 m/s over a time interval of 0.5 s. The average acceleration is:

$$\bar{a} = \frac{\Delta v}{\Delta t} = \frac{v - v_0}{t - t_0} = \frac{1.0\text{m/s} - 0.1\text{m/s}}{0.5\text{s} - 0\text{s}} = 1.8\text{m/s}^2. \quad (4.15)$$

4.10.1 Acceleration as the Bending of a Curve

Let's say that the car we filmed (figure 4.1) was accelerating. After filming it, we put the film in a projector and start viewing it. The projector shows us the frames in the order $i = 0, 1, 2, \ldots$, with elapsed time Δt between each one, so we view the motion at the same speed it was filmed.

As we watch the car accelerate away from the origin, suppose I suddenly stop the projector at an arbitrary frame i so that now we see the car frozen in time at some moment $t_i = i \cdot \Delta t$. Now we can determine the car's average acceleration from just this frame, the previous one, and the next one.

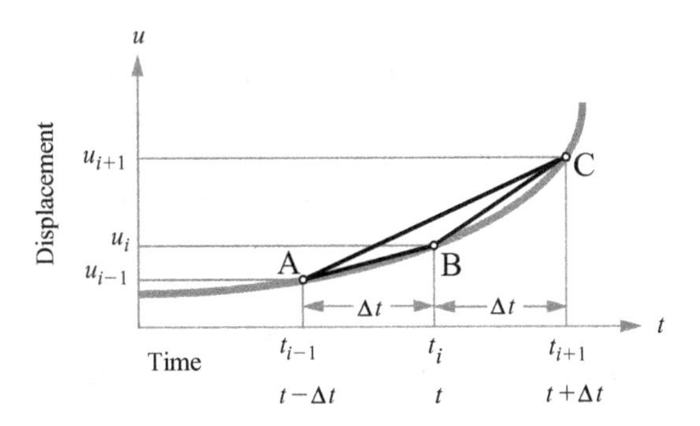

Figure 4.2
Displacement of a car accelerating from a stop.

Suppose B is the displacement of the car at the moment I stopped the film. The previous frame was frame $i - 1$, at time $t_{i-1} = t_i - \Delta t$. Call A the displacement of the car at that moment. The next frame we would see if we started the projector again is frame $i + 1$ at time $t_{i+1} = t_i + \Delta t$. Call C the displacement of the car at that moment. Figure 4.2 shows a graph of the car's displacements A, B, and C at the moments t_{i-1}, t_i, and t_{i+1}.

If we let u_i be the displacement at B, then the displacement of the car at points A and C would be u_{i-1} and u_{i+1}, respectively. Now, the differences between these displacements can be named as follows:

Backward difference	$\Delta u_{AB} = u_i - u_{i-1}$	Displacement from A to B
Forward difference	$\Delta u_{BC} = u_{i+1} - u_i$	Displacement from B to C
Central difference	$\Delta u_{AC} = u_{i+1} - u_{i-1}$	Displacement from A to C

If we relate these differences to the elapsed time between the appropriate points, we can figure out the average velocity of the car during the three frames:

Backward velocity	$\bar{v}_{AB} = \dfrac{u_i - u_{i-1}}{t_i - t_{i-1}}$	Average velocity from A to B
Forward velocity	$\bar{v}_{BC} = \dfrac{u_{i+1} - u_i}{t_{i+1} - t_i}$	Average velocity from B to C
Central velocity	$\bar{v}_{AC} = \dfrac{u_{i+1} - u_{i-1}}{t_{i+1} - t_{i-1}}$	Average velocity from A to C

So we have three slopes corresponding to the average velocity between the three points (figure 4.2).[3]

Figure 4.2 shows that the backward velocity is a shallower slope than the forward velocity: this shows that the car must be accelerating. In fact, the average acceleration is just the difference between these two slopes divided by the elapsed time. Recall from equation (4.14) that average acceleration is the difference of two velocities over time, and velocities are slopes on a graph of displacement vs. time, so acceleration is the amount of change in the slope of a curve over time.

Intuitively, the instantaneous acceleration at point B (figure 4.2) is the amount that the curve bends at that particular point. The more a curve is bent, the greater must be the acceleration along it. Don't forget this. It'll come in very handy when we consider the vibration of strings, reeds, bars, and membranes.

4.10.2 Estimating Instantaneous Acceleration

We've seen that we need just one observation to measure an object's displacement from its origin, and two to measure its average velocity. But it takes three observations to measure its average acceleration.

We can estimate the instantaneous acceleration of an object where we have three observations separated by a finite time difference. (By "finite" I mean not infinite and not infinitesimal.) Referring again to figure 4.2, if we have three displacements $u_{t-\Delta t}$, u_t, and $u_{t+\Delta t}$ at points A, B, and C, separated by the time interval Δt, then the acceleration at point B is approximately

$$a_B \approx \frac{u_{t+\Delta t} - 2u_t + u_{t-\Delta t}}{\Delta t^2}. \qquad \textit{Second-Order Central Difference Approximation} \quad (4.16)$$

The origins of (4.16) and why it is an approximation and not an equality, go beyond the scope of this book. But this approximation will come in very handy when we study the vibration of strings and air.

4.11 Relating Displacement, Velocity, Acceleration, and Time

Having developed the concepts of displacement, velocity, acceleration, and time separately, we can now combine them to understand all aspects of the motion of an object traveling with constant acceleration along a straight line. To simplify things a bit, assume that the object starts accelerating from the origin $s_0 = 0$, so now the displacement $\Delta s = s - s_0 = s$.

Since we're only considering *constant* acceleration here, the average acceleration equals the instantaneous acceleration, that is, $\bar{a} = a$.

4.11.1 Velocity under Constant Acceleration

Suppose the car depicted in figure 4.2 has initial velocity v_0 and constant acceleration a, and we want to know its final velocity v after elapsed time t. Since equation (4.14) relates all these variables, $a = (v - v_0)/t$, we just have to solve (4.14) for v to get the final velocity:

$$v = at + v_0 \quad \text{for constant acceleration.} \qquad (4.17)$$

4.11.2 Displacement under Constant Acceleration

We can get the displacement of the car by solving (4.10), $\bar{v} = (s - s_0)/(t - t_0) = s/t$, for displacement s:

$$s = t\bar{v} \quad \text{for constant acceleration,} \qquad (4.18)$$

but in order to solve this, we must know what the average velocity $\bar{v}$ is. Well, we have the final velocity v and the initial velocity v_0, so clearly the average velocity $\bar{v}$ must be the average of these two velocities:

$$\bar{v} = \frac{v + v_0}{2} \quad \text{for constant acceleration.} \tag{4.19}$$

(Note that (4.19) only applies when the acceleration is constant.) Now we can determine the displacement of the car by substituting (4.19) into (4.18) to get

$$s = t\bar{v} = \frac{t(v + v_0)}{2}. \tag{4.20}$$

With equations (4.14), (4.17), and (4.20) we have solutions for acceleration, velocity, and displacement of an object when acceleration is constant. By suitable choice of terms, we can use these directly or combine them to solve any problem involving constant acceleration. For instance, none of these equations directly deals with finding displacement when only acceleration, time, and initial velocity are known. But we can find it easily enough. Start with (4.20) and substitute into it the value of v from (4.17):

$$\begin{aligned} s &= \frac{t(v + v_0)}{2} \\ &= \frac{t[(at + v_0) + v_0]}{2} \\ &= v_0 t + \frac{at^2}{2} \end{aligned} \tag{4.21}$$

for constant acceleration. Using (4.21), we only need initial velocity, acceleration, and time to determine displacement. Equation (4.21) has the interesting property that it can tell us the displacement even when there is no acceleration. The first term $v_0 t$ gives the displacement if the acceleration is zero and velocity remains constant at v_0, and the term $at^2/2$ gives the additional displacement that results from nonzero acceleration.

One other combination of these variables will prove useful later. First, solving (4.17) for t yields $t = (v - v_0)/a$, and then using this definition for t in (4.20) yields

$$\begin{aligned} s = t\bar{v} &= \frac{v - v_0}{a} \cdot \frac{v + v_0}{2} \\ &= \frac{v^2 - v_0^2}{2a}. \end{aligned} \tag{4.22}$$

This yields displacement if we don't know time but do know acceleration and the initial and final velocities. Finally, solving for v^2 yields

$$v^2 = 2as + v_0^2. \tag{4.23}$$

4.12 Newton's Laws of Motion

Suppose a small rocket plus its propellant has 1 kg of mass. The mass of the rocket's propellant is a tiny fraction of the mass of the rocket, so we can neglect the fact that, as it burns, the rocket contains less mass through time. Now send it into deep space so as to effectively eliminate friction and the effects of gravity on its movement. When the rocket's engine is ignited, it supplies a constant force and the rocket moves away in a straight line. As its propellant is expelled, the mass of the rocket decreases ever so slightly, but it is such a small change that for our purposes the rocket's mass remains virtually the same. Since we know the mass of the rocket, we can measure the *force* that the engine applies by measuring the rocket's acceleration per unit of time according to the equation

$$F = ma, \qquad \textit{Newton's Second Law of Motion} \quad (4.24)$$

where force is F, acceleration is a, and mass is m. Equation (4.24) is known as Newton's second law of motion.

If a mass weighing 1 kg is accelerated by one meter per second per second (1 m/s^2), then the strength of the force is said to be 1 newton (N).

We can derive additional information about acceleration by rearranging (4.24) as $a = F/m$. This says that acceleration increases as F increases and shrinks as m increases. For example, if the propellant constituted a substantial amount of the mass of the rocket, then as the propellant was expelled, the rocket's mass would decrease and its rate of acceleration would correspondingly increase.

We can relate the concepts of force, mass, and motion as follows. When the rocket engine has expelled all its propellant, its acceleration will become zero. But because there is virtually no friction or other force in deep space, the rocket continues traveling indefinitely in the same direction at its final velocity.

This is known as *Newton's first law of motion,* which can be stated as follows:

An object continues in a state of rest or motion at a constant speed along a straight line unless compelled to change that state by a net force.

"Net force" means the sum of all forces acting on the object. Now, a greater force is required to change the direction of an object with greater inertia. Newton's first law is sometimes also called the *law of inertia* when expressed as follows:

Inertia is the natural tendency of an object to remain at rest or in motion at a constant speed along a straight line.

The mass of an object is a quantitative measure of inertia.

Newton's third law of motion is often stated as follows:

For every action, there is an equal but opposite reaction.

Here "action" means force and "reaction" means force in opposition. For example, suppose an astronaut suspended in space pushes against the side of a satellite with force F. The satellite pushes back with a force $-F$, that is, with a force equal in magnitude but opposite in direction. After the force is expended and they are no longer touching, the satellite and astronaut move away from each other at a rate proportional to their relative masses.

4.13 Types of Force

Force is an action in a particular direction upon an object, such as a push or a pull. The net force on an object is the combination of all forces upon it. The effects of force can be seen when an object accelerates, decelerates, twists, or deforms. Force is measured in units of newton. Types of force include gravity, friction, air resistance, turning (as in a screw), pressure, normal force, buoyancy, and tension.

Forces can be categorized as to whether they are contact forces or noncontact forces. Gravitational, electrical, and magnetic forces are examples of the latter because they can be effective whether the forcing object and the forced object are touching or not.

4.13.1 Weight

Weight is the force exerted by gravity. The force of gravity F_g on a mass m is

$$F_g = mg, \qquad \textit{Force of Gravity} \quad (4.25)$$

which we know from (4.24) is expressed in terms of acceleration. Acceleration due to gravity at sea level is about $g = 9.8 \text{ m/s}^2$. So, according to (4.24), if we have a mass of 1 kg, for example, the force of gravity exerted on it at sea level will be

$$f = m \cdot a$$

$$F_g = 1 \cdot 9.8 = 9.8 \text{ N}.$$

4.13.2 Normal Force

In mathematics, *normal* means perpendicular to a plane. So a *normal force* is one that is perpendicular to surfaces that are in contact. For example, a weighing scale exerts a force opposite to gravity until the spring force of the scale and the force of gravity balance, and the scale supports the object. The normal force is a function of the electrical forces between charged particles within the atoms of the compressed springs. We measure the weight of a body by observing the strength of the normal force, indicated by the amount the springs are deformed. If the supporting surface is inclined, or if it is accelerating, the normal force will not necessarily correspond to the weight of an object.

4.13.3 Frictional Force

Suppose we must push a heavy wooden box across a rough concrete floor. At first, even if we push hard, it might not budge because the box is pushing back with an equal but opposite *static frictional force*. Once in motion, its opposition to movement is called the *sliding frictional force*, or *kinetic frictional force*.

The static frictional force is often greater than the corresponding sliding frictional force. Less force is needed to keep it moving than is needed to start it moving.

If a car is driven with its emergency brake set, it's hard to get it rolling because the brakes grab, but once it attains some speed, the engine seems to have less work to do. If it slows to a stop, at some point the more powerful static frictional force takes over and the car abruptly halts.

This is a good explanation of how a violin string vibrates under the influence of a bow. If the bow is stationary, a powerful static frictional force sticks the bow and string together. As the bow moves, it drags the string with it until the elastic force of the string overcomes the static frictional force. The lesser sliding frictional force takes over as the string glides back opposite to the direction of the bow. When the elastic force is spent, the string slows. As it slows, the more powerful static frictional force kicks in again and entrains the string with the movement of the bow.

Friction is a nonlinear force because it does not vary uniformly with the velocity of the object but tends to increase at low velocities. The forces of friction are a result of atomic-level interactions between the sliding surfaces.

4.13.4 Tension

The tension of a guitar string can be thought of as a force that seeks to pull the two ends of the instrument together. Indeed, in older guitars the strings sometimes bend the neck, pulling the nut toward the bridge in a shallow bow. Alternatively, we can think of tension as the tendency of the string to be pulled apart. One end of the string applies a force T to the guitar, and as dictated by Newton's third law, the guitar applies a reaction force $-T$ to the string. The same is true of the other end of the string; hence the tension tends to pull the string apart. Tension is also a result of atomic-level forces.

4.14 Work and Energy

If I apply a force F to lift something a distance s off the floor, then I have performed *work* to counteract gravity. If I depress a string on a guitar in order to pluck it, I have similarly performed work to counteract the string's force of tension.

Work is the force applied to move an object times the distance it is moved.

Mathematically,

$$W = Fs\,. \qquad \textit{Work} \quad (4.26)$$

If there is no distance covered ($s = 0$), then no work is done. Thus, if I try to lift a piano and can't budge it, I may exhaust myself, but I've done no work.

Force and distance are measured in newtons and meters, respectively, and since work is the product of these two, it is measured in *newton-meters*. Fortunately, this rather unwieldy unit for work has been given a simpler name in the SI system: the *joule* (J), named after James Joule (1818–1889) for his research on work and energy.

4.14.1 Kinetic Energy

Energy is the ability to do work. When a force performs work on an object, the result is a change in the kinetic energy of the object. The work done by the net forces on an object equals the change in the kinetic energy of the object.

Because of Newton's second law, we know that when a constant net force F is applied to an object of mass m, it experiences acceleration according to $a = F/m$. As a result, the object's speed changes from an initial value v_0 to a final value v over a time interval. We can relate Newton's second law to speed by way of (4.23), $v^2 = 2as + v_0^2$, as follows. First, because of (4.26) and because $F = ma$, we can write

$$W = Fs = mas.$$

Note that both (4.23) and the above equation contain the term as. Solving (4.23) for as gives $as = \frac{1}{2}(v^2 - v_0^2)$, and substituting this into $Fs = mas$ gives

$$W = Fs = \frac{1}{2}mv^2 - \frac{1}{2}mv_0^2. \quad (4.27)$$

Equation (4.27) relates work to the difference between two terms, each of which has the form $\frac{1}{2}$(mass)(speed)^2. This quantity is called *kinetic energy.* The kinetic energy of an object with mass m and speed v is given by

$$E = \frac{1}{2}mv^2. \quad \textit{Kinetic Energy} \ (4.28)$$

By (4.27) and (4.28), work is just the difference between an initial kinetic energy and a final kinetic energy. That is, for some initial kinetic energy E_0 and final kinetic energy E, $W = E - E_0$. Because work is just the difference between two kinetic energies, work and kinetic energy are expressed in the same SI unit: the joule (J).

From (4.28) we see that

Kinetic energy is proportional to the square of velocity.

For instance, when a car's velocity doubles, its kinetic energy quadruples. Suppose it is going 30 kilometers per hour and takes 30 meters after braking to come to a complete stop (that being the distance it takes to completely dissipate the motion energy into heat). Then, if its speed doubles to 60 kilometers per hour (assuming the same road conditions), it will take four times as long to stop (120 meters) because the car has four times the amount of kinetic energy to dissipate.[4]

There are many forms of energy, including electrical, thermal, chemical, radiant, nuclear, and mechanical. Acoustical energy is a kind of mechanical energy.

4.14.2 Potential Energy

Kinetic energy is measured by its mass and velocity, as in equation (4.28). But an object may also possess *potential energy* simply by virtue of its position. Like kinetic energy, potential energy represents the ability to do work, but only potentially. Once potential energy is enabled to perform work, it becomes kinetic energy. For example, an object suspended in the earth's gravitational field has *gravitational potential energy* by virtue of its position with respect to the earth. The greater the height, the greater its potential to do work if it were to be released and allowed to drop. If it is allowed to fall, its potential energy changes to kinetic energy in proportion to the height of its drop. Recalling that the force of gravity $F_g = mg$, we can say that the potential of an object to do work because of the force of gravity is

$$W_g = mgh, \qquad \textit{Gravitational Work} \quad (4.29)$$

where h is the height of the object. Note that whether a ball rolls down a hill or falls vertically, the work done by gravity would be the same, because only a change in vertical distance can be attributed to the force of gravity; any other motion would have to be attributed to another force.

We can define the gravitational potential energy as

$$E_p = mgh. \qquad \textit{Gravitational Potential Energy} \quad (4.30)$$

There are many other forms of potential energy besides gravity. A stretched or compressed spring has *elastic potential energy*. A string under tension has *tensile potential energy*.

4.15 Internal and External Forces

Forces are either internal or external. External forces can increase or decrease the available energy in a system because the force comes from outside the system. If the external force is positive, the system's energy will increase; if it is negative, its energy will decrease. External forces include applied force, normal force, tensional force, frictional force, and air resistance. The system experiencing the force may increase or decrease either or both kinetic or potential energy. For instance, friction always acts as a negative force to reduce the total energy in a system.

Internal forces cannot change the total energy of a system, but they can change kinetic energy into potential energy, and vice versa. Internal forces include gravitational force, magnetic force, electrical force, tensile force, and spring force. For instance, when the force of gravity displaces an object from a high location to a lower one, some of its potential energy is transformed into kinetic energy, but the total energy remains the same. Its movement under the influence of gravity will undoubtedly include friction and other external forces, but the change due to the internal force of gravity (or other internal force) does not itself change the total amount of energy.

4.16 The Work-Energy Theorem

The *total mechanical energy* E of a system is the sum of its potential and kinetic energy:

$$E = E_k + E_p. \qquad \textit{Total Mechanical Energy} \quad (4.31)$$

An external force applied to an object can change the total mechanical energy if work is done (that is, if there is displacement, and the displacement is directly related to the applied force). If the work is positive, energy is added to the system; if it is negative, energy is taken away. The gain or loss in energy may be either kinetic or potential energy, or both. For instance, a rocket traveling upward gains gravitational potential energy; if it is accelerating, it is also gaining kinetic energy. Work done in these circumstances equals the change in mechanical energy, both potential and kinetic.

When work is performed on an object only by internal forces (such as springs or gravity) the total mechanical energy of the system is unchanged. But the energy changes form: some kinetic energy will be converted to potential energy, or vice versa. For example, a marble rolling down the inside of a bowl loses potential energy as gravity draws it downward, but its kinetic energy correspondingly increases as it drops. This is reversed when the marble climbs the far side of the bowl. Similarly, in the moment that a vibrating piano string is at rest at its point of maximum stretch, its energy is only potential; but when the tension force starts to pull it back, this potential energy is converted to kinetic energy. Where there is no change in total mechanical energy, the total mechanical energy is said to be *conserved.*

4.17 Conservative and Nonconservative Forces

Another way to classify kinds of force is whether they dissipate energy or not. *Conservative forces,* as the name implies, do not dissipate energy, whereas *nonconservative forces* do. Conservative forces store energy that can be retrieved later. For instance, if I roll a ball up a slope, I increase its potential energy. If later it rolls back down again, it does so by converting the previously stored potential energy to kinetic energy. Thus, gravity is a conservative force. Other examples of conservative forces include the elastic force of springs, momentum, and the electrical force between charged particles, because these forms of energy can be stored and recovered. A wound spring can unwind; a charged battery can discharge through a circuit; momentum given to a ball by a toss can be delivered into the hands of the person catching it; and so on.

Nonconservative forces dissipate or transmit energy. Nonconservative forces include frictional forces, viscous forces such as air resistance, and propulsive forces.

As with gravity, potential energy as well as kinetic energy can be associated with all conservative forces. Kinetic energy of motion can be converted into potential energy of position, for instance, when a moving object coasts up a hill, and can then be converted back to kinetic energy when it rolls down again. While kinetic energy E_k and potential energy E_p may be interconverted or transformed into each other, total energy is preserved, and $E = E_k + E_p$ according to the principle of conservation of mechanical energy, provided no work is done by nonconservative forces.

Note that when a mass is lifted against gravity, the potential energy increases regardless of the direction in which the mass is raised (so long as it is upward). And when the stored potential energy is released by lowering the mass, the path downward does not matter: potential energy due to gravity is measured only by the difference in height. *If the work done is independent of the path the motion takes, the force is conservative.*

Another way to describe conservative forces is to consider a car on a hilly closed loop. Discounting friction, the potential energy gained going up exactly balances the potential energy going down when the car gets back to the start. Discounting friction, *no net work is done by a conservative force on a closed loop.*

Only kinetic energy can be associated with nonconservative forces. For nonconservative forces, the work done depends on the path the motion takes: the longer the path a sliding object takes, the greater the force due to friction. Because no work can be stored in such a force, potential energy is not defined for it. The energy may dissipate, for instance, as heat, or be transmitted, for instance, via sound waves.

The energy a musical instrument receives from a performer, such as when a string is struck on a piano, is a nonconservative, propelling force because energy leaves the performer. The instrument receiving the energy seeks to return to its original energy level by dissipating it. But not all energy is immediately dissipated; some energy is stored in conservative forces, which work to vibrate the string. The energy that dissipates from the string enters the sound board, is transmitted into the air, and arrives at our eardrums.

As with this musical example, both conservative and nonconservative forces typically combine in everyday situations to produce a net force on objects. The total work W done by this net force is the sum of conservative work W_c and nonconservative work W_{nc}, and

$$W = W_c + W_{nc}.$$

According to the work-energy theorem, the work done by the net forces on an object equals the change in the kinetic energy of the object, so we can also write

$$E_k = W_c + W_{nc}.$$

4.18 Power

Power is work done per unit of time. Thus, it also measures the rate at which energy is transferred or transformed. The average power $\overline{P}$ is the ratio of work W to time t:

$$\overline{P} = \frac{W}{t} \quad \text{or} \quad \overline{P} = \frac{E_k}{t}. \qquad \textit{Average Power} \ (4.32)$$

The SI unit of power is J/s = W (joules/second = watt) in honor of James Watt (1736–1819).

For instance, if it takes 1 second to transfer 1 joule of energy to lift a body up 1 m, then the power expended is 1 watt. A 100 W lightbulb uses in 1 second the same amount of energy as would be required to lift a mass that is pulled to earth by 1 newton of force 100 m up in the air in 1 second.

4.19 Power of Vibrating Systems

Kinetic energy tends to dissipate from where there is more to where there is less. So to perpetuate a vibration requires a way to replenish its energy. This fact leads to an important distinction between different kinds of musical instruments.

4.19.1 Classes of Musical Instruments

There are two classes of musical instruments:

- *Sustaining instruments* receive energy continuously from the player and produce a continuous output. The performer of an Australian didgeri-do performs circular breathing, and a pipe organ has a motorized wind chest to supply a sustaining tone. Wind instruments can sustain for the duration of a player's breath, and practiced string players can sustain tones indefinitely by periodically reversing bow direction.
- *Nonsustaining instruments* receive energy from the player only at note onset. Their sound lasts until the energy is dissipated. Examples include the piano, guitar, banjo, and most percussion. Dissipation by natural frictional forces causes the sound to die away slowly; the player can usually stop the note more rapidly if desired by increasing the rate of dissipation, for example, by resting the hand on a vibrating string.

4.19.2 Efficiency

Efficiency is the ratio of useful power output p_o to total power input p_i. To express efficiency e as a percentage, we can write

$$e = 100 \cdot \frac{p_o}{p_i}.$$

Efficiency determines, among other things, the ease with which a brass or wind instrument can be made to speak. For example, a trumpet is very efficient when the vibrating frequency of the performer's lips matches one of the trumpet's natural vibrating modes. It is very inefficient at all other frequencies, which helps produce a stable tone.

On the other hand, a piano sounding board is designed to have about the same efficiency for every frequency so that no one frequency is favored over any other. Good loudspeaker systems are also designed to have the same efficiency at all frequencies so they don't favor one frequency over another.

These ideas are developed more fully in volume 2, chapter 6.

4.19.3 Power of Musical Instruments

Most musical instruments are not as powerful as a 100 W lightbulb. Here are some examples:

Orchestra	75 W
Percussion	1–10 W
Trombone fortissimo	6 W
Piano	1 W
Violin	0.1 W
Violin pianissimo	0.001 W

Why, then, are audio power amplifiers built to generate on the order of 300 to 600 W per channel? It's because loudspeakers are very inefficient, on the order of 1–10 percent, so that they will not color the sound by emphasizing selected frequencies.

4.20 Wave Propagation

Waves propagate in a medium by displacing differences in force or pressure from one place to another. The crest in the rope in figure 4.3 moves along the rope by displacing the force that shook it. An acoustical wave such as the one shown in figure 1.2 propagates in air by displacing pressure differences from one space to another.

There are three ways that waves can propagate through a medium. The different ways relate the direction of motion of the particles *in* the medium to the direction of wave motion *through* the medium:

- *Transverse* Like surface waves in water, the direction of motion that creates the wave is perpendicular to the direction of wave motion. If we tie a rope to a wall at one end and shake the other end, we might see shapes as shown in figure 4.3.
- *Longitudinal* Like sound waves in air and under water, the direction of motion that creates the wave is the same as the wave motion. Figure 4.4a shows a spring at rest. In 4.4b its left end is given a momentary shove right, creating a compressed region. In 4.4c it is given a momentary shove left, creating a stretched region while the compressed region continues to propagate to the right. In 4.4d the left end of the spring is returned to its initial position while the compressed and stretched regions continue to move to the right.
- *Torsional* The direction of motion that creates the wave rotates about the axis of wave motion. Putting a medium under twisting stress creates torsional waves. Figure 7.9 shows a Shive wave machine, which is an example of torsional wave propagation. Torsional wave motion proceeds down the central wire that connects all the machine's transverse bars.

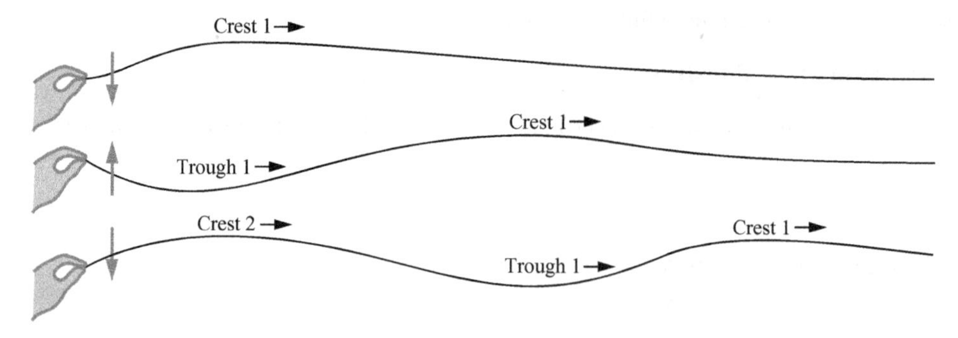

Figure 4.3
Transverse waves.

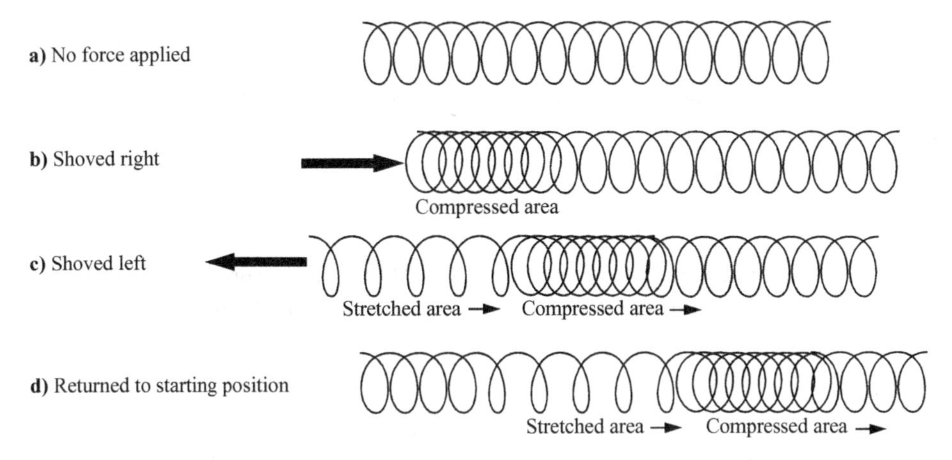

Figure 4.4
Longitudinal waves.

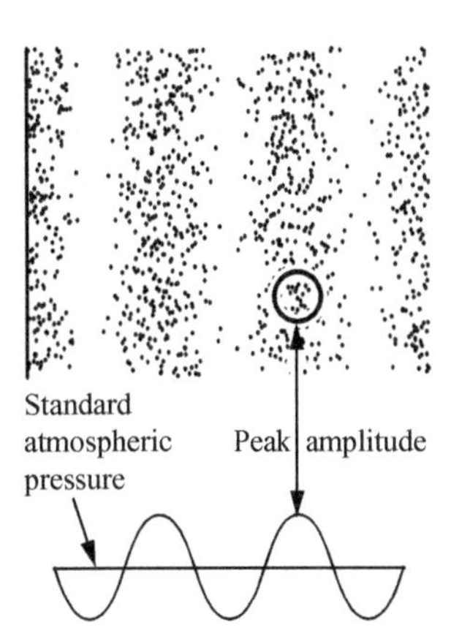

Figure 4.5
Wave amplitude.

4.21 Amplitude and Pressure

The *amplitude* of a wave is the distance from its peak height to its point of zero displacement or equilibrium (figure 4.5). For a sound wave, amplitude is the difference between the wave's peak pressure and standard atmospheric pressure. Sound amplitude is usually measured as *sound pressure level* (SPL), which is the difference between the greatest pressure in a wave and standard atmospheric pressure.

If you dive deeply into a swimming pool, you experience pressure against your eardrums. Pressure is the force applied by the molecules of water pressing perpendicularly on the surface area of your eardrum. Physicists would say that the force of the water is applied normal to the surface of your eardrum. A normal force F is distinguished from other forces by adding the symbol $\perp$ as a subscript. So, in general, pressure p is the amount of force applied normal to a surface $F_{\perp}$ divided by the area a over which it is applied:

$$p = \frac{F_{\perp}}{a}. \qquad \textit{Pressure} \quad (4.33)$$

Pressure is measured in newtons per square meter. The SI unit of pressure is the *pascal* (Pa), named after the French scientist/mathematician Blaise Pascal (1623–1662). So 1 Pa = 1 N/m^2.

4.22 Intensity

Energy from the motion of sound waves flows through the eardrums and into the inner ear, where it registers as sound. Intensity I is the energy E per unit of time t that is flowing across a surface of unit area a:

$$I = \frac{E/t}{a^2}. \qquad (4.34)$$

According to equation (4.32), $P = E/t$. So we can say $I = P/a^2$, that is, intensity I is the power flowing across the surface of area a. The standard area unit is 1 m^2, so intensity is measured in W/m^2. We must also take into account the direction the energy is flowing relative to the surface it is flowing across:

$$I_{\perp} = \frac{P}{m^2}, \qquad (4.35)$$

where $I_{\perp}$ is intensity flowing normal to the surface.

4.23 Inverse Square Law

In the absence of any barriers, sound has a spherical radiation pattern. To compare sound intensities at varying distances from a source (figure 4.6), we must make spherical measurements.

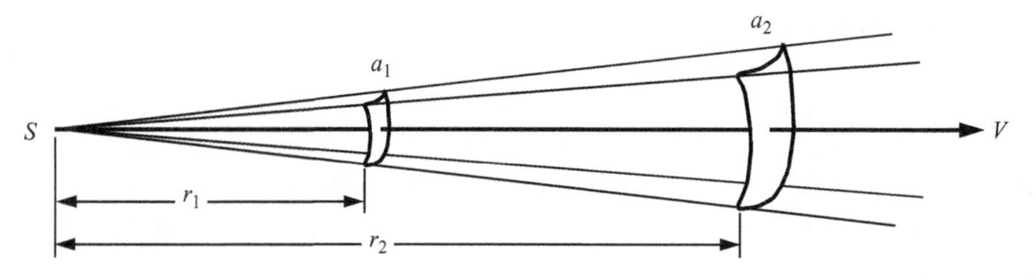

Figure 4.6
Comparison of spherical surfaces.

Let P be the power of a wave at distance r_1 propagating along direction V. This means that an amount of energy P is flowing through surface a_1 each second. If no energy is lost, then the same power will flow through a_2 each second as well, and $P/a_1 = P/a_2$. Since the areas of surfaces a_1 and a_2 are proportional to the squares of their distances from the source S, the intensity I varies inversely as the square of the distance to the source, and

$$\frac{I_2}{I_1} = \left(\frac{r_1}{r_2}\right)^2. \quad (4.36)$$

4.24 Measuring Sound Intensity

Just as the range of frequencies we can hear is limited, so is our perception of sound intensity. The threshold of hearing is the minimum amount of sound intensity required for a sinusoid to be detected by an average listener in a noiseless environment. The limit of hearing (also called the threshold of pain) is the intensity above which sound is registered as (possibly damaging) pain by most of us.

Perception of loudness is not as straightforward as perception of pitch. While loudness is primarily affected by intensity, it is affected also by other perceptual and acoustic factors, especially frequency. We are generally less sensitive to very low and very high frequencies (see section 6.5). For the ear's most sensitive frequencies, around 1000 Hz, the range between the threshold of hearing and the limit of hearing is staggeringly large:

- Sound intensity at the *threshold of hearing* at a frequency of 1 kHz is approximately $t_h = 1 \cdot 10^{-12}$ W/m^2 for a very sensitive listener.
- Sound intensity at the *limit of hearing* at a frequency of 1 kHz is approximately $l_h = 1 \cdot 10^0$ W/m^2.

Thus, the range of sound intensities our ears can register at 1 kHz is on the order of 10^{12}, which is about 1 trillion to 1. No other sense faculty has this range of sensitivity.

4.24.1 Sound Intensity Scale

To establish a sound intensity scale, we must determine its boundaries and the gradations into which it is divided. It makes sense to use the threshold of hearing and the limit of hearing as the lower and upper boundaries of the scale. Expressing the range of hearing as a ratio shows the extent of the scale:

$$\frac{l_h}{t_h} = \frac{10^0}{10^{-12}} = 10^{12}\ \text{W/m}^2. \quad \textit{Intensity Range of Hearing}\ (4.37)$$

But it is awkward to talk meaningfully about ratios with a range of 1 trillion to 1. It would be easier if we could measure sound intensities using a small set of values that could be mapped to the wide range of intensities. This is similar to the approach I took to represent pitch, where the semitones of the equal-tempered scale provided a simple mapping between linear pitch and exponential frequency.

Using the exponent of the powers of 10 as the units of the sound intensity scale would allow us to represent the enormous dynamic range of perceived sound intensity simply with the numbers 0 to 12.

We can use the log function to extract the exponent of a quantity. For instance, $\log_{10}10^{-12} = -12$. So we can extract the exponents in (4.37) by writing

$$\log_{10}\frac{l_h}{t_h} = \log_{10}\frac{10^0 \text{ W/m}^2}{10^{-12} \text{ W/m}^2} = 12 \text{ bel.} \qquad \textit{The Bel Scale} \quad (4.38)$$

The sound intensity scale developed this way is called the *bel,* invented by engineers at Bell Telephone Laboratories in the 1920s and named in honor of Alexander Graham Bell (1847–1922).

Because we are measuring log ratios, the size of the bel increases with increasing differences in intensity. For example, if $I = 10$ W/m^2 and $I' = 100$ W/m^2,

$$\log_{10}\frac{I'}{I} = \log_{10}\frac{100}{10} = 1 \text{ bel .}$$

If $I = 10$ W/m^2 and $I' = 1000$ W/m^2,

$$\log_{10}\frac{I'}{I} = \log_{10}\frac{1000}{10} = 2 \text{ bel .}$$

If $I = 10$ W/m^2 and $I' = 10{,}000$ W/m^2,

$$\log_{10}\frac{I'}{I} = \log_{10}\frac{10{,}000}{10} = 3 \text{ bel} \ldots.$$

The bel scale covers the entire audible range of sound intensities with just a dozen integer values, so we have satisfied an important design criterion. In fact, we have satisfied it a little too well: the range 0–12 is too coarse-grained for practical work. The preferred unit is the *decibel* (dB), which is ten times the resolution of a bel:

$$10\log_{10}\frac{l_h}{t_h} = 10\log_{10}\frac{10^0 \text{ W/m}^2}{10^{-12} \text{ W/m}^2} = 120 \text{ dB.} \qquad \textit{The Decibel Scale} \quad (4.39)$$

Perhaps you've heard that the intensity range of hearing is 120 dB. Now you know where that number came from.

We can generalize the decibel scale to compare two arbitrary intensities. The *sound intensity level in decibels* of a sound with intensity I' is defined as

$$10\log_{10}\frac{I'}{I}, \qquad \textit{The Decibel} \quad (4.40)$$

for some reference intensity I. We can use (4.40) to make a comparison of two relative intensities. For instance, if $I = 10^{-2}$ W/m^2 and $I' = 10^{-1}$ W/m^2, then the intensity of I' is greater than the reference I by $10 \log_{10} 10^{-1-(-2)}$ dB $= 10$ dB.

We can use (4.40) to make a loudness comparison against either t_h or l_h, depending upon whether we are measuring up from silence or down from the limit of pain. Sound intensity meters commonly measure up from silence by setting $I = t_h$ in equation (4.40). A very quiet room might have a sound intensity level of about 40 dB. Continuous exposure to a sound intensity level of over 90 dB can be harmful to hearing. So the useful range of musical intensities is from about 45 dB to 95 dB, with peaks ranging upward to a maximum of 120 dB.

Why was 10 chosen as the base? Why not pick 2 as the base, as we did for relating pitch to frequency (see equation (2.1))? One reason is that there is no obvious loudness equivalent to the interval of the octave. We don't always interpret a doubling of intensity as twice as loud. Also, powers of 10 give a much more compact scale to work with than would powers of 2.

To obtain an absolute intensity level from a decibel level requires reversing the previous process:

1. Convert decibels to bels.
2. Make the resulting value an exponent of 10.
3. Make it proportional to the reference against which it was originally measured, that is, t_h or l_h.

For example,

$$y \text{ W/m}^2 = \frac{10^{(x \text{ dB})/10}}{t_h}. \qquad \textit{Decibel-to-Intensity Conversion} \quad (4.41)$$

The decibel scale is used to measure sound level and is also used in sound recording and communications. Another scale based on the same logarithmic principle is the Richter scale, used to measure the intensity of earthquakes. Variants of the decibel are used to measure power, sound pressure, voltage, or intensity.

4.24.2 Loudness in Recording Equipment

Measurements of ambient sound, such as in a concert hall or factory, are typically measured up from the threshold of hearing. In contrast, recording engineers usually want to measure down from the limit of the loudest sound they can record without distortion on their recording equipment. Let l_r be the *limit of recording,* louder than which the recorder would distort. Now let $I = l_r$ be the reference loudness in (4.40). Then the loudness of the sound we want to compare is I'. The loudest sound we could record without distortion is $I' = l_r$. The decibel value corresponding to this is $10 \log(l_r/l_r) = 0$ dB. For any softer sound, $I' < l_r$, and the corresponding dB value will be negative. For this reason, the level meters on recorders are measured in negative dB values (figure 4.7). Approximate musical loudness levels are given for comparison. A value of 0 dB on such a scale means the recorded sound is very close to saturating the recording medium. Negative dB values indicate softer levels.

It is customary for the designers of recording equipment to leave 10 dB or so for *head room* at the top of the scale as insurance against any sound's being distorted. So, typically the reference

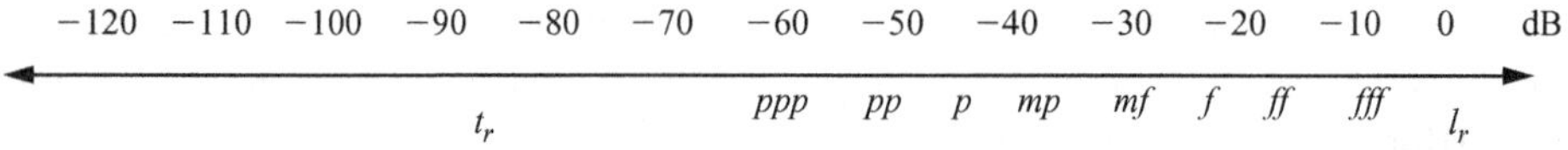

−120 −110 −100 −90 −80 −70 −60 −50 −40 −30 −20 −10 0 dB

ppp pp p mp mf f ff fff

t_r l_r

Figure 4.7
Loudness levels, measuring down from the limit of hearing.

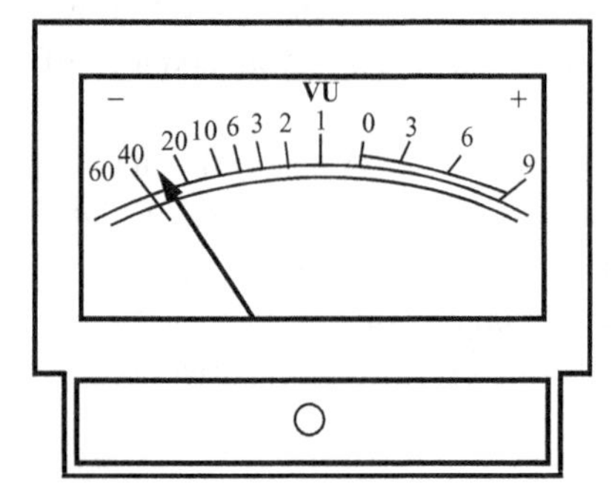

Figure 4.8
VU meter.

intensity $I = l_r - 10$ dB. This is why level meters on recorders show some positive dB values above 0 dB. The top positive value is where distortion would actually begin. Any sound with loudness above 0 dB is in danger of being distorted.

Recording equipment typically does not have the same wide dynamic range as human hearing. Let the *threshold of recording* t_r be the level below which the noise floor of the recorder's electronics overshadows the recorded sound. Often, the practical limit is on the order of –90 to –100 dB. Below that the noise floor of the recorder's electronics is louder than the recorded sound. Figure 4.7 reflects these considerations. Figure 4.8 shows an example volume unit (VU) meter with a scale that goes to –60 dB.

Because the decibel scale is logarithmic times 10, we know we can make a sound ten times louder by increasing its intensity by 10 dB. We can work out the decibel values corresponding to doubling or halving the intensity as follows. We know from algebra that $\log(i/j) = \log i - \log j$ (see appendix A). So, for instance, $\log(2/1) = \log 2 - \log 1$. Now $\log 2 \approx 0.3$ and $\log 1 = 0$, and $10 \cdot (0.3 - 0) = 3$ decibels per doubling of intensity.[5] Thus, if we double the intensity level of a sound, we raise its loudness by approximately 3 dB.

4.24.3 Sound Pressure

Unfortunately, measuring energy flow is not always convenient or even possible. Intensity is only meaningful when energy flows through an area. Yet there are occasions when energy is present but not flowing. For example, consider a thick steel pipe tightly sealed at both ends with steel caps, containing a battery-operated radio. Virtually all the energy from the radio sound reflects off the ends of the pipe and virtually no sound escapes. The resulting waveforms inside the pipe are called standing waves. Since virtually all energy that the radio's loudspeaker pumps into the air in the pipe is returned to it by the captured air, virtually no energy flows, and there is virtually no measurable intensity.

However, average pressure variation remains meaningful even with standing waves because we can still measure the pressure difference inside the pipe between the pressure crests and troughs. Since it is also generally easier to build equipment to measure pressure differences than it is to measure intensity, it would be very convenient if we could find a way to relate sound pressure to sound intensity.

Relating Sound Pressure to Intensity The relation of intensity I of a sinusoid to the value of the average pressure variation Δp in air is

$$I = \frac{\Delta p^2}{V\delta}, \tag{4.42}$$

where V is the velocity of sound in air and δ is the density of air. The pressure variation Δp is one half the pressure difference from peak to trough of the wave (that is, it is the distance from the mean to the extreme), measured in pascals (N/m^2).

Equation (4.42) says that intensity is proportional to the square of the pressure variation. This is the most important part of this equation. It also says that increasing the velocity of sound in the medium or increasing the density of the medium would decrease the intensity. However, since we're almost always dealing with air at or near standard atmospheric pressure, these factors can be neglected. The most important item is the square relationship between pressure variation and intensity.

Using standard values for V and δ, if we take the threshold of hearing as $I = 10^{-12}$ W/m^2 (intensity), then under normal atmospheric conditions, the average pressure variation corresponding to this intensity will be $2 \cdot 10^{-5}$ N/m.

So we can relate average pressure variation to intensity. But how can we actually measure the average pressure variation of a sound? There are a number of approaches we can take.

Measuring Sound Pressure We could measure *air particle displacement,* that is, the distance air molecules are pushed aside, but that is difficult to measure because of the random thermal motion of gas molecules, and we would have to be able to resolve to a distance of 1 micrometer or less just to observe them. We could measure *air particle velocity,* but that has the same problem.

By far the easiest way to measure the strength of a sound is to measure the average pressure change over a large area. Since pressure is force per unit of area, all we have to do is sample a large enough area to get a force we can register on even relatively crude measuring devices. So we define *sound pressure level* (SPL) as the average pressure variation per unit area.

Standard atmospheric pressure A is 10^5 Pa, roughly 14.7 pounds per square inch. Pressure level changes caused by sound waves are very small in comparison, ranging from about 0.1 Pa at the threshold of hearing up to 1 Pa at the limit of hearing. In comparison to atmospheric pressure, sound pressure is minuscule.

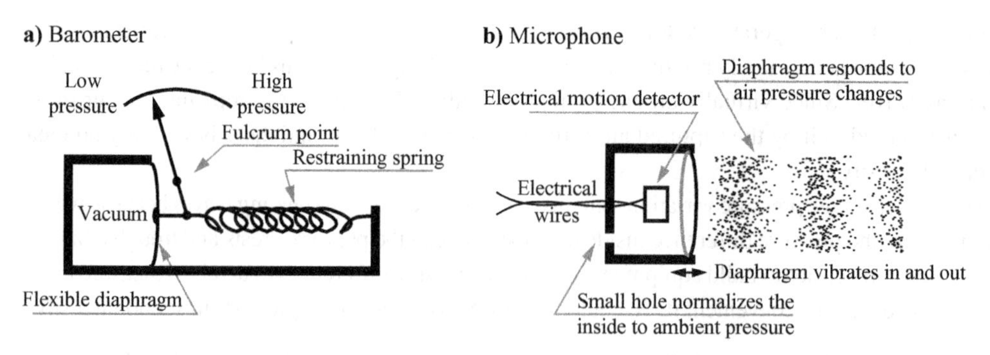

Figure 4.9
Simplified barometer and microphone.

The Microphone The easiest way to measure sound pressure level is with a microphone. Like barometers, *microphones* measure air pressure variation. Whereas a *barometer* encloses a vacuum and can therefore measure absolute pressure changes, a microphone encloses a small volume of air at the prevailing atmospheric pressure level and records the relative pressure changes of passing sound waves. A small flexible diaphragm on the end of the microphone enclosure is displaced by high and low pressure wavefronts impinging on it from a passing sound wave. The amount of displacement can be measured electrically by a number of different techniques. Figure 4.9a shows a simplified barometer, and figure 4.9b shows a simplified microphone.

Given how vanishingly small the pressure fluctuations of sound are, how can a relatively massive microphone diaphragm be displaced by such tiny forces? The answer is, we make the diaphram large so that it encounters more of the sound's force field. But this poses other design problems. If we make it too large, the diaphram becomes heavy and unresponsive. So we make it as thin as possible to reduce its mass. But then it becomes fragile. So we choose materials that are strong but flexible. This is the domain of mechanical engineering.

4.24.4 Proximity Effect

There's another important and practical microphone design problem that arises from the geometry of waveforms. Waves expand from a sound source spherically under normal conditions; what happens when a curved wave front encounters a flat microphone diaphragm?

If the diaphragm is near the sound source, and the wavelength is sufficiently short (and therefore the frequency is high), the diameter of the diaphragm is large in comparison to the diameter of the spherically expanding wave. As shown in figure 4.10a, the diaphragm cuts across several high and low pressure areas. The high pressure areas cancel the low pressure areas for a diaphragm in this position, so it receives little net energy.

The diaphragm in figure 4.10b is relatively far from the source and does not cut across pressure areas. It experiences uniform high and low pressure across its whole surface at this frequency and distance, thereby receiving maximum net energy from the passing wave.

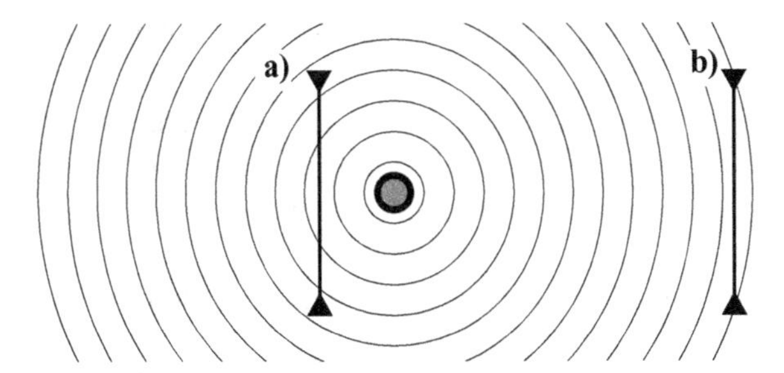

Figure 4.10
Proximity effect.

At a sufficient distance from the source, sound waves tend to flatten out into sheets (plane waves). The farther the receiver is from the source, for a same-sized receiver, the more plane the wave front becomes. The region near the source where the curvature of the wave front is significant to the receiver is the *near field*, and beyond where the curvature stops being significant is the *far field* (see section 7.6).

The significance of this phenomenon, the *proximity effect,* is that as the distance from microphone to spherically radiating sound source decreases, the high frequency sensitivity of the microphone decreases. As a speaker moves closer to a microphone, the timbre of the speaker's voice becomes warmer. That's the proximity effect in action. The proximity effect can be reduced, if necessary, by choosing a microphone with a smaller diaphragm. It can also be fixed by moving the microphone farther away, but this can cause it to start picking up undesirable room noise. This is the domain of audio engineering.

4.25 Summary

The Système International d'Unités (SI) is used to represent basic physical proportions, matter, distance, dimension, and time.

Periodicity is the amount of time elapsed between the start and end of a single event. Frequency is the number of events occurring in a single elapsed time interval. Periodicity and frequency are each other's inverses.

Mass is the quantity of matter contained in an object. Matter is anything that occupies space and exhibits inertia. Inertia is the tendency of a body to impede acceleration. Mass is an intrinsic quality of matter, unchanged by such things as location. Weight, the force of gravity acting on an object, depends upon the object's location.

Density measures how tightly packed together the material in a body is. We can distinguish linear distance, area distance, and volumetric distance.

Displacement indicates distance from a starting point or origin. The ratio of distance to time is average speed. Velocity indicates distance per time interval as well as linear direction. Instantaneous velocity is the ratio of two small but nonzero values of distance and time. We take the limit of the ratio as the time interval goes toward zero.

Average acceleration is change in velocity measured in meters per second per second. Instantaneous acceleration is the ratio of two small but nonzero values of velocity and time. We take the limit of the ratio as the time interval goes toward zero. When plotted, the instantaneous acceleration at a point is just the amount that the curve bends at that particular point.

If we know three of the four motion variables—displacement, velocity, acceleration, and time—we can always find the fourth by algebraic substitution.

Newton's first law of motion states that, because of inertia, an object continues in a state of rest or motion at a constant speed along a straight line unless compelled to change that state by a net force. The mass of an object is a quantitative measure of inertia. Newton's second law of motion equates force to mass times acceleration. Newton's third law of motion states that for every action, there is an equal but opposite reaction.

Force is an action in a particular direction upon an object. Contact forces include friction, air resistance, turning (as in a screw), pressure, normal force, buoyancy, and tension. Noncontact forces include gravitational, electrical, and magnetic forces.

Work is the force applied to move an object times the distance it is moved. Energy is the ability to do work. When a force performs work on an object, the result is a change in the kinetic energy of the object. Kinetic energy of an object is proportional to the square of the object's velocity.

Objects may possess potential energy by virtue of position. Forms of potential energy include gravitational, elastic, and tensile energy.

External forces can increase or decrease the available energy in a system. They include applied force, normal force, tensional force, frictional force, and air resistance. Internal forces cannot change the total energy of a system, but they can change kinetic energy into potential energy, and vice versa. Internal forces include gravitational force, magnetic force, electrical force, tensile force, and spring force.

The total mechanical energy of a system is the sum of its potential and kinetic energy. When work is performed on an object only by internal forces, the total mechanical energy of the system is unchanged but the energy changes form.

Conservative forces store energy that can be retrieved later. Nonconservative forces dissipate or transmit energy. Nonconservative forces include frictional forces, viscous forces such as air resistance, and propulsive forces. No net work is done by a conservative force on a closed loop. Only kinetic energy can be associated with nonconservative forces. For nonconservative forces, the work done depends on the path the motion takes. Conservative and nonconservative forces combine in everyday situations to produce a net force.

Power is work done per unit of time. Thus, it also measures the rate at which energy is transferred or transformed. Kinetic energy tends to dissipate from where there is more to where there is less. To perpetuate a vibration requires constantly replenishing lost energy. There are sustaining and nonsustaining musical instruments, differentiated by whether there is a constantly renewable source of energy to drive the instrument's vibration.

Efficiency is the ratio of useful power output to total power input.

Waves propagate in a medium by displacing differences in force or pressure from one place to another. The movement can be transverse, longitudinal, or torsional.

The amplitude of a wave is the distance from its peak height to its point of zero displacement or equilibrium. For a sound wave, amplitude is the difference between the wave's peak pressure and standard atmospheric pressure. Sound amplitude is usually measured as sound pressure level, which is the difference between the greatest pressure in a wave and standard atmospheric pressure. Pressure is the amount of force applied normal to a surface divided by the area over which it is applied.

Energy from the motion of sound waves flows through the eardrums and into the inner ear, where it registers as sound. Intensity is the energy per unit of time (power) that is flowing across a surface of unit area. Sound has a spherical radiation pattern if not blocked. Intensity varies inversely as the square of the distance from the source.

The ear detects sound intensity between the threshold of hearing and the limit of hearing. The decibel is 10 times the logarithm base 10 of the ratio of two intensities. The decibel scale covers the entire audible range of sound intensities with 120 values.

Measurements of ambient sound, such as in a concert hall or factory, are typically measured up from the threshold of hearing. In contrast, recording engineers usually measure down from the limit of the loudest sound they can record without distortion.

Sound intensity is typically less useful as a measure than sound pressure because there are pressure differences even in standing waves that we can measure. Although we could measure particle displacement or velocity, it's easier to measure average pressure variation per unit area, called sound pressure level (SPL). A microphone measures relative pressure variation from one side of a diaphragm to the other.

5 Geometrical Basis of Sound

Geometry is frozen music.
—*Goethe*

5.1 Circular Motion and Simple Harmonic Motion

Suppose a pendulum swings back and forth above a turntable. The turntable has a marker, such as a small cone, placed on its surface (figure 5.1). The cone moves with *uniform circular motion* because a motor drives it in a circle at a constant speed. Now adjust the length of the pendulum so that it makes one full swing in the same time that the turntable makes one complete revolution, and release the pendulum at exactly the same moment the cone moves under it so that the two movements are synchronized. With the two motions so aligned, if we look directly edge-on at the turntable, the pendulum and the cone seem to have exactly the same left/right motion even though we know that the pendulum moves in a line while the turntable moves circularly. Intuitively, it looks like circular motion and simple harmonic motion are in some way equivalent if seen from the right vantage point. This train of thought suggests that we can use the geometry of circles to study simple harmonic motion and wave behavior.

5.2 Rotational Motion

Circular motion and simple harmonic motion are closely related. In fact, to understand circular motion is to understand sine waves, which are the basis of all musical sound. This section reviews information provided by geometry and trigonometry about circular motion.

5.2.1 Angular Displacement

The center of a rigid rotating body, such as a turntable, defines its *axis of rotation* as a point around which circular motion revolves. The angle through which the rigid body rotates about its axis of rotation is its *angular displacement*. Suppose a turntable rotates from an initial angle θ_0 to a final angle of θ_f. We say the turntable sweeps out the angle θ, defined as

$$\theta = \theta_f - \theta_0. \qquad \textit{Angular Displacement} \quad (5.1)$$

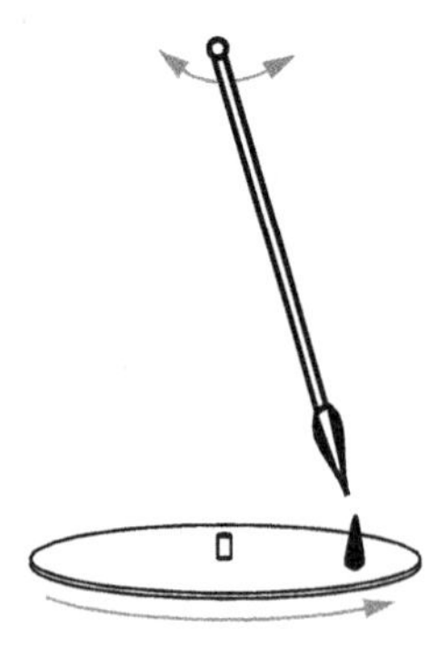

Figure 5.1
Pendulum and turntable.

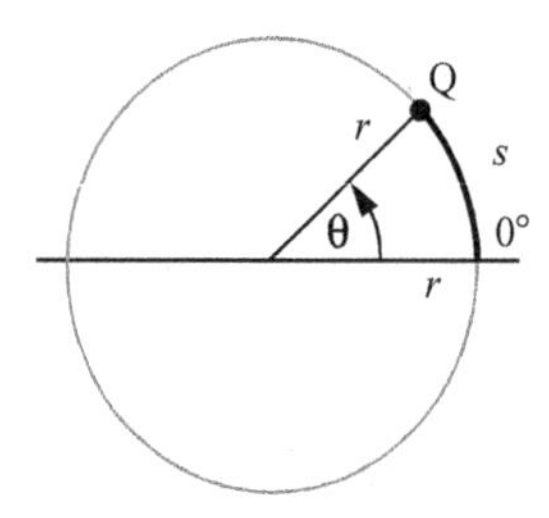

Figure 5.2
Radian measure.

Rotatable objects can turn either clockwise or counterclockwise.

Counterclockwise angular displacement is taken to be positive, and clockwise angular displacement is taken to be negative.

Thus, θ indicates counterclockwise rotation, and $-\theta$ indicates clockwise rotation.

5.2.2 Radians

It is common to use degrees to measure angular displacement or to refer to entire revolutions. One *revolution* returns a turntable to its initial position and equals 360°.

Suppose a turntable sweeps out an angle θ as shown in figure 5.2. As it does so, point Q traces out an arc of length s. Clearly, the length of s grows if either its radius r or the angle θ grows. In fact, we can show with elementary geometry that

$$\theta = \frac{\text{Arc length}}{\text{Radius}} = \frac{s}{r}. \tag{5.2}$$

When $s/r = 1$, that is, when the arc length is the same as the radius, the angle θ is equal to 1 *radian* (rad). Since both s and r are measures of distance, their ratio is a dimensionless number (because

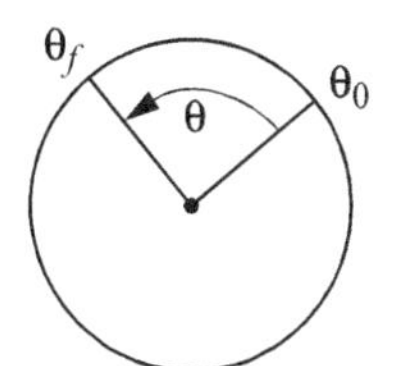

Figure 5.3
Angular rotation.

the units in the numerator and denominator cancel out). A dimensionless number is a "pure number" unencumbered with physical significance.

If the point Q sweeps out one entire revolution of radius r, its angular displacement will be $\theta = 2\pi$ and its arc length s will equal the circumference of the circle, $2\pi r$.

Since one revolution equals 360°, we can equate degrees and radians. If $\theta = 360°$, then $s = 2\pi r$ and

$$\frac{s}{r} = \frac{2\pi r}{r} = 2\pi \text{ rad}, \tag{5.3}$$

and 2π rad = 360°. Solving for rad, we see that one radian is

$$\text{rad} = \frac{360°}{2\pi} \cong 57.3°. \qquad \textit{Radian} \tag{5.4}$$

This constant, the radian measure, allows us to use simple integers and ratios of integers to specify useful divisions of a circle.[1] For example, the circumference of the circle is 2π radians, and a half circle (180°) is one half of that, exactly π radians. Similarly, one quarter of the circumference is $\pi/2$ radians, which is therefore 90°, the size of a right angle.

When angles are stated in radians, the constant π tends to drop out from equations, greatly simplifying calculations. Radian measure also simplifies calculation of the length of an arc. Solving (5.2) for s yields

$$s = r\theta, \qquad \textit{Length of an Arc} \tag{5.5}$$

so we can get the length of s simply by multiplying the radius of its circle by the arc's angle in radians.[2]

5.2.3 Angular Velocity

Suppose a turntable starts at angle θ_0 and rotates to angle θ_f (figure 5.3). Then its angular displacement is $\theta = \theta_f - \theta_0$. Further, suppose the turntable performs this rotation in t seconds. Then its *angular velocity* is the angular displacement θ divided by elapsed time t:

$$\omega \equiv \frac{\theta}{t}, \qquad \textit{Angular Velocity} \tag{5.6}$$

which we measure in SI units of rad/s.

Angular velocity is the rate at which angular displacement changes.

Compare (5.6) to linear velocity, which is the rate at which linear displacement changes. Counterclockwise angular velocities are positive, whereas clockwise angular velocities are negative. In (5.6) the symbol ≡ means "defined as." I use it to signify that I am defining ω to have a particular meaning, namely, θ/t. Later, when I use ω, it will carry this significance.

Here's another way to calculate angular displacement. Suppose the turntable shown in figure 5.7 is set so that the cone is at its rightmost position, aligned with the x-axis. Then we start the turntable and start a timer at time $t = 0$. The turntable rotates counterclockwise at a constant rate of ω rad/s, moving through angle θ in time t. Since the turntable rotates at a uniform speed, the size of the angle θ grows at a constant rate. Therefore, the angular displacement θ at time t is the angular velocity times the elapsed time t:

$$\theta = \omega t. \qquad \textit{Angular Displacement with elapsed time} \quad (5.7)$$

5.2.4 Angular Acceleration

If the turntable shown in figure 5.7 starts rotating with angular velocity ω_0 and ends at time t with angular velocity ω_f, the change in angular velocity is $\omega = \omega_f - \omega_0$. If the change is not zero, the turntable exhibits *angular acceleration* α, which is *change* in angular velocity ω through time t:

$$\alpha = \lim_{\Delta t \to 0} \frac{\Delta\omega}{\Delta t}, \qquad \textit{Angular Acceleration} \quad (5.8)$$

measured in SI units of (rad/s)/s = rad/s^2.

Angular acceleration is the rate at which angular velocity changes.

5.2.5 Rotational Speed

If a bicycle's wheel is turning once per second at a constant rate, and the tire's radius $r = 0.5$ m, how fast is the bicycle going? If the circumference of the wheel is $c = 2\pi r = 3.14$ m, then the velocity of the bicycle must be about 3.14 m/s. Every point on the circumference of the tire is also traveling at 3.14 m/s. Thus, for some radius r and some period of time T,[3] the *rotational speed* of a point on a circle is

$$v = \frac{2\pi r}{T}. \qquad \textit{Rotational Speed} \quad (5.9)$$

5.2.6 Centripetal Acceleration

Speed doesn't imply direction, but velocity does. As a point on the circle travels, its direction changes moment by moment. So, even though the speed of a point on the circle remains uniform, its velocity changes from instant to instant because its direction changes.

Figure 5.4a shows a circle rotating through points p_1 and p_2. The velocity at these points can be drawn as vectors, v_1 and v_2, representing the linear velocity of each point. The difference of the two vectors is the change in velocity $\Delta v = v_2 - v_1$. The difference of two vectors can be shown by putting

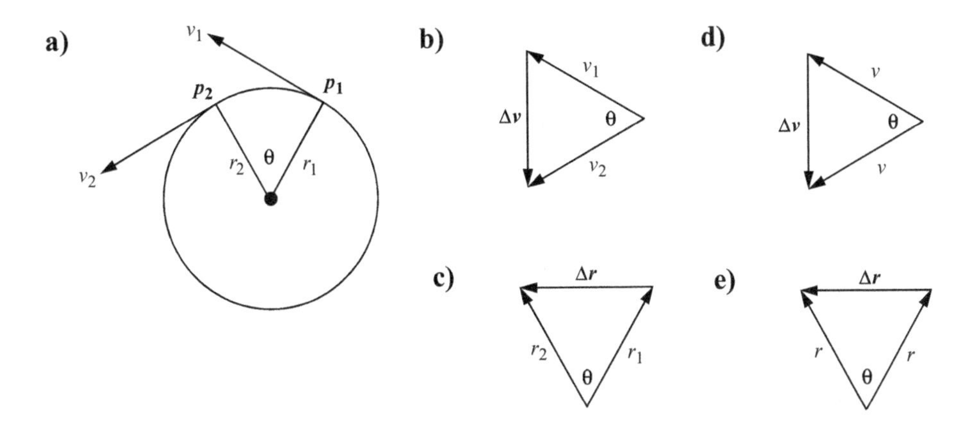

Figure 5.4
Uniform circular motion.

their bases together and measuring the distance between their tips (figure 5.4b). Similarly, the vector distance between p_1 and p_2 is $\Delta r = r_2 - r_1$ (figure 5.4c). Since the length of $v_1 = v_2$ and the length of $r_1 = r_2$, triangle $r_1 r_2 \Delta r$ and triangle $v_1 v_2 \Delta v$ are both isosceles triangles.

Let's simplify things a bit. Since $v_1 = v_2$, let's define $v = v_1 = v_2$, and since $r_1 = r_2$, let's define $r = r_1 = r_2$ (figures 5.4d and 5.4e). Note that the isosceles triangles in 5.4d and 5.4e have the same angle θ. So they are similar. From geometry we know that for similar triangles,

$$\frac{\Delta v}{v} = \frac{\Delta r}{r}. \tag{5.10}$$

For the next step, we can make a simplifying assumption. First, let $\Delta t = t_2 - t_1$, the time it takes for p_1 to get to p_2. Now, for small angles θ,

$$\Delta r \approx v \cdot \Delta t. \tag{5.11}$$

That is, Δr is approximately equal to $v \cdot \Delta t$ for small angles θ. Properly speaking, the length we should use is the arc of the circle between p_1 and p_2 because that's the distance the point will actually be traveling. But for small angles, the difference between the length of the arc from p_1 and p_2 and the length of the chord from p_1 and p_2 can be ignored. Being able to ignore this will greatly simplify what follows.

If we substitute (5.11) into (5.10), we derive the acceleration of the point on the circle as follows:

$$\frac{\Delta v}{v} = \frac{v \cdot \Delta t}{r}$$

$$\Delta v = \frac{v^2 \Delta t}{r}$$

$$\frac{\Delta v}{\Delta t} = \frac{v^2}{r}.$$

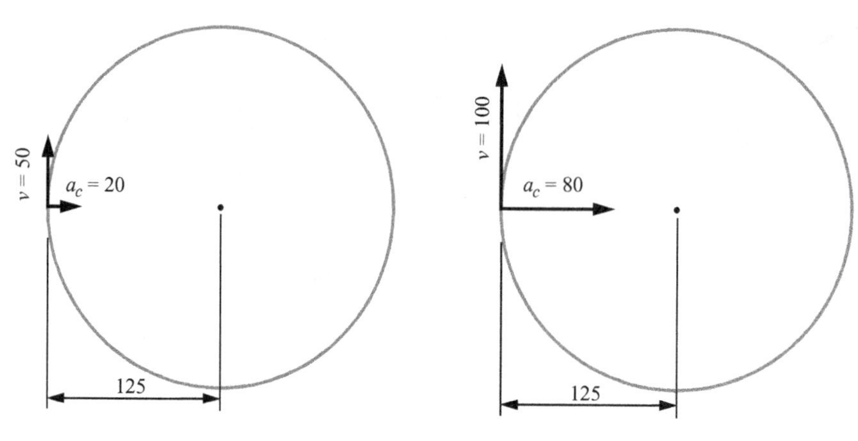

Figure 5.5
Centripetal acceleration.

The ratio $\Delta v/\Delta t$ is acceleration because it represents change in velocity over time. This is called *centripetal acceleration* because the direction of the bending force is always toward the center of the circle (see figure 5.5). It is defined as

$$a_c = \frac{v^2}{r}, \qquad \textit{Centripetal Acceleration} \quad (5.12)$$

where a_c is centripetal acceleration, v is velocity, and t is time.

Suppose we can control a rocket in deep space and want it to turn in a circle around a point with radius r. To get it to turn, we would have to ignite one rocket on its tail to propel it forward with a force proportional to v and ignite another pointing sideways with a force proportional to a_c. Figure 5.5a shows that for $v = 50$ and $r = 125$, a_c must be $50^2/125 = 20$. Figure 5.5b shows that if v is doubled to 100 for the same r, then a_c must quadruple to 80 in order for the rocket to maintain a circle of the same size.

5.2.7 Tangential Velocity

On a merry-go-round the circular motion pushes riders away from the center, and pushes them harder, the further from the center they are. But is the direction of the push *radial,* directly away from the center? Setting an object on a turntable, we can spin it at some angular velocity ω sufficient to make it fly off. Suppose it flies off at point Q (figure 5.6). We would observe that the object's angular velocity is instantly converted into some linear velocity in a direction *tangent* to the point where it flew off.[4] This is understandable because

Circular motion is the result of a centripetal force applied at right angles to the instantaneous velocity.

If we suddenly eliminate the centripetal force, the remaining linear velocity is all that is left, and the object flies off in whatever direction it was last aimed. In figure 5.6 the velocity of the

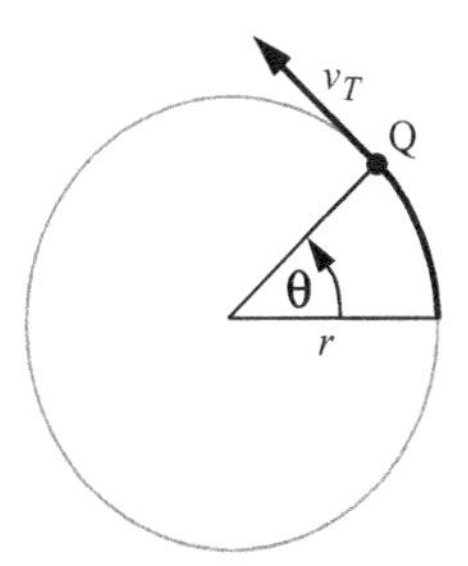

Figure 5.6
Tangential speed.

object at point Q is shown by a vector v_T anchored on Q and drawn tangent to the circle. The vector v_T indicates the *tangential velocity* of the object at point Q corresponding to its linear velocity.

Clearly, the object is subject to tangential velocity even when it is still on the turntable because this represents its linear velocity at each moment in time. Velocity implies both speed and direction, but the vector v_T is constantly changing direction as it progresses around the circle. So the magnitude of the vector is just its length (without regard to which direction it points) and corresponds to its speed.

Intuitively, we can tell that the tangential speed v_T must be related to the turntable's angular velocity $\omega = \theta/t$ as well as to its radius r because an increase in either would tend to give more velocity to the object. But how can we express this?

Recall that (5.5) relates angular displacement θ and radius r to the arc length s by $s = r\theta$, and that (5.6) relates angular velocity to angular displacement and time by $\omega = \theta/t$. If we introduce (5.6) into (5.5), the result combines angular velocity and radius, as we require. Dividing both sides of (5.5) by time t, we obtain

$$\frac{s}{t} = \frac{r\theta}{t} = r \cdot \frac{\theta}{t} \text{ rad/s.} \qquad (5.13)$$

The right-hand side now has a term θ/t in it. Since angular velocity $\omega = \theta/t$, (5.13) can be rewritten as

$$\frac{s}{t} = r\omega \text{ rad/s.}$$

Since s measures arc length, the ratio s/t expresses the speed of a point on the circle. Thus, *tangential speed* is defined as

$$v_T = \frac{s}{t} = r\omega \text{ rad/s.} \qquad \textit{Tangential Speed} \ (5.14)$$

We must use units of rad/s because this equation was derived from (5.5), which defines radian measure. When an object is thrown off a turntable, its tangential speed is converted into tangential velocity because then it has a particular direction, namely, tangent to its last point of contact.

5.2.8 Period and Frequency

As the cone on the turntable in figure 5.1 completes one revolution, the corresponding simple harmonic motion completes one back-and-forth cycle. The period T of this cycle clearly depends upon the angular velocity ω of the circle. Since by (5.6), $\omega = \theta/t$, and the circle completes one revolution of $\theta = 2\pi$ radians in $t = T$ seconds, we can relate angular velocity to period T as follows:

$$\omega = \frac{\theta}{t} = \frac{2\pi}{T},$$

and so

$$T = \frac{2\pi}{\omega} \qquad \textit{Period related to angular velocity} \quad (5.15)$$

Since frequency $f = 1/T$, we can relate the angular velocity to frequency:

$$\omega = 2\pi\frac{1}{T} = 2\pi f.$$

Relating angular velocity to frequency in this way will be so useful in subsequent chapters that it deserves being repeated:

$$\omega = 2\pi f. \qquad \textit{Radian Velocity} \quad (5.16)$$

In this book, when I write ω, I will almost always mean its definition $2\pi f$. Solving (5.16) for f, we derive the definition of frequency:

$$f = \frac{\omega}{2\pi}. \qquad \textit{Frequency related to angular velocity} \quad (5.17)$$

This definition says that frequency is the ratio of the angular velocity, $\omega = \theta/t$ (see equation (5.6)), to the arc length of a circle. The greater the angular velocity, the more often it completes a full circle, hence the higher its frequency.

5.3 Projection of Circular Motion

Figure 5.7 shows a spring/mass system vibrating vertically next to a turntable that has a cone mounted on its edge. By appropriate choices of rotational speed of the turntable, elasticity of the spring, and weight of the mass, the motion of the shadows of the cone and mass can be synchronized on a screen behind them. This suggests that the simple harmonic motion of a pendulum or a weighted spring can be related to uniform circular motion via projection.

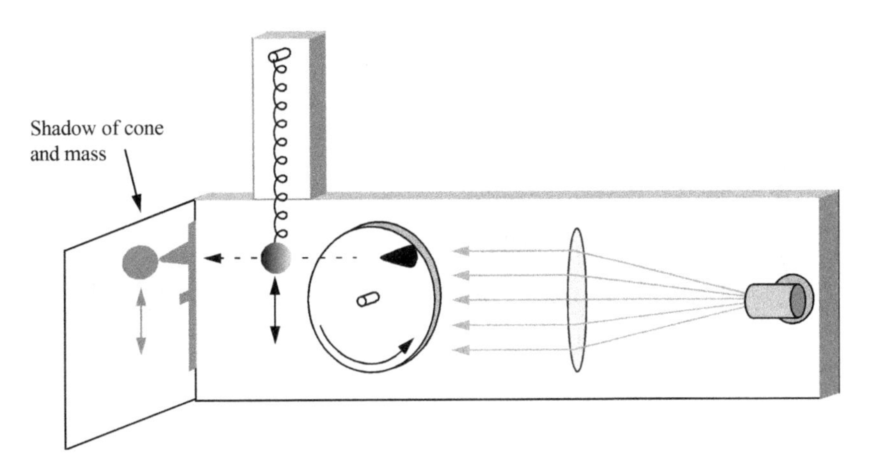

Figure 5.7
Simple harmonic motion as the projection of uniform circular motion.

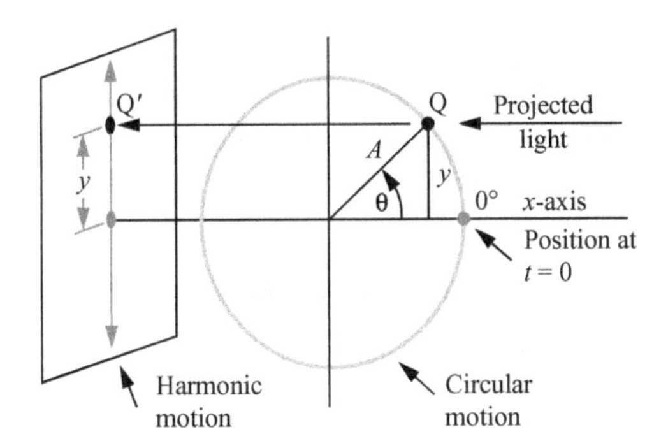

Figure 5.8
Front view of turntable.

Figure 5.8 shows the turntable and screen from figure 5.7 with the cone at point Q. Since light shines across the circle parallel to the x-axis, point Q′, which is the shadow of Q, appears on the screen at the same displacement above the x-axis.

The displacement of points Q and Q′ from the x-axis is y, the projection of the radius A onto the y-axis.

Elementary trigonometry shows that the radius A, its angle θ, and the value of y are connected by the *sine relation* (see appendix A).

$$y = A \sin\theta = A \cdot \frac{y}{A}. \qquad \textit{Sine Relation} \quad (5.18)$$

Equation (5.18) relates the height y of the triangle, and hence the height of its projection on the screen, to the radius A and its angle θ. The sine relation allows us to reconcile circular motion with

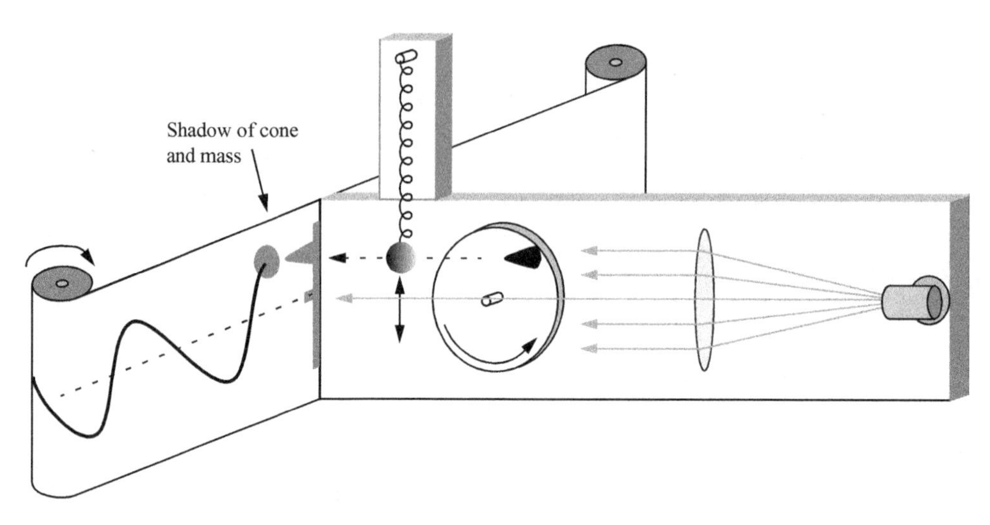

Figure 5.9
Simple harmonic, uniform circular, and sinusoidal motion.

simple harmonic motion. In order to see how the vertical displacement y changes, figure 5.9 adds a strip of film to record the position of the mass and cone through time, allowing us to see the sinusoidal motion of the spring/mass system together with the motion of the turntable. Mathematically and intuitively, it should be clear now that

Simple harmonic motion and the projection of uniform circular motion are the same.

5.3.1 Relating Displacement of Simple Harmonic Motion to Time

Since, by (5.6), $\theta = \omega t$, we can relate the vertical displacement y of the cone's shadow at time t as follows:

$$y = A \sin \theta = A \sin \omega t, \tag{5.19}$$

where A is the radius of the turntable, θ is the turntable's angular displacement, t is time, and ω is angular velocity.

The expression ωt in (5.19) determines the rotational position of the turntable at time t; taking the sine of that rotational position determines the height of the vertical displacement y; multiplying the vertical displacement by A scales the displacement for the size of the turntable.

Equation (5.19) shows the identity of simple harmonic motion and circular motion and provides a way to determine the displacement of a sinusoidal wave at any time t. We see that

The projection of simple harmonic motion through time generates sinusoidal motion.

The term A in (5.19) can be interpreted either as the radius of a circle or as the amplitude of the corresponding simple harmonic motion because this value determines both attributes.

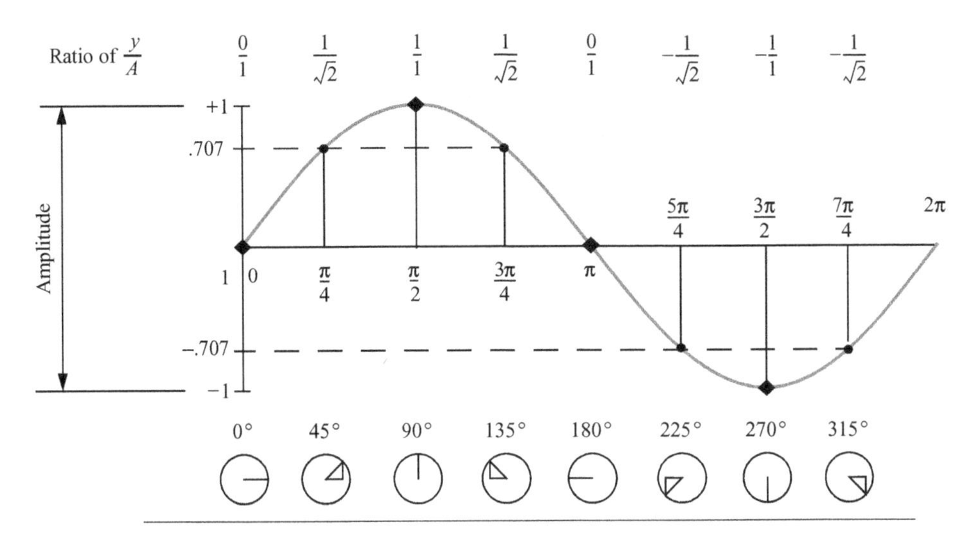

Figure 5.10
Constructing a sine wave.

5.4 Constructing a Sinusoid

A simple way to generate a sine wave is to plot a few selected points of (5.18) and connect the points with a smooth line. Figure 5.10 shows eight values of $A \sin \theta$ every 45° as θ makes one complete revolution. The y-axis shows the corresponding values of $A \cdot (y/A)$ for radius $A = 1$. A circle of radius 1 is a *unit circle*. It is convenient to set A to 1 in order to keep the example simple, but it can be any value. Notice that y takes on values in the range –1 to 1 as θ varies.

- When the angle θ is 0° or 180°, the displacement of $y = 0$ and $\sin 0° = \sin 180° = y/A = 0/1 = 0$.
- When θ is 90°, $y = 1$ and $\sin 90° = y/A = 1/1 = 1$.
- When θ is 270°, $y = -1$ and $\sin 270° = y/A = -1/1 = -1$.

These *cardinal points* are marked with diamonds in figure 5.10.

- At 45°, triangle Axy in figure 5.8 becomes an isosceles right triangle, and by elementary geometry,

$$y = \frac{A}{\sqrt{2}} = \frac{A}{1.414...}.$$

Plugging this value into the sine relation yields the formula

$$\sin\frac{y}{A} = \sin\frac{A/\sqrt{2}}{A} = \sin\frac{1}{\sqrt{2}} = \sin 45° = 0.707....$$

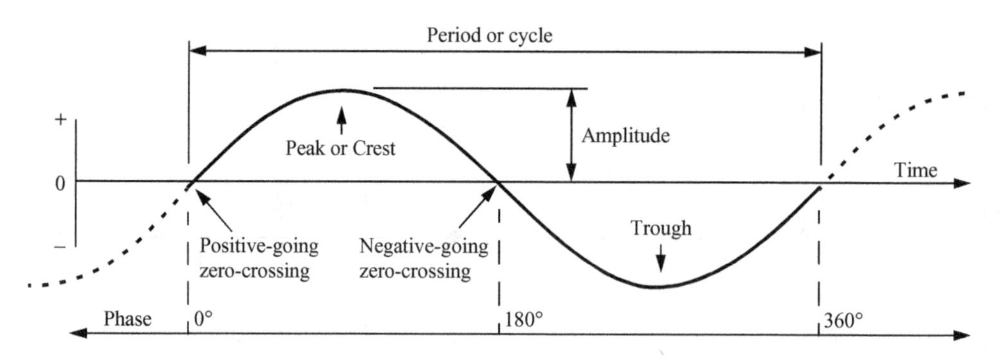

Figure 5.11
Anatomy of a sine wave.

- Similar reasoning establishes the values at 135°, 225°, and 315° (figure 5.10).

So as θ goes from 0 to 360°, sin θ exhibits one period of sinusoidal motion.

5.4.1 Anatomy of a Sinusoid

The landmarks of the sinusoidal wave are shown in figure 5.11. The *y*-axis shows the amplitude, which is proportional to a corresponding circular radius *A*. The *x*-axis shows the phases of the sine wave's various notable features, such as where it crosses the *x*-axis (zero crossings), and its *crests* and *troughs*. We speak of the "phases of the moon" in the same sense: *phase* describes the characteristic points reached periodically each time a wave repeats. The *period,* or *cycle,* of a sine wave is one complete movement through all its phases, corresponding to one complete revolution of a corresponding circle.

It will often be more convenient to show the passage of time on the *x*-axis rather than the size of the angle θ. Solving (5.7) for *t* yields

$$t = \frac{\theta}{\omega}, \tag{5.20}$$

which shows that *time is directly proportional to angular displacement* θ. This means the *x*-axis can either measure elapsed time or elapsed phase.

Since frequency is the reciprocal of time, $f = 1/t$, (5.20) can be rearranged:

$$f = \frac{\omega}{\theta}, \tag{5.21}$$

which shows that *frequency is directly proportional to angular velocity* ω. The greater the angular velocity, the more rapidly the turntable turns.

If the *x*-axis shows elapsed time, we are measuring frequency; if the *x*-axis shows elapsed phase, we are measuring periodicity.

Sinusoids, like circles, have no beginning and no end, so the period of a sine wave can start anywhere. Conventionally, sine wave periods are usually regarded as beginning at a positive-going zero crossing (figure 5.11) and extending until just before the next positive-going zero crossing. But we could just as well measure the period from crest to crest, or from trough to trough, by suitable choice of phase offset.

5.4.2 Phase Offset

Equation (5.19) requires that the turntable start at its 0° position, which is when point Q in figure 5.8 is aligned with its positive x-axis. In this position, the vertical displacement $y = 0$ because $\sin 0 = 0$. If we wish to be able to start the turntable at any orientation, we must introduce a way to specify its starting position in (5.19). If we don't start with $\theta = 0$, y will have a nonzero initial value.

Let's define a constant ϕ, which is the *phase angle* (or *phase offset* or *phase shift*) of the turntable's starting position. The vertical displacement of the cone's shadow at time t with phase offset ϕ can then be written as

$$y = A\sin(\omega t + \phi), \tag{5.22}$$

where ϕ defines a constant offset from 0°. It can take on any positive or negative real value. For instance, suppose we set $\phi = \pi/2$. Note in figure 5.10 that $\sin(\pi/2) = 1$. Then at time $t = 0$,

$$y = A\sin(\omega t + \pi/2) = A\sin(\pi/2) = A.$$

This means that at $t = 0$ the turntable starts rotating with the cone positioned at the top of the turntable, which is rotated 90° counterclockwise from the previous starting position.

5.4.3 Wavelength

The physical length of a waveform period, its *wavelength,* depends upon the medium through which the wave is traveling and its frequency. In air, sound waves travel at about 340 m/s (approximately 1100 feet per second) at a temperature of 20°C (see section 7.4).

So a frequency of 1 kHz in air has a wavelength of approximately

$$\frac{1\,\text{second}}{1000\text{ periods}} \cdot \frac{340\text{ meters}}{1\,\text{second}} \cong 0.34 \text{ meters per period,}$$

or

$$\frac{1\,\text{second}}{1000\text{ periods}} \cdot \frac{1100\text{ feet}}{1\,\text{second}} \cong 1.1 \text{ feet per period.}$$

Note how these three measurements are interrelated.

	is a measure of . . .	is measured in . . .	unit
Periodicity	duration	seconds/cycle	seconds
Frequency	how rapid or how often	cycles/second	hertz
Wavelength	length	meters/period	meter

5.4.4 Velocity of Simple Harmonic Motion

How can we characterize the velocity of an object moving in simple harmonic motion when both the direction and speed of such an object change through time as the object vibrates back and forth? Since harmonic motion is the projection of circular motion, we should be able to understand the velocity of harmonic motion by thinking more about tangential velocity.

Figure 5.12 shows the projection of tangential velocity v_T of an object on the edge of a turntable. By a combination of geometry and trigonometry (see appendix A), we see that the velocity v of the shadow that is projected on the screen is just the y-axis component of the vector v_T, that is,

$$v = v_T \cos \theta, \tag{5.23}$$

where $\theta = \omega t$.

Recall from (5.14) that the tangential velocity v_T is related to the angular velocity ω by $v_T = r\omega$. Let's substitute amplitude A for radius r, so now $v_T = A\omega$. Substituting $A\omega$ for v_T in (5.23), we obtain the *velocity of simple harmonic motion:*

$$v = A\omega \cos \theta = A\omega \cos \omega t. \tag{5.24}$$

This tells us that even though an object on a rotating circle moves with uniform circular motion, the velocity of its corresponding simple harmonic motion is *not* uniform. The velocity constantly varies between maximum and minimum values through time sinusoidally. When θ equals exactly 90° or 270°, velocity is exactly 0, and the object in simple harmonic motion is momentarily stationary. Velocity is positive maximum when θ equals 0, and at that point it equals

$$v = A\omega. \qquad \textit{Maximum Velocity of Simple Harmonic Motion} \tag{5.25}$$

Velocity is negative maximum when θ equals 180°.

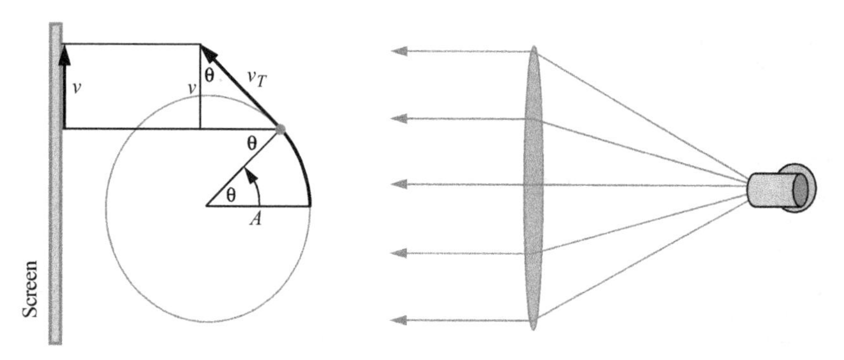

Figure 5.12
Projection of tangential velocity.

5.5 Energy of Waveforms

Equation (5.25) says that the velocity of an object vibrating in simple harmonic motion is proportional to both the amplitude and the angular velocity of the corresponding unit circle. In other words, simple harmonic motion—the projection of circular motion—will have higher velocity either if the corresponding circular motion has a longer radius or if that radius turns faster. This suggests that a mass moving in simple harmonic motion would have greater momentum if either its amplitude or its frequency were increased.

In section 4.14, kinetic energy was shown as the product of the mass m of an object times the square of its velocity v, or $E = \frac{1}{2}mv^2$. I used an automotive metaphor to show that doubling a car's speed quadruples its energy. Now let's apply this understanding to a molecule of air zipping in and out of someone's ear as part of a sound wave impinging on their eardrum.

If the amplitude of a wave doubles while the frequency remains the same, the particle must cover *twice the distance in the same amount of time* (via one period of doubled amplitude). Or, if the frequency of the wave doubles, the particle must cover *twice the distance in the same amount of time* (via two periods at the original amplitude). In either case, the energy of the molecule of air has quadrupled because the velocity of its simple harmonic motion has doubled.

If the wave in figure 5.13a is stretched out, it has the length shown in figure 5.13d. The wave in 5.13b, with twice the amplitude of the wave in 5.13a, has the length shown in 5.13e. Wave 5.13c has the same amplitude and twice the frequency of wave 5.13a, and its length also equals that shown in 5.13e. Since the duration T of all three waves (5.13a, 5.13b, and 5.13c) is the same, but the length of waves 5.13b and 5.13c is twice that of wave 5.13a, clearly waves 5.13b and 5.13c have twice the speed of wave 5.13a. So we see that a wave's energy depends on both its amplitude and its frequency.

Consider a point on the turntable in figure 5.12. If the turntable's radius is A, it has circumference $d = 2\pi A$. Since, by equation (4.8), velocity is $v = d/t$, the circular velocity of the point is $v = 2\pi A/t$,

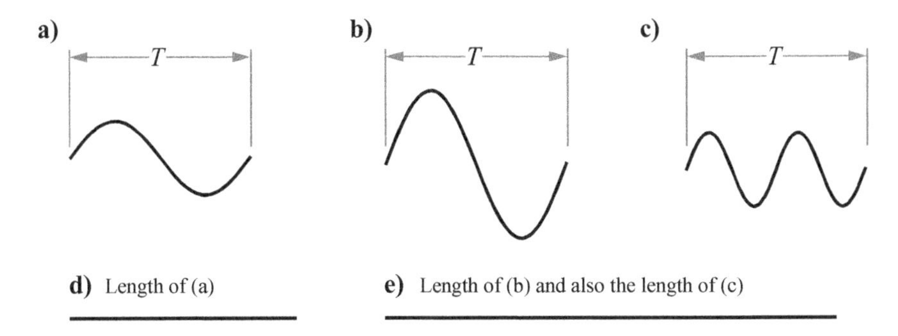

Figure 5.13
Path lengths.

which also can be written as $v = 2\pi A \cdot (1/t)$. Since, by equation (4.1), the frequency of rotation is $f = 1/t$, we can also write

$$v = 2\pi Af. \qquad \textit{Rotational Velocity} \quad (5.26)$$

Taking $E = \frac{1}{2}mv^2$ from equation (4.28) and substituting v from (5.26) yields

$$E = \frac{1}{2}m(2\pi Af)^2, \qquad \textit{Rotational Energy} \quad (5.27)$$

which confirms that wave energy depends upon both frequency and amplitude.

5.5.1 Measuring the Energy of Waveforms

Peak Pressure Level Perhaps the easiest way to measure the strength of a waveform is to examine how its *maxima* and *minima*—its highest and lowest points—relate to the ambient pressure level. *Peak pressure level* of a sound wave is the difference between the ambient pressure level and the magnitude of either the maximum or minimum pressure level of the sound wave, whichever is greater:

$$l_p = \max(|l_+|, |l_-|) - l_a , \qquad \textit{Peak Pressure Level} \quad (5.28)$$

where l_p is peak pressure level, l_a is ambient pressure level, l_+ is the highest peak, and l_- is the deepest trough. The operator $|\ldots|$ gives the magnitude of the enclosed expression, and the function $\max(a, b, \ldots)$ chooses the greatest value of its arguments. Figure 5.14 shows the peak pressure level.

Peak-to-Peak Pressure Level Every sound recording device has some limit beyond which it can no longer accurately represent the strength of the waveform being recorded, and waveforms with peaks greater than the limit are distorted (see section 4.24.2). Modern recorders often contain volume level meters that measure the strength of the recorded waveform based on (5.28) to help the recordist prevent distortion. The peak-to-peak pressure level of a waveform is the magnitude of the distance between l_+ and l_-:

$$l_{pp} = |l_+ - l_-| \qquad \textit{Peak-to-Peak Pressure Level} \quad (5.29)$$

Why Average Pressure Level Doesn't Work Peak-to-peak level shows the limits of a waveform's amplitude, but it does not always provide the best information about a waveform's strength. For example, a recording that is mostly silence except for a brief tone burst may have a large peak

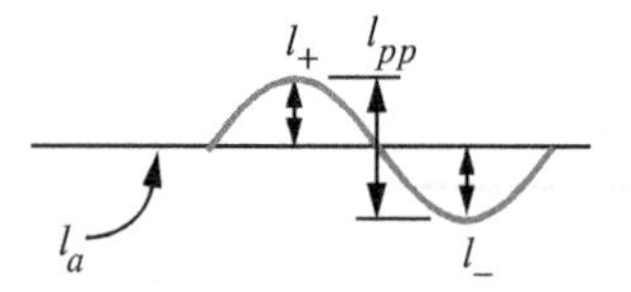

Figure 5.14
Peak pressure level.

amplitude if the tone burst is loud, but there is little energy in the waveform over its total duration because it is mostly silent.

One might try to get a clear picture of a waveform's strength by averaging the waveform's pressure over time, hoping to smooth out the peaks. But sound waveforms are usually evenly balanced above and below ambient pressure, so in general $l_+ - l_- \cong 0$. Therefore, the mean value of most sounds is typically close to zero, and so average pressure is not a useful way to measure the strength of a waveform.

RMS Level Ideally, it would be useful to observe the power contained in the waveform because power is the energy in the waveform through time. But all we can easily measure with a microphone is the waveform's pressure fluctuations. How can we derive a measure of energy from pressure? The key lies in recalling that there is a square relation between amplitude and energy.

The average value of cos *t* over one full period is 0.0. The peak amplitude $l_p = |l_+| = |l_-|$ of the cosine is 1.0. The peak-to-peak amplitude is $l_{pp} = 2.0$ (figure 5.15).

Let $s(t) = \cos t$. There is a trigonometric identity (see volume 2, appendix A.4.1) that says

$$\cos a \, \cos b = \frac{\cos(a-b) + \cos(a+b)}{2}.$$

If we square $s(t)$, then

$$\begin{aligned} s^2(t) &= \cos^2 t \\ &= \cos t \cos t \\ &= \frac{\cos(t-t) + \cos(t+t)}{2}. \\ &= \frac{1 + \cos 2t}{2} \end{aligned} \tag{5.30}$$

So, by (5.30), $s^2(t)$ is a cosine wave at twice the frequency, offset by 1, and then divided by 2 (figure 5.16). This is what the original cosine waveform, shown in figure 5.15, looks like when squared. Note that all values are now positive. The peak value is still 1.0. Its *mean value* is 0.5.

Now let's take the mean value for this squared waveform (0.5) and undo the effects of the squaring operation. The square root of the mean value is $0.5^{1/2} \cong 0.707$. This is the *root mean squared*

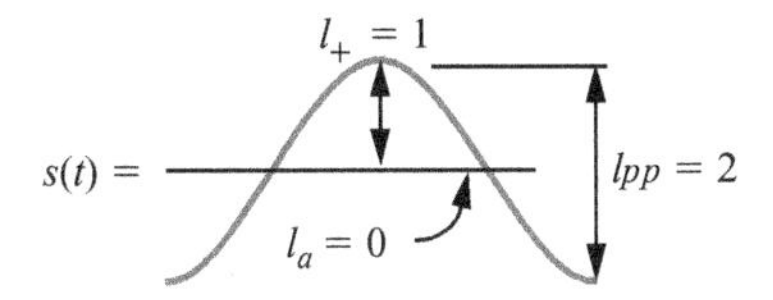

Figure 5.15
RMS level.

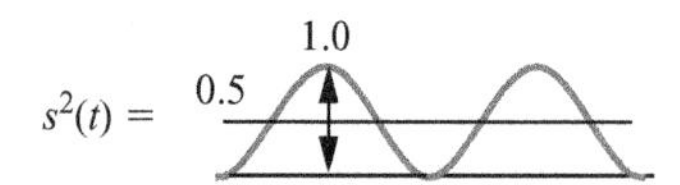

Figure 5.16
RMS cosine.

(RMS) value of the waveform. So the RMS amplitude of $s(t) \cong 0.707$. This allows us to say something useful about the average energy of a sinusoid knowing only its amplitude. The relation of the amplitudes is as follows:

Average	0
RMS	0.71
Peak	1
Peak-to-peak	2

In general, if $s(t)$ is a sinusoid with peak amplitude A, then its RMS amplitude is $A/\sqrt{2}$.

Because we used a sinusoid to derive RMS amplitude, this measure is only valid for sinusoids. In particular, it is not valid for time-varying waveforms. This, of course, leaves out all the interesting real-world audio waveforms we'd like to measure with it. Nonetheless, this definition of RMS is widely used in practice[5] because, I suppose, it's better than nothing. But there are more sophisticated techniques to overcome this difficulty and find the true RMS value of arbitrary waveforms (see volume 2, chapter 1).

Sound Pressure Level Although the decibel scale was developed for sound intensity, we can adapt it to measure sound pressure level. Equation (4.40) defined decibels of sound intensity level (dB SIL) as

$$y \text{ dB SIL} = 10\log_{10}\frac{I'}{I}, \qquad \textit{dB SIL} \quad (5.31)$$

where I is a reference intensity, and I' is the intensity being measured. Recalling that intensity is proportional to the square of amplitude, we can define decibels of *sound pressure level* (*dB SPL*) as

$$10 \cdot \log_{10}\left(\frac{A'}{A}\right)^2$$

$$2 \cdot 10 \cdot \log_{10}\frac{A'}{A}$$

and

$$y \text{ dB SPL} = 20\log_{10}\frac{A'}{A}, \qquad \textit{dB SPL} \quad (5.32)$$

where A is a reference amplitude, and A' is the amplitude being measured.

Decibels of sound pressure level (SPL) correspond to twice the equivalent decibels of sound intensity level (SIL). Where a doubling of intensity corresponds to an increase of 3 dB SIL, a doubling of pressure corresponds to an increase of 6 dB SPL. An intensity ratio of 10:1 equals 10 dB SIL and 20 dB SPL.

5.6 Summary

Uniform circular motion is circular movement at a constant speed. Simple harmonic motion is the projection of circular motion. Angular displacement is the angle through which a rigid body rotates about its axis of rotation. Counterclockwise angular displacement is taken to be positive and clockwise angular displacement to be negative. The angle formed by a radius and an arc the length of the radius is called a radian. Measuring angles with radians simplifies many calculations. Angular velocity is the rate at which angular displacement changes. Angular acceleration is the rate at which angular velocity changes.

By Newton's laws, objects tend to travel in a straight line. To travel in a circular path, an object must experience centripetal acceleration to overcome the object's tendency to travel linearly. Circular motion is linear velocity forward constrained by centripetal force toward a center. There is no such thing as centrifugal force.

Simple harmonic motion of a pendulum or a weighted spring can be related to uniform circular motion via projection. Simple harmonic motion is the same as the projection of uniform circular motion. The projection of simple harmonic motion through time generates sinusoidal motion.

An object on a rotating circle moves with uniform circular motion, but the velocity of its corresponding simple harmonic motion constantly varies between maximum and minimum values through time sinusoidally. The speed of an object vibrating in simple harmonic motion is proportional to both the amplitude and the angular velocity of the corresponding unit circle.

Peak pressure level of a sound wave is the difference between the ambient pressure level and the magnitude of either the maximum or minimum pressure level of the sound wave, whichever is greater. The peak-to-peak pressure level of a waveform is the magnitude of the distance between its lowest and highest point. The root mean squared (RMS) value of a waveform is a useful measure of energy in a sinusoid, calculated by squaring the waveform to derive its mean value and then taking the square root of the mean value to determine the RMS value. Technically, this operation is valid only for sinusoids.

Since it's easier to measure pressure variations in air than sound intensity, we adapt the decibel of sound intensity level (dB SIL) to the decibel of sound pressure (dB SPL) by doubling the dB SIL value.

6 Psychophysical Basis of Sound

Pongileoni's bowing and the scraping of the anonymous fiddlers had shaken the air in the great hall, had set the glass of the windows looking on to it vibrating: and this in turn had shaken the air in Lord Edwards' apartment on the further side. The shaking air rattled Lord Edwards' membrana typani; the interlocked malleus, incus, and stirrup bones were set in motion so as to agitate the membrane of the oval window and raise an infinitesimal storm in the fluid of the labyrinth. The hairy endings of the auditory nerve shuddered like weeds in a rough sea; a vast number of obscure miracles were performed in the brain, and Lord Edwards ecstatically whispered 'Bach!'
—Aldous Huxley, *Point Counter Point*

The length of strings is not the direct and immediate reason behind the forms [ratios] of musical intervals, nor is their tension, nor their thickness, but rather, the ratios of the numbers of vibrations and impacts of air waves that go to strike our eardrum.
—Galileo Galilei, "Two New Sciences"

We must distinguish carefully the ratios that our ears really perceive from those that the sounds expressed as numbers include.
—Leonhard Euler, "Conjecture sur la raison de quelques dissonances généralement reçues dans la musique"

What has been said of sonorous bodies should be applied equally to the fibres which carpet the bottom of the ear; these fibres are so many sonorous bodies, to which the air transmits its vibration, and from which the perception of sounds and harmony is carried to the soul.
—Jean-Philippe Rameau, "Generation Harmonique"

6.1 Signaling Systems

Suppose you and I were about to play a duet. In order to start, I might signal you by saying, "Ready? One, two, three. . . ."

For there to be a signal, there must be a source, a receiver, time, distance, and a medium—in this case, air—which spans the distance and connects the source to the receiver. Altogether, this constitutes a *signaling system*. A *signal* is a physically detectable quantity such as the pressure of an acoustical wave that traverses a signaling system. More generally, a signal is a description of how any one parameter varies with any other parameter. A *system* is any function that produces an output signal based on an input signal.

Acoustics is the study of signals and signaling systems where the medium is air. The full treatment of acoustics covers these elements:

- *Source*—how sounds are created, including the mechanics of vibrating systems of all kinds. If only musical sources are considered, the subject is musical instrument acoustics.
- *Medium*—how sound behaves in air, including how sound is transmitted through air (spreading) and what happens to it along the way if it encounters obstacles such as walls (absorption and scattering) that cause transmission losses. Scattering happens, for example, when a sound strikes a wall: some is reflected, the rest is transmitted through the wall. Room acoustics is the study of sound transmission in rooms. Interference from other sources produces an ambient noise level that may block or degrade the reception of a signal by a receiver.
- *Receiver*—how we hear. If this entire subject is acoustics, and what goes on between our ears is psychology, then the subject of how we hear is *psychoacoustics*. The question of how objective measures of sensory stimuli relate to the subjective experience is the concern of *psychophysics*.

The properties of the receiving end of the human sonic signaling system are covered in this chapter, and the properties of sources and media are covered in chapter 7.

6.2 The Ear

Consider the problems that our hearing helps us to solve. The ears detect, analyze, and classify biologically interesting sounds: they compile spectral and temporal information of incoming signals, parse them into various sources, localize these sources in time and space, and construct a model of the *auditory scene* that surrounds us.[1] In a lecture I once heard, Albert Bregman characterized our hearing faculty as follows: Suppose we scraped out two minor indentations (representing our ear canals) on the edge of a vast lake (representing our sonic environment) and installed two floats in them (corresponding to our ear drums). Suppose that by simply observing how the waves moved the floats up and down, we were somehow able to understand everything that was happening in the lake—to correctly identify boats going by, to note their position, to distinguish boats from fish, wind, reflections, and so forth. Our hearing effectively carries out all these functions and others with little more than this kind of arrangement.

The auditory system is attuned not only to listen to certain sounds but to ignore sounds that are not biologically relevant. Such sounds include ambient noises and the effects of sound reflection, refraction, and diffusion, which together can produce ugly distortions of a sound source. If we notice these secondary signals at all, it is to use them constructively to characterize our acoustic environment. No information is wasted by the auditory system. If we add to this the fact that our audition is also capable of carrying us away in transports of rapture when we hear music that moves us, this is an extraordinary faculty indeed.

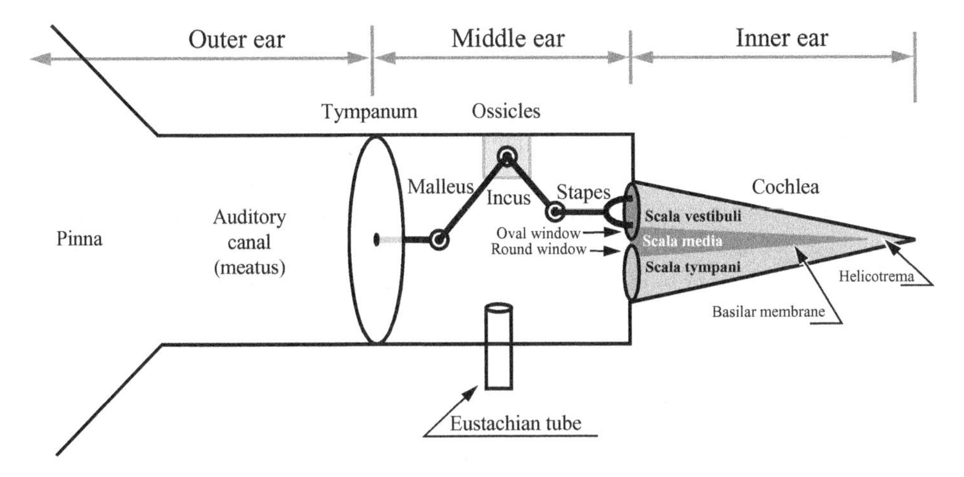

Figure 6.1
Schematic diagram of the human ear.

Figure 6.1 is a simplified drawing of a cross-section of one human ear showing the outer ear, middle ear, and inner ear.

6.2.1 Outer Ear

The outer ear consists of the *pinna* (the part that sticks out from your head), the *auditory canal (meatus),* and the *eardrum (tympanum).*

The funnel-shaped pinna helps collect sound from the environment. Its shape modifies the arriving frequency information depending on the direction of the sound source, imprinting directional clues that we use to identify the location of the source. Even the shape of the head and torso and the distance between the ears influence how we identify direction. Frequencies around 3000 Hz are transferred most efficiently by the meatus, and this is the frequency range of our greatest hearing sensitivity.

The tympanum is bent by the force of arriving sound and transmits the motion to the middle ear. Although it vibrates most easily between 1 and 3.5 kHz, it transmits sound to the inner ear over the entire audible frequency range.[2] It has a conical shape, a highly detailed and fibrous structure, and an angular placement in the ear canal. The function of the eardrum and the middle ear is to provide mechanical advantage to resolve the mismatch between the density of air in the outer ear and the fluid of the inner ear. Without this impedance matching, very little acoustical energy would be absorbed by the inner ear and hearing would be severely limited. It is still largely a mystery how the tympanum accomplishes this task over such a wide frequency range.

6.2.2 Middle Ear

The middle ear is the chamber immediately behind the tympanum. It is connected to the throat by the Eustachian tube, which allows air pressure behind the tympanum to normalize to external air pressure. A mechanical linkage system couples vibrations arriving at the tympanum to the inner ear, consisting of three tiny bones called *ossicles,* known as the *hammer (incus),* the *anvil (malleus),* and the *stirrup (stapes),* named for their shapes. The hammer is attached to the tympanum, and the stapes is connected to the oval window leading to the inner ear.

Since the outer ear is in air but the inner ear is in fluid, the density difference between them would allow very little energy from the air to penetrate into the inner ear were it not for the leverage provided by the ossicles. For instance, go to a swimming pool and have a friend talk to you as you put your head under water. You might still be able to hear your friend as your head goes under, but the sound is weak and muffled. Most of the sound from your friend's voice bounces off the water, back into the air, because of the difference in density between the two media. The middle ear passes along sound energy to the inner ear by providing a mechanical leverage of about 25 to 1 using the ossicles to move the denser inner ear fluid. That is, the middle ear *matches the impedance* of air to the inner ear fluid (see volume 2, chapter 8). Most of the mechanical energy present in the tympanum is transmitted efficiently to the inner ear, although the outer ear and middle ear transfer frequencies in the range of 1–3.5 kHz about 50 times more efficiently than frequencies outside this range.[3]

Acoustic Reflex and Temporary Threshold Shift The middle ear also has a few small muscles that can temporarily protect the inner ear from intense sounds. The *stapedius* muscle reduces the mobility of the ossicles by pulling the stapes to the side. These muscles are activated by the bilateral *acoustic reflex* within about 10–20 ms of when sound pressure exceeds 90–100 dB. The acoustic reflex provides about 20 dB of protection. However, the response time for this reflex is about 30–40 ms *after* the sound has started, and full protection takes up to about 150 ms longer. The route from the auditory nerves to the stapedius is hardwired in the brain, so the acoustic reflex is ordinarily below cognitive control (although some individuals, including the author, can voluntarily activate it). Thus for explosions and gunfire, damage to the ear can take place before these natural protections come into play. This strongly suggests the use of artificial noise suppression via ear plugs where explosive sounds are a possibility.

The *tensor tympani* muscle is attached to the malleus and increases the tension on the eardrum as part of a more general acoustic reflex to loud sounds that can take as long as 1 or 2 seconds. These systems are not fail-safe protections. Extended exposure to loud sounds (in excess of 100 dB or so) fatigues these muscles, reexposing the inner ear to punishing sound levels and risking hearing damage. However, our ears do not alert us to this condition. Instead, another longer-term protective mechanism comes into play, the *temporary threshold shift* (TTS),[4] whereby our hearing gradually adjusts to ongoing elevated intensity levels, and we lose the sense that the sound is too loud. If the hair cells are not allowed to recover through periods of relative quiet, they gradually lose their ability to respond, and they die, resulting in permanent hearing loss, or *permanent threshold shift.* In addition to damaging the auditory mechanism, noise may contribute to loss of sleep, tension, headaches, reduced vision, sexual impotence, heart disease, and even mental illness (Cohen, Anticaglia, and Jones 1970). The moral:

Too much noise is bad for you!

6.2.3 Inner Ear

The stapes connects to the *oval window* at one end of the *cochlea,* a fluid-filled tube that connects to the *auditory nerve*. The cochlea is a coiled double tube, connected at the center. Figure 6.1 shows it uncurled. One end of the double tube is the oval window, the other end is a *round window,* which is also covered with a membrane. The oval window side of the cochlea is the *scala vestibuli*. The round window side of the cochlea is the *scala tympani*. At the apex of the cochlea, these two scala are connected by a narrow aperture, the *helicotrema*. The two scala are filled with *perilymph,* which is similar to cerebral spinal fluid. As the oval window is vibrated by the stapes, the perilymph moves back and forth. The membrane over the round window is pushed in and out in a complementary motion.

The scala vestibuli and scala tympani enclose the *scala media,* filled with *endolymph,* similar to intracellular fluid. Within the scala media is the *organ of Corti,* which is the receptor organ for hearing. It rests on part of the membranous labyrinth, the *basilar membrane*.

6.2.4 Basilar Membrane

The basilar membrane runs down the center of the cochlea. About 30,000 hairlike receptor units called *hair cells (cilia),* are attached to it along its length. On the other end, the hair cells are anchored to the more stable *tectorial membrane*. The hair cells connect the two membranes along their entire length. There is a row of *inner hair cells* and several rows of *outer hair cells.* The inner hair cells provide most of the afferent information to higher neural centers.

The basilar membrane vibrates under the pressure of the perilymph in response to sound. It is thinner, stiffer, and narrower at the base of the cochlea than at the apex. Imagine a guitar string that is thicker at one end than the other: the thin end will vibrate more readily at high frequencies than the thicker end. Thus, for a pure tone of given frequency, only one relatively narrow region on the basilar membrane vibrates sympathetically. Low frequencies vibrate the perilymph most intensely at the apex of the basilar membrane, and high frequencies vibrate it most intensely near the oval window. Thus, the position along the basilar membrane encodes frequency for the auditory nerve. In the language of psychoacoustics, the basilar membrane transforms *frequency to place*. According to ideas originally put forth by G. S. Ohm and Helmholtz, the basilar membrane was thought of as a kind of spectrum analyzer that maps frequency to position. Figure 6.2 shows a map relating frequency to position along the basilar membrane. As the basilar membrane is vibrated, the hair cells are sheared back and forth between the tectorial membrane and the basilar membrane. Hair cells receiving significant movement trigger an electrical signal that is transmitted to a nerve lying under the organ of Corti. These neurons transmit signals back along the auditory nerve to the brain stem.

Figure 6.2 shows that about half of the basilar membrane is used to encode frequencies between 25 Hz and about 1.6 kHz. All the remaining frequencies in the range of human hearing—from 1.6 kHz to about 20 kHz—fit into the remaining half of the area. Perhaps not surprisingly, we have greater difficulty discriminating higher pitches than lower ones.

Place Theory If the frequency of a tone doubles, the position of maximum displacement along the basilar membrane moves toward the oval window by a constant amount. This suggests that the

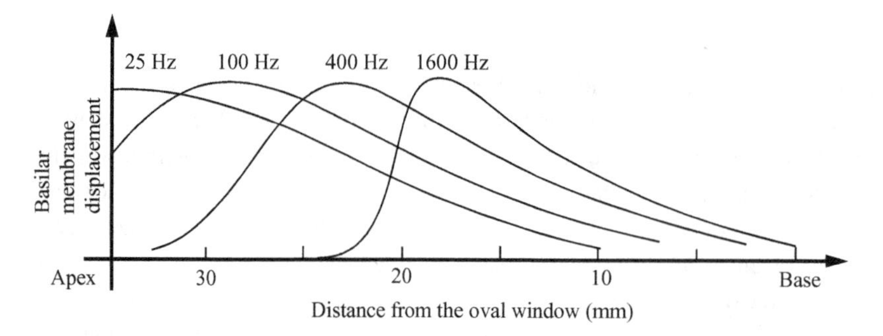

Figure 6.2
Frequency response of the basilar membrane. (Adapted from Békésy 1960.)

basilar membrane encodes frequency ratios, not frequency differences. Here is physiological evidence of the logarithmic relation between pitch and frequency: the basilar membrane uses a logarithmic encoding for pitch. This observation, the *place theory* of pitch, holds that there is a direct relation between the frequency presented to the basilar membrane and the place along its length that is displaced most strongly. More generally, the place theory holds that there is a *tonotopic mapping* between the basilar membrane and an associated region of the auditory cortex that performs frequency discrimination based on the topology of the basilar membrane.

Frequency Sharpening But there is at least one problem with the frequency-to-place theory. The curves in figure 6.2 suggest that our ability to discriminate between two close frequencies should be much poorer than it actually is. In fact, our hearing does a much better job than one would predict from the passive mechanics of the basilar membrane. Kachar et al. (1986) discovered a possible explanation. They observed through video microscopy that outer hair cells change length in response to nerve stimulation. Ashmore (1987) stimulated a single outer hair cell and observed its length change substantially. The effect persisted at frequencies into the kilohertz range. Current thinking is that outer hair cells help to sharpen the tuning of the basilar membrane by affecting how it vibrates, directing and focusing the responsiveness of the inner hair cells. It seems that sound analysis in the cochlea is influenced by a dynamic neurophysiological feedback process.

6.3 Psychoacoustics and Psychophysics

The aim of this section is to develop a simple model of the hearing system. The psychologically relevant characteristics of music include pitch, loudness, timbre, duration, amplitude envelope, spectral envelope, consonance, volume, rhythm, vibrato, and sound location information.

Psychoacoustics is the science of how we perceive sound. An interdisciplinary field, it draws upon physics, biology, psychology, engineering, and music. Psychoacoustics starts with the basic subjective attributes of sound as we perceive it and seeks to understand the ways these perceptions relate to each other.

Psychophysics focuses just on the crossover point where physics leaves off and psychology begins—where the objectively observable stops and the subjective starts. Its aim is to develop metrics that relate the external physical variables of sound (the Φ variables) to the internal psychoacoustic variables (the Ψ variables).[5] For example, the Φ intensity of a sound can be quantified easily by direct measurement (see section 4.24). The corresponding Ψ variable is loudness. The idea that the Ψ variables could be quantified was first suggested by G. T. Fechner in the 1860s (Allen and Neely 1997).

6.3.1 Science and Perception

Several problems exist in developing objective measures of our perceptions of sound and music.

- *Subjectivity* Objective measurement is a cornerstone of the scientific method, but perception of music and sound is subjective and not directly available to objective measurement. For example, it would be nice to have an objective measure that relates Φ sound intensity to Ψ loudness. But we can no more directly apply objective measurements to subjective states than we can develop a thermometer for happiness. Subjective states are only indirectly available for objective observation.
- *Nonlinearity* Ψ variables are often not linearly proportional to their corresponding Φ variable. Pitch and loudness are cases in point.
- *Nonorthogonality* Ψ variables often influence each other in quixotic and counterintuitive ways. For instance, Φ frequency clearly has a major impact on Ψ pitch perception, but Φ sound intensity also has an impact on Ψ pitch. In two-dimensional Cartesian space, the x and y dimensions are orthogonal and x and y can vary independently. Pitch and loudness are nonorthogonal.

6.3.2 Science Is Limited

Psychoacoustic research must rely on experimental methods that externalize the inner experiences of listeners. We can use such information to construct models of how human hearing functions. But there are many problems with this approach. For example, there is the problem of reconciling differing results due to conflicting experimental methodology. Suppose we conduct loudness experiments using noise bursts as stimuli; how do we relate our results to another experiment that used pure tones? How do we relate either of these to an experiment that used orchestral instruments? This is like surveying the ocean by sampling its depth in only a few places. What if in one experiment we ask subjects to evaluate how "agreeable" a musical interval is, but in another we ask how "consonant" the interval is? How shall we reconcile such semantic differences in experimental design? Psychologists thus face a problem not unlike that described in the ancient tale of the blind men and the elephant.[6] Anyone following the progress of science must live with the suspense of an unsolved mystery. To those who are not a part of the conversation, scientific discourse can be very much like tuning into a heated talk radio program—in Greek.

6.3.3 Science Is Messy

The ideas developed by science that seem effective usually result in a body of explanatory literature that describes the mind-set, or *paradigm,* that these ideas represent. Upheavals in this mind-set occur at unpredictable intervals when new, more expressive models of the subject emerge. The valid kernels of truth within the old paradigm (if they exist) are incorporated as a component of the new paradigm. However, it is not always the case that a new paradigm is simpler than the old; it may assert the importance of previously ignored or undiscovered elements, thereby actually complicating matters.

Sometimes the discarded elements of old mind-sets persist long after they are shown to be limited or erroneous. For whatever reason—social convenience or aesthetics—they linger on. An example of this phenomenon is the so-called psychophysical law that claims that the relationship between Φ intensity and Ψ intensity is always logarithmic. By this "law" the multiplication of Φ intensity by some amount purportedly always produces a corresponding addition to the perceived Ψ intensity. This concept is often associated with Weber's law,[7] which says that as the intensity of a stimulus increases, the ability to detect a difference between two levels of the stimulus decreases. In fact, I tacitly referred to this when I described the motivation for constructing the decibel scale in chapter 4.

Unfortunately, the rationale behind the decibel scale as a measure of loudness is inadequate at least in part because it ignores the fact that our hearing varies in its sensitivity to different frequency and intensity ranges. Decibel measure isn't used anymore to measure Ψ intensity, but it is still valuable as a measure of Φ intensity in engineering disciplines, for instance, in designing and using recording equipment.

I suppose we could come up with a crude metric of the complexity of a subject by tallying up all the partial explanations and conflicting theories that are currently extant about it, and then multiplying that by the number of years scientists have been studying the problem. The development of a scientific model of human hearing has been under way for at least 140 years, since the early work of Fechner, and we are still nowhere near having an established body of laws. By this measure alone we can see that the auditory system is hugely complex, containing redundancies, contradictions, and even deceptions.

Some things, such as the outer limits of loudness and pitch, are by now well established. However, though I try to restrict this discussion to just the settled facts, there is no mistaking this territory for the comforts of home. The best advice I can offer to the interested reader is to buy a radio and start learning Greek!

6.4 Pitch

Pitch is the subjective Ψ variable corresponding most closely to the objective Φ variable frequency. Pitch is sometimes called the *response pattern* to frequency. But there's no simple equality between them. While our sense of pitch is roughly proportional to frequency, it is also influenced by frequency range, loudness, and the presence or absence of other frequencies. Another difference is that pitch is limited to our range of hearing (17 Hz to 17 kHz) but frequency is unlimited.

A commonly quoted definition of pitch given by the American National Standards Institute (ANSI 1999) says, "Pitch is that auditory attribute of sound according to which sounds can be ordered on a scale from low to high." Unfortunately, stipulating precisely what "that auditory attribute" is turns out to be surprisingly complex.

A sound is *pitched* if its wave shape is highly redundant through time. Otherwise we hear *noise*. Even a pitched tone must have a certain minimum duration for its pitch to be perceived; otherwise it is heard as a click. Tones with rich harmonic spectra will appear to have a more definite pitch than sinusoids, simpler harmonic spectra, or inharmonic spectra. Very complex inharmonic spectra may appear to have several pitches. In the case of large bells, the fundamental, or *hum note,* is not the same as the perceived pitch of the instrument, the *strike note.*

6.4.1 Pitch Perception

G. S. Ohm (1843) first put forward a theory that the ear derives pitch by performing Fourier analysis on acoustical signals (see volume 2, chapter 3). Ohm's theory, sometimes called *Ohm's law of acoustics,* which he developed just after Fourier's original work, was perhaps the first place theory of pitch. One of the predictions of this theory is that the ear should be relatively insensitive to phase information, which has been shown generally to be true.

But place theory fails to account for how the ear organizes frequency components into tones instead of hearing all frequencies as unique pitches. Also, because of the nature of the Fourier transform, place theory implies a one-to-one correspondence between frequencies in the acoustical signal and pitches that the ear should detect. But we sometimes hear phantom pitches where there is no energy in the signal. How can that be?

The Missing Fundamental The place theory of Ohm hit a major stumbling block with an experiment performed by August Seebeck (1841). Suppose I play two tones for you: one is a pure sinusoid, the other is pitched but complex (having many harmonics). You can adjust the pitch of the pure tone with a knob. Your job is to adjust the pitch of the pure tone to match the pitch of the complex tone. It is virtually certain that you will adjust the frequency of the pure tone to the fundamental frequency of the complex tone *even if there is no measurable energy at the fundamental frequency* (see section 2.8.1).

Suppose the partials of the complex tone are 300, 400, and 500 Hz. You will most likely distinctly hear a "fundamental" at 100 Hz, the greatest common factor of the overtones. You will *not* hear an inharmonic tone with fundamental at 300 Hz. So convinced are our ears of the ubiquitous phenomenon of a fundamental with harmonics at integer multiples that even if there is no fundamental, our hearing is hardwired to invent one. This means that Ohm's theory, which requires a one-to-one correspondence between frequencies and pitch, runs into the contradiction of a pitch with no corresponding frequency.

The phenomenon of the missing fundamental is what enables us to hear satisfying music come from the tiny speaker of a transistor radio: our hearing invents the fundamentals that the speaker can't reproduce.

Periodicity Theory The explanation that Seebeck provided as a substitute for Ohm's place theory came to be called *periodicity theory.* It was developed further in the 1940s by Schouten, Ritsma, and Cardozo (1962). This theory supposes that the neural signals from the cochlea to the brain encode timing information related to the phase of the acoustical signal and that the brain has some means of measuring time intervals.

Periodicity theory notes that the combination of several high harmonics can sum to create a waveform with prominent time domain features whose period is the same as that of their common fundamental. This way, a pitch period-measuring capability in the brain would get more or less the same information from a tone with or without a fundamental.

Periodicity theory also explains why amplifying the electrical activity in the auditory nerve results in an electrical signal similar to the acoustical signal presented to the ear.

However, it is neurologically impossible for neurons to fire more rapidly than about 1 ms, called the *absolute refractory period.* So periodicity theory runs into trouble for pitches above 1000 Hz. Another difficulty is that periodicity theory would lead us to expect the ear to be quite sensitive to the phase of the harmonics in complex tones. However, place theory—that the ear largely ignores phase—agrees very well with experiments. Ohm had suggested that perception of sound depends only on the distribution of energy among partials and does not depend upon differences of phase. The physical demonstration of this was considered a major accomplishment of Helmholtz, and the theory was effectively unchallenged for a century.

Beyond the Peripheral Theories Clearly, place and periodicity theories have merit and also liabilities. Both suppose that acoustical processing occurs in the periphery of the auditory system: the basilar membrane and lower nerve centers of the auditory cortex. So these theories are called jointly the *peripheral theories.*

The main drawbacks of the peripheral theories are (for periodicity theory) sensitivity to the phase relationship between partials, and (for place theory) the impossibility of explaining the missing fundamental in spectral terms.

Also, experiments with *dichotic* signals (where different information is sent to each ear separately via headphones) have demonstrated a necessary role for the brain in pitch detection. Houtsma and Goldstein (1972), for example, demonstrated that we still manage to hear a missing fundamental even if some harmonics are sent to one ear via headphones and different harmonics of the same fundamental are sent to the other ear. This shows that the brain must be the agent that combines the harmonics to determine the fundamental, because if this were handled peripherally, we would hear different pitches in each ear rather than a single fused tone.

Difficulties with peripheral theories and experiments with dichotic signals led researchers to *central processing theories* of pitch perception that emphasize central processing conducted in the brain (Goldstein 1973; Wightman 1973; Terhardt 1974). These theories presume that pattern-matching systems in the brain search for order in the components arriving from the peripheral auditory system. Pattern matching accounts well for our ability to detect the fundamental of stretched harmonics of a piano tone, and to dig harmonic information out of intensely noisy signals.

Assuming higher neural processing for pitch perception also helps explain the fact that we can learn pitch discrimination. When I taught solfeggio and sight-singing in college, I observed over the course of the semester that students' capacity to discriminate and categorize pitch improved, sometimes dramatically. (See section 9.22 for a discussion of self-learning neural systems.)

Pitch perception remains one of psychoacoustics' longest-running controversies, with an unbelievable number of competing theories. Perhaps the theoretical difficulties are a consequence of the importance of pitch perception to survival. A faculty this critical to life can't be entrusted to only one adaptation; redundancy and competitive analysis in both the periphery and the brain are required.

6.4.2 Range and Quality of Pitch Sensation

When I indicated that the range of hearing is 20 Hz to 20 kHz, that was just to throw out some round numbers that are easy to remember. In fact, the boundaries are fuzzy and vary enormously with age, gender, and life experience.

At the top end, a young person in good health might be able to hear up to 17 kHz or so. Adults lose the top end until, nearing old age, it might be down to around 12 kHz for women and 5 kHz for men.

Pitch discrimination drops off above about 5 kHz for all of us, which perhaps explains why few musical instruments are designed to intone beyond that range. The highest note on the piano is C8, 4186 Hz.

At the low end, sounds below about 30 Hz become progressively harder to hear as having a pitch. Below that frequency, we start to *feel* sound as physical impact. The lowest note on the piano is A0, 27.5 Hz.

The range of finest perception both in terms of pitch and loudness is between 1 kHz and 4 kHz, which coincidently is where most speech information occurs.

6.4.3 Just Noticeable Difference of Pitch

Two important attributes of a ruler are its length and the fineness or precision of measurements that can be made with it.[8] If the range from lowest to highest frequency the ear can hear corresponds to the length of some kind of ruler, then to what perceptual quality does the precision of measurement correspond?

If the difference between two pitches is not noticeable, we judge them subjectively to be the same, whether they are physically the same or not. Recall that Euler wrote, "The sense of hearing is accustomed to identify with a single ratio, all the ratios which are only slightly different from it, so that the difference between them be almost imperceptible."

Effectively, pitches must differ by a minimum threshold for us to distinguish them. This threshold is the *just noticeable difference* (JND) of pitch. The *pitch JND* is the measure of sensitivity of the ear to changes in pitch. It is sometimes called the pitch *difference limen,* or pitch *DL*. How well the ear can distinguish between adjacent pitches determines the precision of our hearing.

6.4.4 The Weber-Fechner and Stevens Laws

Interestingly, the size of the pitch JND is not constant. The JND of high frequencies covers a larger span of frequencies than the JND of low frequencies. Attributed to Ernst Weber, the JND is a classic psychophysical invention that has been applied not just to the senses (e.g., color and taste) but even to the price of houses. In general, Weber observed that the greater the magnitude of a stimulus, the greater must be the change in that stimulus before any difference is detected.

Weber's law correctly predicts that the just noticeable difference of pitch grows with increasing magnitude (greater magnitude means, in this case, higher frequency). If we call the size of the JND ΔI and the magnitude of a comparison stimulus I, Weber's law says that

$$\frac{\Delta I}{I} = k, \qquad \textit{Just Noticeable Difference (JND)} \quad (6.1)$$

where k is a constant of proportionality. The parameter k takes on different values for different sensory stimuli.

Gustav T. Fechner (1801–1887) based his work on Weber's JND but refined it by suggesting that for many percepts (including pitch and loudness), a *geometric* increase in Φ magnitude is perceived as an *arithmetic* increase in Ψ magnitude. Thus, Ψ magnitude increases in proportion to the logarithm of the Φ magnitude, and large changes in Φ are compressed into smaller changes in Ψ. Fechner's law can be expressed as

$$\frac{\Psi}{\log \Phi} = k, \qquad \textit{Weber-Fechner Law} \quad (6.2)$$

where Ψ is the magnitude of the sensation, Φ is the magnitude of the stimulus, and k is the constant of proportionality. It was Fechner's work that led to the theoretical underpinnings of the decibel.

Experiments have shown that the Weber-Fechner law works better for some stimuli than others and generally works best for stimuli of medium intensity. Stevens (1962) generalized the Weber-Fechner law so it could be applied more widely. He suggested that Ψ magnitude increases in proportion to the Φ magnitude raised to a power:

$$\frac{\Psi}{\Phi^p} = k, \qquad \textit{Stevens Law} \quad (6.3)$$

where Ψ is the magnitude of the sensation, Φ is the magnitude of the stimulus, p is its exponent, and k is a constant of proportionality.

A logarithm and a power function can be made to resemble each other if the exponent is between 0 and 1. For example, compare curves 1 and 2 in figure 6.3a. Curve 1 is a power law approximation of the Weber-Fechner log curve 2. Since even in the best of circumstances we can only estimate

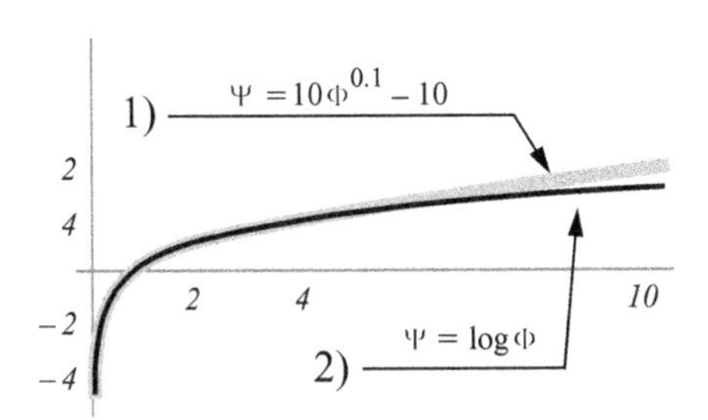

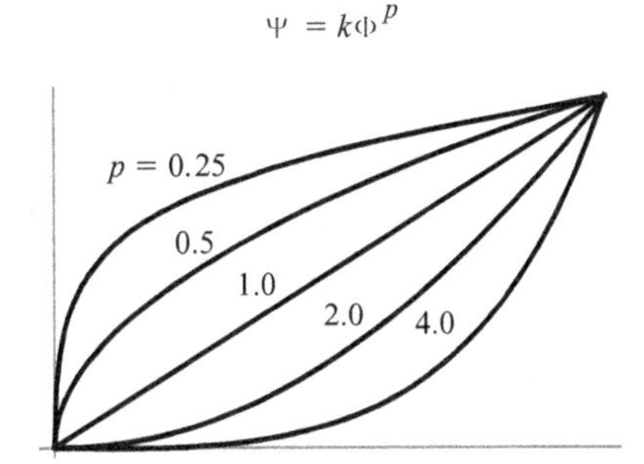

Figure 6.3
Comparison of the Weber-Fechner law and the Stevens law functions.

the Ψ/Φ relation (because Ψ is subjective and we can't objectively measure it), the two approaches are reasonably interchangeable. However, for stimuli such as the apparent length of an event, the degree of compression between Φ and Ψ domains is less than the Weber-Fechner law would predict and may be better modeled with a power exponent greater than 1. In these cases, Φ changes can produce equal or even larger changes in Ψ. In cases like this, the Stevens law, illustrated in figure 6.3b, provides a richer range of mappings to experimental data and has been widely used.

6.4.5 Determining Pitch JND

How are such metrics established experimentally? For example the pitch JND can be determined as follows. Suppose I play a sequence of two sinusoids, both with the same loudness. The first tone has a constant pitch; the second tone has a small vibrato. This allows you to tell the two tones apart. As the subject of the experiment, you must tell me each time whether the pitch of the second tone is "above" or "below" the first. (Saying, "the same" is not an option.)

This process is a simplified version of the experimental method called *two-alternative forced-choice (2AFC)*. If the difference between the tones is large, your judgments will tend to be categorical. But where the difference is slight, your answers will become increasingly arbitrary, and when the frequencies are too close to be distinguished, your answers will be effectively random (right approximately half the time). The experimenter examines your responses, looking for the range over which your responses transition from 100 percent correct to random (50 percent correct). The midpoint of this transition zone, around 75 percent, is taken to be the JND at that frequency.

Such a method could tell us the JND of any frequency, but only for sinusoids (because that's all we tested with). What about other sounds—sustained sounds, short sounds, sounds with varying pitch, sounds with steady pitch or quickly varying pitch, simple sinusoids vs. complex tones? If complex (read: musically interesting) tones are used, which complex tones shall we compare? All these parameters (and more) will have an effect on the pitch JND we end up measuring.

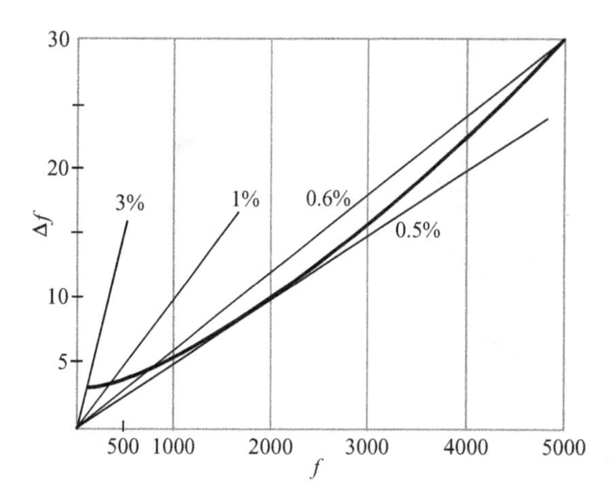

Figure 6.4
Just noticeable difference for pitch. (Adapted from Roederer 1973.)

Psychophysicists have traditionally taken a bottom-up approach to such questions. If they can get a theory right for simple steady-state sinusoids, they figure they can use it later to explain more complex phenomena. I must say, as a musician, I am always disappointed by this approach because it seems that the elementary results of psychophysics are almost uselessly simplistic in realistic musical situations. On the other hand, correct but limited knowledge is better than none (and is certainly a big improvement on erroneous information or superstition).

The JND of pitch has been found experimentally to depend not just upon frequency but also upon intensity and duration as well as the rapidity of frequency change. The heavy line in figure 6.4 shows the pitch JND for constant-intensity (80 dB) sinusoids whose frequency was slowly and continuously modulated up and down. The light lines show several JND thresholds for reference: 0.5 percent and 0.6 percent, 1 percent and 3 percent. We observe that the heavy line mostly lies between 0.5 percent and 0.6 percent.

The figure shows frequency f on the x-axis and the corresponding detectable frequency difference Δf on the y-axis. The ratio of $\Delta f/f$, sometimes called the *frequency resolution* of the ear, shows the pitch JND for frequencies between about 30 and 5000 Hz. The closer this line is to the x-axis, the smaller is the JND. For example, we don't seem to notice a difference of less than ±5 Hz around 1 kHz; thus the JND, expressed as a percentage, is 0.5 percent at 1 kHz. We also don't seem to notice a difference of less than ±30 Hz for tones around 5 kHz, or 0.6 percent.

Note also that

- The low and high ends have wider JNDs, and the bottom end is worse than the top end.
- The most acute region is from 1 to 3 kHz, where the JND is about 0.5 percent of the frequency. For reference, that's about one twelfth of a semitone, or 8.3 cents (see section 3.4).
- Rapidly changing frequency fluctuations can produce JNDs up to 30 times as small.

- Shorter-duration tones produce larger JNDs.
- Frequency resolution of the ear is relatively independent of sound intensity.

JNDs also depend a great deal upon the individuals tested and their degree of musical training as well as upon the methods used to measure the JNDs.

6.4.6 Interval Perception

I suggested that we could compare pitch JND to the tick marks on a ruler, but, like all analogies, this one has its limits. Wouldn't it be convenient if our ears measured pitch difference as the number of JNDs between pitches? Alas, it is not so, and it really can't be if we think about it.

While pitch JND gives us an understanding of pitch similarities, the JND provides no information about how we judge pitch differences. The only thing that JND knowledge contributes to this subject is that pitches lying inside a JND are experienced as the same while pitches lying outside a JND are experienced as different, but JND says nothing about the quality of that difference. We must address this question separately.

Suppose I play a pair of sinusoids, one fixed, the other beginning in unison with it but diverging from it by slowly gliding up in frequency. We might hypothesize that just as the difference between two points constantly increases as the points diverge in space, so too the ear should experience a constantly increasing difference in frequency as a constantly increasing difference in pitch.

This hypothesis is partly true. We hear the tone height of the pitch that is gliding up continue to grow. However, ever more widely separated pitches do not always sound increasingly different, as one would expect if the only thing the ears paid attention to was the number of JNDs between pitches. Instead, as the distance reaches a doubling in frequency, the tones begin to sound alike again, as they did when they were in unison. This perception repeats at each subsequent frequency doubling, an effect called octave equivalence (see section 2.3.3). The equivalence is felt so strongly that virtually all musical scales around the world are organized around the 2:1 ratio of the octave, and pitches related by octaves are virtually always given the same name.

Interestingly, the octave, as a physical frequency ratio of 2:1, always corresponds exactly to the subjective pitch difference of an octave, making the octave a rare instance where objective and subjective measurements seem to match exactly. Perhaps this symmetry between object and subject is why pitch is the element of hearing most heavily relied upon to convey musical information.

All this suggests that pitch is more than a one-dimensional sense of high and low. Révész (1954) developed a *two-component theory of tone*, suggesting that there are at least two principal interlocking structures in pitch:

- The linear span of JND pitch differences from the bottom to the top of our hearing range, which he called *tone height*.
- The circular span of interval differences within the compass of each octave, which he called *chroma*. Chroma refers to the position of a tone within an octave.

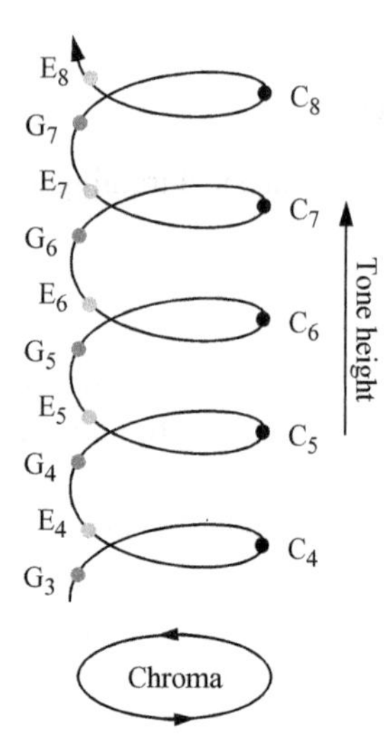

Figure 6.5
Tone height and chroma.

We can reconcile the concepts of tone height and chroma in two dimensions (Shepard 1982). Figure 6.5 represents tone height along the *y*-axis and chroma as an angle on a circle in the *x*-axis and *z*-axis. The combination results in a helix. The movement of a sinusoid from C4 to C5, for instance, is represented as a movement upward in tone height but as a return to the starting angle in chroma.

Octave equivalence is perhaps just a very strong instance of *interval affinity.* Similar intervals are highly identifiable—a trait much exploited by musicians. Fourths show "fourthness" and fifths show "fifthness" regardless of their orientation in pitch space. Understanding the musical qualities of affine intervals is one of the subjects of harmony theory, which in turn is one of the subdisciplines of music theory.

There remains the problem of how to actually construct useful musical scales out of the continuum of available pitches within the chroma. Ordinarily, musicians select a small subset of intervals from the chroma, and these become the *pitch classes* of the scale. When the pitch classes are replicated across each octave, they become the pitches of the available *pitch space,* or gamut. In the West, the scale has 12 chromatic pitches. We can visualize the pitch space of the equal-tempered scale as shown in figure 6.6, which is a projection of figure 6.5 along the *y*-axis.[9] Because humans can hear ten or so octaves, the spiral shows ten revolutions, where the outer ones are the lower octaves. The 12 lines radiating out are the 12 chromatic pitch classes of the Western scale. The set of points where the lines intersect the spiral form the gamut of pitches of the equal-tempered scale (see chapter 3).

To get a feeling for chroma and tone height, perform the following experiment on a piano: starting from a low tone, play a sequence of major seventh intervals (separated by 11 semitones) up the keyboard. For instance, C1, B1, A♯2, A3, G♯4, G5, and so on. You may hear an ambiguous

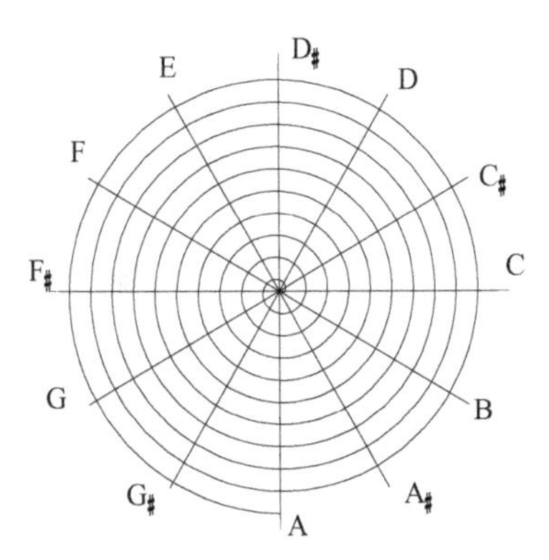

Figure 6.6
Chromatic pitch space.

effect: though the pitch rises by sevenths, you might also be able to hear the sequence as though it were *decreasing by semitones*. While the tones of a major seventh interval are relatively far apart in terms of tone height, they are close together in terms of chroma. Hence, if you focus on tone height, you hear the sequence ascend. If you focus on chroma, you hear it descend. Roger Shepard noticed this effect in his early research, which led ultimately to his famous illusion.

6.4.7 Shepard Scale Illusion

Shepard (1964) wanted to test Révész's theory of tone height and chroma. If he could suppress one of the two effects and the other effect still persisted, that would demonstrate that they are separate perceptual attributes of pitch. In particular, if Shepard could suppress the sense of tone height, chroma should be all that is left. The helix in figure 6.5 would collapse into a circle, and pitch judgments would also become circular. He devised a demonstration of pitch circularity in 1964 that proved Révész's theory. It has come to be known as the Shepard scale illusion or Shepard tone demonstration.

A set of ten sinusoids at octave intervals is played as shown in figure 6.7. The frequencies glide up smoothly together, rising continuously in pitch (in some versions, they rise by semitone steps). The intensity of the low and high sinusoids is increasingly diminished, so the ear mostly hears the sinusoids in the middle frequencies (implemented as a Gaussian-shaped intensity contour). As the top sinusoid goes off the top end of the hearing range, it gradually drops below the threshold of hearing and a new sinusoid is introduced from below. The whole effect is rather like the visual illusion of a barber pole in motion, or the impossible staircase of the visual artist M. C. Escher—constantly rising, never getting anywhere (figure 6.8).[10]

The equation for creating the original Shepard tone illusion based on movement by semitones is given by

$$F(t,c) = F_{min} \cdot 2^{(ct_{max}+t)/t_{max}}, \qquad (6.4)$$

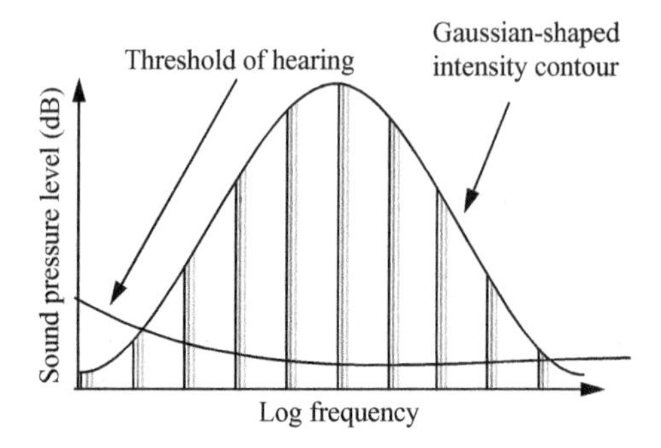

Figure 6.7
Shepard scale illusion.

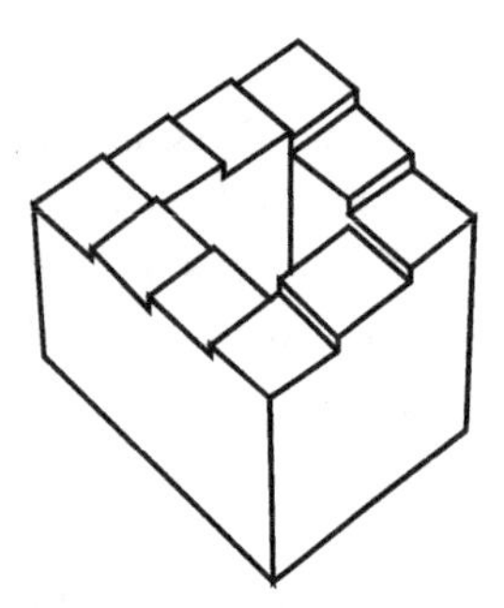

Figure 6.8
Impossible staircase.

where $F(t, c)$ is the frequency of the partial c of tone t, t_{max} is 12 (because this version of the effect is based on the chromatic scale), and F_{min} is the frequency of the lowest partial of the lowest tone. The range of t is $0 \leq t < t_{max}$, and the range of c is $0 \leq c < N$, where N is the number of partials to be generated. Shepard used $N = 10$. To create the first set of partials, set $t = 0$ and evaluate (6.4) for all c. For the next step, increment t by 1, and evaluate for all c again; repeat for all t. It is also necessary to adjust the loudness of each partial to achieve the contour shown in figure 6.7. This step and the modification to variables t and t_{max} in equation (6.4) can be used to effect a smooth glissando. Note that the Gaussian envelope shape is given in log frequency so that equal pitch intervals occupy a uniform distance along the frequency axis.

6.5 Loudness

Loudness is the subjective Ψ variable corresponding most closely to the objective Φ variable intensity. Loudness is sometimes called the response pattern to intensity. But there is no simple equality between them. While our sense of loudness is roughly proportional to intensity, it is also influenced by frequency range and the presence or absence of other frequencies. Another difference is that loudness is limited to the distance between our threshold of hearing t_h (10^{-12} W/m^2) and the limit of hearing l_h (10^0 W/m^2) but intensity is unlimited (see section 4.24).

The *loudness JND* is the amount by which the intensity of a sound must change in order for the ear to register a difference in loudness. The size of the loudness JND is approximately proportional to the intensity of the sound: the louder the sound, the greater must be the change in its loudness before the change in loudness is registered. (This is a restatement of the Weber-Fechner law for loudness.) However, the loudness JND varies substantially with frequency and intensity range, so there is no simple linear relation. Also, there is no loudness equivalent to the octave, that is, judging a sound to be "twice as loud" shows a much greater deviation among subjects than does judging a pitch to be "an octave higher."

6.5.1 Relating Pitch and Loudness

As mentioned, the rationale of the decibel scale assumes our perception of loudness to be independent of all other percepts such as frequency, but it is not. Because of the mechanical advantage the ear gives to frequencies of 1–3.5 kHz, tones in this range are perceived as louder than tones of equal intensity in other ranges.

Since loudness depends upon both intensity and frequency, a loudness scale properly requires three dimensions: the independent Φ variables frequency f and intensity I and the dependent Ψ variable loudness L. Since we're dealing with perceptual variables here, we must explicitly test for every relation we want to measure. Thus, if we want to know when loudnesses are equal, we must develop a metric for the equality of two loudnesses at different frequencies. If we want to compare loudness differences, we must develop a metric for the difference of two loudnesses at equal frequencies. The first metric is the *phon,* a measure of equal loudness. The second metric is the *sone,* a measure of comparative loudness. Together, they allow us to account for the ear's varying sensitivity to frequency and intensity.

6.5.2 The Phon Scale

The phon scale identifies equal loudnesses across all perceivable frequencies and intensities. It consists of a set of *equal loudness contours* that relate intensity in one region of frequency to the intensity required to achieve equal loudness in other regions of frequency. By definition, any frequency at the threshold of hearing is exactly 0 phon.

The phon is defined as identical to dB SIL at 1000 Hz from the threshold of hearing to the limit of hearing. Thus, at 1000 Hz the threshold of hearing, $t_h = 10^{-12}$ W/m^2, is defined as 0 phon, and a level of 120 phons equals 120 dB SIL above the 0 phon reference (recall that dB SIL expresses a ratio of two intensities). For example, at 1000 Hz a sinusoid with 10 dB SIL has a loudness of 10 phons, a 20 dB SIL sinusoid has a loudness of 20 phons, and so on.

Having defined the phon scale at 1000 Hz as identical to dB SIL, we now extend the phon scale to frequencies other than 1000 Hz. We do this by comparing sinusoids at various frequencies and intensities to a set of reference intensities at 1000 Hz. In general, for a sinusoid with frequency f, we want to know what intensity I is required so that it will have the same loudness L as a sinusoid at 1000 Hz. Let ε be the criterion of equal loudness. Then for some frequency f and loudness L, we want to solve the relation $I = \varepsilon(L, f)$, which tells us what intensity I is required for a sinusoid to achieve loudness L at frequency f.

Ordinarily, the phon scale is evaluated at 10 phon increments from 0 to 120 phons. An approximation to this set of curves is shown in figure 6.9. It shows the contours of equal loudness for sinusoids that were first established experimentally by Fletcher and Munson (1933). We see that in general low frequencies must have greater intensity in order to have the same loudness as frequencies around 1000 Hz. This is especially true when low frequencies also have low intensity. The same is also true of high frequencies but with somewhat less exaggeration. The curves in figure 6.9 are adapted from those recommended by the International Standards Organization (ISO 226, 1987).

The equal loudness curves shown in figure 6.9 are also called *equal loudness contours* because they can be thought of as delineating curves of equal elevation above the two-dimensional frequency/intensity plane. Imagine figure 6.9 as a three-dimensional map that we are looking straight down on. Greater phon levels rise up toward us like a 3-D relief map.

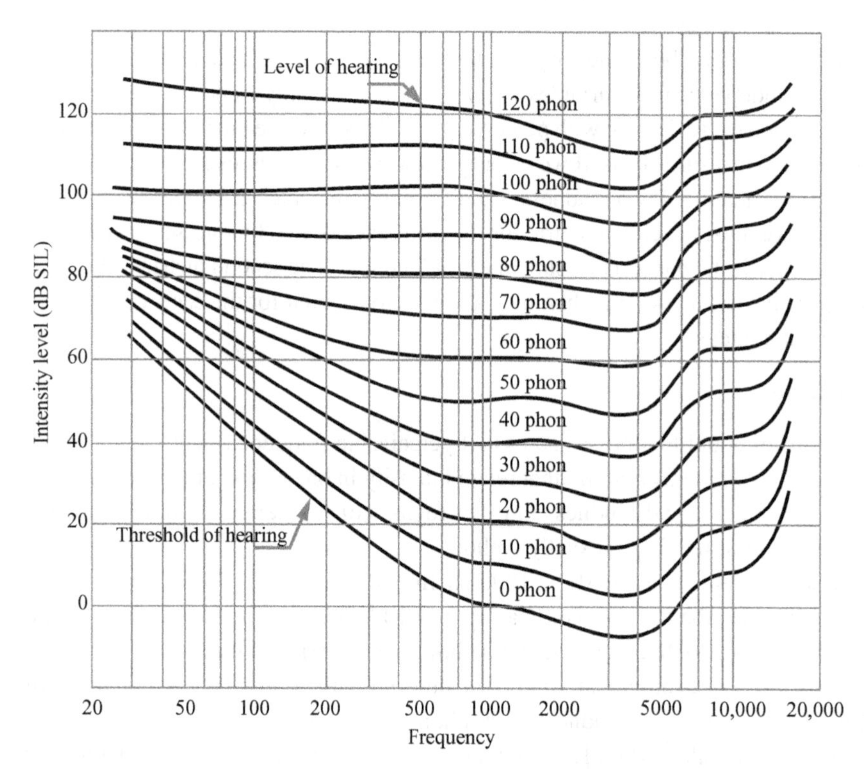

Figure 6.9
Equal loudness contours. (Fletcher and Munson 1933.)

Here's a practical application for the phon scale. Suppose we record a symphony orchestra performing with intensities between 60 and 95 dB SIL. If we play it back at a lower intensity, say, 40–75 dB SIL, it sounds tinny, lacking in bass and treble. Reproduced at lower intensity, low and high frequencies of low intensity receive greater attenuation in our perception because of the ear's lack of sensitivity at these frequencies. If we compensate by boosting the bass and treble according to the equal loudness curves, we can restore something like the original balance of intensities. Some audio amplifiers come equipped with a so-called loudness knob that applies an approximation of the above curves for different listening levels.

Sound level meters approximate the loudness corresponding to the intensity of sound. They usually include switchable weighting networks, which are filters applied to the input signal that mimic the Fletcher-Munson curves, attenuating frequencies where our hearing is less sensitive. In this way the response of the instrument can be made to provide a rough approximation of the perceived loudness of a sound. The meters typically come with A, B, and C weighting networks, which are simplified inverse functions of the 40, 70, and 100 phon curves, respectively (Stevens 1961; ISO 1975).

6.5.3 Threshold of Hearing

Perhaps the most salient equal loudness curve is the threshold of hearing. An approximation to the threshold of hearing is given by Terhardt (1979) as

$$T_q(f) = 3.64\left(\frac{f}{1000}\right)^{-0.8} - 6.5e^{-0.6(f/1000-3.3)^2} + 10^{-3}\left(\frac{f}{1000}\right)^4. \tag{6.5}$$

This complicated-looking function is an approximation of the threshold of hearing for a young adult with acute hearing. When plotted for f in the range of human hearing, it produces the graph shown in figure 6.10. This curve can be used to determine the maximum allowable energy level for noise and distortion that can be added by a recording system before it is noticed as distortion

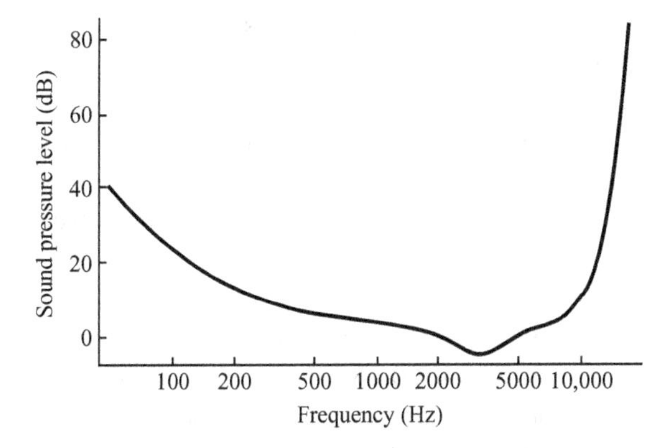

Figure 6.10
Threshold of hearing.

by a sensitive listener. This has great relevance for the design of audio systems in general and is a crucial metric for perceptual audio coders, such as MPEG audio, including the well-known MP3 audio coding format.

A final note on the phon scale: remember that it is a measure of *equal* loudness. It can only answer the question: Is the loudness of two frequencies equal? It does not tell us about loudness *differences*. For instance, a doubling of loudness in phons does not necessarily result in a sound's being heard as twice as loud. To compare proportional loudness requires the sone scale.

6.5.4 The Sone Scale

We can characterize the ratio of two sinusoids with different intensities at the same frequency with the sone scale. One sone is defined as the loudness of a 1 kHz tone at 40 dB SIL. This is the reference loudness of the sone scale. This also means that 1 sone = 40 phons. A sound that is judged to be twice as loud as the reference has a loudness of 2 sones, a sound that is judged to be half as loud as the reference has a loudness of 0.5 sone, and so on. For example, the average listener hears a 1 kHz sinusoid at 50 dB as about twice as loud as a 1 kHz sinusoid at 40 dB. Hence, the 50 dB 1 kHz sinusoid has a loudness of 2 sones.

Loudness in sones L_s can be related to loudness in phons L_p as follows:

$$L_s = 2^{(L_p - 40)/10}. \qquad \textit{Phon/Sone Conversion} \ (6.6)$$

For 1000 Hz tones, the Ψ variable L_s relates to the Φ variable sound intensity roughly following a power law:

$$L_s \cong kp^{0.6}, \qquad \textit{Sones and Intensity} \ (6.7)$$

where p is the pressure in pascals, and k depends on frequency. These two equations, based on the work of Stevens (1956), indicate that loudness doubles for a 10 dB increase in SPL. The calibration of the sone scale is controversial because of the difficulty subjects have in identifying loudness ratios with certainty. For instance, Warren (1970) found a doubling of loudness for a 6 dB increase in SPL. Thus, (6.7) should not be taken too literally.

It should be clear by now that the relation between the Ψ and Φ domains is anything but simple. What is especially remarkable to me is that musicians are able to navigate the complexities of all these nonlinear, nonorthogonal relations with ease, balancing intonation and loudness to achieve precisely calibrated sonic effects. More astonishing yet is that naive listeners are effortlessly able to sort it all out.

6.5.5 Pitch Shift with Loudness Change

Another example of the nonorthogonality of Ψ variables is that loudness has an impact on pitch. If the intensity of a 100 Hz tone is increased from 40 dB to 100 dB SPL, the pitch decreases by about 10 percent. At 500 Hz, the pitch changes by about 2 percent for the same increase in SPL. Try taking headphones on and off while listening to music with the volume as loud as is comfortable. You will probably be able to hear the pitch shift as you put them on and off.

6.6 Frequency Domain Masking

When two sinusoids are presented simultaneously, the fainter sinusoid can be rendered inaudible, or *masked,* by the louder sinusoid if the fainter one lies within a certain frequency range of the louder one. Figure 6.11 shows a 1 kHz 60 dB sinusoid, called the *masker,* which has the effect of raising the threshold of hearing in its vicinity. The skirts to either side of the 1000 Hz tone indicate the just noticeable level required for a test tone to be audible in that range. The dashed line indicates the threshold of hearing in the absence of the masker. For example, a 1.5 kHz sinusoid at 35 dB or a 750 Hz sinusoid at 35 dB would be inaudible in this case. Frequencies above the masker frequency are more strongly masked than those below it (Zwicker and Fastl 1990).

The masker will mask any signal that lies below the masking threshold, not just sinusoids. In particular, any artifacts of a recording process, background noise, and so on, will all be masked so long as they are below the threshold.

6.6.1 Temporal Masking

Masking also occurs when tones are played in succession. This is called *temporal masking*. There are three possibilities:

- *Forward masking* Even after a sound ends, its effect on the threshold of hearing lingers for a while. The threshold of a test signal following the masker is impaired for a period of time. Forward masking can last as long as 100 to 200 ms. The relative loudness of the masker and test signal and their precise timing affect the audibility of the test signal.

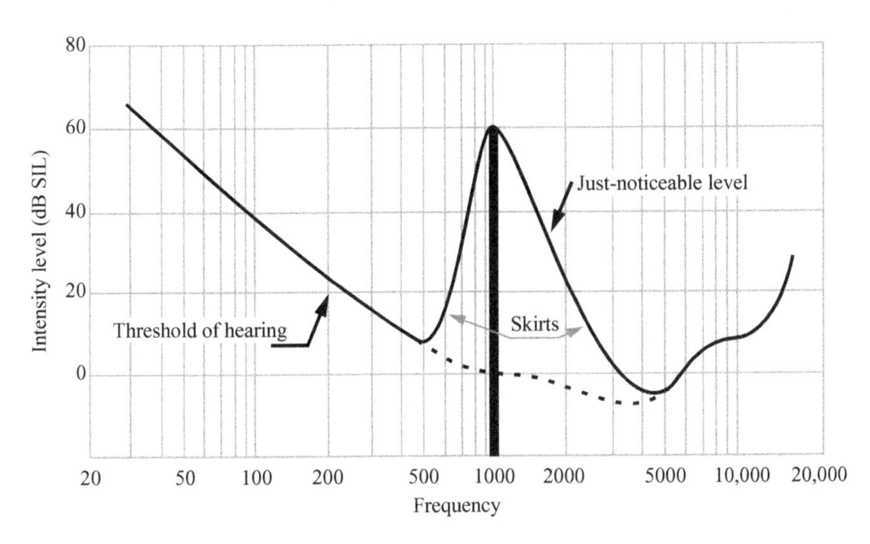

Figure 6.11
Frequency domain masking.

- *Simultaneous masking* This occurs when the masker and the test signal are presented at the same time, and is identical to frequency domain masking.
- *Backward masking* This occurs when a masker influences the audibility of a fainter test signal that *precedes* it. While this might seem at first to require prescience on the part of the ear, it can be explained by realizing that sound perception is actually integrated over a time interval preceding the moment of recognition. The time interval is generally regarded as being on the order of 200 ms. Fainter sounds lying within this interval are subject to some degree of masking regardless of their order of arrival. The amount of masking diminishes the more the test signal precedes the masker. It is also affected by the relative loudness of the two signals.

Approximate durations of forward and backward masking are suggested by figure 6.12 (Zwicker and Fastl 1990). The curve indicates the level that a short tone burst must have in order to be just noticeable in the presence of a relatively long masker of 200 ms duration. The *y*-axis shows intensity level expressed in dB above the just noticeable level of the test signal by itself, that is, 0 dB is the reference intensity of the just noticeable level of the test signal alone. The *x*-axis indicates the onset time of the test signal relative to the masker signal, which begins at time 0.

We can see that the effectiveness of backward masking decreases sharply as the test signal increasingly precedes the masker. Backward masking is generally thought to be effective only up to about 5 ms.

When musicians are supposed to strike a note at the same instant, they rarely manage to do so. But because of temporal masking, our ears are forgiving, perceiving the onset as simultaneous so long as the attacks lie within the temporal masking intervals. For instance, if a softer tone attacks up to 5 ms earlier than a louder one, it is masked by the louder one, so we tend to hear the two tones as simultaneous.

Some modern perceptual coders such as MPEG audio divide the audio stream into packets in order to transmit it more efficiently. Temporal masking makes it possible for the audio packets to still sound seamless so long as the temporal masking boundaries are not exceeded. To succeed, encoders like

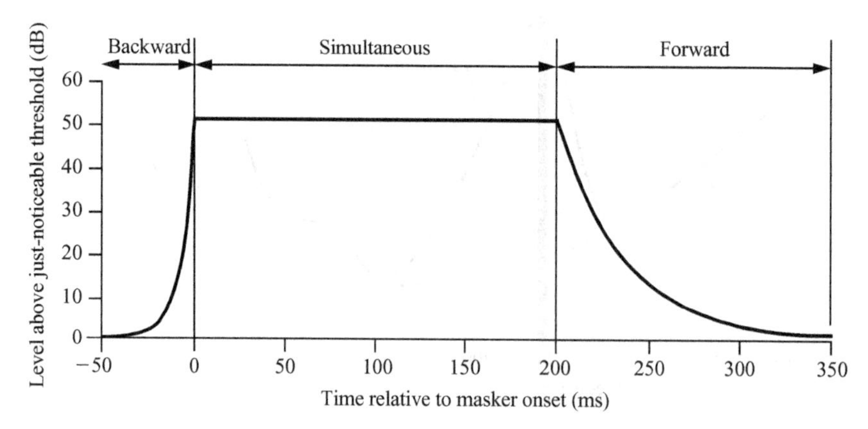

Figure 6.12
Forward and backward masking.

MPEG audio must be able to provide temporal resolution under 5 ms to ensure that events that are supposed to be heard as simultaneous are actually perceived that way (Bosi and Goldberg 2003).

6.7 Beats

When sinusoids of slightly different pitch f_1 and f_2 are sounded together, the phase difference changes through time so that they sometimes reinforce and sometimes cancel at a rate of $\Delta f = f_2 - f_1$ (see volume 2, chapter 2). The amplitude of the sum of the two waves modulates at a rate equal to the difference between their frequencies. Such slow, periodic fluctuations in amplitude are called beats. Figure 6.13 shows the beating that results when two sinusoids with a frequency ratio of 91/100 Hz are added together.

When the ear hears two pure tones of slightly different frequency, the combination produces a sensation of audible beats at the *difference frequency* Δf. This is heard as a kind of fluttering or wavering of the amplitude of the combined sound. The musical term for this effect is *tremolo*. If Δf is greater than about 10 Hz, the tremolo effect disappears and the tone becomes rough-sounding and unpleasant, that is, *dissonant*. If Δf keeps growing beyond about 20 Hz, the ear starts hearing two distinct, but still rough, tones. As Δf keeps growing, the roughness eventually goes away somewhere near a major third, and we simply experience two separate tones. This effect is best described by reference to the theory of critical bands (see section 6.9).

6.7.1 Tonal Fusion

Beats are often used by musicians as an aid in tuning their instruments because these qualitative changes supply additional information along with the ear's pitch perception. Figure 6.14 provides a graphical representation of tonal fusion and perception of beat frequencies.

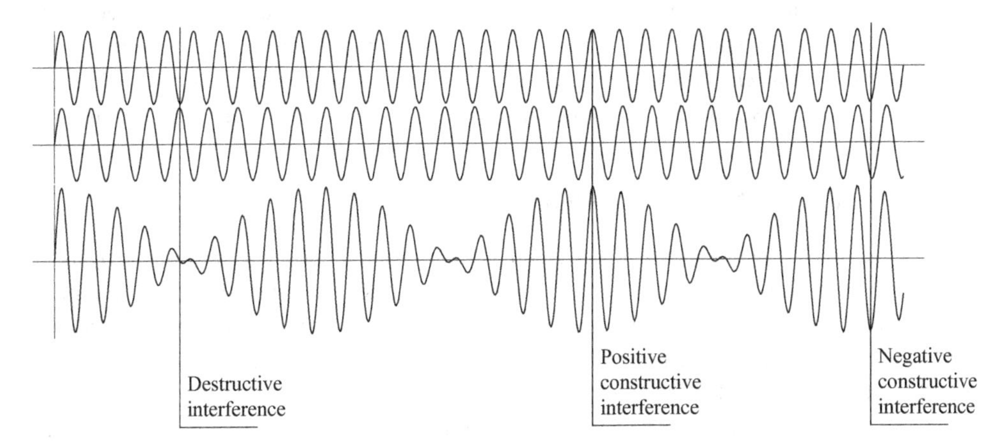

Figure 6.13
Beats.

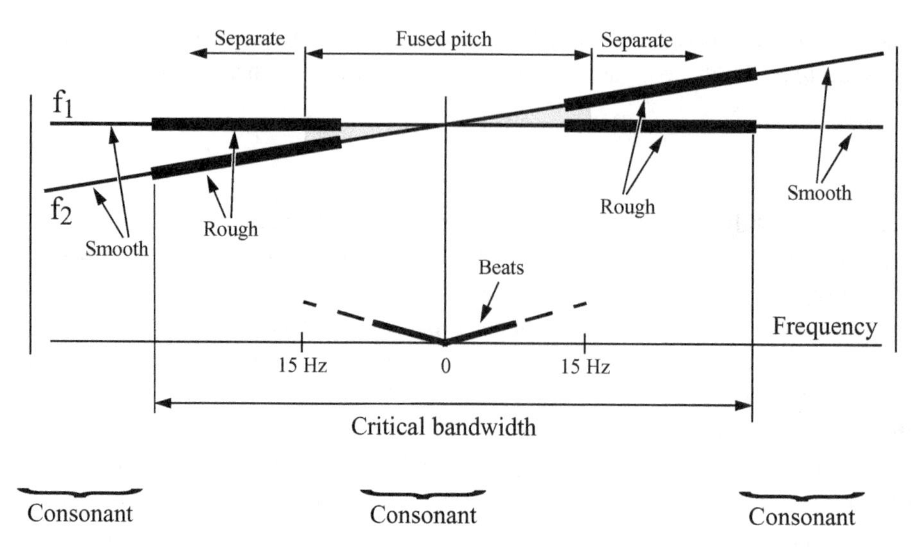

Figure 6.14
Tone fusion and the perception of beat frequencies. (Adapted from Roederer 1993.)

The beat phenomenon arises from pure tones that are very nearly in unison, called *first-order beats*. The ear hears the Φ effect created by the amplitude envelope of the two tones and also experiences a Ψ effect from neural processing.

Beats may also be heard between pure tones that are very nearly an octave, fifth, or fourth apart. These are called *second-order beats.* However, this beating results only from the effects of neural processing.

When frequencies of two sinusoids are within 15 Hz of each other, the ear tends to hear just one pitch. The two sinusoids lose their separate perceptual identity, and we hear a single fused pitch. Carl Stumpf (1848–1936) studied the circumstances under which tones appear to be fused. He defined *tonal fusion (tonverschmelzung)* as the effect of hearing two tones not as a sum but as a whole, or unity (Stumpf 1883/1890). He found tonal fusion to be most pronounced in the consonant intervals (unison, octave, and fifth) and less pronounced in the increasingly dissonant intervals.

6.7.2 Tonal Fusion and Music Composition

Tonal fusion was evidently of concern to J. S. Bach. He sought to compose pleasing music by using consonant intervals, and he wanted to project a polyphonic musical style, where multiple independent musical lines are separately discernible. But the most consonant intervals have a tendency towards tonal fusion, and tonal fusion destroys the sense of polyphony by making separate voices appear as one. David Huron (1991) conducted a statistical analysis of Bach's music and concluded that while Bach preferred consonant intervals, he avoided consonant intervals to the extent that they promoted tonal fusion so as not to compromise polyphony.

Tonal fusion was used explicitly by composer Maurice Ravel in his composition *Bolero*. The repetitive melody of this work is passed around in the key of C to various instruments over the course of its 17 minutes' duration. When it's the French horn's turn, Ravel adds a piccolo playing the melody transposed strictly at the twelfth (up an octave and a fifth), in the key of G, matching the third French horn harmonic. He also adds another piccolo playing the melody transposed strictly at the seventeenth (up two octaves and a third), in the key of E, matching the fifth French horn harmonic. The tone color of the piccolos and French horn fuse into a single unique hybrid timbre (Slonimsky 1948, 187–188).

Ravel may have been inspired by the design of *mutation stops* in French organs. Mutation stops are ranks of organ pipes that sound at a pitch other than the unison or octave. When played with regular stops, they alter, or *mutate,* the timbre of the regular stop. *Nazard* is the French name for a mutation stop that sounds at the twelfth; the *tierce* sounds at the seventeenth.

Pitch is only one factor that can cause tones to fuse; spectral content and articulation (such as tremolo, vibrato depth, and rate) are also factors. John Chowning presented a striking example of tone fusion based on vibrato in his work on the singing voice (see volume 2, chapter 9).

6.8 Combination Tones

The violinist, theorist, and composer Giuseppe Tartini famously noted in 1754 that when two loud, pure tones are sounded together, a third is sometimes also heard at the difference frequency, $\Delta f = f_u - f_l$, where f_u and f_l are the upper and lower frequencies.[11] For example, 2100 Hz and 2000 Hz produce a difference tone of 100 Hz. This effect can be demonstrated easily by having two pennywhistle players or soprano recorder players stand near each other playing very high-pitched tones very loudly. The players and anyone sufficiently close to them will hear low-pitched rough tones at the difference frequency. This phenomenon is called *difference tones,* sometimes also Tartini's tones.

Helmholtz claimed to have discovered a tone at $\Delta f_s = f_u + f_l$ that he called *sum tones*. Mathematical theory strongly suggests sum tones exist, but they are so hard to hear that it is an open question as to whether they have ever been perceived experimentally. We know now that the reason sum tones are hard to hear has to do with masking.

Difference tones and sum tones are called generally *combination tones*. It is easy, but incorrect, to suppose that this phenomenon is related to beats. Beats cannot explain sum tones because beats only arise in the difference of two frequencies. Also, the effect of beats dissappears as the two frequencies diverge beyond a minor third, whereas for combination tones, Δf need not be small to be quite audible. Finally, if the tones are presented one to each ear, beats are still discernible but combination tones are not.

Helmholtz (1863, app. 12) conjectured that we hear combination tones because of nonlinear processing of loud signals in the ear. He supposed that the strength of the tones was forcing the excursion of the tympanum and other elements of the middle ear beyond their region of linear elasticity, thereby distorting the sound in the ear. Nonlinear systems can respond to vibration by generating signals not actually present in the stimuli (see volume 2, chapter 9). The square of the sum of two signals, $\sin^2(a + b)$, which is a quadratic nonlinear expression, includes tones at $a + b$ and $a - b$ (see volume 2, appendix).

Studies by Guinan and Peake (1967) have shown that nonlinear effects in the middle ear cannot by themselves explain combination tones. Current theory favors an effect within the cochlea for combination tones, although dynamic feedback paths from the auditory cortex may explain some other distortion products that have been observed (B. Moore 1997). There is still a high level of theoretical ambiguity in this subject.

6.9 Critical Bands

Fletcher (1940) unified many of the phenomena described in the sections on frequency domain masking, temporal masking, and beats with a concept that he called *critical bands*. These can be thought of as channels of frequency-selective psychoacoustic processing that affect our perception of pitch, loudness, and masking of frequency components lying within a critical frequency distance of one another. This insight eventually led to the psychoacoustic encoding of sound and the introduction of the MPEG audio encoding standard. MPEG takes advantage of the effects that critical bands have on hearing.

6.9.1 Critical Bands and Loudness

Zwicker and Feldtkeller (1955) provided an elegant demonstration of critical bands based on a loudness effect. They played a narrowband noise signal containing all frequencies between 980 to 1020 Hz. The bandwidth of the signal was 40 Hz, and its band center was 1000 Hz (figure 6.15a). Then, keeping the band center at 1000 Hz, and keeping the total intensity constant, they gradually

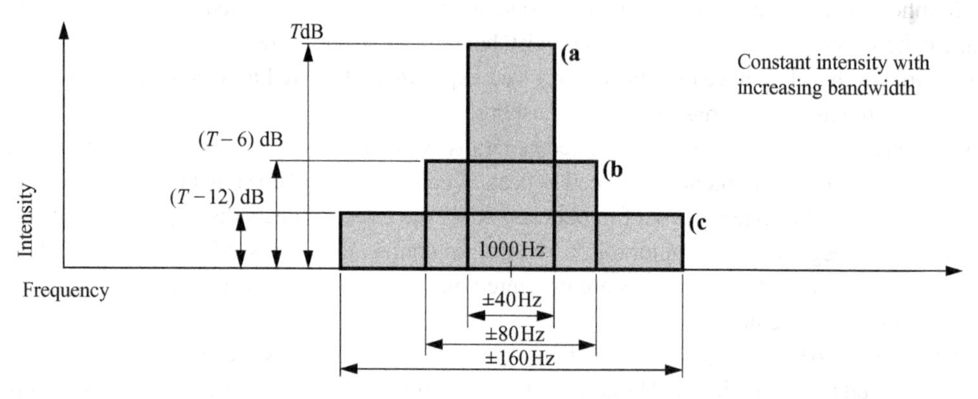

Figure 6.15
Critical bands and loudness.

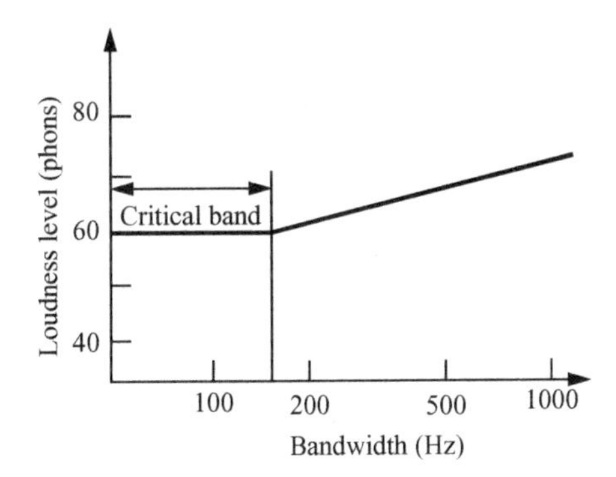

Figure 6.16
Effect of critical bands on loudness.

increased the bandwidth, spreading the same energy over a larger and larger frequency range (figures 6.15b and 6.15c). One might expect that each of these signals would be heard as equally loud because each contains the same total energy. And, indeed, subjects reported that the loudness remained constant . . . but only up to a certain bandwidth, after which perceived loudness began to increase *even though there was no increase in total energy.* With the band center at 1000 Hz, loudness began to increase when the bandwidth exceeded about 160 Hz.

So when the noise bandwidth was kept narrower than a critical threshold (160 Hz bandwidth at 1000 Hz band center frequency), the researchers got the expected effect: subjects reported that noise bands of varying bandwidth and constant intensity all sounded equally loud. But when bandwidth exceeded the critical threshold, subjects reported increasing loudness, even though total energy in the noise spectrum remained constant. Figure 6.16 shows how they observed loudness to increase after the bandwidth of the noise grew beyond the width of a critical band.

To explain this effect, Zwicker and Feldtkeller theorized that the ear lumps together the loudness of components that lie within the same critical band. Loudness increases when significant energy spills into more than one critical band. Thus, within the critical band, loudness is a function of the spectral width and spectral intensity. But once the bandwidth of the noise is broader than a critical threshold, all that matters is the spectral intensity. In order to understand this effect, it is necessary to do another experiment with masking.

6.9.2 Critical Bands and Masking

Suppose I play a sinusoid with frequency f_s and a wideband noise source with a band center f_c that is distant in frequency. You adjust the loudness of the sinusoid so that you can just barely hear it over the noise. Let's call this your *just noticeable loudness threshold,* T_0 (figure 6.17a). Now, keeping its spectral amplitude the same, if I move the noise signal's center frequency so it is the same as the sinusoid (figure 6.17b), the noise masks the sinusoid, and you can no longer hear it. Now

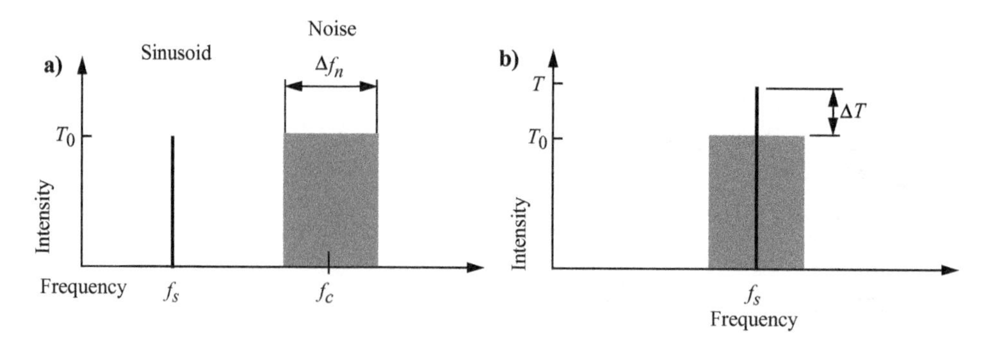

Figure 6.17
Sinusoid with wideband noise signal.

suppose I allow you to raise the level of the sinusoid to some level T so you can hear it above the noise. The difference between the thresholds, $\Delta T = T - T_0$, is the amount by which the noise signal masks the sinusoid.

Keeping its amplitude the same, if I now increase the bandwidth Δf_n of the noise, the sinusoid will again become inaudible. You must make the sinusoid even louder (by increasing T) before you can hear it again. Therefore the amount of masking ΔT increases as the bandwidth of the masking signal increases.

However—and this is the interesting part—beyond a critical threshold *increases in the noise bandwidth Δf_n no longer increase the amount of masking*. No further increases in T are required, no matter how much broader the bandwidth of the noise signal becomes.

Fletcher, whose experiment this is, suspected that this effect occurred because of a neurophysiological structure in the ear. He suggested that areas of the basilar membrane responded together to selected frequency ranges, the critical bands. The bandwidth of the noise signal where it ceased to further increase the just noticeable loudness threshold of the sinusoid was taken as the width of a critical band, centered on that frequency.

6.9.3 Critical Bands and Pitch

Plomp and Levelt (1965) and Greenwood (1961a) suggested that a pitch-based relation exists between consonance and critical bands. They believed that pitches of sinusoids separated by less than a critical band give rise to the effects described in the section on beats, including tonal fusion, whereas sinusoids that are separated enough to resolve into two distinct critical band regions on the basilar membrane give rise to the perception of two distinct tones. Experiments showed that *sinusoids appear most dissonant at approximately 40 percent of a critical band.* Sinusoids both closer and farther than that pitch distance become less dissonant. The dissonant sensation does not occur in the region of tone fusion and also does not occur where the difference in frequency exceeds the width of a critical band.

6.9.4 MP3 and Critical Bands

MP3, a component of the MPEG standard (see volume 2, chapter 10), is a practical application of masking due to critical bands. MP3 is an extension of the technology for encoding digital audio, known as pulse-code modulation, or PCM (see volume 2, chapter 1). MP3 has created a revolution in music distribution because audio can be transmitted and stored much more efficiently in MP3 format than with PCM while maintaining satisfactory sound quality.

Both MP3 and PCM encoding rely on a psychoacoustical model—a set of judgments about what we can and can't hear—to determine what information in the signal should be encoded. PCM encoding uses a relatively weak psychoacoustical model:

- Frequencies above 22.5 kHz are not encoded because they are above the range of human hearing.
- Sounds louder or softer than certain limits are also not encoded.

The MP3 psychoacoustical model inherits the PCM model. (In fact, the input to an MP3 encoder is a PCM-encoded audio signal.) But the MP3 psychoacoustical model also includes criteria about human temporal and spectral masking, and so it can encode more efficiently than PCM.

MP3 encoding takes place in two principal stages:

1. A psychoacoustical model of the critical bands identifies irrelevant frequency components—those that would not be perceived because of temporal and spectral masking effects. Masked components are encoded with less detail than is employed for unmasked components, thereby simplifying the spectrum of the encoded signal. Although simplifying the encoding of the masked components distorts them, the distortion isn't noticeable because the components are masked. (The amount of simplification, and hence the amount of distortion, must be constantly monitored and adjusted to be sure that the distortion introduced by this process never exceeds the masking threshold.)
2. The simplified spectrum is then subjected to additional steps to remove redundant information in the signal and put it in the most compressed representation possible (see section 9.15).

The result is an encoding of sound that can be transmitted with less effort or stored in less space than is required for PCM. Savings of between 12 and 20 times that required for PCM audio are possible without substantial degradation of sound quality.

In order to recreate the audio signal, an MP3 decoder is required. The MP3 decoder restores the simplified spectrum from the compressed representation encoded by step 2 above. However, the decoder cannot reverse the simplified encoding of the masked components in step 1 because that information was discarded. Because MP3 can't recover exactly the signal presented to its encoder, it is a *lossy encoding* or *lossy compression* scheme.

Strictly speaking, only the simplification of masked components in step 1 is lossy. The redundant information removed in step 2 is recovered completely in the decoding stage, so step 2 performs lossless encoding or lossless compression. Technically, PCM audio encoding is also

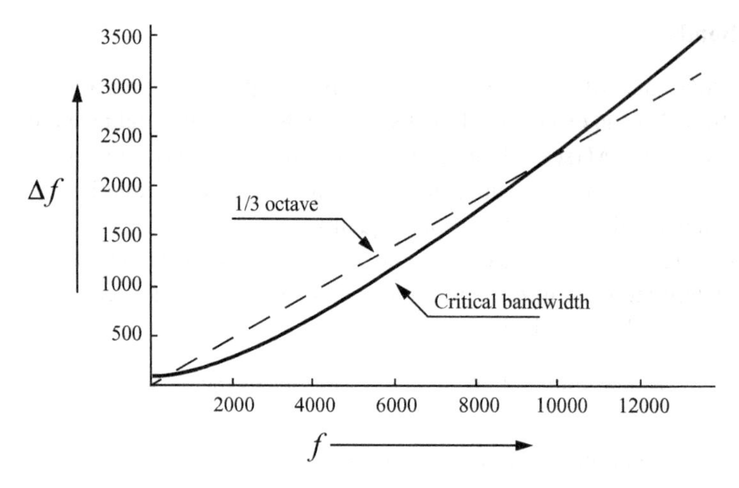

Figure 6.18
Critical bandwidth vs. center frequency.

lossy: because of the frequency and amplitude limits it imposes, it doesn't recover exactly the signal presented to its encoder. However, because there is no simplification of masked components, PCM is less lossy than MP3.

6.9.5 Measuring Critical Bands

Although critical bandwidth estimates vary substantially depending upon the type of experiment used to measure them, they average about one third of an octave for most of the audible range (but are greater at low frequencies). A semitone interval is $\sqrt[12]{2}$, so one third of an octave would be $\sqrt[3]{2} \approx 1.26 \approx 5/4$, or slightly over a major third. Thus, the ear appears to take into account the stimulus of neurons as far away as one third of an octave in order to determine the loudness of a sound.

Figure 6.18 shows how critical bandwidth varies with frequency for a typical listener (Zwicker, Flottorp, and Stevens 1957). The dashed line indicates one third of an octave. For instance, at 10 kHz, one third of an octave is between 2 and 2.5 kHz.

Critical bandwidth remains fairly constant up to about 500 Hz, then grows by about 20 percent of frequency thereafter. A reasonable approximation of the critical bandwidth is given by Zwicker and Fastl (1990) as

$$BW_c(f) = 25 + 75\left[1 + 1.4\left(\frac{f}{1000}\right)^2\right]^{0.69} \text{ Hz}, \qquad \textit{Critical Bandwidth} \quad (6.8)$$

where $BW_c(f)$ is the critical bandwidth at frequency f.

Although the critical bands are continuous, it is sometimes useful to think of the ear as comprising a discrete set of bandpass filters that obey (6.8). Using this approach, it is common to divide the ear's spectrum into 24 discrete critical bands, as shown in table 6.1 (Zwicker 1961). This

Table 6.1
Bark Scale: Critical Bandwidth, Center Frequency, and Critical Band Rate

Bark No.	Center Frequency	Critical Bandwidth	Lower Band Edge	Bark No.	Center Frequency	Critical Bandwidth	Lower Band Edge
0	50	80	0	13	2,150	100	2,000
1	150	100	100	14	2,500	380	2,320
2	250	100	200	15	2,900	450	2,700
3	350	100	300	16	3,400	550	3,150
4	450	110	400	17	4,000	700	3,700
5	570	120	510	18	4,800	900	4,400
6	700	140	630	19	5,800	1,100	5,300
7	840	150	770	20	7,000	1,300	6,400
8	1,000	160	920	21	8,500	1,800	7,700
9	1,170	190	1,080	22	10,500	2,500	9,500
10	1,370	210	1,270	23	13,500	3,500	12,000
11	1,600	240	1,480	24	19,500		15,500
12	1,850	280	1,720				

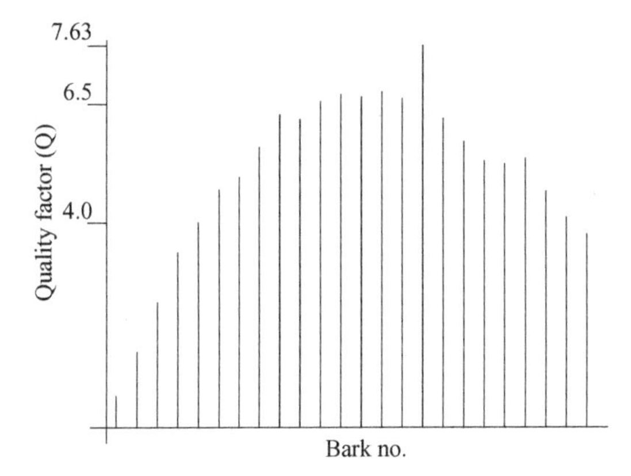

Figure 6.19
Quality factor of critical bands.

numbered list of discrete critical bands is the *bark scale*. The bark scale encodes the center frequency and bandwidth of each numbered critical band.

6.9.6 Quality Factor of Critical Bands

Table 6.1 shows that the bandwidth of the critical bands increases in relatively constant proportion as the center frequency of the band increases. The ratio of the center frequency to the bandwidth of a bandpass filter is its *quality factor,* often abbreviated Q. Figure 6.19 shows the Q for each of

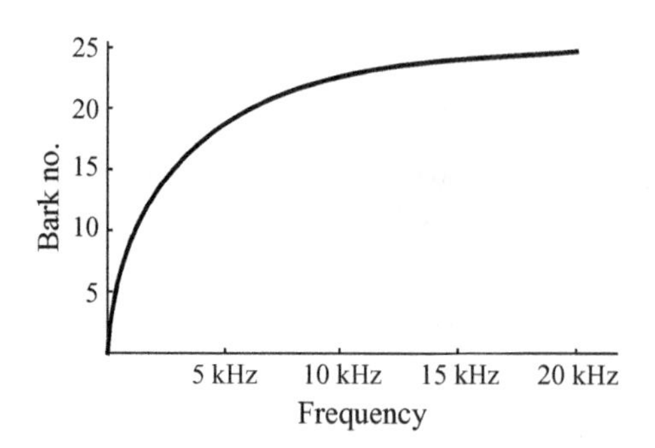

Figure 6.20
Frequency to bark function.

the critical bands. The narrower the bandwidth, the higher the Q, and the stronger will be its resonance when stimulated by a signal whose frequency lies within the band (see section 8.9.6). Like the critical bands, a *constant Q* filter varies its bandwidth as a function of the center frequency, keeping a constant ratio between them. Note in figure 6.19 that the Q of most bands is fairly constant in the range from 4 to 6, especially in the center range of hearing. Thus, the critical bands can be viewed as similar to constant Q bandpass filters.

The center frequencies and bandwidths in table 6.1 are only samples of the continuous frequency response of the ear. In reality, the auditory effects of critical bands are formed around the frequencies of the signals the ear hears and are not associated with a specific fixed filter bank in the ear.

The bark number for a frequency in Hz can be obtained with the following equation (Zwicker and Fastl 1990):

$$z(f) = 13\,\text{atan}(0.00076f) + 3.5\,\text{atan}\left(\frac{f}{7500}\right)^2. \qquad \textit{Bark Number} \ (6.9)$$

Figure 6.20 shows (6.9) plotted for the range of 20 Hz to 20 kHz.

6.9.7 Critical Bandwidth and Pitch JND

The curve of critical bandwidth vs. center frequency in figure 6.18 is very close to the same shape over the same range as the curve for pitch JND in figure 6.4. While the pitch JND spreads from about 3 to 30 Hz in 5 kHz, critical bandwidth goes from about 100 to 900 Hz over the same range. Thus, critical bands are proportionally about 30 times wider than the pitch JND at the same frequency.

6.10 Duration

How long does it take for us to identify the pitch of a tone? How long does it take to determine the loudness of a sound?

6.10.1 Effect of Duration on Pitch

Tones with quick onset times (such as vibraphone or marimba) have a clicking or percussive attack that is essentially a broadband noise. Very brief tones (under 10 ms) sound like clicks no matter what timbre the tone has. As the tone lengthens beyond about 15 ms, and if the tone's onset becomes more gradual, pitch perception solidifies up to about 30 ms. Pitch perception becomes stronger as the tone continues growing in length regardless of onset time.

Tones of greater complexity do not necessarily take longer to recognize. The ear can identify many pitches simultaneously in nearly constant rate time. While it can be shown that pitch depends on duration, this dependence is for extremely short tones only. This allows us, for instance, to follow extremely rapid polyphonic musical passages with sufficient accuracy to enjoy the experience. Music as we know it would be radically different if pitch were substantially dependent on duration.

Under optimal conditions we establish a sensation of pitch about 4–8 cycles after tone onset. The conventional wisdom is that the attack noise masks the underlying periodic vibration of the instrument and that this masking delays our pitch recognition. However, I have a different experience to report.

I once developed a pitch-tracking computer system for an electronic violin built by Max Mathews. While developing the system, I spent many hours listening to, and analyzing, violin tones. I observed that I was generally able to identify the correct pitch well before any significant periodic information was available in the signal. This suggested that my ear was using the characteristic broadband noise in the violin's attack transient to help identify the pitch. Spectral and temporal qualities of the attack noise may provide additional early clues to the correct pitch, perhaps through a cognitive learned response.

6.10.2 Critical Bands and Acoustical Uncertainty

Generally, the more precise we wish to be about the exact frequency of a sound we are hearing, the longer we must listen to it. Let's say that the smallest frequency difference we can discriminate is $\Delta f = f - f_0$ and that the required duration over which we must listen in order to resolve this frequency difference is $\Delta t = t - t_0$. Then we can say that the *acoustical uncertainty* is

$$\Delta f \Delta t = k, \qquad \textit{Acoustical Uncertainty} \quad (6.10)$$

where k is a constant that relates achievable frequency resolution to required temporal resolution, and vice versa.

Under optimum conditions $k \approx 0.1$ for the auditory system (Majernick and Kaluzny 1979). This means that to achieve frequency resolution of 0.1 Hz, our ears require about 1 second of the stimulus under optimum conditions. If all we have is 0.1 second of stimulus, our ears can achieve frequency resolution of about 1 Hz under optimum conditions. To achieve finer frequency resolution (smaller Δf) requires a correspondingly larger time interval (larger Δt). Thus, k represents the fundamental limit on our ability to know the precise frequency of a signal within a precise time interval.

Uncertainty plays many important roles in the mathematics of music (see section 9.15, equation (9.19); and volume 2, chapters 3 and 10).

Clearly, we want our ears to have the finest frequency precision possible (small Δf) with the shortest response time possible (small Δt). But the basilar membrane is governed by (6.10). So increasing its frequency precision would necessarily require us to lessen its temporal precision. However, our hearing neatly sidesteps the limitations of (6.10) by using critical bands. The ear divides up the audio spectrum into frequency bands each of which has a relatively broad Δf. Hence, within each band, Δt can be relatively small. That way we get good frequency resolution without suffering poor temporal resolution.

The trade-off is that the critical bands provide relatively poor pitch discrimination by themselves. That the pitch JND is about 30 times finer than the width of a critical band suggests just how greatly aided we are by critical bands and the dynamic feedback processes, described in the section on frequency sharpening, that refine our sense of pitch within a critical band.

6.10.3 Loudness and Duration

The acoustical uncertainty principle applies for loudness as well. The ear averages over a duration of about 200 ms to determine the loudness of a sound. Because of this fact, sounds that are shorter than 200 ms must be proportionately more intense to appear to have the same loudness as sounds that are longer than this threshold. Put another way, loudness is proportional to duration up to about 200 ms. More precisely, loudness grows by 10 dB as duration grows by a factor of 10, up to 200 ms. This correlation is even stronger for broadband sounds and extends up to about 1 second. An important consequence of these facts is that our ears lack a means to protect themselves naturally against impulsive high-intensity sounds, such as gunfire (see the section on acoustic reflex and temporary threshold shift).

6.11 Consonance and Dissonance

Consonance was defined in chapter 3 as tones that sound well together. But what process governs our perception of consonance? There are many theories of consonance. They generally fall into one or more of the following categories:

- *Cultural theories* examine social, cultural, and stylistic norms.
- *Acoustic theories* look at the physical properties of acoustical signals, such as properties of musical instruments and scale systems.
- *Psychophysical theories* look at how the neurophysical structure of the ear may affect consonance.
- *Cognitive theories* examine learning, expectation, and categorical perception.

As an example of a cognitive theory, a dissonant sound may be heard as consonant if it is preceded by many sounds that are even more dissonant. While learning how to write sixteenth-century chorale harmonizations in the style of J. S. Bach, I experienced a shift in my expectation of consonance as my ear acclimated to this antique style. I came to appreciate why his contemporaries found some of Bach's chorale settings shocking, whereas to a modern listener they can seem bland.

Theories of consonance stretch back at least to Galileo Galilei. Plomp and Levelt (1965) quote Galileo (1638) as follows: "Agreeable consonances are pairs of tones which strike the ear with a certain regularity; this regularity consists in the fact that the pulses delivered by the two tones, in the same interval of time, shall be commensurable in number, so as not to keep the ear drum in perpetual torment." The relatively large number of extant theories of consonance goes far beyond what can be summarized here. Instead, I develop a simple psychophysical model based on critical bands to give an idea of the subject.

If two sine tones of 1000 Hz and 1015 Hz are played, we do not hear two distinct tones. Instead, we tend to hear one fused pitch with a 15 Hz beat frequency and attendant roughness (see section 6.7). This is because the critical bandwidth for 1000 Hz is about 160 Hz, and the two tones lie close within the same critical band. If we let "dissonance" define this roughness, then "consonance" defines its absence. In terms of figure 6.14,

- Frequencies differing by less than a JND of pitch form a *perfect consonance,* or *unison.*
- Frequencies differing by more than a critical band form a consonance.
- Frequencies differing by between 5 percent and 50 percent of a critical band are the most dissonant.

Greenwood (1961b) was the first to observe a relation between critical bandwidth and judgments of consonance and dissonance. He analyzed data collected by Mayer in 1894 and compared it to the estimated size of critical bands. Sounding intervals with tuning forks, Mayer had asked listeners to identify the smallest interval for which no dissonance was perceived. Greenwood's plot of Mayer's data suggested that the dissonance disappears when the distance between pure tones is greater than or equal to the size of a critical band. Linking dissonance to position within critical bands is called *tonotopic dissonance.*

We can expand this observation into a simple psychophysical metric for the consonance or dissonance of any two complex tones by counting how many of their partials land together in critical bands (and discounting any that lie within a JND or do not share a common critical band). The hypothesis is that the more the partials of the two tones fall within the 5 percent to 50 percent critical band range, the more dissonant the two tones should be (Plomp and Levelt 1965).

The following is an illustration of this approach.

1. Start with two complex tones that form a musical interval, say, a perfect fourth.
2. Count the number of dissonant partials *d*. For each partial of the lower tone p_l, count how many partials of the upper tone p_u form an interval that is within a critical band of p_l. Use equation (6.8) to compute the critical bandwidth for each harmonic. If the interval is small enough to fall within the same pitch JND, exclude it from the count of dissonances because it is perceived as a unison and hence is consonant.

3. As we go up in frequency, at some point the successive harmonics of each tone will begin to fall within the same critical band because the critical bands widen with increasing frequency but the harmonics do not. Stop counting when the harmonics of either tone by themselves begin to fall within one critical band, because if the ear uses a method like this, we presume it would have to stop here as well.

Figure 6.21 shows the count of dissonances for $p_l = 220$ Hz with p_u set in turn to the 12 equal-tempered semitone intervals from unison to one octave above p_l, based on the critical band function given in equation (6.8).

If we list the intervals from figure 6.21 in order of increasing dissonance, the results are as shown in table 6.2. The order of the shaded intervals agrees with the ones in table 3.5, which is ordered according to standard Western cultural norms of musical consonance. So this approach is fine up to a point. The rest of the orderings in table 6.2 are arguable. The minor seventh and major second are predicted to be more consonant than the major third, for example. Also, it does not seem right that the tritone should have the same consonance as the major third. This result agrees with Terhardt (1974), who wrote that consonance is "only slightly and indirectly correlated with musical intervals. Thus, psychoacoustic consonance cannot be considered as the basis of the sense of musical intervals."

While better computational estimates of consonance are available (e.g., Kameoka and Kuriyagawa 1969a; 1969b), the sheer number of competing theories of consonance extant in the world today suggests that the only way forward is to perform (sigh) more research.

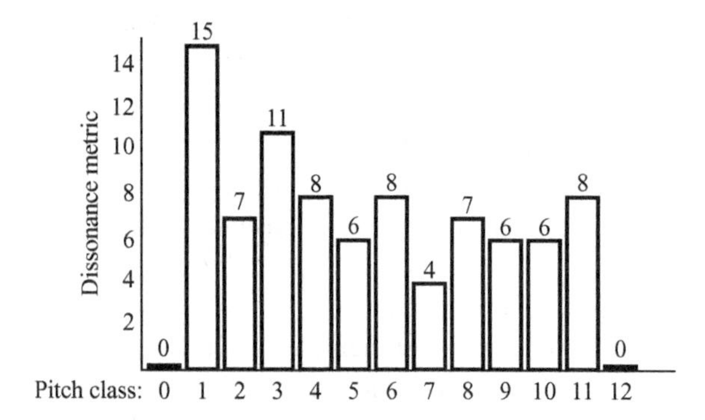

Figure 6.21
Dissonance metric of equal-tempered intervals based on critical bands.

Table 6.2
Intervals in Increasing Order of Dissonance

Interval	Unison	Octave	Fifth	Fourth	M. 6th	m. 7th	m. 6th	M. 2d	M. 3d	Tritone	M. 7th	m. 3d	m. 2d
Pitch class	0	12	7	5	9	10	8	2	4	6	11	3	1
Dissonance	0	0	4	6	6	6	7	7	8	8	8	11	15

6.12 Localization

How is it that we can so easily tell which direction a sound is coming from? In placing a sound in space, we extract psychophysical cues from arriving sounds based on

- The geometry of the outer ear and the placement, size, and shape of the pinnae and ear canal
- The geometry and orientation of the head, chest, and shoulders
- The distance of the ears above the ground

We add to these psychophysical cues a cognitive framework that includes

- Understandings about the acoustical properties of the sound source
- Basic acoustical facts about sound transmission in air
- Information about the known acoustical environment
- Information about our orientation in space in six degrees (up/down, left/right, forward/backward, pitch, yaw, and roll)

Incredibly, for each sound source in the environment, our hearing automatically and instantaneously creates a psychological image of the sound with its direction and distance encoded so that we register it subjectively as an object in space/time, together with the nature of the acoustical environment that it lies within. Pretty amazing. But that's not all. We can also tell

- Whether the sound is coming from above or below
- Its rate of relative motion
- Its rate of relative acceleration

and much, much more.

6.12.1 Angular Cues

Angular cues tell the direction of a sound on the horizontal plane. John Strutt (1907), the third Lord Rayleigh, a pioneer in spatial hearing research, theorized about the cues the ear uses to determine the angle of an incident sound. He began by noting that if a sound source is located to one side of the receiver, the sound energy received at the closer (*ipsilateral*) ear will be more intense than at the further (*contralateral*) ear because sound must travel a longer distance to reach the contralateral ear, and intensity decreases as the square of distance.

6.12.2 Interaural Level Difference

He also noted that sound traveling to the contralateral ear must navigate around the head. He knew that high frequencies are attenuated relatively more than low frequencies when they diffract around an object (see section 7.11). The sound heard at the ipsilateral ear will be brighter than at the contralateral ear because the head shadows the contralateral ear.

Rayleigh reasoned that by comparing the difference in intensity level, especially of high-frequency sounds received by the ears, our hearing should be able to tell the direction of the sound. Rayleigh grouped the intensity cue and the diffraction cue together and called them jointly the *interaural level difference* (ILD).

Interaural level difference is small for wavelengths less than about four times the diameter of the human head (averaging approximately 17 cm). So this cue shouldn't work for frequencies below about 500 Hz. But diffraction by the head increases rapidly with increasing frequency, and above about 3000 Hz, Rayleigh figured that head shadowing should cause a 20–30 dB drop in level at the contralateral ear, making this a very effective angular cue in this frequency range.

Rayleigh realized that his theory implied that directional sensitivity should vanish for sounds that contain no energy above about 500 Hz. But when he experimented on this, he was surprised to discover that he could determine the direction of pure tones even as low as 128 Hz. So he went back to the old drawing board.

6.12.3 Interaural Time Difference

Rayleigh then considered the possibility that hearing is sensitive to the difference in phase between signals arriving at each ear because of the greater time that must elapse for the signal to arrive at the contralateral ear. Rayleigh called this the *interaural time difference* (ITD) (Strutt 1907). He used a simplified trigonometric approximation of head shape to calculate the different lengths of the sound paths to the two ears. His first simplification was to model the head as a sphere with radius r. Next, he considered only plane waves, whose rays arrive at the ears in parallel. He characterized the direction of sound arrival, the *angle of azimuth,* as follows. He drew a radius from the center of the head forward through the nose as the zero-degree reference (figure 6.22a). He drew another radius at the angle of the incident plane wave. The angle of these two radii is the azimuth angle z. With this simplified model the difference in the length of the straight-line path to the two ears in terms of azimuth is $2r \sin z$. For example, if $z = 0°$, $\sin 0 = 0$ and there is no delay difference for sounds arriving directly from the front (or back). If, however, $z = 90°$, then $\sin 90° = 1$,

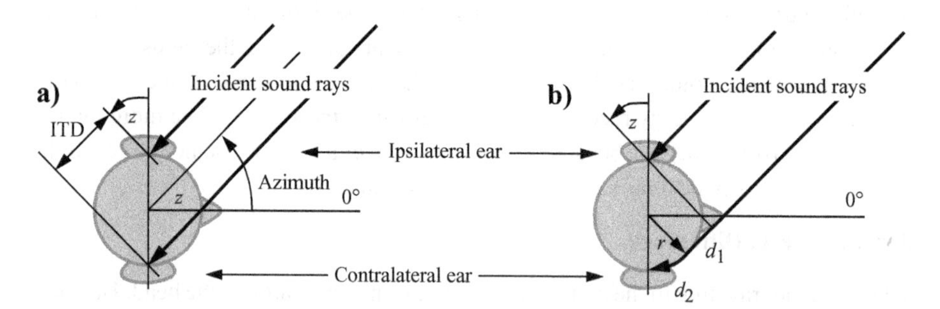

Figure 6.22
ITD, spherical head.

and the sound (which is now coming at the head directly from the left side) experiences a delay of approximately the head diameter to reach the other ear. If we assume sound travels at speed c, this corresponds to an ITD of $2r \sin(z)/c$.

There are obvious difficulties with this analysis, including the fact that it calculates the sound as traveling through the head. Figure 6.22a, shows an incident sound ray traveling through the head to arrive at the contralateral ear. Still, for small angles of azimuth, this is not a bad approximation. We can improve it slightly as in figure 6.22b by calculating the delay to the far ear along a ray that arrives at a tangent point on the side of the head (line length d_1), then arcs around by diffraction to the ear (arc length d_2). With equation (5.2) for arc length, the path length will be $d_1 + d_2 = r \sin z + rz$. Converting path length to ITD, we have

$$\text{ITD} = \frac{r \sin z + rz}{c}, \qquad |z| < 90^\circ. \qquad \textit{Interaural Time Difference (ITD)} \quad (6.11)$$

Equation (6.11) only works for azimuth angles whose magnitude is less than 90° because beyond 90° the rays arriving from behind the head would be closer to the source, and as the angle approached 180°, the arriving sound rays would be time-aligned again from behind the head.

While (6.11) is an improvement, it is still not a good estimate for the large class of people whose heads are not precisely spherical. Also, most people's ears are not on a diameter through the center of the head but are somewhat back from it.

Hearing is also sensitive to the onset of sounds, and we also use the onset time difference between the ears, the *lateral onset cue,* to help establish direction of arrival.

6.12.4 Problems with ITD

The ear is only sensitive to ITD for frequencies whose wavelength is less than half the distance between the ears because above this frequency the effect becomes ambiguous. To see why, let's assume the diameter of the average head to be $d \approx 0.175$ m. When exactly half a wavelength spans the distance between the ears ($\lambda = 2d$), the ear registers the same pressure differences at both ears regardless of which direction the sound is coming from. If our ears were sensitive to ITD for waves in the range $d \le \lambda \le 2d$, we would hear an apparent source location on the *opposite* side from the true direction of arrival. Perhaps anticipating the potential for adaptive catastrophe here, our evolutionary intelligence wisely bred this capability out of us. Taking the speed of sound at standard temperature and pressure to be 340 m/s, the frequency corresponding to $\lambda = 2d$ is $(340/0.175)/2 = 971$ Hz, which, not surprisingly, is about where our ears stop paying attention to ITD cues. Our hearing is most sensitive to ITD around 500 Hz. At that frequency, experiments show we have a JND of azimuth Δz near the forward direction of between 1° and 2°. Using (6.11) to calculate the ITD for $\Delta z = 2°$ yields the astonishingly small figure of 18 microseconds. Given the comparatively sluggish synaptic delay time of about 1 millisecond for average neurons, it seems incredible that our ears are capable of measuring such small delays with such precision, but they do.

6.12.5 Duplex Theory

ILD and ITD together are known as the *duplex theory* because these two effects are complementary. Our hearing is responsive to ILD cues from 500 Hz upward becoming reliably strong above about 3000 Hz, so ILD cues are best for high frequencies. On the other hand, ITD cues are strongest below about 1000 Hz. For frequencies around 2 kHz, where neither cue works well, our localization is not very good (Stevens and Newman 1936).

6.12.6 Anatomical Transfer Function

A major difficulty of the duplex theory of ILD and ITD cues is that it implies there should be regions where we experience front/back and top/bottom source direction ambiguity. The most obvious case where the theory predicts this should occur is for sounds on the median plane (see figure 6.24). Sounds arriving from any position on the median plane have ITD and ILD of 0 because they are equally centered between the ears. Therefore, we should have no clue as to the elevation of a sound. But we can clearly distinguish sounds from above and below. Furthermore, identical ILD and ITD cues are supposed to be produced by sounds at positions a, b, c, and d in figure 6.23. The duplex theory implies that the region of ambiguity forms the surface of a *cone of confusion* whose apex is the ear. There should be as many cones of confusion as there are ITD/ILD cues. But the ear, being ignorant of this scientific difficulty, is quite able to distinguish these cues. So something is missing from the duplex theory.

Researchers eventually noticed the large flaps of skin (pinnae) that stick out from the sides of our heads. They discovered that our hearing is very sensitive to the way the spectrum of arriving sounds is modified by the sound shadowing and scattering effects of the pinnae as well as the head, shoulders, and torso. All these parts of the body cause sounds coming from different directions to be filtered differently on their way to the eardrum in a manner that is highly predictable by our hearing (after all, the shape of our bodies is well known to us). These direction-dependent cues, variously called the *anatomical transfer function* (ATF) or the *head-related transfer function* (HRTF), are the essential cues for discriminating front/back and elevation of sound sources and also play a role in discriminating lateral cues. Think about it. Why are our pinnae always *behind* our ear canals? So that we can

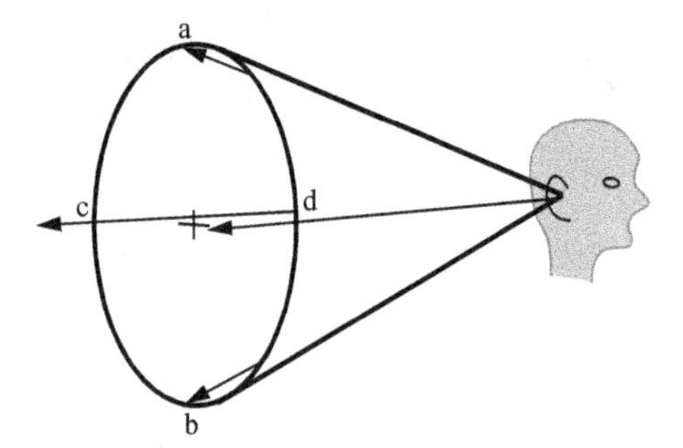

Figure 6.23
Cone of confusion.

tell sounds in front from sounds in back. Sounds arriving from behind the head are subject to more diffraction than sounds coming from in front because the pinnae block the direct path for signals behind the head. Similarly, we can tell up from down by diffraction effects caused by our anatomy. Thus, spectral modifications caused by ATF are important cues to locate sounds in space.

6.13 Externalization

ATF also solves another problem. If I play you a stereo signal through loudspeakers, you hear the sound coming from the general direction of the speakers, that is, outside your head. But if I play you the same sound over headphones, you generally experience the sound inside your head. What's the difference between these two presentation modes? Your ATF. Headphones bypass the filtering applied by your head, pinnae, shoulders, and torso to incoming signals, depriving you of a sense of direction for arriving sounds, and your hearing seems to conclude that the sound therefore must be coming from inside your head. If I simulated your ATF cues by filtering the sounds I send to your headphones based on measurements of your ATF, you would hear the sounds outside your head again.

6.13.1 Measuring ATF

I could determine your own particular ATF experimentally as follows. First, I ask you to sit in a stable position and (very carefully) insert tiny microphones into your ear canals. While you sit still, I beam a click or short noise burst from a small loudspeaker at your head from a constant distance and from all angles of azimuth z and elevation ϕ around you. I record the signals received by the microphones in your ear canals, which show how the shape of the waveform is changed by the scattering/shadowing properties of your body on its way to your ears. I then apply well-known signal-processing techniques to the recordings (see volume 2, chapter 3), resulting in a set of spectra describing your ATF for all measured angles of azimuth and elevation.

With this information, I can give you the illusion that you are hearing a sound coming from any azimuth z and elevation ϕ of my choice. All I have to do is select the spectrum corresponding to the direction I want, and filter a recorded sound with that spectrum and play it for you over headphones. The illusion even works with loudspeakers under favorable conditions.

The illusion even works pretty well if I substitute someone else's ATF for yours. These ideas are the technological foundation for better-quality 3-D audio surround systems. They are able to skip the tedious part of having to measure each individual's ATF by using the ATFs of people who are good ATF "donors," that is, whose own response patterns are characteristic of those of many other individuals.

6.13.2 Head Movement and Spectral Context

There are still unresolved issues, even with ATF theory. I said that we use spectral modifications caused by ATF to help locate sounds. But how can we know a priori whether the spectral features of a signal we hear are due to ATF-induced filtering or are just built-in aspects of the source signal's

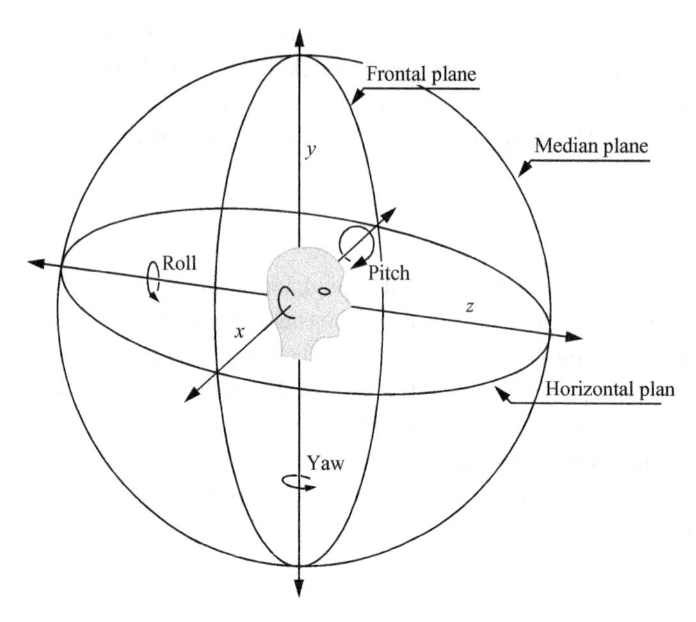

Figure 6.24
Three-dimensional coordinates.

original spectrum? In fact, studies show we really can't tell the difference, that is, unless we can turn our heads, which immediately clarifies whether a sound is really coming from that direction or whether it just happens to match a positional spectral cue. If the source also moves, our discrimination improves even more.

The effectiveness of 3-D audio surround systems is greatly improved if the system can compensate for head movement. The listener wears a head-tracking system that allows the experimenter to monitor all six degrees of freedom: pitch, roll, and yaw plus the standard $\{x, y, z\}$ Cartesian coordinates of position in space (figure 6.24). Additionally, if the sound spectrum is well known to us, or we can observe the position of the sound source visually, we can discern whether spectral features are due to the sound source or to our own ATF.

6.13.3 The Precedence Effect

When we hear a sound in a natural environment, we hear not only the signal that travels through the air directly to our ears but also its many echoes reflected off nearby surfaces. These echoes superimpose an incoherent jumble of delayed and scaled copies on top of the direct signal, which is then delivered to our ears to sort out.

Performance of the ITD cue degrades in highly reverberant environments because it depends upon receiving coherent phase information between the two ears. The ILD cue fares somewhat better with reverberation because it only looks for level differences. But it is still subject to confusion in rooms with strong standing wave patterns where intensities of sound are subject to local minima and maxima determined by room geometry. For example, if I play a 500 Hz continuous sinusoid over a loudspeaker in a room and have you walk around in it, you'll get different impressions of the location of the speaker at different positions in the room because of the standing waves.

The saving grace in all this is that, generally, the direct path to our ears from the source is shorter than any of the reflected paths; thus we hear sounds traveling on the direct path first. Our hearing is keenly aware of this fact and gleans as much directional information as possible from the onset of the sound before the reflections begin to arrive. This is the *precedence effect* (Wallach, Newman, and Rosenzweig 1949; Haas 1951). In general, sound location is perceived to be in the direction from which the first signal arrives, so long as the strongest echoes arrive within about 35 ms, they are spectrally and temporally similar, and they are not much louder than the direct signal.

Under these circumstances, echoes are suppressed by the precedence effect. The precedence effect can be demonstrated with a stereo audio system with loudspeakers separated by a few meters. Set the controls to monophonic reproduction so that the same signal is sent to both speakers. First, stand in front of the speakers exactly on the midline between them while playing some music or speech. You will hear the signal in front of you, somewhere between the speakers. If you then move toward one of the speakers by a meter or so, suddenly you will believe that only the nearby speaker is making any sound—it's as though the distant one was switched off. But the other speaker is clearly still contributing loudness and spaciousness to what you hear, which you can demonstrate if you have an accomplice unplug the far loudspeaker. You will notice a reduction in overall loudness and a reduction in spaciousness. But if it is plugged back in, you still can't hear the sound as arriving from it, and your sense of the direction stays with the local speaker.

6.13.4 The Trade-off between Time and Intensity

ITD and ILD seem to be processed by the brain separately before being combined at a higher level with other cues to model lateral position of sound. This can be exploited within a certain range to play off ITD and ILD cues against each other.

If you stand in the median plane between two loudspeakers (called the sweet spot in the audio literature) fed with a stereo sound signal, you will hear the stereophonic sound field in front of you. If you move too far to one side, the sound field collapses and the precedence effect reinforces the percept that the nearby speaker is the location of the sound source. However, to a certain extent, this can be compensated for by boosting the intensity of the far speaker until it overcomes the ITD cue, and you can restore again the sensation of being in the sweet spot. The ear apparently weighs the ratio of ITD and ILD cues to determine lateralization.

6.13.5 Distance Cues

How do we know how far away a sound source is? Suppose I set up two loudspeakers in a room behind an acoustically transparent but visually opaque screen. The first speaker is 3 meters in front of you and I play a sound at intensity I. Suppose I then switch to a second speaker at twice the distance and play the same sound with the same intensity I. You'd have no trouble telling which was the closer sound source: because of the inverse square law, the intensity of the direct signal arriving from the far speaker is $I_d = I/4$; therefore you hear the second speaker as farther away.

But suppose we do a second experiment where I secretly increase the intensity of the far speaker to $4I$, so that now $I_d = 4I/4 = I$, and repeat the procedure. Though the inverse square law cue is now gone, you will still correctly tell me which speaker is the far one and will perhaps also mention that I appear to have made the far one louder. How did you figure that out?

For every sound, your hearing judges not just the intensity of the direct signal I_d but also the ratio of the direct signal intensity to the attendant reverberant signal intensity R as a cue for distance. In the first experiment, we're pumping the same intensity I into the room from either speaker; therefore the average reverberant intensity in the room is R no matter which speaker plays. Reverberant energy is distributed uniformly throughout the room quickly after a sound starts. But meanwhile the direct signal intensity went from I in the close speaker to $I/4$ in the far one. Thus, your ear judged that

$$\frac{I}{R} > \frac{I/4}{R}$$

and reasoned that if the reverberation intensity *stayed the same* but the direct signal intensity *went down,* then the second speaker must be farther away.

However, in the second experiment, the intensity in the room goes from I to $4I$. Therefore the amount of reverberation in the room likewise goes from R to $4R$. But meanwhile the intensity of the direct signal that you heard remained the same. (Because I quadrupled the intensity of the distant speaker, the direct signal strength you experience from either speaker is identical.) Thus, your ear judged that

$$\frac{I}{R} > \frac{I}{4R}$$

and reasoned that if the direct signal intensity *remains the same* but the reverberant intensity *increases,* the sound must be both farther away *and* louder.

We can confirm that your hearing is factoring reverberation into its cue for distance by repeating this experiment in an *anechoic chamber.* As the name implies, it is a room that is so padded that it produces no echoes, depriving you of the reverberation cue. This time you would experience the second experiment as ambiguous and wouldn't be able to tell which speaker was farther away.

Another distance cue is based on the fact that high frequencies are absorbed more quickly by air than low frequencies. The greater the distance, the more the high frequencies in a signal are attenuated. The effect is more exaggerated with greater humidity. So even in a large space without echoes—like a flat desert—you can still tell relative distance because your hearing has a built-in sense of how much air absorbs high frequencies.

6.14 Timbre

"Timbre," a word borrowed from French, is sometimes defined as "sound color." Here's the ANSI (1999) definition: "Timbre is that attribute of auditory sensation in terms of which a listener can judge that two sounds similarly presented and having the same loudness and pitch are dissimilar. . . . Timbre depends primarily upon the spectrum of the stimulus, but it also depends upon the waveform, the sound pressure, the frequency location of the spectrum, and the temporal characteristics of the stimulus."

The first sentence of the ANSI definition is an example of the common but not very helpful tendency to define timbre by what it is *not*. According to such negative definitions, timbre is what's left over after pitch, loudness, and duration are accounted for. Let's call this approach the residue theory of timbre. The second sentence is a bit more helpful.

Musicians have a highly developed, though informal, description of timbre. Musically, timbre can refer to the features of a tone that serve to identify the instrumental source, such as oboe or violin, or the instrumental family, such as woodwinds or strings. Alternatively, timbre can denote the semantic quality of a musical tone, such as dark, dull, bright, or shrill. The most useful terms are those that relate to measurable phenomena, such as *sharpness,* the ratio of high frequency energy to total energy. Sharpness can be thought of as the "center of gravity" of the spectral envelope of a sound (Bismarck 1974). *Roughness* characterizes tones or noises that contain frequency or amplitude modulations between about 20 and 200 Hz (see section 6.7; and volume 2, chapter 9).

Positive theories of timbre have only recently begun to arise. The theoretical difficulties stem partly from the multidimensional complexity of timbre (Licklider 1951; Plomp 1970) and partly from the bias toward viewing timbre as the residue of pitch, loudness, and duration.

Modern psychoacoustical research into timbre has sought to understand

- What are the principal perceptual structures the auditory system uses to determine timbre? In other words, what timbral effects are we sensitive to, and in what order of precedence?
- How does the auditory system categorize and order timbre? In other words, does the ear have a natural taxonomy of sounds?

Research has shown that the two most significant perceptual structures of timbre are spectral energy distribution and evolution of spectral energy distribution over time. Thus, timbre consists primarily of the static and dynamic properties of a sound's spectrum, leaving aside pitch, loudness, and duration. Timbre identification has been shown to depend a great deal on spectral evolution.

Although this definition is still basically a residual definition of timbre, at least it suggests a way to take a small step forward. Suppose we take a collection of instrument tones and normalize them so that they all have the same perceived pitch, loudness, and duration. Then any remaining differences between the tones would be, by definition, their timbre. We could then do experiments on the normalized collection to study how subjects experience the differences between the tones and try to understand from this how the auditory system organizes and categorizes timbre.

Such research was carried out by John Grey (1975).[13] Although his entire experiment goes beyond the scope of this book, in brief, he recorded a collection of standard orchestral instrument tones and performed a set of experiments to normalize them for pitch, loudness, and duration. Therefore, by definition, the normalized instrument tones differed only in timbre. Subjects were then played each possible pairing of these tones at random and asked after each pair how dissimilar they were on a scale of 1 to 10. The experiment generated thousands of perceptual dissimilarity judgments from a multitude of subjects. These judgments allowed Grey to construct a multidimensional constellation of the orchestral instruments where the distance between all the instruments is proportional to how dissimilar each tone is felt to be from all the others. It is important to note that the experiment included no hypothesis a priori; all Grey started with were the dissimilarity judgments of his subjects.

The next step was to determine what features of the sound stimuli might best account for the differences his subjects heard. He used multidimensional scaling (MDS) techniques (Kruskal 1964) to reduce the number of dimensions of the dissimilarity judgments into a set of distances in three-dimensional space. Figure 6.25 shows a view of the dissimilarity judgments expressed as distances in three dimensions. Abbreviations for the instrument tones are O1, O2, oboes; C1, C2, clarinets; X1, X2, X3, saxophones; EH, English horn; FH, French horn; S1, S2, S3, strings; TP, trumpet; TM, trombone; FL, flute; BN, bassoon.

It is important to note that the data specify only relative distances between data points. Grey examined the data in one, two, and three dimensions in all possible rotations and decided that the three-dimensional orientation shown in figure 6.25 offered the best possibilities for explanation of timbre differences.

In this rotation, Grey noted the *y*-axis relates to the *spectral energy distribution*. On the one extreme, the french horn (FH) and strings (S3) have relatively narrow spectral bandwidth (fewer harmonics) with most energy concentrated in the lowest harmonics. At the other extreme, the trombone (TM) has a very wide spectral bandwidth (many harmonics) with energy more evenly distributed among them all.

The *x*-axis relates to *temporal energy distribution,* specifically to how partials align during attack and decay. At one extreme, higher harmonics of the woodwinds enter and exit simultaneously with the low ones at onset and termination of a note. At the other extreme, the higher harmonics of strings, brass, flute, and bassoon tend to enter after the lower harmonics and exit more quickly than the lower ones.

The *x*-axis also expresses musical instrument family partitioning. The woodwinds appear on the far left, the brass in the middle, and the strings on the far right. The exceptions to this pattern are the clustering of bassoon with the brass, and flute with the strings.

Grey also interpreted the *z*-axis in terms of temporal patterns. At one extreme, the strings, flute, clarinets, saxophone (X1, X2), and oboe (O1) display initial high-frequency low-amplitude energy, most often inharmonic, during the attack segment. The tones at the other extreme, including brass, bassoon, and English horn, either have low-frequency inharmonicity or at least no high-frequency initial energy in the attack.

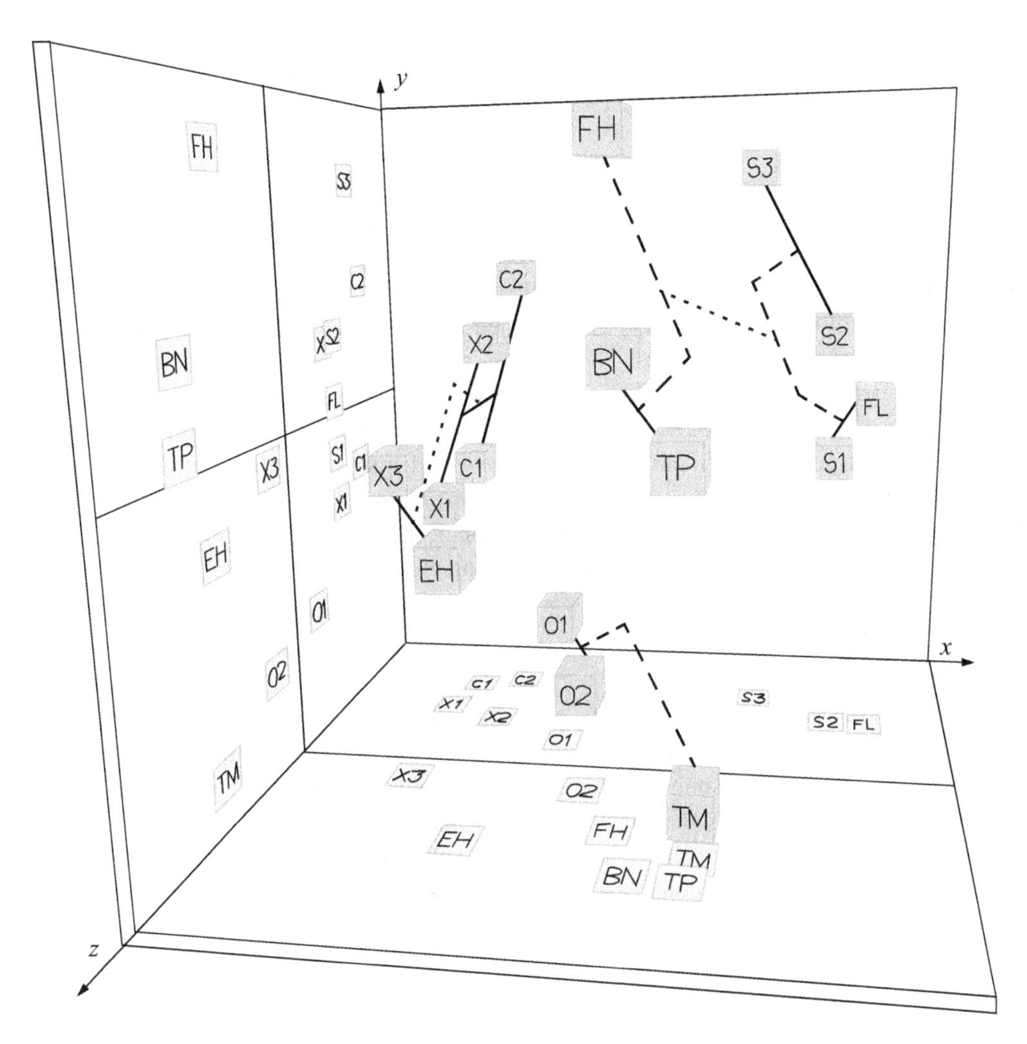

Figure 6.25
Three-dimensional hierarchical clustering analysis of timbre similarities. (Adapted from Grey 1975.)

Grey also performed a clustering analysis of the data. The solid lines in figure 6.25 indicate the strongest clustering, followed by dashed and then dotted lines. For instance, string S1 is most like the flute, and string S2 is most like string S3. Also, the group {S1, FL} is more like the group {S2, S3} than the group {FH, BN, TP}. Last, the group {S1, FL} is more like group {FH, BN, TP} than anything except group {S2, S3}.

The composer Henry Cowell (1930) wrote,

> If tone-qualities were arranged in order, and a notation found for them, it would be of assistance to composer and performer alike. . . . Tone-quality thus becomes one of the elements in the composition itself

> and ceases to be only a matter of performance. . . . Progress in the field of new or graduated tone-qualities in composition has been greatly hindered by lack of notation, as it has been justly felt that if music demanding new tonal values were set down in present notation, the desired effect would be likely to be entirely lost in the performance. (34)

Figure 6.25 provides composers with interesting information on the use of timbre as an organizing principle in composition. For instance, to reinforce the independence of two melodic lines, composers could choose timbres that are far apart in the figure. One can entertain such ideas as transposable timbre by moving gradually from mellow to bright instruments. Timbres with more bite tend to stand out in an ensemble because their onset transients tend to contain a higher percentage of total energy.

Grey's research is but one study with a very narrow focus: it covers only 16 specific sounds at one pitch, one duration, and one loudness level. We know little about the spaces between these timbres, let alone the possible maps of timbre space that would arise using other control parameters. The orchestral instruments themselves are not constant in timbre at different pitches, durations, and loudnesses. For example, the timbre of the clarinet demonstrates a wide range of effects across its pitch range. Nonetheless, Grey has provided a tempting glimpse.

6.15 Summary

The most salient aspects of musical sound are pitch, loudness, duration, sound location, and timbre. Psychometric scales such as decibels and phons were developed to give some objectivity to subjective judgments. While such units provide a common language for discussing the auditory abilities of a population of listeners, they should not be considered to be in the same league as physical measurements. Nonetheless, subjective judgments can help to expose a coherent subjective structure if collected over a sufficient number of data points, and this structure can in turn be related to various acoustical parameters such as amplitude and frequency.

6.16 Suggested Reading

Allen, J. B., and S. T. Neely. 1997. "Modeling the Relation Between the Intensity JND and Loudness for Pure Tones and Wide-Band Noise." *Journal of the Acoustical Society of America* 102 (December): 3628–3646.

Bosi, Marina, and Richard E. Goldberg. 2003. *Introduction to Digital Audio Coding Standards.* Dordrecht, The Netherlands: Kluwer.

Green, D. M. 1976. *An Introduction to Hearing.* Hillsdale, N. J.: Erlbaum.

Moore, Brian. 1986. *Frequency Selectivity in Hearing.* San Diego: Academic Press.

———. 1995. *Hearing.* San Diego: Academic Press.

———. 1997. *An Introduction to the Psychology of Hearing.* 4th ed. San Diego: Academic Press.

Pickles, J. O. 1988. *An Introduction to the Physiology of Hearing.* 2d ed. San Diego: Academic Press.

Plomp, R. 1976. *Aspects of Tone Sensation.* San Diego: Academic Press.

Tobias, J. V., ed. 1970/1972. *Foundations of Modern Auditory Theory.* 2 vols. San Diego: Academic Press.

Yost, William A. 2000. *Fundamentals of Hearing: An Introduction.* 4th ed. San Diego: Academic Press.

7 Introduction to Acoustics

A physicist who looks back over the history of his subject is struck by the prominent place that was originally occupied by musical acoustics. In fact it was one of the important sources of information about the nature of the physical world and a prime source of intellectual stimulation.
—Arthur H. Benade, *Trumpet Acoustics*

7.1 Sound and Signal

Having focused on the listener in the previous chapter, I now focus on the medium and consider how sound travels.

The sounds we hear correspond to pressure disturbances in the medium we are immersed in—air or water. In chapter 6 I mentioned that sound implies a source, a medium, and a receiver. This raises the age-old question: If a tree falls in the forest and there's no one to hear it, did it make a sound?

One could argue that pressure disturbances in air are not sound until a subject experiences them, but this seems academic. A way out of the difficulty is to differentiate between a sound and a signal: A sound has meaning by the information it conveys from a source to the receiver; a signal is a sound that conveys such information. Therefore, an unheard sound is not a signal, but it is still a sound. Practically speaking, just as there is no harm in talking about a sunset (even though it's the Earth that turns), it's okay to discuss the propagation of sounds and signals without regard to a source or receiver so long as we're aware of the potential for contradiction.

7.2 A Simple Transmission Model

One simple model of the transmission of signals that takes into account the receiver, source, and medium is

Receiver	Source	Medium
Observed sound =	Original sound –	Transmission loss,

where the transmission losses are spreading, absorption, and scattering of sound on the path from source to receiver.

For sound transmission to carry information (for it to be a signal), it must be detectable at the receiver, whether the ear or a microphone. Ordinarily, this requires that

- The sound must be above the receiver's threshold of sensitivity and within its frequency range.
- The sound must be greater in strength than the ambient noise, that is, the *signal to noise ratio* must be greater than 1.

Otherwise the signal is considered to be buried in the background noise. For signals meeting these detection criteria, we can get a rough prediction of whether a signal can be heard by relating the ambient noise level to the sound intensity level:

$$\text{Signal to noise ratio} = \frac{\text{Observed intensity level}}{\text{Ambient noise intensity level}}.$$

If the result is greater than 1, it's a relatively safe assumption that it can be detected (although this must remain a rough estimate because for hearing, detectability is potentially affected by masking).

7.3 How Vibrations Travel in Air

The speed of sound is a function of how quickly a medium can transport energy by wave motion. This in turn depends upon the physical properties of the medium.

When energy is injected into a medium, it seeks to return to its lowest energy level by radiating the energy away. In an elastic medium such as air, energy can be radiated either by heat convection (heat exchange between adjacent molecules) or by wave motion, or both. When a sound wave travels through a gas, the regions that are compressed become slightly warmer, and the regions that are expanded become slightly cooler. But the wavelength for most audible sound is relatively large in comparison to the rate that heat flows through air, so no appreciable heat flows from a condensation to an adjacent rarefaction. So most of the work in sound transmission happens because of pressure changes rather than thermal convection.

A system that performs work without heat flowing into or out of it is *adiabatic*. Under normal atmospheric conditions, and for most audio frequencies of interest to humans, energy propagation by convection is much slower than energy propagation by sound wave, so air is considered an adiabatic medium.

In the medium's undisturbed state, molecules of fluid media such as air collide randomly in three dimensions under the forces of thermally induced motion. Average particle speed is proportional to temperature.

We can usefully think of air as an *ideal gas,* representing its molecules as a collection of perfectly hard spheres that collide but otherwise have no interaction with each other. An ideal gas stores all its energy in the *translational velocity* of the particles (that is, the particle speed). The random motion of the molecules is the air's *internal energy* or *microscopic energy,* as distinct from the air's *macroscopic energy,* which characterizes the large-scale motion of an air mass as a whole. Sound and wind are forms of macroscopic energy.

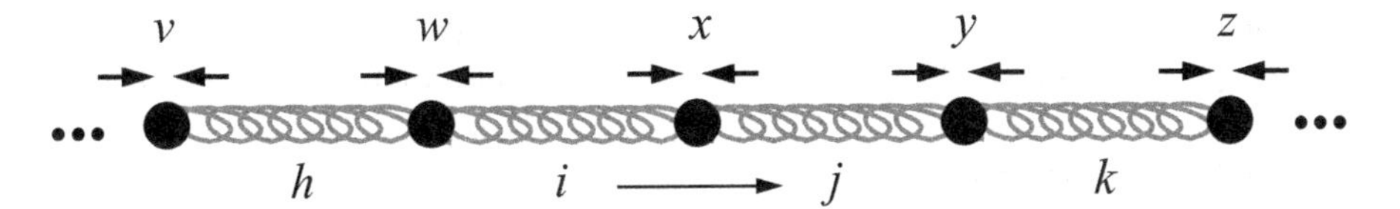

Figure 7.1
Idealized one-dimensional representation of air.

Imagine a packet of air such as that enclosed by your lungs.

- If you move the air packet (by breathing in or out), energy is transferred from your lung muscles to the air and stored in the momentum of the air particles—an *inertial property* of air.
- If you compress the air packet (by holding your breath and squeezing your chest and diaphragm), energy is stored as heat—an *elastic property* of air.

In either case the energy is *stored* in the air's momentum or compression because all the energy (except that lost to friction) will be released again when the air packet decelerates or the air packet decompresses. Understanding wave propagation depends on understanding how the medium's inertial and elastic properties interact.

Figure 7.1 presents an idealized one-dimensional representation of air, in which small packets of air molecules are represented as balls $\{v, w, x, y, z\}$ between springs $\{h, i, j, k\}$. In the beginning, the springs are all pressing with even force upon the balls, the forces balance, and there is no movement, corresponding to the ambient background air pressure.[1]

If I give ball x an instantaneous shove right, it further compresses spring j and expands i. The force from x to y grows while the force from x to w shrinks. Therefore, first w and y are drawn to the right, then v and z, and so on. The movement of segment $\{x, y, z\}$ is a *compression wave,* and the movement of $\{x, w, v\}$ is an *expansion wave.*

If stiffer springs are used, or if the springs are more compressed, the force displacing x would be conducted to y and w faster. Therefore, speed of wave propagation goes up with increasing stiffness. Just as the stiffness of a spring is raised as the pressure on it grows, so the stiffness of a gas is increased by raising the pressure P it undergoes.

Gases are compressible to the extent that they can easily convert pressure into internal energy, and different gases have different energy-storing capacities. A gas with a higher *heat capacity ratio* γ is like a spring with greater inherent stiffness: it compresses less easily, that is, it requires more force to compress, and therefore it can store more energy per unit of volume than a more compressible gas with lower γ. So the elastic properties of a gas are its pressure P and its ability to store heat γ. Increasing either P or γ (or both) increases wave propagation speed.

Now for the inertial properties. If the mass of the balls in figure 7.1 were increased (and the springs were left unchanged), the balls would have more inertia, and the force displacing x would be conducted to y and w more slowly. The inertial property of a gas is its *density* ρ, defined as its mass m per unit of volume V, or $\rho = m/V$. Increasing ρ decreases wave propagation speed.

The *phase of matter* has a huge impact on the inertial and elastic properties of different media. In general, solids have greater stiffness than liquids, which have greater stiffness than gases. For this reason, longitudinal sound waves travel faster in solids than in liquids or gases. One might think that because of gases' relatively small mass per unit volume, the speed of sound would be faster in gases. But the stiffness is so much greater in liquids and solids that, in general, for speed of sound c,

$$c_{\text{solid}} > c_{\text{liquid}} > c_{\text{gas}}.$$

7.4 Speed of Sound

To summarize the foregoing,

- Increasing elastic properties P or γ (or both) *increases* wave propagation speed.
- Increasing the inertial property ρ *decreases* wave propagation speed.

Combining these observations, we can say that the *speed of sound* c_s in a gas is proportional to the ratio of its elastic and inertial properties:

$$c_s \propto \frac{\text{elasticity}}{\text{density}} \propto \frac{\gamma P}{\rho}. \qquad (7.1)$$

Since energy is proportional to the square of velocity, we can rewrite (7.1) as a proper equality:

$$c_s^2 = \frac{\gamma P}{\rho},$$

and thus the speed of sound is

$$c_s = \sqrt{\gamma \frac{P}{\rho}}. \qquad \textit{Speed of Sound } (7.2)$$

All we need now is to find appropriate values for P, γ, and ρ for air in order to determine the speed of sound in air, but to find them requires a few additional discoveries.

7.4.1 Heat Capacity

Heat is energy that flows from a higher-temperature object to a lower-temperature object. Because it is a kind of energy, its unit is the joule (J), the same unit used for work, kinetic energy, and potential energy.

Heat that flows originates in the internal energy of the hotter substance. Internal energy is the sum of the molecular kinetic energy (the random kinetic motion of the molecules), molecular potential energy (forces acting within and between molecules), and other forms of energy. The internal energy of a substance is not called heat unless it is flowing.

The amount of heat needed to raise the temperature of a substance by a certain amount—its *heat capacity*—depends upon the kind of substance and upon the mass of the substance. The heat capacity Q of materials can be shown to be directly proportional to the change in temperature ΔT and the amount of mass m, so that $Q \propto m\,\Delta T$. Adding a constant of proportionality c, the *specific heat capacity,* allows us to determine the heat capacity of a specific material, $Q = cm\,\Delta T$. The value of c must be determined experimentally for each specific material. Solving for c, we have a way to determine the heat capacity of specific materials:

$$c = \frac{Q}{m\,\Delta T}. \qquad \textit{Specific Heat Capacity} \quad (7.3)$$

From (7.3), the SI unit for specific heat capacity is J/(kg · C°). For example, the specific heat capacity of copper is 387 J/(kg · C°), and the specific heat capacity of water (at 15°C) is 4186 J/(kg · C°).

The specific heat capacity of gases is different depending upon whether the gas is measured with *constant pressure* or *constant volume*. This distinction is usually not important for solids and liquids but can be significant for gases, such as air. The specific heat capacity measured holding pressure constant is called c_p, and the specific heat capacity measured holding volume constant is called c_v. For example, the constant pressure specific heat capacity c_p of oxygen is 912 J/(kg · C°), and the constant volume specific heat capacity c_v of oxygen is 651 J/(kg · C°).

The importance of this distinction may not at first be obvious, but it turns out to be crucial for correctly calculating the speed of sound. Newton first analyzed the speed of sound in *The Principia*. His analysis was correct, but the predicted result was far smaller than measured values. This problem dogged theorists for the better part of a century and set back the progress of acoustics until the difference between c_v and c_p was discovered, and the mystery was solved.

7.4.2 Heat Capacity Ratio

The ratio of c_p/c_v, the heat capacity ratio, characterizes the inherent molecular springiness of a gas:

$$\gamma = \frac{c_p}{c_v}. \qquad \textit{Heat Capacity Ratio} \quad (7.4)$$

It is the ratio of the specific heat capacity of a gas at constant pressure to the specific heat capacity at constant volume.

If we compress a gas, we add to its internal energy, causing its temperature to rise. The compressibility of a gas depends on how its particles accommodate change of heat energy. This in turn determines the ratio of the change in heat energy to the change in temperature.

For an ideal gas, c_p exactly equals c_v so that $\gamma = c_p/c_v = 1.0$. This means that the specific heat capacity is the same whether we hold pressure or velocity constant. If we double the pressure on an ideal gas, the volume is halved.

If γ is greater than 1.0 (because $c_p > c_v$), the gas is not ideal. Nonideal gases store energy in the translational velocity as well as the *rotational velocity* and *vibrational velocity* of the particles. For nearly diatomic gases such as air, γ is 7/5 = 1.40.

7.4.3 Mass Density

Having considered the elastic properties of air, we must next look at its inertial properties before we can establish the physical basis of the speed of sound. Wave propagation is slower in more dense media because denser particles accelerate less quickly for the same applied force, therefore they communicate their force to their neighbors less quickly.

The *mass density* ρ of an undisturbed gas is its mass *m* per unit of volume *V*:

$$\rho = \frac{m}{V}, \qquad \textit{Mass Density} \quad (7.5)$$

where mass is the quantity of matter contained in an object, and matter is anything that occupies space and exhibits inertia. The SI unit of mass density is kg/m³. For example, the mass density of helium is 0.179 kg/m³, and the mass density of air is 1.29 kg/m³ (Beranek 1986). To determine the speed of sound, we must determine the average molecular mass of air, which is composed of numerous different gases.

Air is composed of about 78 percent nitrogen (N_2), 21 percent oxygen (O_2), 0.9 percent argon (Ar), and 0.03 percent carbon dioxide (CO_2) by mass, and its *average molecular mass* is the sum of the products of the various atomic masses times their percentages. So to determine the average mass of air, we must first determine the atomic mass of the individual gases, which we do as follows.

The *mole* is the SI base unit for expressing the amount of a substance measured in molecules. Amedeo Avogadro (1776–1856) discovered that 12 grams of carbon-12 contains 6.022×10^{23} atoms. This number is known as *Avogadro's number* N_A. One mole (abbreviated mol) of a substance contains as many particles (atoms or molecules) as N_A.

Since 1 mol has the same number of atoms regardless of what the substance is, the difference in mass between substances is due to the difference in their molecular weights. For example, 6.022×10^{23} atoms of carbon-12 weigh 12 g, and the same number of nitrogen atoms weigh 28.013 g. Table 7.1 shows calculations of average molecular mass for air. If 6.022×10^{23} particles of air weigh 28.87 g, one particle weighs

$$m = \frac{28.87\,\text{g}}{\text{mol}} \cdot \frac{1\ \text{mol}}{6.022 \times 10^{23}\ \text{particles}} = \frac{4.79 \times 10^{-23}\ \text{g}}{\text{particle}}, \qquad \textit{Average Mass, Atom of Air} \quad (7.6)$$

or 4.79×10^{-26} kg per air particle.

Table 7.1
Average Molecular Mass of Air

Element	Percent		Atomic Mass		g/mol
N_2	78.08	×	28.013	=	21.87
O_2	20.95	×	31.998	=	6.70
Ar	0.934	×	29.948	=	0.27
CO_2	0.031	×	44.010	=	0.01
					28.87

7.4.4 Pressure, Volume, and Temperature

Pressure P is force per unit area, measured in atmospheres. An *atmosphere* (atm) is defined as the average atmospheric pressure at sea level, with a standardized value of 101,325 Pa (pascal), a little over 10^5 N/m^2, or about 14.7 pounds per square inch (see section 4.21). Measurements of air density are made by reference to *standard temperature and pressure* (STP), defined as 1 atm of pressure at 0°C, or 273.15 Kelvin.

The pressure fluctuations of sound waves are very small in comparison to standard atmospheric pressure. Sound pressure level (SPL) ranges from about 0.1 Pa at the threshold of hearing up to about 1 Pa at the limit of hearing. This corresponds to a fluctuation of between 10^{-7} N/atm and 10^{-5} N/atm.

If we have a volume of a fixed size, and we add more molecules of gas to it (for example, by pumping more air into a tire), the pressure increases. When the volume and temperature of an ideal gas are kept constant, doubling the number of molecules of air doubles the pressure. So pressure is proportional to the number of molecules, or equivalently, to the number of moles n of the gas, so we can write $P \propto n$.

If we have a volume of variable size, and we add more molecules of gas to it (for example, by pumping more air into a balloon), the volume increases. When the pressure and temperature of an ideal gas are kept constant, doubling the number of molecules of air doubles the volume. So volume is also proportional to the number of moles n, and we can write $V \propto n$.

If we have a fixed number of molecules of air in a certain volume and we reduce the volume, the pressure increases, as happens when pressing down on the plunger of a bicycle pump with the outlet closed. When the number of molecules and the temperature of an ideal gas are kept constant, halving the volume doubles the pressure because the same number of molecules now occupy half the space. So for an ideal gas, pressure and volume are reciprocal:

$$P \propto \frac{1}{V}, \quad \text{or} \quad PV \propto 1.$$

The final factor we must consider is temperature T. When volume is kept constant, raising the temperature of an ideal gas raises the pressure. And when the pressure is kept constant, raising the temperature of an ideal gas increases the volume. So temperature affects both pressure and volume directly, and we can write $PV \propto T$.

Combining these four proportionalities, we can write $PV \propto nT$. We can rewrite this as an equation by inserting a proportionality constant R, called the *universal gas constant*. The result is the *ideal gas law:*

$$PV = nRT, \qquad \textit{Ideal Gas Law} \quad (7.7)$$

where n is the number of moles of gas present, T is absolute temperature in Kelvin, and R is a magic number worked out experimentally in the late 1700s and codified in Boyle's law, Charles's law, and Avogadro's hypothesis. A good value for it is 8.314351 J/(mol · K).

Equation (7.7) expresses the ideal gas law in terms of moles n, but using Avogadro's number, it is easy to express this in terms of the total number of particles N. The total number of particles N is just the number of moles n times the number of particles per mole, N_A, in other words, $nN_A = N$. Substituting into (7.7), we have

$$PV = nRT = nN_A\left(\frac{R}{N_A}\right)T = N\left(\frac{R}{N_A}\right)T.$$

The constant term R/N_A is *Boltzmann's constant*,[2] usually represented by the symbol k:

$$k = \frac{R}{N_A} = \frac{8.31\ \text{J/(mol} \cdot \text{K)}}{6.022 \times 10^{23}\ \text{atoms/mol}} = 1.38 \times 10^{-23}\ \text{J/K}. \qquad \textit{Boltzmann's Constant}\ (7.8)$$

Substituting k into the ideal gas law, we have

$$PV = NkT. \qquad \textit{Ideal Gas Law using Boltzmann's Constant}\ (7.9)$$

Notice that the values of (7.7) and (7.9) are expressed only in units of pressure, volume, quantity of gas, and temperature. From these we can calculate the speed of sound in air.

7.4.5 Calculating the Speed of Sound

Solving (7.9) for pressure, we get

$$P = \frac{N}{V}kT. \qquad (7.10)$$

This can be rewritten in terms of mass density ρ and molar mass m because $N/V = \rho/m$ (a consequence of equation (7.5)). Substituting this equality into (7.10), we get

$$P = \frac{\rho}{m}kT. \qquad (7.11)$$

A rearrangement of (7.11) yields

$$\frac{P}{\rho} = \frac{kT}{m}. \qquad (7.12)$$

Now, recall equation (7.2) for the speed of sound:

$$c_s = \sqrt{\gamma\frac{P}{\rho}}.$$

Substituting (7.12) into (7.2) yields

$$c_s = \sqrt{\gamma\frac{kT}{m}}. \qquad (7.13)$$

We have an equation for the speed of sound based only on easily measurable quantities: temperature, compressibility, and mass. Plugging in the values $\gamma = 7/5$, $m = 4.79 \times 10^{-26}$ kg/m^3, and $k = 1.38 \times 10^{-23}$ J/K yields

$$c_s = \sqrt{\gamma \cdot \frac{kT}{m}} = \sqrt{\frac{7}{5} \cdot \frac{1.38 \times 10^{-23} \cdot T}{4.79 \times 10^{-26}}} = 20.0833\sqrt{T}. \qquad \textit{Speed of Sound} \ (7.14)$$

Temperature (in Kelvin) is the only nonconstant term for the speed of sound in air at standard atmospheric pressure. To use the more familiar Celsius scale, substitute $T = 273.15 = 0°\text{C}$. For example, the speed of sound at standard temperature and pressure is

$$20.0833\sqrt{T} = 20.0833 \cdot 16.52725 \cong 331.9 \text{ m/s}, \qquad \textit{Speed of Sound at STP} \ (7.15)$$

or a little over 1000 ft/s or about 1 ft/ms.

Air is *nondispersive,* meaning that c_s does not change with frequency, as light does in glass, for example.

7.4.6 Universal Wave Equation

Now that we have an analytic means to determine the speed of sound, we can use it to relate the period of a wave directly to its frequency. The speed of sound c is

$$c = f\lambda \text{ m/s}, \qquad \textit{Universal Wave Equation} \ (7.16)$$

where f is frequency in cycles per second, and λ is wavelength in meters. Knowing any two of these properties allows us to find the third:

$$c = f\lambda, \qquad f = \frac{c}{\lambda}, \qquad \lambda = \frac{c}{f}.$$

For example by (7.15), the frequency of a wavelength of 10 m is $f = 33.19$ Hz. The wavelength of 1000 Hz in air at STP is $\lambda = 0.3319$ m, or about 1 ft.

7.5 Pressure Waves

Sound is propagated through a medium such as air by *longitudinal waves,* where particle motion, wave motion, and energy flow are all in the same direction. Longitudinal waves are also called *pressure waves* (or *P-waves*) because wave propagation is carried by pressure differences in the medium. A two-dimensional representation of a longitudinal pressure wave is shown in figure 7.2. The figure can be thought of as a density contour map of the medium where the darker areas, or *compressions* C, have greater density than the lighter areas, or *rarefactions* R. The ordering of densities is always $R < \rho_o < C$, where ρ_o is density of air at STP.

As previously discussed, when a packet of air is compressed, the molecules store the energy as heat. When a packet of air is accelerated, the molecules store the energy as inertia. If we think about

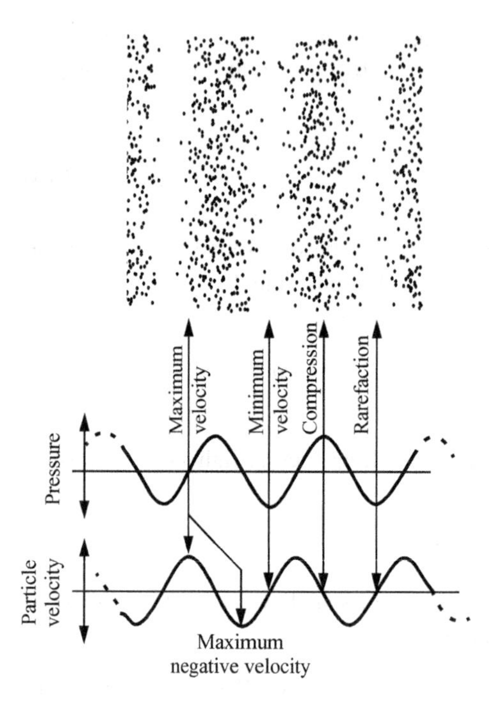

Figure 7.2
Longitudinal pressure wave.

figure 7.2 as a snapshot of a sound waveform in time, we see that the wave travels by *alternating inertia storage and heat storage through time.* At any moment, the most compressed/rarefied air packets (hence the ones with the most/least heat) have no momentum, whereas the air packets with the most velocity (hence the ones with the most momentum) have no compression.

7.6 Sound Radiation Models

Sound propagates with a spherical radiation pattern in free space of uniform density,[3] and the intensity of a signal at some distance will generally be inversely proportional to the square of the distance. But the actual distribution of transmitted sound energy depends upon other factors, including the *radiation pattern* of its source, which is a function of the efficiency of sound propagation in three dimensions. For instance, we experience the loudest signal from a violin if we directly face its top; it sounds quieter if we face its side.

Additionally, sound can be reflected, refracted, or absorbed on its way from a source to a receiver by objects encountered along the way. For instance, we experience an even quieter signal from a violin by standing directly behind the violinist, in the player's *acoustical shadow.* What arrives at our ears is determined by these factors (among others) in combination.

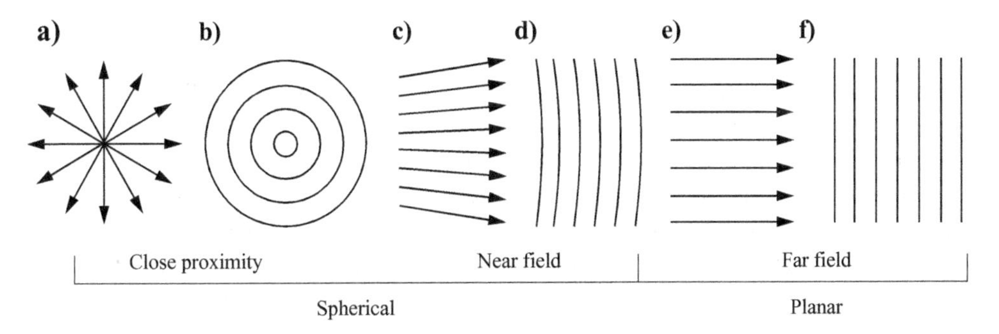

Figure 7.3
Graphical models of the transmission of sound.

The transmission of sound can be modeled as either waves or rays. Figures 7.3a and 7.3b show how these models represent two-dimensional waves in close proximity of a point source radiating in all directions. Near the source, we see great curvature of the wave fronts. In three dimensions, wave fronts emerge in layers of pressure areas from the source in a spherical pattern. In figures 7.3c and 7.3d, we are viewing sound at intermediate proximity from its source, in the near field or the *Fresnel zone*. Equivalently, we are viewing a magnified portion of the sound through a small aperture nearby its source. Though the total 2-D signal path is still circular, our view of it is so limited that what we see begins to look more like parallel rays or parallel wave fronts. Figures 7.3e and 7.3f show the sound at great distance from its source, in the far field, or *Fraunhofer region*. Equivalently, we are viewing at extreme magnification through a very small aperture nearby the source. We see so little curvature that for all intents and purposes the rays and wave fronts are parallel. In this region, the radiation pattern is independent of distance. These are called *plane waves,* although strictly speaking they are only geometrically planar in the limit at an infinite distance from the source. In three dimensions, plane waves pass by like sheets of pressure areas from the direction of the source.

Portraying transmission of sound as rays helps us visualize the direction and strength of sound propagation, and portraying them as waves helps us visualize wavelength. Both approaches are just models of the underlying physical phenomena, which we use to help make sense of what we experience. We can adopt whichever perspective helps us understand, and switch back and forth between models at will.

At what point do we go from near field to far field, from effectively spherical to effectively planar transmission? Three factors are significant: the distance from the source s_0, the area of the aperture a through which we observe the waveform pass, and the wavelength λ of the waveform (see section 4.24.4). An observation is termed far field if the distance from the source is much greater than double the area a of the aperture divided by the wavelength λ, called the *Rayleigh distance:*

$$s_0 \gg \frac{2a}{\lambda}. \qquad \textit{Rayleigh Distance} \ (7.17)$$

7.7 Superposition and Interference

Wave interference occurs when two or more waves act simultaneously on a medium. When such waves pass through each other in an ideal medium, the resulting disturbance at any point in the medium can be found simply by adding the individual displacements that each wave would have caused by itself. This is the *principle of superposition.*

Where the interfering waves have the same sign, the sum of their displacements will be larger than either wave by itself, resulting in *constructive interference*. Thus, for example, when the crest of one wave is superposed upon the crest of another, they interfere constructively. The same goes for a trough superposed upon a trough.

Where the interfering waves have opposite sign, the sum of their displacements will be smaller than either wave by itself, resulting in *destructive interference*. This occurs, for example, when a crest is superposed upon a trough, or vice versa. For examples of constructive and destructive interference, see figure 6.13.

Where two waves of opposite sign and equal magnitude coincide, they cancel, resulting in no displacement of the medium. A listener at that position would hear silence. This is essentially the principle behind noise-canceling headphones. A microphone mounted near the ear detects an incident sound wave, and an electronic circuit creates an inverse waveform matching the incident sound wave, and plays it through the loudspeaker in the headphones so that the incident sound is canceled by the inverse waveform when it reaches the ear.

7.8 Reflection

Interpreted as rays, sound obeys Newton's laws of motion because sound rays continue in a state of motion "at a constant speed along a straight line, unless compelled to change that state by a net force." Reflection acts as a net force upon a wave to deflect its direction.

Reflection of sound waves occurs only where the speed of sound changes, which, according to (7.2), happens only where the density or elasticity of the medium changes. Reflection occurs at the boundaries between media with different densities and elasticities, for instance, where sound strikes a wall. But reflections can also occur within the same medium where its density or elasticity changes. For example, the enormous impulsive sound wave created by a bolt of lightning in the clouds would be heard as a single clap, like a sonic boom, except for the numerous reflections caused by the turbulent differences in pressure and temperature within the storm. We hear the effect of these reflections as rolling thunder.

Sound reflection is just like light reflection: the angle of the incident ray and the reflected ray lie on the same plane, and the angles of incidence θ_i and reflection θ_r are equal (figure 7.4a),

$$\theta_r = \theta_i. \qquad \textit{Law of Reflection} \quad (7.18)$$

Specular reflection, which is reflection from smooth, relatively plane surfaces, creates a phantom source equidistant to the perpendicular of the reflecting surface (figure 7.4b). Nonspecular

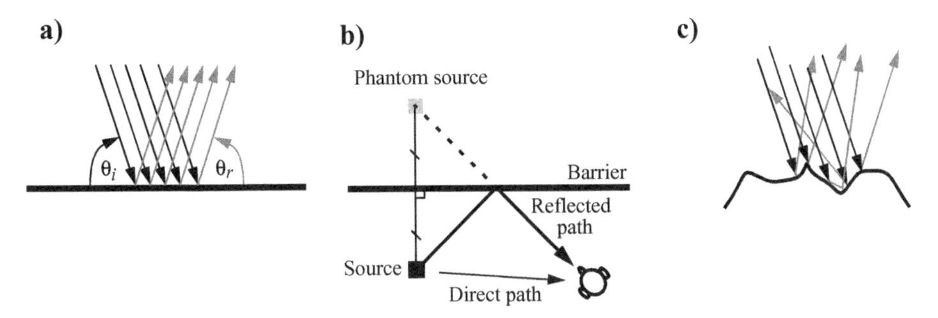

Figure 7.4
Phantom reflection source.

reflection (figure 7.4c) creates dispersion, scattering, or diffuse reflection. If the reflecting surface is sufficiently diffuse, no phantom source is created.

Looking at reflection from the wave perspective, we'd say that each local point on the reflecting surface emits a new spherically spreading wave front in response to the incident wave. The direction in which this new reflected wave front is propagated is constrained by

- The local geometry of the surface it strikes
- The pressure it experiences from other local wave fronts

This characterizes any reflection, specular or not. For a plane wave striking a plane surface, the hemispheres radiated by each point form a coherent wave front that resembles the impinging wave front but traveling in a new direction. For nonspecular reflection, we must examine the way each hemisphere is constrained by its local surface and the influences of nearby hemispheres in response to their own local conditions.

A wave front will experience *scattering* if the dimensions of the object it encounters are small in comparison to its wavelength. A wave front will experience *reflection* if the dimensions of the object it encounters are large in comparison to its wavelength. For example, a table top 1 m in diameter will tend to scatter wavelengths larger than 1 m and reflect smaller wavelengths.

Assuming that the direct path between source and receiver is not blocked, the first sound we hear from a source is always the direct signal because it travels the shortest distance between source and receiver. Since there is strong evolutionary survival value in knowing the direction from which a sound originates, our hearing suppresses reflected copies arriving after the direct signal from other directions (see section 6.13.3). But the precedence effect only works for about 35 ms after the arrival of the direct signal. Reflections arriving after that are experienced as distinct *echoes*. Reflections within the precedence interval are experienced as lending spaciousness to the sound. If there are so many reflections that we cannot distinguish them, we hear them as *reverberation*.

7.8.1 Determining Distance from Reflections

Reflected energy can be used to measure distance to a reflective object because some energy usually returns to the source. This is the way radar works. The same technique also works for sound: for example, some cameras are equipped with a distance-finding device consisting of an element that makes a highly directional clicking sound, and a microphone. The click emitted by the device reflects off the object the camera is aimed at, and some of the energy returns to the device's microphone. The device measures the elapsed time and uses equation (7.16), the universal wave equation, to calculate the distance to the object.

Since sound can penetrate opaque objects, sound reflection can be used to map underground rock strata. Using *acoustic pulse reflectometry,* geologists track the reflections of a shock wave transmitted through the earth by setting off a small explosion and recording the echoes (figure 7.5). By examining the delay and amplitude of the most prominent reflections at various microphones, and knowing the average speed of sound in the earth, geologists can infer the depth and location of strata of different densities.

The same principle has been applied to deriving the bore profile of tubular musical instruments such as winds and brass, and the tracheal tubes of humans and animals, using a noninvasive technique (Ware and Aki 1969). Though these instruments can in principle be measured with calipers, in practice complications such as side holes and the inaccessibility of interior tubes limit the accuracy of this approach. For the voice, a noninvasive approach to measuring the tracheal tubes of live subjects is invaluable. With acoustic pulse reflectometry, an impulse of sound is injected into the instrument. Reflections arise in response to the impulse where the bore diameter changes. The more rapid the change in bore diameter, the larger the reflection it causes. The resulting impulse response function is measured by a microphone. The impulse response is converted into a cross-sectional area as a function of axial distance from the impulse source, providing the desired bore profile.[4]

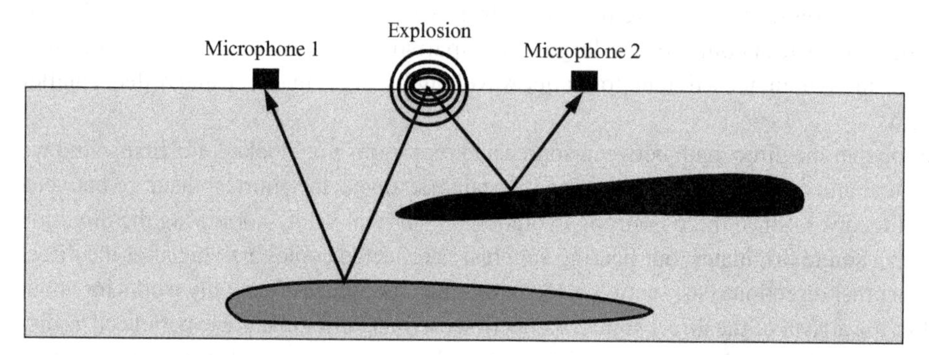

Figure 7.5
Acoustic pulse reflectometry.

7.8.2 The Old Rope Trick

We can model sound reflection in transverse waves by attaching a rope to a wall. Holding the free end, if you snap the rope, an impulsive wave travels down its length. When the wave reaches the wall, some of the wave energy reflects back towards your hand while some is transmitted into the wall. The returning wave experiences *phase reversal,* meaning that the returning pulse travels on the opposite side of the rope: if the outgoing pulse travels above the rope, the reflection returns below the rope (figure 7.6a). This is how sound travels in strings with rigid terminations.

Attaching the end of the rope to a lightweight thread and the thread to a wall makes the rope act as though it were free at the wall end (figure 7.6b). The reflected wave remains on the same side of the rope both coming and going, so there is no phase reversal.

When the rope's end is fixed, as in figure 7.6a, we can think of the rope's reflected wave as coming from an imaginary inverted source on the other side of the wall (figure 7.7). If the rope is fixed to the wall, the wave's displacement clearly must be zero when it reaches the wall (assuming the wall is inflexible). We would achieve the same effect if we had an identical rope on the other side of the wall that was shaken the opposite (inverted) direction. The imaginary inverted wave arriving from the imaginary source would exactly cancel the real source when the two motions meet at the wall, also providing zero displacement.

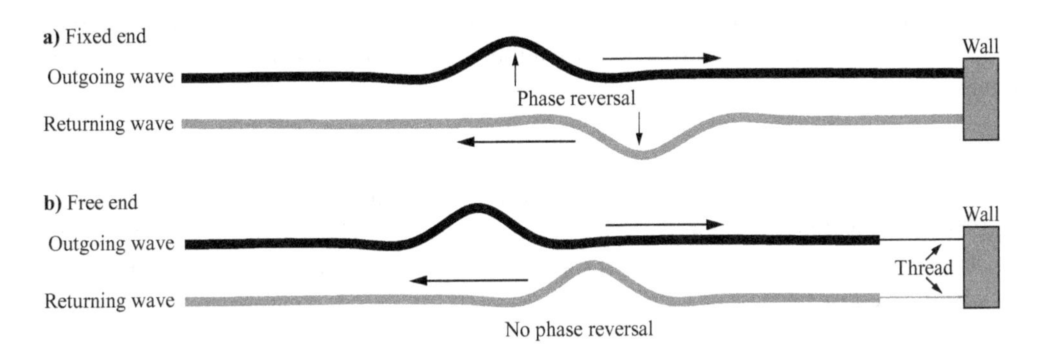

Figure 7.6
Reflection of an impulsive wave on a rope.

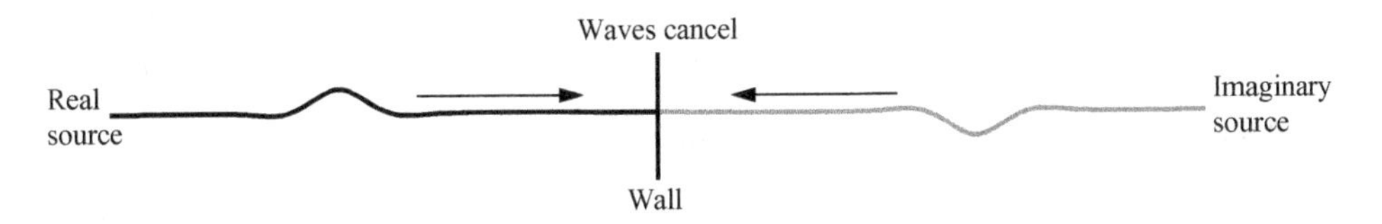

Figure 7.7
Rope with fixed end.

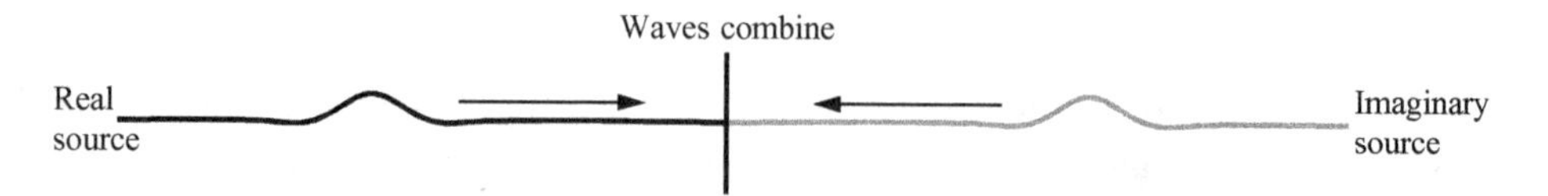

Figure 7.8
Rope with free end.

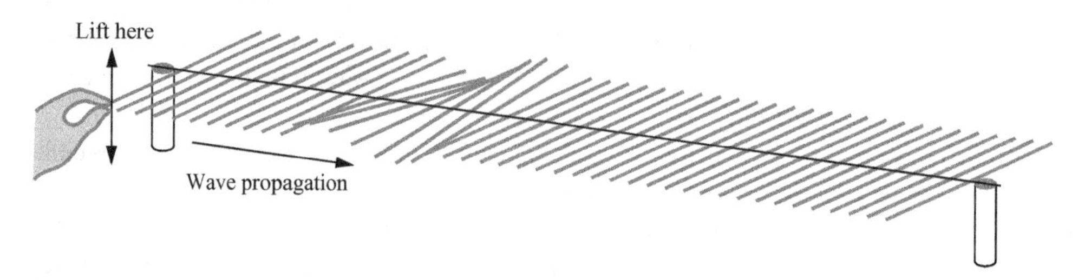

Figure 7.9
Shive wave machine.

However, when the rope's end is free, as in figure 7.6b, the end snaps like a whip, momentarily doubling its displacement. We can model this as an uninverted imaginary wave arriving from the imaginary source, so the waves add when they meet, providing twice the displacement (figure 7.8).

7.8.3 Shive Wave Machine

Most natural wave motion occurs too fast or is otherwise too subtle to follow easily with the eye. In the 1950s, John N. Shive developed a wave machine at Bell Telephone Laboratories that clearly reveals transverse wave motion. An array of stiff steel rods are attached crosswise at regular intervals to a wire (figure 7.9). Because of the relatively large inertia of the rods compared to the elasticity of the wire, a wave takes several seconds to travel from one end of the array to the other. If the tips are painted with phosphorescent paint and the apparatus is viewed under black light, only the tips are visible, and one sees an array of dots moving up and down in transverse wave motion. This can be used to replicate the experiments with the ropes. With the far end free, as in figure 7.9, a positive impulse sent down the wave machine results in a positive reflected wave, as in figure 7.6b. With the rod at the far end clamped so it can't move up or down, a negative wave is reflected, as in figure 7.6a.

7.8.4 Reflection and Transmission at Media Boundaries

The Shive wave machine can also be used to examine what happens when waves cross the boundary between two media with different speeds of sound. The crossbars at the right end of the wave machine shown in figure 7.10 have been shortened so that the speed of wave propagation along this section of the machine is doubled.[5] Call the slower speed of wave propagation in the long bars

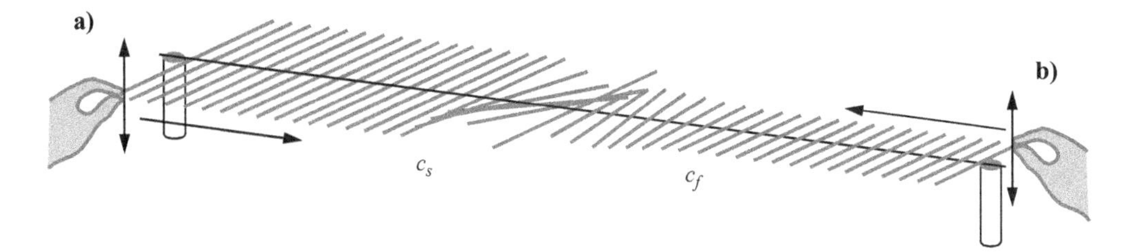

Figure 7.10
Reflection and transmission at a boundary.

c_s and the faster speed in the short bars c_f. The figure shows a wave crossing over the boundary, one half still in the slow section, one half in the fast section.

We can compare what happens when a wave crosses from a slower to a faster medium (send an impulse from (a) in figure 7.10) and from a faster to a slower medium (send an impulse from (b) in the figure).

- If, as in figure 7.9, the bars of the Shive machine do not change length, then $c_s/c_f = 1$ and the speed of propagation remains unchanged. All energy is transmitted along its entire length, none is reflected until it reaches the end.
- If, as in figure 7.10, the bars change length, then $c_s/c_f \neq 1$. Because there is a difference in the speed of propagation, some energy is transmitted and some is reflected.
- The initial disturbance that moves toward the barrier is the *incident wave,* the *reflected wave* is returned from the boundary, and the *transmitted wave* passes through the boundary.
- The sum of the reflected and transmitted energy always equals the total original energy (apart from that lost to friction or sound).
- Where wave motion goes from the slower medium into the faster medium, as in figure 7.10a, the returning wave does not experience a phase reversal, like the rope attached to a string.
- Where wave motion goes from the faster medium into the slower medium, as in figure 7.10b, the returning wave experiences a phase reversal, like the rope attached to a wall.
- In no case is the phase of the transmitted wave reversed.
- The frequency of the wave is preserved across the boundary.

We can account for these phenomena as follows. Suppose there is a boundary where medium 1 has speed of propagation c_1 on one side and medium 2 has speed of propagation c_2 on the other. Assume that the incident wave always starts in medium 1. Then the amplitude coefficient of the reflected wave A_r will be

$$A_r = \frac{c_2 - c_1}{c_1 + c_2}. \qquad \textit{Reflection} \ (7.19)$$

The amplitude coefficient of the transmitted wave A_t will be

$$A_t = \frac{2c_2}{c_1 + c_2}. \qquad \textit{Transmission} \quad (7.20)$$

For example, if the speed in medium 1 is twice that of medium 2, so that $c_1/c_2 = 2/1$, and the incident amplitude is equal to 1, then by (7.19) the amplitude of the reflected wave will be –1/3 and by (7.20) the amplitude of the transmitted wave will be 2/3. This is equivalent to starting the incident wave from (b) in figure 7.10.

If we reverse this so that $c_1/c_2 = 1/2$, then reflected amplitude is 1/3 and transmitted amplitude is 4/3. This is equivalent to starting the incident wave from (a) in figure 7.10.

If $c_1/c_2 = 1/1$, reflected amplitude is 0 and transmitted amplitude is 1. We can use these equations to determine the behavior of the rope described in section 7.8.2:

- In the limit, if the second medium's speed of propagation is zero, $c_1/c_2 = 1/0$, then reflected amplitude is –1 and transmitted amplitude is 0. This is the case for the rope with fixed end and the Shive wave machine in figure 7.9 with its end clamped—all the energy is reflected and comes back inverted.
- If the second medium's speed of propagation is infinite, the reflected amplitude is 1 (not inverted) and transmitted amplitude is 2. This corresponds to the rope with free end and the Shive wave machine with its end unclamped.

Remember, total energy is conserved in the system, but amplitude adjusts according to the mass and elasticity of each medium at the boundary.

Reflection of longitudinal waves, such as sound waves in air, can be visualized as follows. Consider a long pipe with a drum head at the left end (figure 7.11a). We can create an impulsive wave by striking

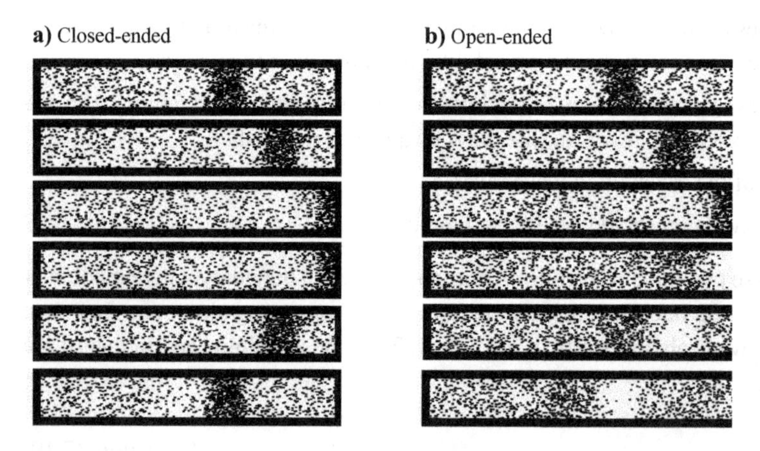

Figure 7.11
Closed-ended and open-ended reflections in a tube.

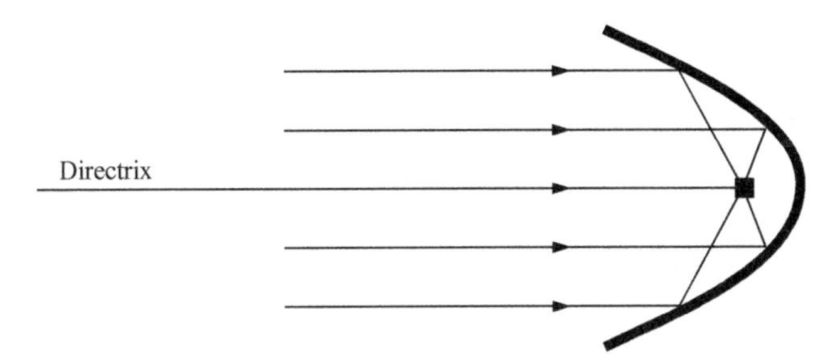

Figure 7.12
Parabola focusing sound on a microphone.

the drum head, which sends a sharp positive pressure wave down the tube. If the tube is closed at its right end, the positive pressure doubles when the impulsive wave strikes it, and the wave is reflected back as a positive pressure wave—the same as described for a free-end rope reflection.

However, if the tube is open at its right end, the wave interacts with the air surrounding the mouth of the tube. When the positive pressure wave exits the tube, it displaces the air, which is at normal atmospheric pressure around the mouth. This displacement is then propagated away from the opening as a high-pressure wave. A new low-pressure area surrounding the mouth is created in its wake. Air from outside and inside the tube is drawn to this new low-pressure zone. The air outside the tube that is drawn back then propagates away from the tube as a low-pressure wave, and the air that is drawn from inside the tube then propagates back up the tube as a low-pressure wave (figure 7.11b). This is the same as described for a fixed-end rope reflection.

When sound reflects from a concave surface, wave intensity is focused exactly as light intensity is focused in a reflecting telescope. If a microphone is placed at the focus of a parabolic dish, sound waves arriving from the same direction as the *directrix* (the line bisecting the reflector) are focused on it because each point on the parabola is equidistant from the focus and the directrix (figure 7.12).

7.8.5 Acoustical Coupling

There are situations where it would be good to have as much energy as possible travel from one medium into another while reducing or eliminating reflections due to the discontinuity at the barrier between the two media. For instance, the boundary at the oval window in the ear has air outside and denser perilymph inside (see section 6.2.3). If sound waves struck the oval window directly, most energy would reflect back out of the ear and little would get inside to enable hearing. The tympanum and the bones of the middle ear provide the inner ear with a mechanical coupling that passes almost all energy into the inner ear for a range of frequencies of greatest biological interest, thereby providing the inner ear with a more intense signal at these frequencies. Such coupling devices are called *transformers*.

Figure 7.13 shows the Shive wave machine adapted to couple energy across a boundary with minimal reflection. The average speed of sound in the central region is the geometric mean of the

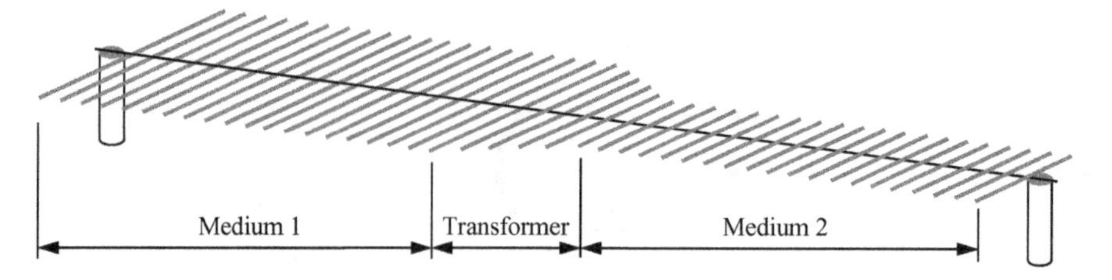

Figure 7.13
Transformer at a boundary.

speed in the two surrounding media. The transformer couples most efficiently for frequencies whose wavelengths are close to four times the length of the transformer. It works progressively less well for other frequencies.

7.9 Refraction

That part of a sound's energy that is transmitted through a boundary between two media enters into the new medium and is subject to *refraction*. Suppose a plane wave strikes a plane surface with a different speed of sound c. If the *angle of incidence* α of the wave is not perpendicular to the new medium, the wave is bent upon entering it and goes off at a different angle β.

There are three cases to consider:

- The speed of sound is the same in the two media: $c_1 = c_2$. Then a sound that strikes the surface of the second medium at an angle of incidence α will enter the new medium at angle β, and $\alpha = \beta$ (figure 7.14a).
- The speed of sound is faster in the first medium: $c_1 > c_2$. The angle of incidence $\alpha > \beta$, resulting in a *focusing effect* (figure 7.14b). The energy is focused because relatively wide angles of incidence result in relatively narrower angles of refraction.
- The speed of sound is faster in the second medium: $c_1 < c_2$. Then $\alpha < \beta$, resulting in a *dispersive effect* (figure 7.14c). The energy is dispersed because relatively narrow angles of incidence result in relatively wider angles of refraction.

In general, for some angle of incidence α, the sound will be refracted into area Q if $c_1 < c_2$ and into area P if $c_1 > c_2$ (figure 7.14d). We see that the angle of incidence α is related to angle of entry β by $\beta = K\alpha$, where $K \propto c_2/c_1$.

The exact relation, known as Snell's law, is

$$\frac{\sin \alpha}{c_1} = \frac{\sin \beta}{c_2}. \qquad \textit{Refraction} \ (7.21)$$

Solving for β, we have

$$\beta = \sin^{-1}\left(\sin\alpha\frac{c_2}{c_1}\right).$$

For example, if $c_1 = 1.0$, $c_2 = 1.25$, and the angle of incidence is $\alpha = 45°$, the angle of entry is $\beta = 62.1°$.

Where an incident sound wave moves from a slower to a faster medium ($c_1 < c_2$), there is a *critical angle* α_{crit} that causes the angle of entry to equal 90°. When $\alpha = \alpha_{crit}$, the energy of the refracted wave (called a *creep wave*) travels along the boundary, and is rapidly attenuated. When $\alpha > \alpha_{crit}$, all incident wave energy is reflected. The value of the critical angle is

$$\alpha_{crit} = \sin^{-1}\left(\frac{c_1}{c_2}\right).$$

7.9.1 Gradient Refraction

Figure 7.14 shows refraction at sharply defined boundaries between media of different densities, but refraction takes place wherever any parameter affecting the speed of sound changes for any reason. Refraction can occur continuously over a gradient, for example.

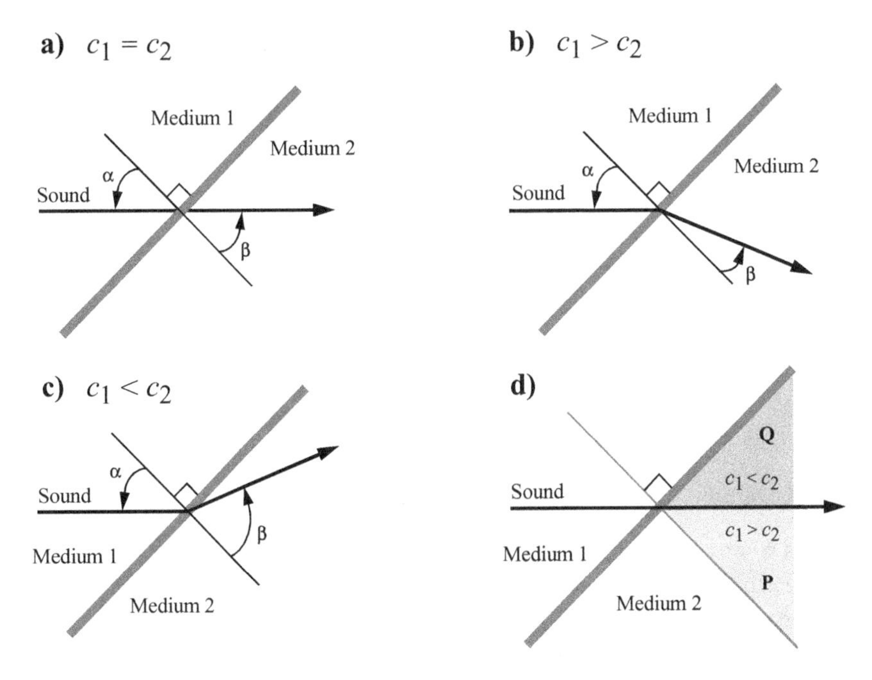

Figure 7.14
Refraction at a barrier.

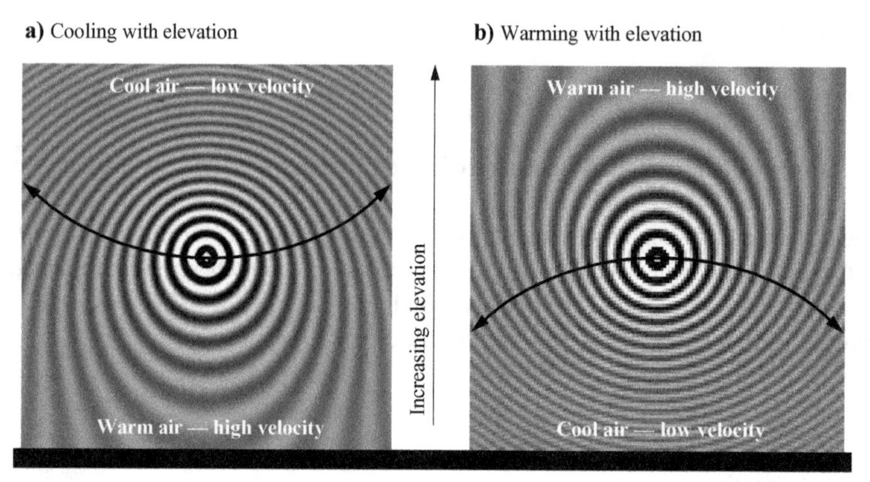

Figure 7.15
Refraction of sound in the atmosphere.

On a still evening with no clouds, the atmosphere near the ground remains warm because of the ground's lingering heat, and the air generally is colder with increasing elevation. As the density increases over a gradient, the speed of sound correspondingly decreases. Figure 7.15a shows a sound source at some elevation above the surface of the earth. The warm air below causes the wave front to speed up, the cool air above causes it to slow down, and the result is that the wave front bends upward with increasing distance from the source.

On a still morning with no clouds, air in the upper atmosphere is warmed by the sun faster than air near the ground, so the atmosphere generally gets warmer with increasing elevation. Figure 7.15b shows the wave front bent downward with increasing distance from the source by the early morning refraction gradient. Sound can often be heard on the surface of the earth at a farther distance when refraction is downward under these weather conditions because the wave fronts travel around the tops of surrounding obstacles. I once lived a half mile from the ocean. During the day, the surf sound was blocked by tall cliffs. But on calm nights a temperature gradient would form, refracting the sound. Every night (reliably within 5 minutes of midnight, for some reason), suddenly, I'd hear the surf.

7.9.2 Land Speed of Sound

Consider what happens when one shouts into the wind to a listener. The *land speed of sound* must be measured with respect to the average wind speed. Thus, if $c = 331$ m/s but the air itself is moving at 10 m/s, the land speed of sound c_l is actually 341 m/s in the direction of the wind.

When wind flows smoothly without vortices or other turbulence, it travels slower at ground level than higher up. The speed of air molecules in contact with the earth must be effectively zero because the earth is stationary with respect to the wind and these air molecules are bound to earth

Figure 7.16
Refraction due to wind speed.

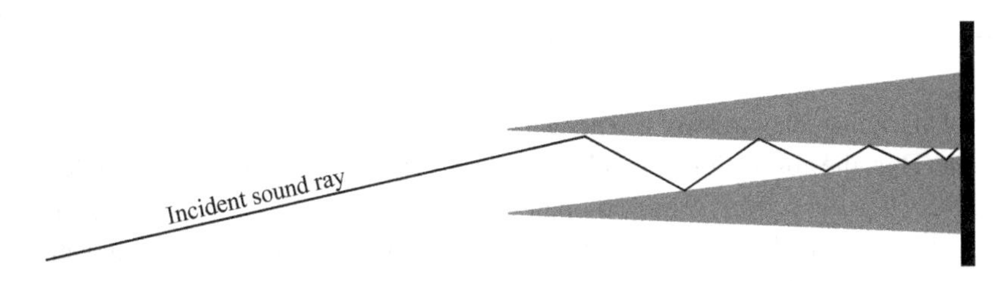

Figure 7.17
Sound absorption panel.

by strong molecular forces. Air molecules above them are progressively less subject to this resistance, with increasing elevation causing the air to move in horizontal sheets, an effect called *laminar flow*. Thus the speed of wind—and hence the land speed of sound—shows a gradient increase with elevation in the direction of the wind. Figure 7.16 shows this effect on two listeners at equal distances downwind (a) and upwind (b) of a sound source. The listener upwind receives less intensity than the downwind listener because upwind sound is refracted up into the sky.

7.10 Absorption

When sound energy is transformed into another kind of energy, such as heat, we say the sound is absorbed. Different materials absorb sound to varying degrees: a cement wall absorbs little and reflects most sound energy; a wood wall absorbs some and reflects some; carpet absorbs most and reflects little.

Air itself absorbs sound energy, depending upon its temperature and relative humidity. Cold, dry conditions favor sound transmission, whereas hot, moist conditions absorb sound, with high frequencies being most subject to attenuation. This means that the intensity of a signal being transmitted through air will actually be less than would be predicted by the inverse square law of distance because the air itself dissipates energy from the sound.

An anechoic chamber is one whose walls absorb all sound, providing no detectable echo or reverberation. Usually the walls are constructed of large wedges of soft, fibrous material on all surfaces. If we envision a sound wave incident upon a wedge as a ray (figure 7.17), we can see that reflection causes the ray to strike the wedge many times, each time transferring some of its energy as heat into the wedge until it is completely absorbed.

Table 7.2
Absorption Coefficients of Various Materials at Various Frequencies

	Frequency (Hz)					
Material	125	250	500	1000	2000	4000
Concrete block	0.36	0.44	0.31	0.29	0.39	0.25
Wood	0.15	0.11	0.10	0.07	0.06	0.07
Carpet	0.08	0.24	0.57	0.69	0.71	0.73
Air	–	–	–	–	0.01	0.02

The best absorber of all is an open window: 100 percent of the energy that goes through it is lost to listeners in the room. Thus the absorption of all other materials is compared to that of a window of equal area, and the *absorption coefficient* of an open window is defined as $a = 1$. If a square meter of carpet absorbs half as much sound as a window of equal size, then the absorption coefficient of the carpet material would be $a = 0.5$.

A surface having an area S and an absorption coefficient a can be said to have *total absorption* $A = Sa$, equal to an open window of area A. Materials vary as to how much sound they absorb in different frequency bands. The absorption coefficients of concrete, wood, carpet, and air are shown in table 7.2. The entry for air assumes a temperature of 20°C, 30 percent relative humidity.

Our ears use attenuation due to air absorption and surface reflection to help identify objects in space in a number of ways:

- In the open, if we hear a sound we recognize, but it sounds muffled, we assume it is far away. The muffling is a consequence of the attenuation of high frequencies with distance through air.
- In a hall the direct signal from a source not only arrives first but is also spectrally the brightest that we hear from that source because it is least subject to attenuation due to air absorption at the walls or in the air. Our ears compare the spectral brightness of the direct signal to the reflected signals as a cue to the location of the sound source (along with other cues such as intensity and time of arrival).

7.11 Diffraction

Since sound waves emanate spherically from a sound source, one might think that if an object blocked the sound, the size of the sound shadow would grow with distance beyond the blocking object (figure 7.18a). Instead, the sound shadow shrinks with distance (especially for low frequencies) and even disappears (figure 7.18b). This is pretty good news if your seat at the concert hall is directly behind a large pole: you'll still be able to hear the music—at least the low frequencies. This tendency of sound to spread out into sound shadows is called *diffraction*.

Diffraction arises in two common situations: apparent bending of waves around small obstacles or past sharp edges, and spreading out of waves past small apertures. *Small* in this case means small in proportion to the wavelength of the passing waveform.

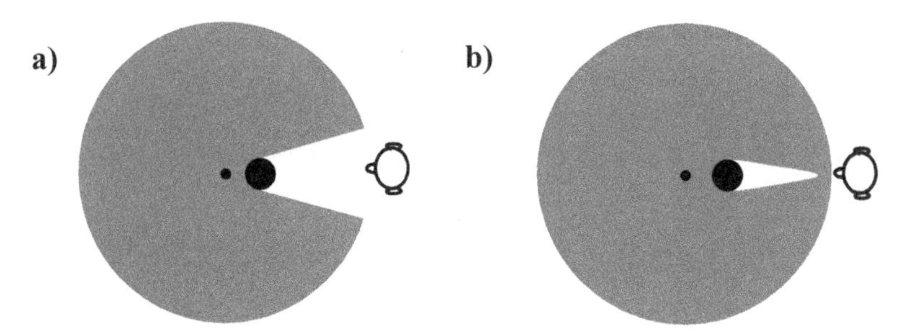

Figure 7.18
Naive expectation vs. actual behavior of sound shadows.

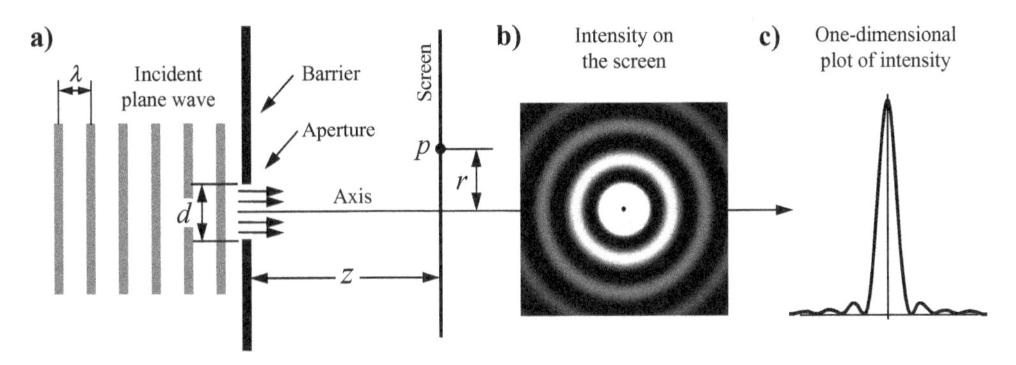

Figure 7.19
Diffraction through an aperture.

It is a lot easier to study diffraction of light than of sound because we can see light. So let's make some simplifying assumptions and consider diffraction of light through an aperture (figure 7.19a). Diffraction of sound works exactly the same way. Assume that the light arrives from such a distance that the wave fronts are virtually parallel, so we can ignore the spherical complexities of waveforms. Now imagine that this wave front impinges on a barrier with a small circular aperture. The light passing through the aperture strikes a screen behind it. Figure 7.19b shows the *diffraction pattern* of light intensity striking the screen for some aperture diameter d, wavelength λ, and distance to screen z. The whiter the area, the more energy it is receiving. Figure 7.19c shows a cross-section of the diffraction pattern with intensity on the y-axis. It is understandable that the light should be most intense on the screen directly opposite the aperture. But what about the fringe areas that also get light energy?

To understand the diffraction pattern, consider how the plane wave front passes through the aperture. According to *Huygens's principle*,[6] the sound wave in the aperture behaves as if *all the points of the wave surface within the aperture were separate radiating sources of sound with the same phase*. This means that every point on the wave surface within the aperture emits vibrations not only directly toward the screen but also in all other directions, hemispherically. Thus all the points on the wave surface within the aperture radiate their energy in such a way that it spreads out uniformly

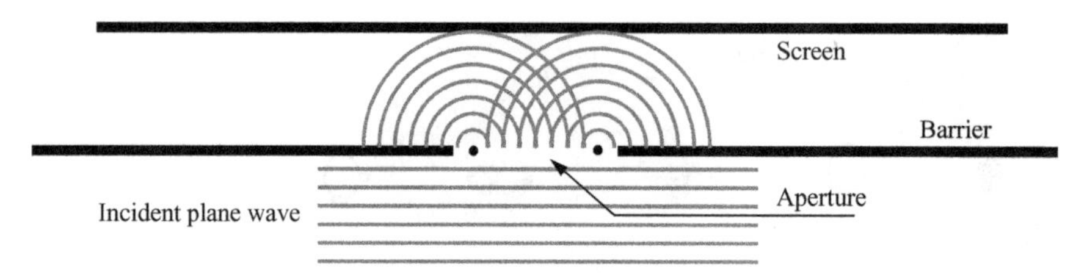

Figure 7.20
Behavior of two points in the aperture according to Huygens.

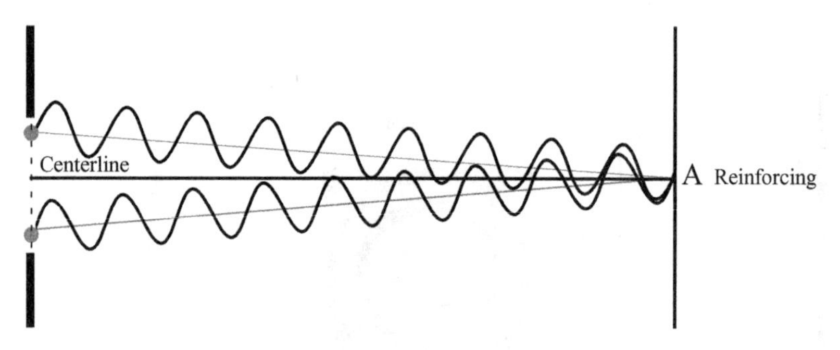

Figure 7.21
Constructive interference.

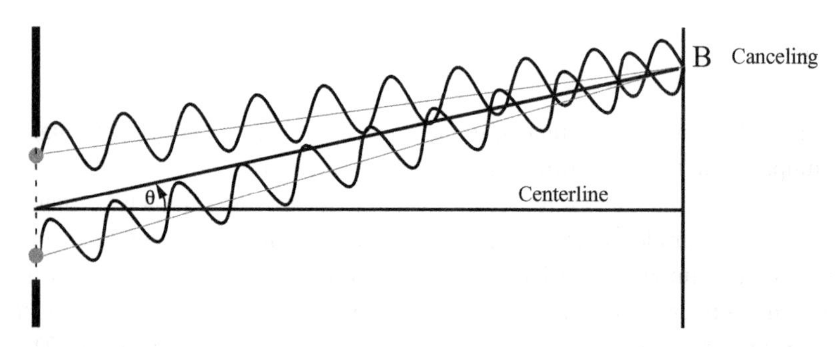

Figure 7.22
Destructive interference.

and hemispherically into the area beyond the barrier. Figure 7.20 shows this behavior for two points within the aperture. (We are assuming the phases of all these separately radiating points are aligned by the plane wave that is driving them from behind.)

Figure 7.21 shows that vibrations emanating from all points on the wave surface reach the center point A in phase and reinforce each other in constructive interference. (Only two points are shown to keep the figure simple, but Huygens' principle holds for an infinite multitude of points interacting this way.) Figure 7.22 shows how the same vibrations at a different angle cancel at point B, resulting in destructive interference. Waves with the same wavelength will sometimes add, sometimes cancel, depending upon θ, the angle of incidence. The same destructive interference shown in figure 7.22 would happen if point B were on the other side of the x-axis at the same distance from the center of the aperture.

If we consider the whole screen, there would be a dark ring with a radius $B - A$. Now imagine a point C at twice the distance B is from A. The phase of the vibrations would once again be constructive as they were at point A, and there would be a white ring on the diffraction pattern. However, it would be fainter because the distance to the screen from the aperture is greater, and light intensity drops with the square of the distance, according to the inverse square law. Thus consecutive concentric rings receive less and less energy until they are insignificant.

The cross-section of the diffraction pattern (figure 7.19c) looks rather sinusoidal, although the values are all positive and it dies away quickly at the edges. In fact, this is a squared sine wave that is scaled so its intensity dies away quickly off-axis.

The approach to diffraction presented here, which considers only plane waves, is due to Joseph Fraunhofer (1787–1826). *Fraunhofer diffraction* involves coherent plane waves incident upon an obstruction. The more general case, *Fresnel diffraction,* is the same except that the curvature of the wave fronts is taken into account. This was first worked out by Augustin-Jean Fresnel (1788–1827). The diffraction pattern shown in figure 7.19b, historically called the Airy disc,[7] corresponds to Fraunhofer diffraction through a circular aperture. Diffraction through other aperture shapes requires different equations.

Before we can construct an equation for diffraction, let's first isolate the factors involved:

- Intensity diminishes as point p is moved further from the axis by increasing r (figure 7.19a). This is simply the inverse square law at work.
- Diffraction grows as the aperture becomes smaller. As the size of the aperture d shrinks, the total amount of energy passed through decreases, and the energy that still gets past is diffracted more strongly. Figure 7.23 shows a cross-section of the Airy disc diffraction pattern as a function of aperture size. The larger the aperture, the more the energy tends to beam (and the more energy gets past

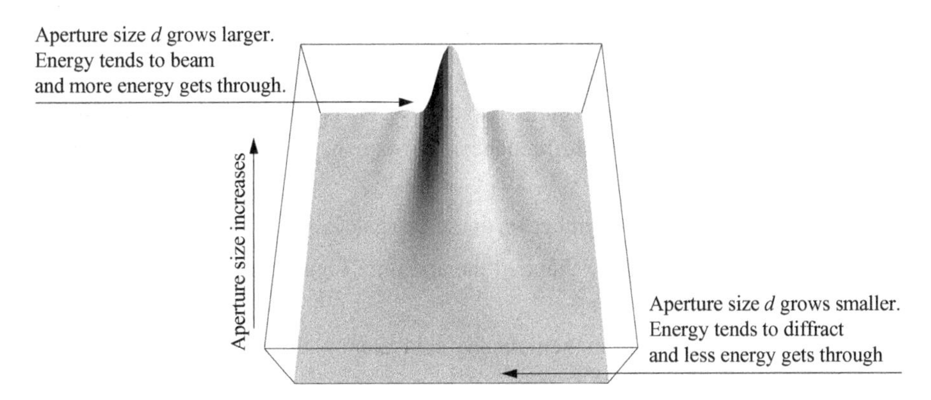

Figure 7.23
Diffraction pattern as a function of aperture size.

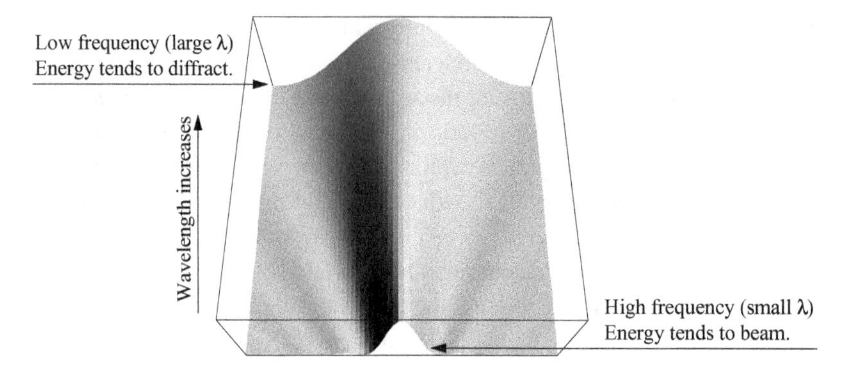

Figure 7.24
Diffraction pattern as a function of wavelength.

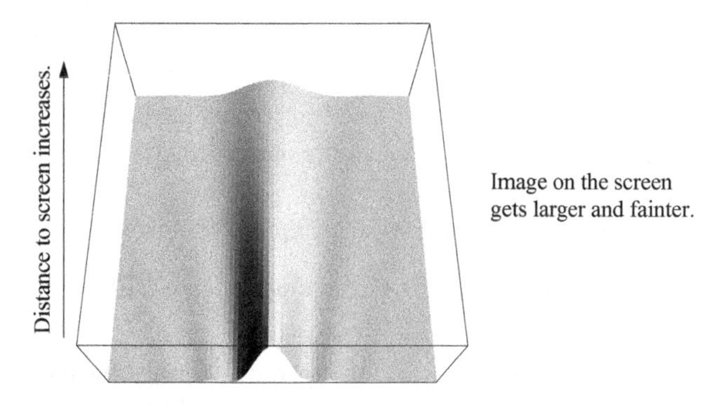

Figure 7.25
Diffraction pattern as a function of distance to screen.

the aperture because it is bigger); the smaller it is, the more the energy spreads out across the screen uniformly (and the less energy gets past the aperture because it is smaller).

- Diffraction is greater, the longer the wavelength. Low frequencies (large λ) tend to diffract, and high frequencies (small λ) tend to beam. Figure 7.24 shows a cross-section of the Airy disc diffraction pattern as a function of wavelength/frequency. As frequency goes from low to high, the energy tends to beam.
- As the distance to the screen z grows, the image on the screen gets larger and fainter (figure 7.25).
- The diffraction pattern changes with the shape of the aperture. Figure 7.26 shows a diffraction pattern made by a rectangular aperture.

Fraunhofer diffraction by an aperture is mathematically equivalent to the Fourier transform of the aperture shape (see volume 2, chapter 3).

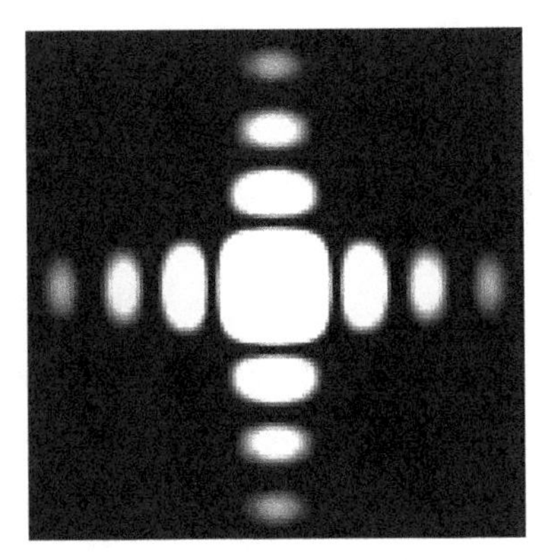

Figure 7.26
Diffraction pattern of a rectangular aperture.

We can engineer an equation for diffraction by putting all these facts together. Referring to the geometry of figure 7.19a, we know that

- Overall intensity and the amount of diffraction are both proportional to diameter d of aperture.
- The diffraction effect is magnified as the distance z from the aperture to the screen grows, while overall intensity simultaneously goes down.
- The amount of diffraction is proportional to wavelength λ.
- As we move away from the axis by a radius distance r, the intensity at point p goes down.
- Overall, the intensity I_a is directly proportional to the amount of sound energy entering the aperture.

Then the intensity I_p of energy at point p on a screen that is distance r from the screen's axis is

$$I_p = I_a\left(\frac{d}{z}\right)^2\left(\frac{\sin\delta}{\delta}\right)^2, \qquad \textit{Diffraction} \quad (7.22)$$

where $\delta = dr/\lambda z$.

The terms d, λ, and z control the distance between the peaks of the diffraction pattern. The peaks will become wider apart if the aperture d is made smaller, if the wavelength λ grows, or if the distance to the screen z grows. Terms d and z also have an effect on the overall intensity: if the aperture d is made larger, more energy is let through; if the distance to the screen z grows, the intensity diminishes.

Why does music played through a loudspeaker sound so different depending upon where one is listening from? On-axis, in front of the speaker, we hear a rich mixture of low and high frequencies; off-axis, the sound gets increasingly muffled until, standing directly behind the speaker, all we hear is low frequencies. Of course, the answer is diffraction. A loudspeaker in an enclosure is subject to exactly the same Fraunhofer diffraction as plane waves passing through an aperture (figure 7.27). We would expect to hear these same diffraction effects, and we do. Figure 7.28 shows

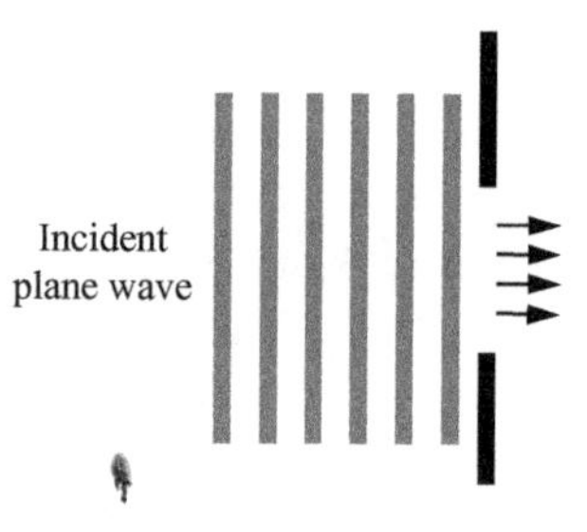

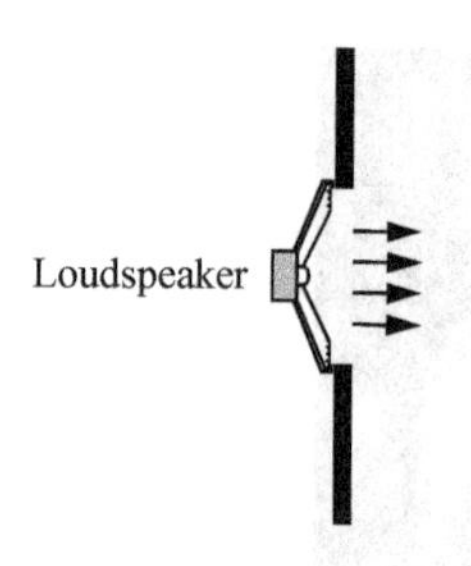

Figure 7.27
Loudspeaker as a Fraunhofer diffractor.

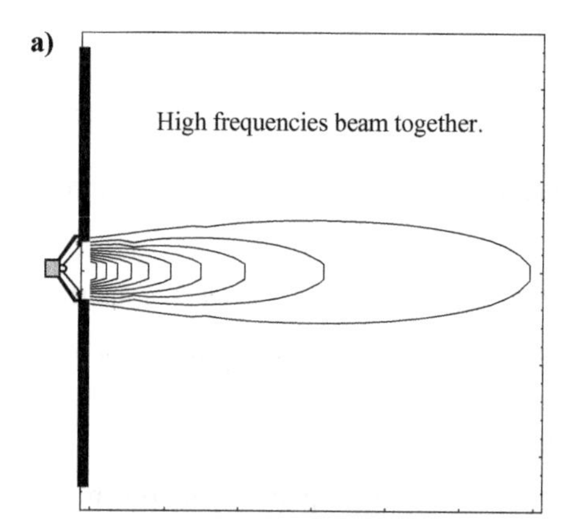

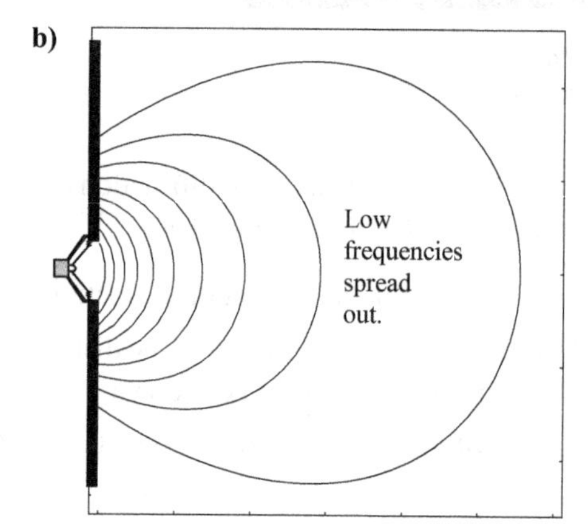

Figure 7.28
Equal-energy contours.

equal-energy contour graphs for high-frequency energy (7.28a) and low-frequency energy (7.28b) from a loudspeaker into a room.

7.12 Doppler Effect

If a train moves past very quickly, we hear the pitch rise as it approaches and then fall as it moves away. Why?

All waves travel at the same speed in a uniform medium, but if the distance between receiver and sound source is shrinking, the more recently emitted waves do not have as far to go to reach the receiver. In the time it takes to produce one wavelength, the source has moved toward the receiver thereby foreshortening the wavelength being emitted in that direction. We hear the foreshortened wavelengths as a higher pitch. The same reasoning can be used to show why the pitch drops for a sound source moving away.

7.12.1 Doppler Shift with Stationary Listener

Figure 7.29 shows how a sound source *S* moving to the right compresses the wavelengths emitted in its direction of travel and lengthens those emitted in the opposite direction. Thus, a stationary listener at point A hears a higher pitch than a listener at point B as a consequence of the movement of *S*.

Assuming the listener is stationary, the equation describing Doppler frequency shift f_d for a moving sound source at frequency f and velocity u is

$$f_d = f \cdot \frac{v_s}{v_s - u}. \qquad \textit{Doppler Shift} \ (7.23)$$

When the velocity of the source $u = 0$ the ratio $v_s/(v_s - u) = 1$ and $f_d = f$, so there is no pitch shift. But for $u > 0$ (corresponding to the source moving toward the listener) the denominator is smaller while the numerator remains constant; therefore the ratio $v_s/(v_s - u) > 1$, causing the Doppler-shifted frequency f_d to go up. If the source is moving away, $u < 0$, and the ratio becomes $v_s/[v_s - (-u)] < 1$, causing the Doppler-shifted frequency f_d to go down. For example, if the source is moving toward us at half the speed of sound, then $f_d = f\ [v_s/(v_s - 0.5v_s)] = 2f$, and we hear the frequency f shifted up exactly one octave (figure 7.30).

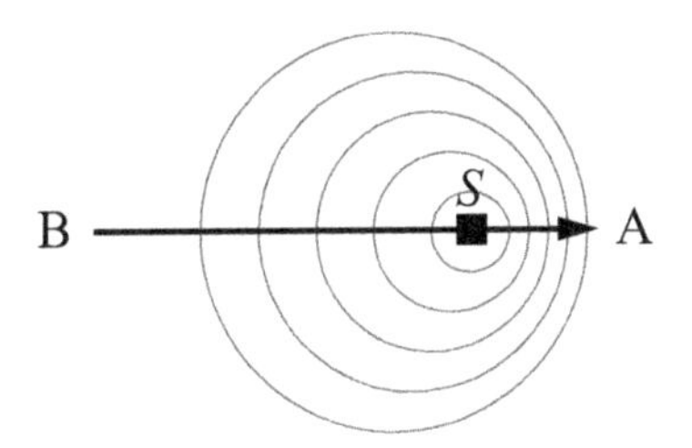

Figure 7.29
Doppler shift.

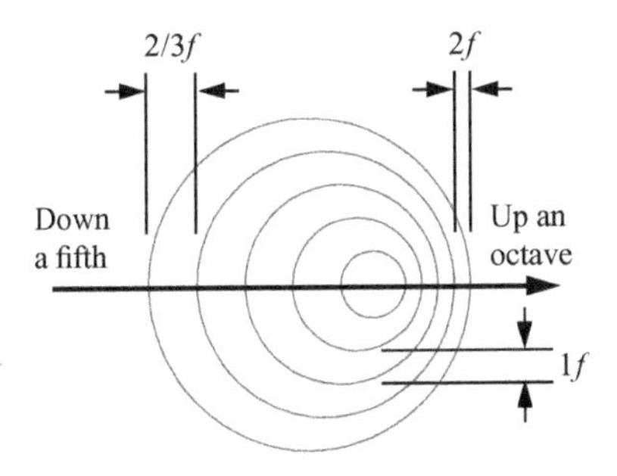

Figure 7.30
Frequency shift heard at half the speed of sound.

We might expect that if the sound source were to move away at half the speed of sound, we would hear a one-octave shift down in frequency, but since $f_d = f[v_s/(v_s - (-0.5v_s))] = f2/3$, the pitch drops only by a fifth.

If the source is traveling away from the listener at the speed of sound, then

$$f_d = \frac{f \cdot v_s}{v_s + v_s} = \frac{f}{2},$$

a drop of an octave (figure 7.31). If the source is traveling toward the listener at the speed of sound, then

$$f_d = \frac{f \cdot v_s}{v_s - v_s} = \infty,$$

so waves emitted exactly in the direction of travel at the speed of sound stack up on top of each other and form a single pulse with infinite frequency (see figure 7.31). A listener standing nearby would first hear a sonic boom, then the sound source shifted down an octave as it flashed past.[8]

What happens if velocity exceeds the speed of sound, that is, $u > v_s$? Theoretically, the listener would hear the sound backward, and only after it had passed by (figure 7.32).

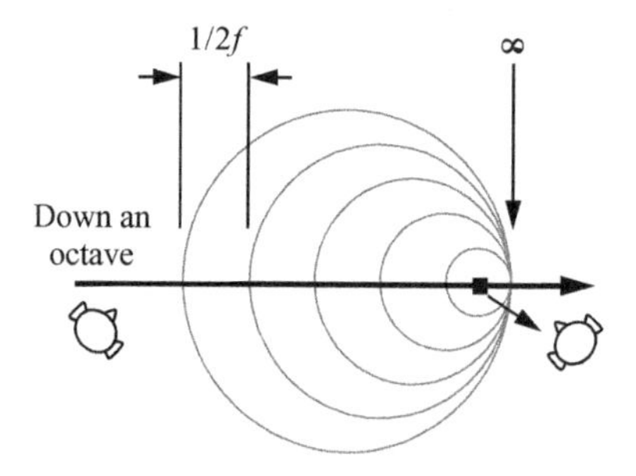

Figure 7.31
Frequency shift heard at the speed of sound.

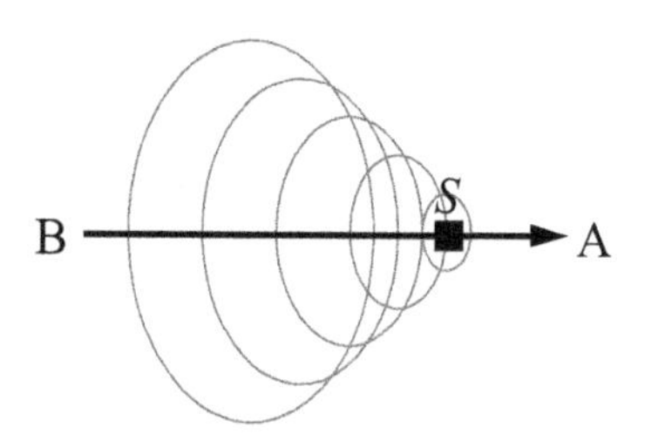

Figure 7.32
Supersonic Doppler shift.

7.12.2 Doppler Shift with Stationary Sound Source

Another possibility for Doppler shift is when the sound source is stationary and the receiver is approaching it. The equation for this case is

$$f_d = f \cdot \frac{v_s + u}{v_s}. \qquad \textit{Doppler Shift, Receiver Moves} \quad (7.24)$$

If the receiver moves away from the source at the speed of sound, then $f_d = f[(v_s - v_s)/v_s] = 0$ because the receiver is traveling at the same speed of the sound; all frequencies are shifted to 0. If the receiver approaches the source at the speed of sound, then $f_d = f[(v_s + v_s)/v_s] = 2f$, for a shift up by an octave.

7.12.3 Doppler Shift with Source and Receiver Moving

If the source and receiver are moving with velocities u_s and u_r, respectively, the equation becomes

$$f_d = f \cdot \frac{v_s + u_r}{v_s - u_s}. \qquad \textit{Doppler Shift, Both Move} \quad (7.25)$$

7.12.4 Two-Dimensional Doppler Shift

Equation (7.25) and all the other Doppler equations in the preceding sections are only accurate in one dimension, that is, where the source and listener are headed either directly away from or toward each other. However, most listeners prefer to stand to one side of speeding trains when observing their Doppler effect. A listener close to the train tracks experiences a sharper swing in pitch as the train passes by than a listener some distance away. What accounts for the difference?

I'll only consider the case where the source travels in a straight line past a listener in two dimensions, as when a train goes past someone standing beside the tracks. However, since paths can usually be broken down into a sequence of linear segments, this approach can be generalized with little effort.

In this case, we can no longer rely on absolute velocity; we must look at the *relative velocity* between source and receiver. How the distance between source and receiver changes through time determines Doppler shift. Figure 7.33 shows a sound source moving on a straight line at some absolute velocity u

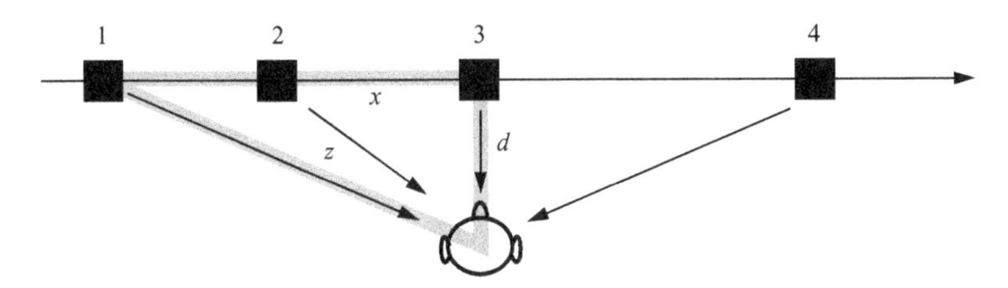

Figure 7.33
Trajectory of a moving sound source.

past a stationary listener at distance d. Clearly, positive Doppler shift is greatest at position 1, less at 2. There is no Doppler shift at position 3 because there is no relative velocity between source and receiver—for a moment they are neither moving closer nor further apart. After position 3, Doppler shift starts going negative, and the negative Doppler shift at position 4 matches the positive Doppler shift at 1. Doppler shift will be maximum at both horizons, and zero at the point of closest approach.

Geometrically, Doppler shift is proportional to the ratio of the arctangent of the lengths of sides x and d, as shown in figure 7.33. The relative velocity $\vec{z}$ from the source to listener in terms of the source velocity u and the distance of the listener from the source's path d can be expressed as

$$\vec{z} = u\frac{2\,\text{atan}(-x/d)}{\pi} - 1,$$

where x is the location of the source along its trajectory (with respect to the point of nearest approach). Setting $u = \vec{z}$ in the Doppler shift equation for a moving source given in (7.23), we have

$$f_d = f \cdot \frac{v_s}{v_s - u\left(\frac{2\,\text{atan}(-x/d)}{\pi} - 1\right)}. \qquad \textit{Doppler Shift in Two Dimensions} \quad (7.26)$$

We can test (7.26) by setting the distance d to 0, so the source heads directly at the listener, and we should get the same Doppler shift results as before. And, indeed, when $d = 0$, $[(2\,\text{atan}(-x/d))/\pi] - 1 = 1$ no matter what x is, so that term drops out of (7.26), and we have just (7.23) again. Figure 7.34 plots the Doppler shift that the listener will hear if a sound source producing a pure tone at 440 Hz goes past at half the speed of sound. The curves show the listener at 1 m, 5 m, 50 m, and 100 m from the closest approach, and the curve is plotted for the span of 100 m on either side of the listener's position. The closer the listener gets to the path, the closer the Doppler shift approaches the one-dimensional case; for instance, at 1 m the frequency is nearly doubled on approach, then drops nearly a fifth (to 293 Hz) when departing.

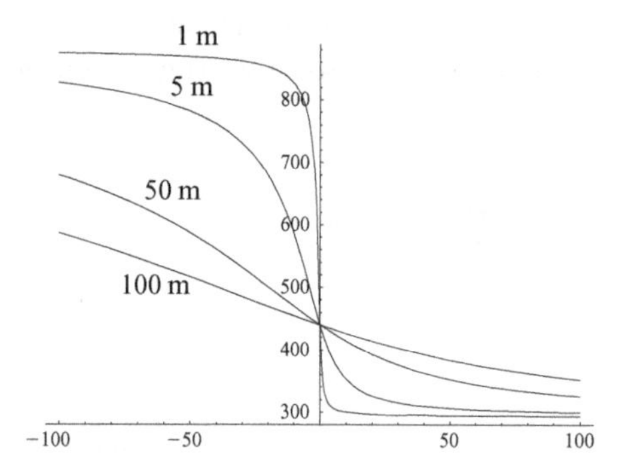

Figure 7.34
Doppler shift curves.

7.13 Room Acoustics

Suppose we pop a balloon in a concert hall and record the result. The sound the room makes in response to this brief impulse of sound is its *impulse response*.

The first few sound paths from source S to receiver R are shown in figure 7.35. The *direct signal D* travels the line-of-sight path from the sound source to the microphone, arriving at time t_D. The next few impulses, *early reflections,* arrive at the microphone after reflecting from nearby surfaces. These include the first-order (one bounce) reflections labeled 1, 2, 3, and 4, and the second-order (two-bounce) reflections labeled 5 and 6. Many other possible paths from source to receiver are not shown. In addition to these paths, there are also reflections from the side walls, from the stairs, and from the stage. Over time, there are so many reflection paths that the sound field in the room ends up composed of plane waves distributed with uniform randomness in all directions.

Figure 7.36 shows an idealized impulse response of a hypothetical room. The original impulse occurs at time t_0. The direct signal arrives at the microphone at time t_D. Depending upon the geometry of the room, the early reflections may occupy the first 10 to 100 ms of the room's reverberation

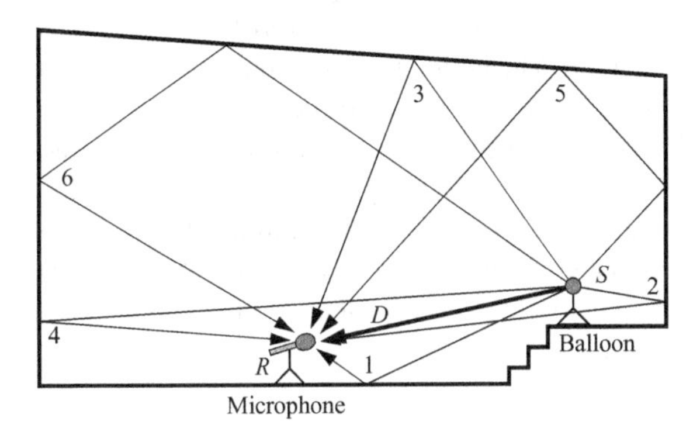

Figure 7.35
Direct sound path and early reflections in a concert hall.

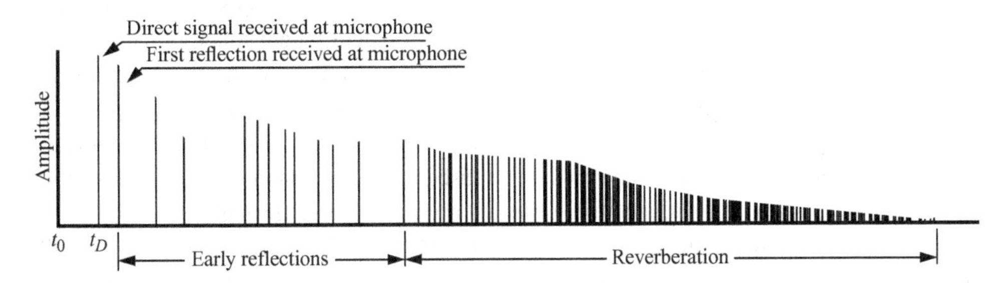

Figure 7.36
Idealized impulse response of a hypothetical room.

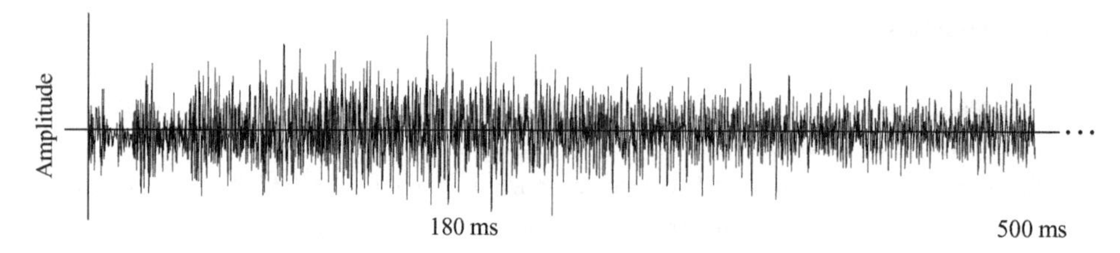

Figure 7.37
Impulse response of a cathedral (0.5 seconds).

time. The time delay of each reflection is proportional to the time it takes the impulse to travel from the sound source to the walls and then to the microphone. The amplitude of each reflection is inversely proportional to the distance traveled, directly proportional to the size of the reflecting surface, and inversely proportional to the material the surface is made of (among other factors).

The remaining reflections, *late echoes* or the *reverberation tail,* are the result of the combinatoric explosion of multiple reflections over time. The sound energy in a good-sounding hall declines approximately in an exponential curve after the source has stopped emitting sound. The shape of the curve is influenced by the position, orientation, and characteristics of the sound source and listener as well as their placement in the room. The notion that the room response can be idealized into distinct sections does not necessarily bear out well in practice.

In addition to reflection, sound distribution is also influenced by spreading, absorption, refraction, and diffraction. Figure 7.37 shows the first half-second of the impulse response of a cathedral. While some early reflections stand out, it is remarkable how quickly and uniformly the density of reflections builds up. The room response shows a uniform, gradual buildup until about 180 ms, then a gradual decline, providing a rich reverberant background. The long reverberation tail (not shown) is audible for about 10 s after the impulse.

7.13.1 Musical Character of Rooms

Music of a particular style is generally designed for a characteristic listening environment and may not sound good if reproduced in an uncharacteristic setting. For example, plainchant (a monophonic vocal style of the Middle Ages in Europe that is made up mostly of long sustained tones and slow tempos) was designed for highly reverberant cathedral spaces where sound lingers for 10 s or longer. Highly intricate and rhythmically active polyphonic music of the Baroque era was designed for halls with reverberation times of about 2–3 s. Plainchant in a Baroque concert hall sounds thin and exposed, and when polyphonic Baroque music is performed in cathedrals, the sound lingering from previous notes interferes with subsequent notes, making it difficult to follow the intricate lines of the music. Architectural taste and musical taste go through cycles of fashion and convenience. A late-romantic symphony by Gustav Mahler, with its focus on sonority, calls for longer reverberation time; a string quartet from any era sounds best in a more modest room; organ music usually calls for cathedral-length reverberation.

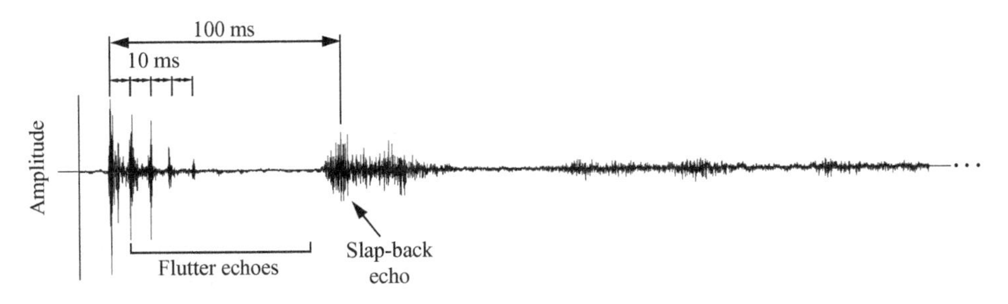

Figure 7.38
Impulse response of a bad sounding room.

For speech, the primary objective is intelligibility rather than a graceful reverberation. Figure 7.38 shows the first half-second of the impulse response of a concrete tunnel with severe acoustical problems. Its evenly spaced early reflections at 10 ms intervals produce a *flutter echo* caused by the parallel walls of the tunnel, which gives an unpleasant shuddering quality to the room response. At 100 ms, a *slap-back echo* from another section of the tunnel reaches the microphone. Since this is outside of the interval masked by the precedence effect (section 6.13.3), it is heard as a separate acoustical event, competing with the direct signal for audition and severely degrading intelligibility.

7.13.2 Reverberation Time

In the 1890s, Wallace Sabine (1868–1919) was presented by Harvard University with the challenge of taming the bad acoustics of a lecture hall. The room was unusable because its reverberation time was excessive, and it also had other problems similar to those shown in figure 7.38. His elegant solution created the foundations of the field of architectural acoustics. Sabine (1921) described the problem with his characteristic lucid prose:

> In the lecture room of Harvard University, . . . a word spoken in an ordinary tone of voice was audible for five and a half seconds afterwards. During this time even a very deliberate speaker would have uttered 12 or 15 succeeding syllables. Thus, the successive enunciations blended into a loud sound, through which and above which it was necessary to hear and distinguish the orderly progression of the speech.

Sabine reasoned that two principal factors competed to determine the reverberation time of a room: its boundary surface area and its internal volume.

- *Surface area*. When sound is reflected from a surface, a great deal of its energy is lost. Some is converted into heat in the wall, and some is transmitted through the wall to the outside. In either case, the sound is absorbed because it is removed from the room (see section 7.10). Increasing the boundary surface area reduces the reverberation time because then the sound has more opportunity to be absorbed.
- *Volume*. The larger the volume of air in a hall, the less opportunity sound has to reflect off the walls and be absorbed. In comparison to the walls, air itself absorbs relatively little energy when transmitting sound (see table 7.2). Increasing the internal volume decreases the rate of sound dissipation and therefore increases reverberation time.

From these considerations, we can see that reverberation time T_R is proportional to the ratio of volume to area:

$$T_R \propto \frac{\text{Internal volume}}{\text{Boundary surface area}}.$$

Suppose we have a bare room with internal volume V, with hard walls that absorb very little sound energy, and one wall contains an open window of area S. All sound that passes through the window to the outside can be said to be absorbed by the window in the sense that it leaves and does not return. The reverberation time T_R is equal to the ratio of the volume V to the area S of the open window:

$$T_R = k \cdot \frac{V}{S}, \qquad (7.27)$$

related by a constant k. When V is in m^3 and S is in m^2, Sabine found that the constant k is 0.161 s/m (see appendix A).

Absorption Not all surfaces absorb sound at the same rate. Rooms lined with carpeting and curtains absorb sound much more readily than do bare walls of brick or concrete.

Sabine modeled the absorption of particular surface materials by comparing them to the ideal absorption of an open window of the same area. Since the open window absorbs all energy that reaches it, he assigned it an absorption coefficient of $\alpha = 1$. A surface material that absorbs half of the incident sound energy has an absorption coefficient of $\alpha = 0.5$. Two square meters of this material would be needed to replace the absorption provided by a window of 1 square meter. From this example, we see that in general, a surface of area S and absorption coefficient α has an absorption

$$A = \alpha S \qquad (7.28)$$

that is equivalent to the absorption of an open window of area A.

Real rooms have a variety of surfaces with different materials. The average absorption $\bar{\alpha}$ of all surfaces is simply the sum of contributions from each surface that reflects sound in the room:

$$\bar{\alpha} = \frac{1}{S} \cdot (\alpha_1 S_1 + \alpha_2 S_2 + \cdots + \alpha_N S_N) = \frac{1}{S} \sum_{i=1}^{N} \alpha_i S_i, \qquad (7.29)$$

where the S_i are the individual surface areas, the α_i are the corresponding absorption coefficients, N is the number of surfaces, and S is the total boundary surface area of the room. Absorption is sometimes expressed in units of metric sabine, the absorption of 1 square meter of open window, named in honor of Wallace Sabine.

Combining volume, surface area and absorption, Sabine's formula for reverberation time is

$$T_R = 0.161 \cdot \frac{V}{\sum_i \alpha_i S_i}. \qquad (7.30)$$

Air Absorption Though the effect is small in comparison to the absorption of surfaces, in a large enough hall the absorption of sound by the air itself must be taken into account. The absorption of air m depends upon temperature and relative humidity and is equal to about 0.012 at 20°C, 30 percent relative humidity. The absorption of air also depends upon the volume V of air the sound must travel through. So we must add a term mV to the denominator of (7.30):

$$T_R = 0.161 \cdot \frac{V}{mV + \sum \alpha_i S_i}.$$ *Sabine's Equation for Reverberation Time* (7.31)

Frequency Response Most surface materials and the air itself tend to absorb high frequencies more readily than low frequencies. Each time the sound strikes a wall, and the farther the sound travels in air, some high-frequency energy is removed from the reflections, providing a low-pass filter effect. The late reflections progressively darken the tone of the reverberation because each reflection absorbs a little more of the remaining high-frequency energy. Our ears use this cue to help us distinguish the sound source from the decaying reverberant sound field and to distinguish newly arriving sound from lingering sound.

Low-frequency sound tends not to be reflected by the walls but passes through them to the outside. Each time the sound strikes a wall, some low-frequency energy is transmitted out of the hall and lost, providing a high-pass filter effect. If too many bass frequencies are lost too quickly, the hall reverberation sounds tinny. Wall material must be quite dense, such as stone or brick, to retard the escape of the lowest frequencies.

7.13.3 Sound Quality of Halls

The combination of low-pass filtering of high frequencies and high-pass filtering of low frequencies means that the reverberation tail contains mostly low-to-mid-range frequencies. The rate at which frequencies in different ranges decay affects the quality of the hall's reverberation. If low-frequency energy lingers too long, the hall sounds "tubby." If high-frequency energy lingers too long, the hall lacks warmth.

Many factors contribute to a good-sounding hall. Studies by Manfred Schroeder (1979) have shown the importance of having sufficient sound reflected to the listeners from the side. Lateral reflections with relative time delays in the range of 25–80 ms add a feeling of pleasant spaciousness.

Beranek (1962) listed 18 subjective attributes that affect the quality of a concert hall. A few of these are intimacy, liveness, warmth, loud-enough direct sound, evenness of reverberant sound throughout the hall, good clarity or definition (the direct signal and early reflections should not be lost in the reverberation), ensemble (players should be able to hear one another easily), and sufficient quiet. To this list can be added no strong echoes, no flutter echoes, no focusing of sound by large concave surfaces, and no sound shadows underneath balconies.

In spite of a century of theory and experimentation, architectural acoustics is far from a science. Major successes and catastrophic failures have been designed by architectural acousticians. Sabine's acoustic design for the Boston Symphony Hall made it one of the premier concert halls in the world. But it was admittedly a combination of good science and good luck. Beranek's acoustic design for Avery Fisher Hall, originally called Philharmonic Hall, in New York was an enormous and costly failure in spite of his extensive research, which included direct measurements of the world's finest concert halls. (The failure was perhaps more a consequence of the fact that the architect failed to take his recommendations fully into account.) Whereas laboratory scientists generally can suffer their failures in the privacy of their laboratories, not so acousticians, who succeed or fail very publicly.[9]

7.14 Summary

Air is adiabatic, and as a consequence, sound travels through air in waves determined mostly by the mass and elastic properties of air molecules. We derived the speed of sound from underlying physical principles and considered types of waves and how sound radiates.

Waves interact in a medium additively through constructive and destructive interference. Sound may be scattered or reflected depending upon the relation between the scale of the object the sound encounters and the frequency content of the sound. Reflections complicate the job our ears have to determine sound location. We can use reflection to determine distance and other properties in an acoustical environment without having to do direct measurements. Reflection occurs at boundaries between media, with or without phase reversal, depending upon whether the sound enters a denser medium. Sound can be transmitted more efficiently by matching the impedance of the two media using transformers.

Sound is also subject to refraction at the boundary between media, depending upon the angle of incidence. If the boundary is continuous, the sound undergoes gradient refraction. When sound energy is transformed into another kind of energy, such as heat, we say the sound is absorbed. The tendency of sound to spread out into sound shadows is called diffraction. Doppler effect is the apparent change in frequency of a sound as a source and a receiver pass each other at relative velocities.

Finally, we examined the acoustics of halls. Wallace Sabine showed that reverberation time depends upon the ratio of its boundary surface area and its internal volume.

7.15 Suggested Reading

Backus, John. 1977. *The Acoustical Foundations of Music.* 2d ed. New York: W. W. Norton.

Benade, Arthur H. 1976. *Fundamentals of Musical Acoustics.* New York: Oxford University Press.

Beranek, Leo. 1986. *Acoustics.* Rev. ed. Melville, N.Y.: American Institute of Physics.

Hall, Donald E. 1980. *Musical Acoustics: An Introduction.* Belmont, Calif.: Wadsworth.

Kinsler, Lawrence E., Austin R. Frey, Alan B. Coppens, and James V. Sanders. 1982. *Fundamentals of Acoustics.* 2d ed. New York: Wiley.

Rossing, Thomas D. 1983. *The Science of Sound.* Reading, Mass.: Addison-Wesley.

8 Vibrating Systems

Philosophy is written in this grand book—I mean the universe—which stands continuously open to our gaze, but which cannot be understood unless one first learns to comprehend the language and to interpret the characters in which it is written. It is written in the language of mathematics, and its characters are triangles, circles and other geometric figures, without which it is humanly impossible to understand a single word of it; without these, one wanders about as in a dark labyrinth.
—Galileo Galilei, *The Assayer*

8.1 Simple Harmonic Motion Revisited

The basis for the production of music and sound lies in the principles of mechanical physics. The physical laws of vibration are highly applicable to music, because they determine not only the sounds instruments make but also how the basilar membrane vibrates in response (see section 6.2.4).

In section 1.2.2 I broached the subject of one-dimensional harmonic motion of a spring and weight system. In chapter 5 I related it to circular motion. Now it's time for a still deeper view. Vibration arises from the interaction of an elastic force and inertia. We saw, for example, that these determine the speed of sound (see section 7.4).

8.1.1 Elasticity

Elasticity is the property of a material that allows it to restore itself to its original shape after being distorted (stretched, compressed, twisted). An elastic material is in *equilibrium* when all forces applied to it sum to zero. If the sum of applied forces is zero and does not change through time, it is in *static equilibrium.*

Suppose I fix one end of a helical spring to a stationary object such as a ceiling, with its other end dangling freely. As I *displace* the free end away from or toward the fixed end, I *distort* the spring's shape. The internal elastic force that pushes or pulls against my hand, seeking to return it to its original shape is its *restoring force.*

8.1.2 Elasticity, Stiffness, and Hooke's Law

Suppose I apply a force to two elastic materials until they are displaced the same amount. If the force needed to achieve the same displacement of the two materials is different in each case, then one material is stiffer than the other. The *stiffness* of an elastic material is the ratio of applied force to the resulting displacement. We can equate the force F required to achieve a displacement x by inventing a mediating constant of proportionality k, which represents the strength of the counterforce applied by the elastic material:

$$\frac{F}{x} = -k\,.$$

The negative sign of k reminds us that the direction of the counterforce seeks to restore the spring to its undeformed state. Solving for F yields

$$F = -kx. \qquad \textit{Hooke's Law} \quad (8.1)$$

This equation[1] relates stiffness, force, and displacement. Stiffness k is sometimes called the *spring constant.* The reciprocal of stiffness is called *compliance.*

Hooke's law is not a fundamental physical law like Newton's laws of motion; it is simply an observation about a common physical phenomenon related to the properties of elastic materials. In particular, the constant k is not a fundamental constant of nature but a value determined experimentally for each material, based on the molecular structure of the material.

8.1.3 Linear and Nonlinear Elasticity

Hooke's law describes a *linear relation* between force, displacement, and spring constant because, when plotted, the spring constant is a straight line with slope $F/x = -k$ (figure 8.1a).

Simple harmonic motion is the term used to describe the vibration of instruments that are governed by linear elasticity because their partials are (for the most part) in harmonic relation, that is, they are integer multiples of the fundamental. These instruments include violins, woodwinds, brass, and tuned percussion instruments.

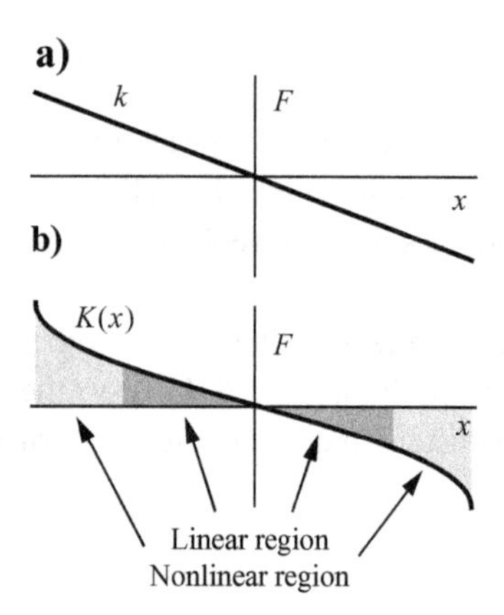

Figure 8.1
Linear and nonlinear elasticity.

A great advantage and a great limitation of Hooke's law is that it does not take into account the *extent* of a material's elasticity. No material is elastic over an infinite range. Although it may respond in a relatively uniform way within a central range, beyond some point it requires much greater force to be deformed further, and eventually many materials bend permanently or break if forced too far. Beyond this central range, we can't speak of a simple spring constant because the force that must be applied to achieve greater displacement does not increase in a straight line: this is a *nonlinear relation*. Beyond this central range, we must construct a curve describing the material's stiffness as a function of displacement: $F/x = -K(x)$ (figure 8.1b). In this case, the amount of restoring force is a nonlinear function of the amount of displacement. All physical materials are to some degree nonlinearly elastic.

The advantage of Hooke's law is that it sheds a great deal of light on the nature of harmonic vibrating systems, which account for a great deal of our acoustical environment. Harmonic systems are also typically easier to understand, mathematically. But it is important to remember that if we study only linear systems, we overlook some of the signature characteristics of musical instruments that result from their nonlinear elasticity, and we won't be able to make sense of highly nonlinear vibrating systems at all. That being said, let's take the easier path and study linear systems first.

8.2 Frequency of Vibrating Systems

Notice that Hooke's law in (8.1) does not include mass but deals only with the elastic properties of objects. Suppose I have a tethered lightweight spring with spring constant k, and I suspend a mass m from its free end. I let the spring stretch to its point of static equilibrium. After it comes to rest, I then displace the spring a distance r by pulling down on it. (I pull it only a small distance so that it remains in its relatively linear elastic range.) Moving it by distance r required me to supply a force $F = kr$ to overcome the spring's stiffness, and the spring now exerts a restoring force of $-kr$.

If I release the mass, it will begin to rise, seeking the spring's point of equilibrium. By Newton's laws of motion, acceleration of the mass is proportional to F/m, so the acceleration will be

$$a = \frac{kr}{m}. \tag{8.2}$$

We now have two equations for acceleration: (8.2) for linear acceleration of a mass on a spring and (5.12) for centripetal acceleration. These can be viewed as equivalent motions (see section 5.1). Therefore we can also equate their accelerations. Doing so, we have

$$\frac{v^2}{r} = \frac{kr}{m}.$$

Note that we have introduced velocity v into the equation. Solving for v yields

$$v = \sqrt{\frac{kr^2}{m}} = r\sqrt{\frac{k}{m}} \tag{8.3}$$

Now, we have two equations for the velocity of vibrating systems: (8.3) and (5.9). Equating them, we have

$$v = \frac{2\pi r}{T} = r\sqrt{\frac{k}{m}}\ .$$

Note that we have introduced periodic time T into the equation. Solving for T yields

$$\frac{2\pi r}{T} = r\sqrt{\frac{k}{m}}$$

$$\frac{2\pi}{T} = \sqrt{\frac{k}{m}}$$

$$\frac{1}{T} = \frac{1}{2\pi}\sqrt{\frac{k}{m}}.$$

Recalling that frequency $f = 1/T$, we can write

$$f = \frac{1}{2\pi}\sqrt{\frac{k}{m}}.$$

Vibrating Frequency (8.4)

Equation (8.4) relates the frequency of a vibrating spring/mass system to its linear spring constant k and its mass m. The equation predicts that the frequency of a vibrating system will double if the spring constant quadruples, and will halve if the mass quadruples.

For a practical example, consider the spring and weight system shown in figures 1.4 and 8.7. To determine its frequency of vibration, we must determine its spring constant, and the amount of mass. We can determine the spring constant by measuring the degree of stretch induced by gravity. Suppose the spring stretches by 0.025 m when loaded with 1 kg. At this point, the elastic force balances the force of gravity, which means $kl = mg$. Solving for k and substituting $m = 1$ and $l = 0.025$, we have

$$k = \frac{mg}{l} = \frac{1 \cdot 9.8}{0.025} = 392 \text{ N/m}.$$

With a mass of 1 kg the vibrational frequency would be

$$f = \frac{1}{2\pi}\sqrt{\frac{k}{m}} = \frac{1}{2 \cdot 3.14}\sqrt{\frac{392}{1}} = 124.4 \text{ Hz}.$$

Doubling the mass to 2 kg drops the frequency to 87.9 Hz.

The method of tuning stringed instruments consists of changing the tension of the strings by stretching them around tuning pegs (rather than adjusting their mass). An increase of tension on the string lowers its elasticity, thereby increasing its vibrating frequency. Since there are practical limits to the elasticity of all materials, it is necessary to trade off mass against elasticity in order to achieve a desired frequency. This is why instrument makers use smaller-diameter strings for higher pitch, because they carry less mass per unit distance.

8.2.1 Radian Frequency and Angular Velocity

We can simplify (8.4) by multiplying both sides by 2π:

$$\omega = 2\pi f = \sqrt{\frac{k}{m}}. \qquad \textit{Angular Frequency} \ (8.5)$$

What does ω signify here? Since ω contains f, it still represents frequency, but by letting ω also include the term 2π, we get a frequency parameter that only involves k and m.

Think of ω as frequency expressed in units of 2π radians. Thus integer values of f measure whole periods of a circular or sinusoidal motion. Parameter ω is called angular velocity or radian frequency, depending on the circumstances, and f is just frequency. For example, if a wheel rotates once per second, it passes through 2π radians each second; therefore its frequency $f = 1$ Hz and its angular velocity $\omega = 2\pi f = 2\pi$ rad/s. If a spring/mass system vibrates in harmonic motion once per second, $f = 1$ and radian frequency $\omega = 2\pi f = 2\pi$. The term *angular velocity* is usually used for circular systems, and the term *radian frequency* is usually used for vibrating systems, but they amount to the same thing.

8.3 Some Simple Vibrating Systems

A simple spring/mass system vibrates in one dimension with one degree of freedom. Below are some other examples of simple vibrating systems. For simplicity, none of these examples takes friction into account.

8.3.1 Pendulum

A simple pendulum (figure 8.2), consisting of a mass m attached to a string of length l, vibrates with *circular harmonic motion* so long as the displacement $x \ll l$.[2] If the mass of the string is much less than m, the frequency of vibration will be

$$f = \frac{1}{2\pi}\sqrt{\frac{g}{l}}, \qquad \textit{Pendulum Frequency} \ (8.6)$$

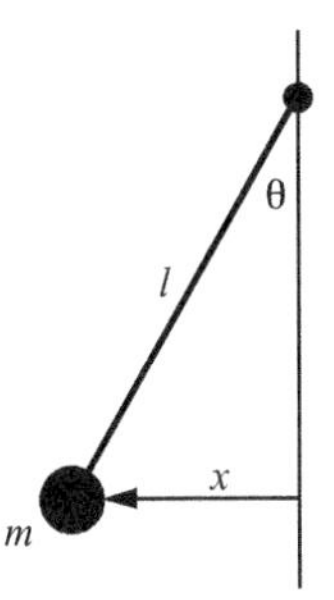

Figure 8.2
Pendulum.

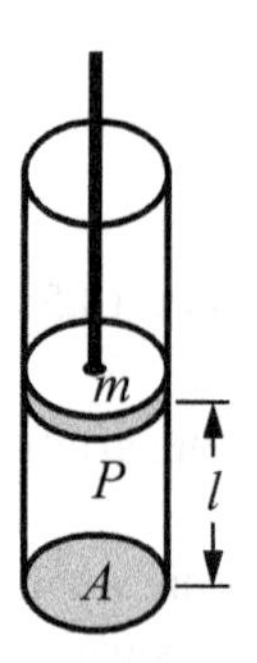

Figure 8.3
Piston.

or, expressed in radian frequency, $\omega = \sqrt{g/l}$. Notice that mass does not appear in this equation. The frequency of a pendulum is strictly a function of length l and gravitational force g.

8.3.2 Piston

Air captured inside a cylindrical tube by a piston of mass m will tend to vibrate at a frequency determined by the mass and the elasticity (otherwise known as compressibility or compliance) of the air (figure 8.3). The compressibility of the air depends upon a number of factors, including the cross-sectional area of the cylinder A, the length of the air column l, the pressure of the gas P, and the heat capacity ratio γ of the gas, which has a value of about 1.4 for air (see section 7.4.2). The spring constant of air is $k = \gamma PA/l$, and the frequency is

$$f = \frac{1}{2\pi}\sqrt{\frac{\gamma PA}{ml}}. \qquad \textit{Piston Frequency} \quad (8.7)$$

Perhaps it is not surprising that frequency should be proportional to inherent molecular elasticity and gas pressure, but the A and l terms may seem a little counterintuitive at first glance. Why does frequency go *up* as the area increases?

To see this, imagine that we replace the air with many very slender springs going from the piston to the bottom of the cylinder. If we increase the length l of the air column, it's as though we add more springs end to end in series (figure 8.4b). Many springs in series are more elastic than one spring by itself. Thus, increasing the length is like adding more springs in series: the compliance goes up, so the frequency goes down.

If we increase the area A of the piston, it's as though we add more springs side by side in parallel (figure 8.4c). Many springs in parallel are stiffer than one spring by itself. Thus, increasing the area is like adding more springs in parallel: the compliance goes down, so the frequency goes up.

8.3.3 Helmholtz Resonator

If air is blown across the mouth of a bottle, the air stream contains many frequencies, but the bottle steals energy (mainly) from just one frequency supplied in the air stream and converts it into simple

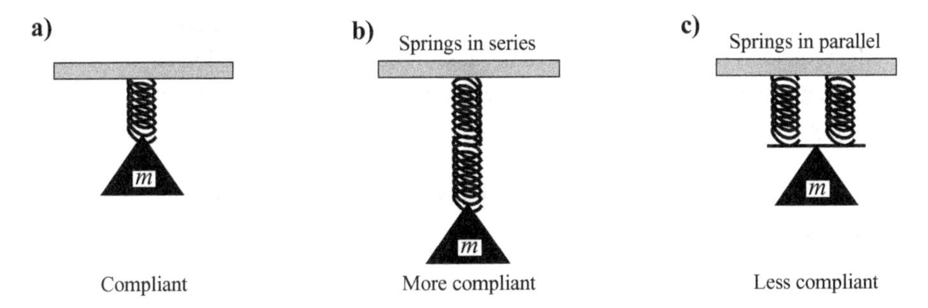

Figure 8.4
Springs in series and parallel.

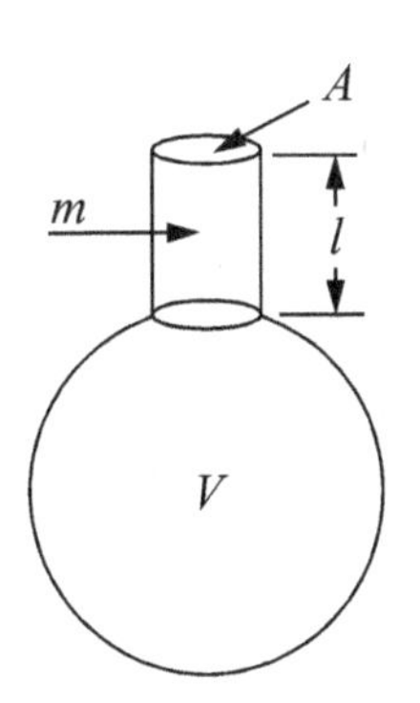

Figure 8.5
Helmholtz resonator.

harmonic motion, which is heard as a breathy tone. Resonance is the tendency of a system to steal energy from, and vibrate sympathetically at, a particular frequency in response to energy supplied at that frequency.

The bottle acts as a Helmholtz resonator,[3] which is a variation on the piston. The air captured in the neck of the bottle constitutes the mass, and the air in the chamber of the bottle constitutes the spring. The frequency of vibration depends upon the compliance of the air in the chamber and the mass of the air in the neck (figure 8.5).

The resonant frequency is approximately

$$f = \frac{c}{2\pi}\sqrt{\frac{A}{lV}}, \qquad \textit{Helmholtz Resonator} \ (8.8)$$

where c is speed of sound, A is the cross-sectional area of the neck, v is bottle volume, and l is the length of the neck. I say "approximately" because the effective length of the neck must be increased a little (an *end correction*) to account for how the air in the tube "recruits" nearby molecules outside the bottle to increase the mass of the air plug. Some end correction must be applied to wind instruments to properly calculate their resonant frequency. Unfortunately, the choice of the end correction scaling term can be rather complicated because the amount of correction required varies depending upon the geometry and proportions of the flange (Benade and Murday 1967; Dalmont, Nederveen, and Joly 2001).

For example, I took a standard Cabernet 750 ml wine bottle with average neck diameter of 19 mm and neck length of 8 cm, drank its contents,[4] then calculated (with somewhat greater difficulty than usual) as follows:

$$c = 331.6 \text{ m/s}.$$

$$A = \pi r^2 = \pi\left(\frac{0.019}{2}\right)^2 \text{ m}^2.$$

$$V = 7.5 \times 10^{-4} \text{ m}^3.$$

Using an end correction of 1.5 times the radius of the neck's opening yielded

$$l = 0.08 + 1.5 \cdot \frac{0.19}{2} \text{ m},$$

for which (8.8) gives a resonant frequency of 105.7 Hz. Experimentally, the resonant frequency was closer to 110 Hz, two octaves below A440, indicating that the end correction of 1.5 was slightly off.

It is perhaps counterintuitive that the frequency of a Helmholtz resonator rises as the area A grows, but the reason is the same as for the piston.

The ducted port loudspeaker enclosure design shown in figure 8.6a is a practical example of a Helmholtz resonator. The port consists of an opening in the side of the loudspeaker enclosure.

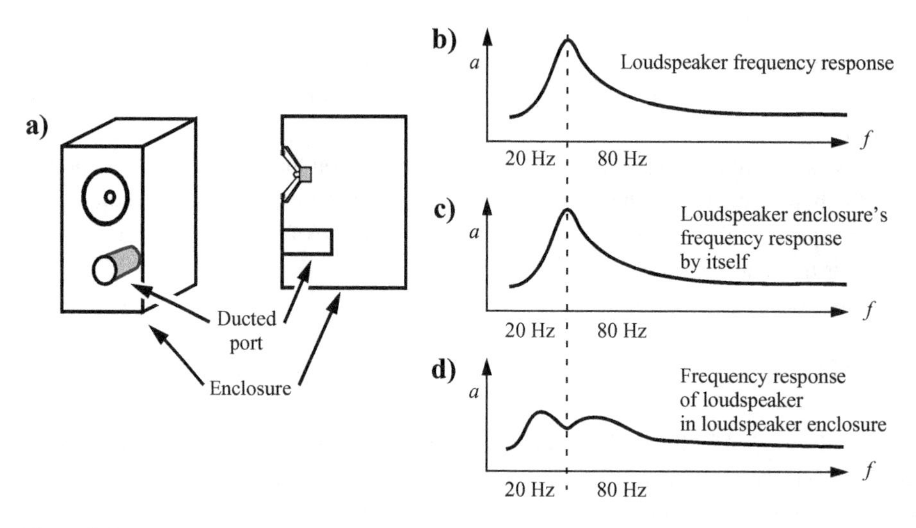

Figure 8.6
Ducted port loudspeaker enclosure.

The duct is a tube inserted into the port, which performs the same function as the tube at the top of a Helmholtz resonator. The loudspeaker enclosure is a cavity that acts like the volume of a Helmholtz resonator.

Ideally, loudspeakers are supposed to be colorless reproducers of other sounds, but since they are themselves essentially a spring/mass system, they have a natural vibrating frequency of their own. Loudspeakers therefore tend to exaggerate the strength of signals that are near their natural vibrating frequency. For high-fidelity speakers, the natural vibrating frequency is often below 100 Hz, resulting in an objectionable "boomy" coloration to bass notes, shown as the peak in the magnitude spectrum plot (figure 8.6b). The purpose of the ducted port enclosure is to compensate for the natural vibrating frequency of the loudspeaker, to even out its response to low frequencies.

The size of the enclosure and the size of the duct are designed so that the air inside the enclosure vibrates at the same frequency as the loudspeaker. When the loudspeaker is sounding at its natural frequency, it causes the air in the enclosure to resonate (figure 8.6c). But, as mentioned, a resonator steals energy at its resonant frequency, thereby bleeding away the excess and providing the loudspeaker system with relatively colorless reproduction at low frequencies (figure 8.6d). Precisely how this stealing of energy takes place is discussed in volume 2, chapter 6.

8.4 The Harmonic Oscillator

The vibrating systems shown in previous sections all arise from the interaction of an elastic force and an inertial force. The elasticity provides a restoring force while the inertia causes the restoring force to overshoot its equilibrium point, thereby extending the vibration. Such systems are called *harmonic oscillators*.

To understand mathematically how vibration arises, let's return to the simplest harmonic oscillator consisting of a mass attached to the end of a lightweight spring, suspended from a crossbar (figure 8.7). We can characterize the vibration by analyzing the forces at work on the harmonic oscillator through time. We combine Hooke's law, which characterizes the spring's restoring force, with Newton's second law of motion, which characterizes the mass's inertial force, and observe how these forces interact to cause a sinusoidal displacement of the mass through time.

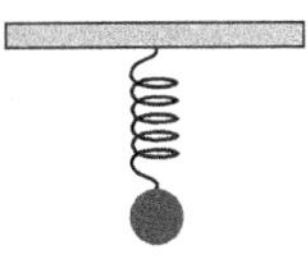

Figure 8.7
Simple spring/mass system.

For a continued discussion of resonance, skip to section 8.9. However, that treatment depends upon the intervening material.

8.4.1 Hooke vs. Newton

When discussing equation (8.1), Hooke's law of linear elasticity, I talked about the *static* force F required to achieve a spring displacement x if the spring has stiffness k. But now we want to examine how the spring force would change if we varied the spring displacement *through time,* so let's consider x as a function of time, $x(t)$. Therefore, we want to study the force

$$F_k = -k \cdot x(t), \tag{8.9}$$

where F_k is the force required to overcome to the spring's stiffness to achieve a displacement x at time t.

By Newton's second law of motion we know that the force required to set a mass in motion is proportional to the mass m times its acceleration a. But here again we want to examine how such a force would change if we varied the mass's acceleration *through time*. So if we consider a as a function of time, $a(t)$, then we want to study the force

$$F_m = m \cdot a(t), \tag{8.10}$$

where F_m is the force required to overcome the mass's inertia m to achieve an acceleration a at time t.

If we apply no external force to a dangling spring/mass system, it will eventually come to rest with the spring displaced downward slightly by the force of gravity on the mass. Where it comes to rest is its point of static equilibrium. A system is in equilibrium when the sum of the forces operating on it is zero. At the static equilibrium point, the force of gravity is exactly opposed by the spring's restoring force. The mass is at rest relative to the spring.

In what follows it will be convenient to eliminate the effects of gravity and friction, which we can do by imagining the spring/mass system vibrating in outer space.[5] Because there is no gravity nearby, we must use a bipolar spring—that is, a spring that provides both a pull when stretched and a push when compressed. Imagine one end of this spring attached to the mass and the other end attached to a very massive object, such as a space station. Let's suppose that the mass is at rest relative to the spring and exerts no force ($F_m = 0$) and the spring exerts no counterforce ($F_k = 0$). Then $F_m = F_k$ because both are zero. This system is in static equilibrium because the sum of the forces equals zero.

But we can show that $F_m = F_k$ even if the mass is vibrating, that is, if the system exhibits *dynamic equilibrium*. A *dynamical system* is one whose state depends upon its previous state (in addition to any other forces acting upon it). For example, suppose I pull down on the weight, stretching the spring an initial displacement x. The restoring force of the spring tugs on the mass with a force proportional to its displacement. In the first infinitesimal moment after I release it, the restoring force attempts to accelerate the mass upward, but the inertia of the mass reacts with a counterforce proportional to its mass. The tendency of a mass to resist change in velocity is its *inertial reactance.* If there were no inertial reactance, the spring would just snap back to its equilibrium point. But instead, during this first infinitesimal moment after release, the elastic force and the inertial reactance still balance, and $F_m = F_k$.

As the inertial reactance gives way, the mass accelerates toward the static equilibrium point. But as it does so, the force applied by the spring diminishes, since there is now less displacement, and the spring tugs on the mass with proportionately less force.

When the mass reaches the static equilibrium point, the restoring force of the spring vanishes. Since this means the restoring force is no longer accelerating the mass, the inertial reactance of the mass also vanishes at this point. Thus here as well, $F_m = F_k$. However, though the mass stops accelerating, its momentum continues to carry it upward, past the static equilibrium point. Now the restoring force begins to oppose the upward movement, causing the mass to decelerate by a proportional amount, and $F_m = F_k$ here as well.

In summary, the restoring force F_k grows with increasing displacement from the equilibrium point. The farther the spring is from equilibrium, the more strenuous is the force it applies to the mass in order to return to equilibrium; but because the mass's inertial reactance opposes it, *the two forces always balance* and are in dynamic equilibrium at all times and in all positions.

8.4.2 Equation of Vibratory Motion

Starting with $F_m = F_k$ and substituting appropriate terms from (8.9) and (8.10) produces $m \cdot a(t) = -k \cdot x(t)$. Expressing this as a dynamic equilibrium yields

$$m \cdot a(t) + k \cdot x(t) = 0, \qquad \textit{Equation of Motion} \quad (8.11)$$

where m is mass, $a(t)$ is acceleration at time t, k is the spring constant, and $x(t)$ is displacement of the mass at time t. Recall that equilibrium means that the sum of applied forces equals zero.

Notwithstanding this intuitive presentation, it would be good to understand how (8.11) can cause an oscillatory vibration because it's not immediately obvious just from looking at it. What we are trying to discover is how displacement x changes as t varies, that is, we want to find an algebraic expression for $x(t)$. Intuition suggests that (8.11) should describe a sinusoidal motion. Actually, to be more precise, it should describe *every possible sinusoidal motion,* because even such a simple spring/weight system is theoretically capable of creating an infinite variety sinusoidal motions with different initial phases, amplitudes, and frequencies. They should all be embodied in (8.11). That all possible sinusoidal motions are indeed embodied in (8.11) is the subject of volume 2, chapter 6.

8.5 Modes of Vibration

The *degrees of freedom* of a vibrating system are determined by how many independent motions the system can make. There are two kinds of motions: translational, which is backward/forward, left/right, and up/down, and rotational, which involves pitch, yaw, and roll.[6]

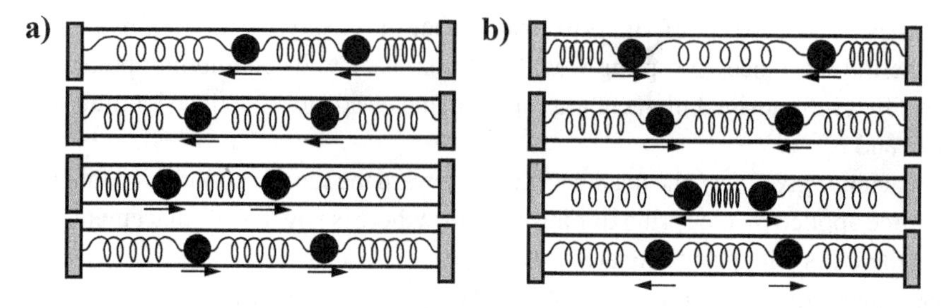

Figure 8.8
System with two degrees of freedom.

A subway has one translational degree of freedom (backward/forward), a car has two (adding left/right), an airplane, three (adding up/down). Adding the rotational motions, an airplane has six degrees of freedom (all three translational and all three rotational motions).

All the vibrating systems described in previous sections have only one degree of freedom. The system shown in figure 8.8, having two weights coupled with springs, has two degrees of freedom. The system in figure 8.8a is just a variation on the simple system in figure 8.7 and exhibits similar harmonic motion. If the total of the mass and spring stiffness of the 8.8a system is the same as that of the 8.7 system, both will vibrate at the same frequency. But even if the mass and spring stiffness of the systems in figures 8.8b and 8.8a are the same, the vibrating frequency of the 8.8b system will be higher because the restoring force from the springs is three times greater. Thus, if the radian frequency for the 8.8a system is $\omega_1 = \sqrt{k/m}$, the radian frequency for the 8.8b system is $\omega_2 = \sqrt{3k/m}$. This method of analysis of vibration was introduced about 1727 by Johann Bernoulli (1667–1748).

These independent vibrational modes are sometimes called *normal modes* or *natural modes.* For each mode, each element of the vibrating system reaches its position of maximum displacement from equilibrium at the same moment. Though the vibrating modes of the 8.8a and 8.8b systems are virtually independent, it is difficult to get a system to vibrate in just one or the other mode without very carefully positioning the balls before releasing them. Ordinarily, the vibration will be a combination of the two modes.

The system shown in figure 8.8 has only two normal modes because it has only two degrees of freedom. A system with N degrees of freedom will have N normal modes. We could add more weights and springs in order to study systems with N degrees of freedom and therefore N modes. Or, we could just increase the number of dimensions of the system from one to two by allowing transverse vibration. The vibrating system in figure 8.9 has four degrees of freedom—the two for the figure 8.8 system plus two more—because now each ball can move in two directions. In general, if the number of masses in a vibrating system is a, and they can each move in b directions, then the number of degrees of freedom $N = ab$.

Each normal mode has its own characteristic frequency made up of some combination of the average contributions of all the masses in the system and the average contributions of the

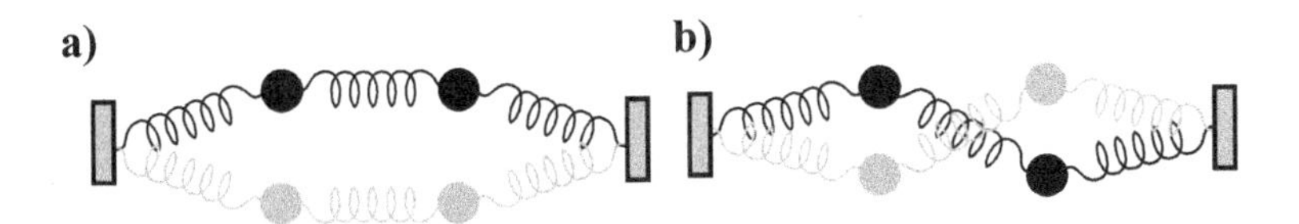

Figure 8.9
Transverse and longitudinal vibration.

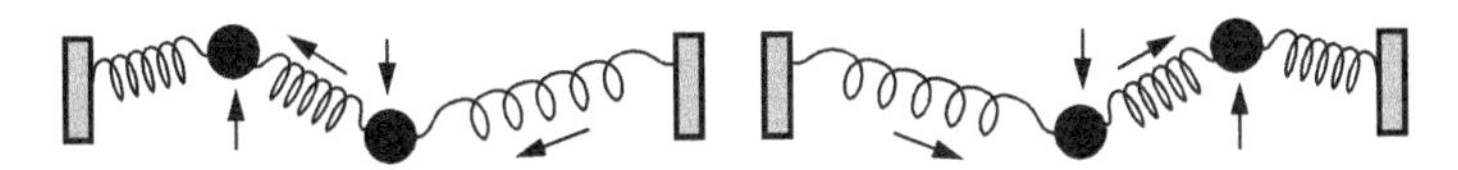

Figure 8.10
Superposition of normal modes.

stiffness in the springs. The resulting motion of the entire system can be characterized as a superposition of all the separate vibrational modes. Figure 8.10 shows the superposition of the normal modes of the systems in figures 8.8a and 8.9b. *Since the modes are virtually independent, and each has its own vibrational frequency, the spectrum of frequencies of the entire system is the linear combination (the sum, or mixture) of each mode.* If sound radiates from such a vibrating system, we hear the sum total of all frequencies of each of the vibrating modes. The strengths of these frequencies is proportional to the amount of energy in each mode, separately.

8.6 A Taxonomy of Vibrating Systems

There are many classification systems of musical instruments, such as the traditional categories of brass, strings, woodwinds, and percussion. Another classification system organizes them as *idiophones* (chimes, cymbals, xylophone, vibraphone, marimba, gongs), *membranophones* (drums), *aerophones* (flutes, oboes, clarinets, trumpets, tubas, whistles, sirens), and *chordophones* (violin, piano, guitar, harpsichord). If we group instruments by the similarity of the fundamental equations governing their vibration, we obtain the simple taxonomy shown in table 8.1.

Tension is the primary restoring force for strings and membranes, and frequency is proportional to tension. Stiffness is the restoring force for bars, air columns, and plates, and frequency is proportional to stiffness.

There are many subgenres for these examples. Bars can be free at both ends or free at only one end. Plates can be clamped at the edge, supported at the edge, supported at the center, or totally free.

Table 8.1
Simple Taxonomy of Musical Instruments

Dimension	Restoring Force	Vibrating Element	Taxonomy
1-D	Tension	Strings	Chordophones
	Stiffness	Bars	1-D idiophones
		Air columns (brass, woodwinds, flutes)	Aerophones
2-D	Tension	Membranes (drums)	Membranophones
	Stiffness	Plates (gongs, cymbals)	2-D idiophones

Bars, plates, and strings can vibrate unhindered or slap against a surface. The saxophone reed is a bar fixed at one end, which slaps against a mouthpiece. Sitar strings slap against a sloping plate attached to the bridge in order to create the characteristic "sizzle" sound. The bottom membrane of a snare drum slaps against an array of coiled wires laid across it to lend it a characteristic "crunch" sound. In all cases, the resulting spectrum contains much more energy in higher partials because of the discontinuity in simple harmonic motion that the slap introduces.

Traditional musical instruments are made from collections of these elements. For example, the essential elements of a saxophone are a bar and an air column; the essential elements of a piano are strings and a plate (the sounding board).

All taxonomies are necessarily reductionist; this one is, too. For example, the strings of a violin actually vibrate in at least four dimensions: up/down, front/back, longitudinal (end to end), and torsional (twisting) vibration. These motions of the strings all affect each other. Also, an important distinction between instruments is whether they are continuously driven (e.g., violins, voice, woodwinds, brass) or impulsively driven (e.g., piano, harpsichord, guitar, percussion). The advantage of this taxonomy is simply that it allows us to group similar instruments together by the basic physical equations that govern their vibration.

8.7 One-Dimensional Vibrating Systems

According to the taxonomy of instruments in table 8.1, the vibration formulas for stringed instruments and bar percussion instruments are closely related. This seems counterintuitive. If they are related mathematically, why do they sound so different? For example, few would mistake the sound of a xylophone for that of a piano, even though the piano's strings are also struck. The piano and other stringed instruments made from long thin wires have largely harmonic spectra, whereas percussion instruments generally have inharmonic spectra.

If the formulas for their vibration are to mean anything, they must account for the vast difference in timbre. The aim of this section is to demonstrate the underlying symmetry of one-dimensional vibrating systems.

8.7.1 Strings

Stringed instruments can be classified by

- How they are sounded:
  Bowed: the violin family
  Plucked: guitar, mandolin, harp, harpsichord
  Struck: piano, hammered dulcimer
- How they select pitch:
  Unstopped strings: harp, piano, harpsichord
  Stopped fretted: guitar, mandolin, lute
  Stopped unfretted: violin family
- Whether their sound can be continuously produced:
  Continuously driven: all bowed strings
  Impulsively driven: all plucked and struck strings

The piano and harpsichord provide an array of strings tuned to consecutive scale degrees, and music is played by selecting the appropriate string. The guitar, lute, violin, and mandolin have a smaller array of strings tuned to nonconsecutive scale degrees, and they provide a fingerboard underneath the strings so the player can sound the pitches in between adjacent strings by stopping off different lengths.

The fingerboard on the *violin family* (violin, viola, cello, and bass viol) is a smooth surface so that any pitch in the continuous pitch space covered by the string may be selected. Guitar, lute, banjo, and mandolin have *frets*—transverse bars across the fingerboard under the strings—so that when stopped by the finger, the length of the stopped string is determined by the fret. Frets provide an improved ability to stop multiple strings with correct intonation.

Continuous pitches may be produced by sliding the finger along the string of a violin, an effect called *glissando.* Sliding the finger along the string of a fretted instrument produces a series of discrete pitches, an effect called *portamento.*

Strings are stretched between rigid supports with a means of adjusting their tension. In virtually every stringed instrument, energy is injected into the string transversely—perpendicular to the string—and transverse motion carries the majority of the energy. Because strings are typically of very low mass and do not displace much air, they are almost always coupled to the air through a sounding board, such as the wooden back of a piano or the body of a violin, mandolin, or guitar. The sounding board allows the energy in the string to be transmitted efficiently into the surrounding air by matching the impedance of the string to the air. Without the sounding board, we would hear very little from a stringed instrument. For example, a strummed unamplified electric guitar is virtually inaudible a few feet away, whereas the sound from an acoustic guitar can fill an auditorium.[7] The difference is that the acoustic guitar matches string impedance to air, and the electric guitar does not (see volume 2, chapter 8).

Bowed instruments produce a continuous tone by replacing energy in the string as it is dissipated. A skilled player can sustain continuity by instantaneously reversing the direction of the bow when the end is reached. Players of impulsively driven instruments, such as the mandolin, can create the illusion of a sustained tone by rapidly replucking the string, an effect called *tremolo.* Bowed instruments can also be plucked, an effect called *pizzicato*.

Unless they are being played with tremolo, all impulsively driven stringed instruments decay gradually to silence from note onset. The rate at which they decay to silence varies enormously. The efficiency with which an instrument radiates energy determines its rate of decay (see section 4.19.2). The banjo is perhaps the most efficient stringed instrument, radiating away all its energy in a few seconds. At the other extreme, the bottom notes of a piano can sustain for several minutes (with the damper pedal down).

In the following subsections I present the vibration of *ideal strings* that are perfectly flexible, have constant mass per unit length, and are connected to massive, nonyielding supports. At first, I will ignore the effects of dissipation on the vibration of strings. However, all stringed instruments depend upon dissipation to carry sound energy into their surroundings where it can be heard. Tension, not stiffness, is the restoring force of the ideal string. Of course, all real strings have some stiffness, and stiffness is the hidden link that relates stringed instruments to bar percussion instruments.

String Modes Strings can be usefully studied as many tiny spring/mass systems concatenated together, similar to those shown in figure 8.9. Since the number of possible vibrating modes is large, for simplicity we consider just the first five modes available when the number of degrees of freedom $N = 5$ (figure 8.11a). Figure 8.11b shows the corresponding first five modes of the infinite number of degrees of freedom of an ideal string ($N = \infty$).

For each mode, the points where the string crosses the equilibrium are called *zero-crossings, points of inflection,* or *nodes* of that mode. Since the strings are fixed at the ends, the ends are nodes as well. Nodes are pivot points around which the string vibrates. For each mode, the points where the string is farthest from equilibrium are called *maxima, points of maximum excursion,* or *antinodes* of that mode.

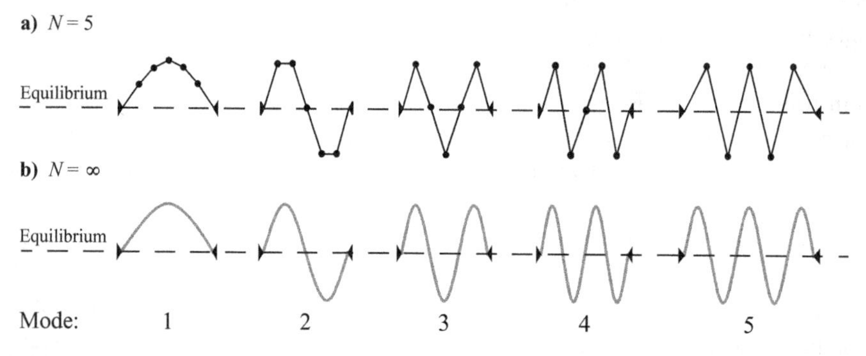

Figure 8.11
Modes of transverse vibration.

Standing Waves and Traveling Waves In an ordinary medium such as air, sound propagates as traveling waves. But the rigid boundaries at the edges of a string cause most energy to be reflected back into the string, and prevent it from radiating away (see section 7.8.4).

The shapes of the modes shown in figure 8.11b are called *standing waves* because the shape of the string remains the same at all moments and only its amplitude changes. (More precisely, the height of the wave is scaled through time in the direction perpendicular to its length.)

The behavior of a standing wave can best be described as the sum of two waves traveling in opposite directions. Imagine two waves, y_1 and y_2, moving through each other from opposite directions along a string. Their combined displacement, $y = y_1 + y_2$, creates a standing wave.

To demonstrate this requires some trigonometry. Let the traveling wave moving to the right be represented as the sinusoid $y_1(x, t) = A\sin(kx + \omega t)$ and the one traveling to the left as $y_2(x, t) = A\sin(kx - \omega t)$, where t is time, ω is radian frequency, x is displacement of the wave from its origin along the direction of travel, k is the rate at which the displacement grows, and A is amplitude.

To see how this represents a traveling wave, we reason as follows. If we set k equal to zero, then $A\sin(kx + \omega t)$ reduces to $A\sin \omega t$, which plots an ordinary sine wave with a zero-crossing at the origin. But if k is nonzero, then as x grows (because the wave is traveling), the zero-crossing at the origin moves away from the origin with velocity k.

Now let's return to the standing wave on a string. To see how two oppositely moving traveling waves combine into one standing wave, consider the following trigonometric identities (see volume 2, appendix):

$$\sin(a + b) = \sin(a)\cos(b) + \cos(a)\sin(b)$$

$$\sin(a - b) = \sin(a)\cos(b) - \cos(a)\sin(b).$$

Suppose we let $a = kx$ and $b = \omega t$; then we can represent the two sinusoids as follows:

$$y_1 = \sin(a)\cos(b) + \cos(a)\sin(b)$$

$$y_2 = \sin(a)\cos(b) - \cos(a)\sin(\mathrm{b}).$$

Summing the two sinusoids, we have

$$\begin{aligned} y = y_1 + y_2 &= 2A\sin(a)\cos(b) \\ &= 2A\sin(kx)\cos(\omega t). \end{aligned} \qquad (8.12)$$

Equation (8.12) shows the product of two sinusoids. Its plot is a standing wave that is the point-by-point sum of the two signals, y_1 and y_2, as they pass through each other. Figure 8.12 shows string mode 4 at several phases and the location of the nodes and antinodes of the string.

Mode Wavelengths Consider the wavelength of the first mode, the fundamental (figure 8.11b). Since this mode outlines half of a sine wave, if the string length is L, then one full period of its

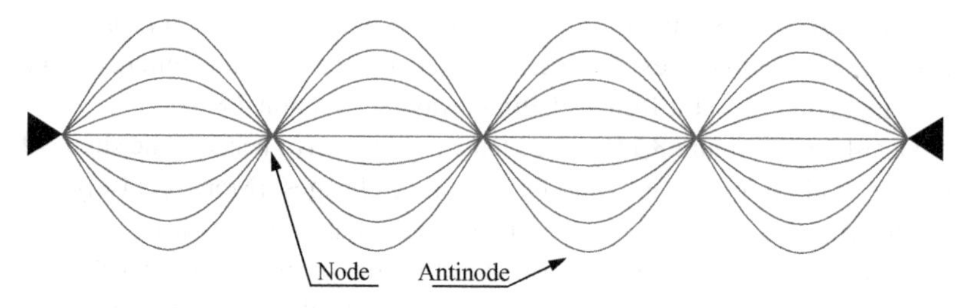

Figure 8.12
String mode 4 as a standing wave.

wavelength $\lambda_1 = 2L$. One full period of mode 2 fits exactly in length L, so we can write $\lambda_2 = L$. And, in general, we can write

$$\lambda_n = \frac{2L}{n}, \qquad \textit{Mode Length} \quad (8.13)$$

where $n = 1, 2, 3, \ldots$.

Mode Frequencies For an ideal string, the velocity of a transverse wave is the same for all modes because the stiffness doesn't increase with the mode number. (This is not true for real strings.) If the velocity of transverse waves on an ideal string is v_t, then we can express the relation between frequency f, wavelength λ, and velocity v_t as $\lambda = v_t/f$, or $f = v_t/\lambda$. Using the definition for λ_n from (8.13), we can express the frequency of mode n as

$$f_n = \frac{nv_t}{2L}. \qquad (8.14)$$

Because the ideal string has no stiffness, v_t depends only on the string's mass per unit length m and its tension T, so that

$$v_t = \sqrt{\frac{T}{m}}.$$

In a string, tension T takes the role of elasticity in a harmonic oscillator. Putting it all together, we can express the frequency of string mode n as

$$f_n = \frac{n}{2L}\sqrt{\frac{T}{m}}, \qquad \textit{String Mode Frequency} \quad (8.15)$$

where $n = 1, 2, 3, \ldots$.[8]

8.7.2 Longitudinal Bars

In the preceding section, we considered the case of the ideal string, which contains tension but no stiffness. The bar vibrating longitudinally, by stretching and shrinking its length, is the other limiting case because it is under no tension; its restoring force is entirely due to its stiffness. Its vibrating frequency equation is very similar to that of the string. The frequency f of mode n is

$$f_n = \frac{n}{2L}\sqrt{\frac{Y}{\rho}}, \qquad \textit{Longitudinal Bar} \quad (8.16)$$

where Y is Young's modulus of elasticity, ρ is the mass density of the material, L is the length of the bar, and $n = 1, 2, 3, \ldots$.

According to (8.16) the modes of the longitudinal bar are in a harmonic frequency series, like strings. The longitudinal vibration modes of a bar are usually very much higher in frequency than the corresponding transverse vibration modes of the same bar. Historically, longitudinally vibrating bars have been used as tuning forks for frequencies above 5000 Hz, where the traditional tuning fork design is no longer satisfactory. Some modern instruments use this vibration mode, for instance, by stroking a steel rod with a rosin-coated cloth to excite longitudinal vibration modes.

Figure 8.13 shows the direction and magnitude of movement of the first three modes of a longitudinal bar.

Young's Modulus The forces needed to stretch a solid object depend upon the following factors (figure 8.14):

- *Amount of stretch* For two identical rods (figure 8.14a), proportionately more force is required to stretch one rod further than the other.

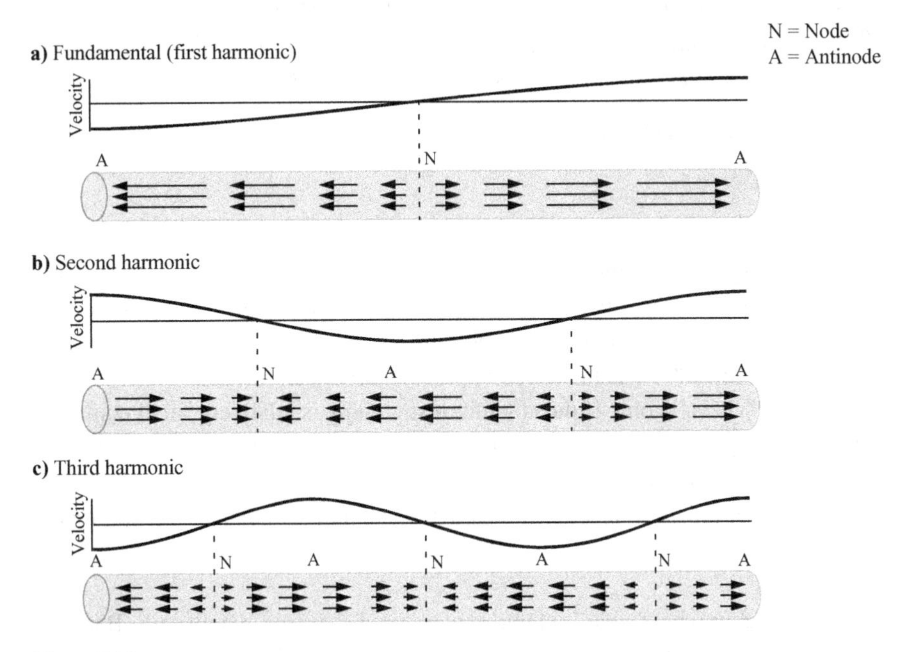

Figure 8.13
Modes of a longitudinal bar. (Adapted from Olson 1952.)

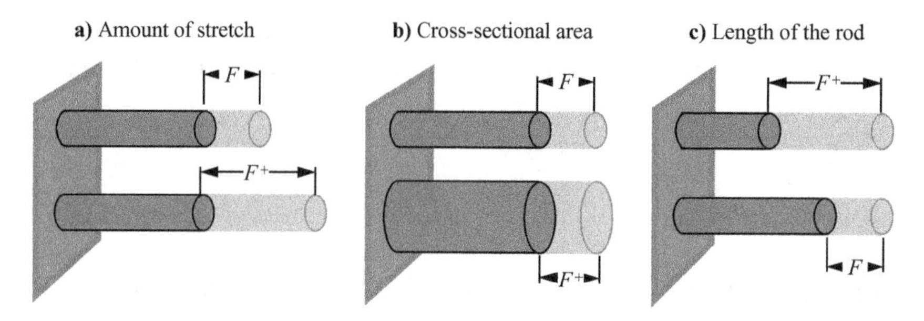

Figure 8.14
Stretching solid rods.

Table 8.2
Young's Modulus for Selected Materials

Polyethylene	0.2–0.7	$\times 10^{10}$	Brass	103–124	$\times 10^{10}$
Wood	0.6–1.0	$\times 10^{10}$	Titanium	110	$\times 10^{10}$
Nylon	2.0–4.0	$\times 10^{10}$	Cast iron	83–190	$\times 10^{10}$
Aluminum	69–79	$\times 10^{10}$	Steel	190–210	$\times 10^{10}$

- *Cross-sectional area* For rods of identical material and length but different cross-section, the amount of force required to stretch the thicker rod will be proportionately greater (figure 8.14b).
- *Length of rod* For rods of identical material and cross-section but different length, the amount of force required to stretch the shorter rod will be proportionately greater (figure 8.14c).

These observations can be combined as follows:

$$F = AY\left(\frac{\Delta L}{L_0}\right), \tag{8.17}$$

where L_0 is the original length of the object, ΔL is the increase in length, A is the cross-sectional area, and Y is a constant of proportionality called *Young's modulus*.[9] Young's modulus is the *ratio of stress to strain* of a material. Its value depends upon the nature of the material. Solving for Y in terms of the units involved shows it is measured in pascals (force per unit area, N/m^2).

Note that equation (8.17) is valid only if the amount of stretching is relatively small compared to the original length of the object because it only applies to linear elasticity (see section 8.1.3).

Table 8.2 is a short list of Young's modulus for various materials. Young's modulus varies a great deal from one sample to the next, depending on the purity of the sample and its manufacturing process.

8.7.3 Transverse Bars

Transverse vibration can occur where a bar is clamped at one end or is free at both ends. Bars free at both ends are used in instruments such as the xylophone, marimba, vibraphone, glockenspiel,

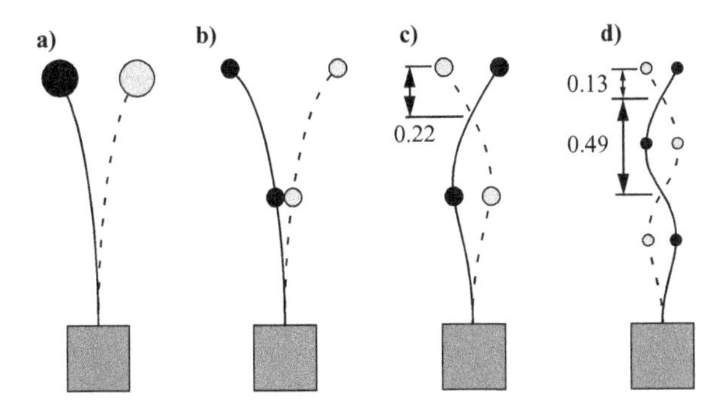

Figure 8.15
Modes of a tuning fork.

and celeste. Bars fixed at one end, also called *cantilever beams,* are the key vibrating elements in the harmonium, accordion, jaw harp, and some organ reed stops.

The vibrating frequency of a longitudinal bar depends upon its length, density, and elasticity, but in transverse vibration frequency also depends on the thickness and cross-sectional shape of the bar because this has a direct effect on the transverse flexibility of the bar. Additionally, transverse bars can twist, creating torsional modes.

Cantilever Beam To study vibration of transverse bars, suppose we take a springy steel wire with relatively little mass, and stick its base into a rigid surface, then attach a mass to the free end (figure 8.15a). When pulled to the side and released, a coherent vibrating movement occurs over the entire length of the spring; hence this is mode 1 vibration, which produces the fundamental frequency. Given a stiffness k and mass m, equation (8.4) determines the vibrating frequency.

Now attach half of the mass at the end and half in the middle of the spring (so that, overall, the mass is the same). Some energy will vibrate mode 1 (figure 8.15b), but some will vibrate mode 2 (figure 8.15c). Because mode 2 flexes the spring much more than mode 1, the spring constant k for mode 2 is higher, making the vibrating frequency of mode 2 a noninteger multiple of the frequency of mode 1, producing a nonharmonic partial. Mode 2 vibration can be about six times higher in frequency than mode 1, corresponding to an increase in the spring constant by a factor of about 18 (since mass is the same overall).

Now we distribute the mass in thirds (figure 8.15d), allowing us to energize mode 3. The amount of flexing that the spring undergoes for mode 3 vibration is even greater, so the spring constant k for mode 3 is even larger. Mode 3's frequency is approximately 18 times higher than mode 1's, corresponding to an increase in k by a factor of about 186.

For every additional mass added to the wire, we more closely approximate an actual bar. Olson (1952) gives the equation for the fundamental frequency of a cantilever beam as

$$f_1 = \frac{0.5596}{L^2}\sqrt{\frac{YK^2}{\rho}}, \qquad \textit{Cantilever Beam} \quad (8.18)$$

Table 8.3
Modes and Frequencies of Fixed/Free Bar

Partial	No. of Nodes	Node Distance from Free End	Partials	Example Frequency
1	0		f_1	33.83
2	1	0.2261	$6.267f_1$	212.00
3	2	0.1321, 0.4999	$17.55f_1$	593.69
4	3	0.0944, 0.3558, 0.6439	$34.39f_1$	1163.36

where L is the length of the bar in meters, ρ is its mass density in g/cm³, Y is Young's modulus, and K is the radius of gyration.

For a bar of rectangular cross-section, Olson gives the *radius of gyration* as $K = a/\sqrt{12}$, where a is the thickness of the bar in the direction of vibration. (Width doesn't matter because we are not considering vibration across the thick side of the rectangle.) For circular cross-section, Olson gives $K = a/2$, where a is the radius of the bar. If the cross-section is hollow, Olson gives

$$K = \frac{\sqrt{a_o^2 + a_i^2}}{2},$$

where a_o is the outside radius of the pipe and a_i is the inside radius. He then gives partial frequencies (table 8.3).

For example, suppose we have a bar 0.5 m long, rectangular in cross-section, made from aluminum that is 10 mm thick. Young's modulus $Y \approx 74 \times 10^9$ Pa for aluminum,[10] the mass density $\rho \approx 2.7 \times 10^3$ kg/m³, thickness $a = 0.01$ m, length $L = 0.5$ m, and because the bar is rectangular, $K = 0.01/\sqrt{12}$. Plugging these values into (8.18) yields a fundamental and partials shown in the last column of table 8.3.

Bar with Free Ends Rossing (1983) supplies the following function for a bar with free ends:

$$f_n = m^2 \frac{\pi K}{8L^2}\sqrt{\frac{Y}{\rho}}, \qquad \textit{Bar with Free Ends} \quad (8.19)$$

where K, L, Y, and ρ are the same as defined in (8.18). The parameter m needs a bit of explaining. Rossing writes, "The frequencies of the modes are in proportion to the squares of the odd integers—almost. The number m begins with 3.0112 and then continues with the simple values 5, 7, 9, . . . , $(2n + 1)$."

We can describe the values for m as follows:

$$m = \begin{cases} 3.0112, n = 1 \\ 2n + 1, n > 1 \end{cases} \quad \text{for} \quad n = \{1, 2, 3, \ldots\}.$$

For example, using the aluminum bar described in the subsection on cantilever beam and plugging those values into (8.19) for $n = 1, 2, 3, 4, 5$ yields the frequencies shown in table 8.4.

Table 8.4
Bar Free at Both Ends

Partial	Frequency	Ratio
1	215.25	1.00
2	593.48	2.75
3	1163.21	5.40
4	1922.86	8.93
5	2872.42	13.34

The ratios of the partials in tables 8.3 and 8.4 are strongly inharmonic. Nonetheless, bars such as these are used for pitched instruments like glockenspiel, chimes, and orchestral bells. This is possible because the higher partials die out quickly (during the initial *clang tone*), leaving the first partial by itself as a relatively pure tone. Also, the inharmonic higher partials of some of these instruments are well beyond the range of human hearing.

Making Transverse Bars Have More Harmonic Spectra The marimba, xylophone, and vibraphone are made from bars free at both ends, suspended over resonating tubes. The bars are thinned in the middle so as to bring the first two partials into a harmonic relation. Here's how it works.

Thinning the middle of a bar has the effect of reducing the stiffness of just its mode 1 vibration. (It also slightly decreases the mass of the bar, which slightly raises its pitch, but the decrease in stiffness is the more important effect.) The result is that the frequency of the first mode is *lowered* relative to the others, which are largely unchanged. Marimba bars are thinned enough so that the relation f_2/f_1 goes from about 2.75/1 to 4/1. Xylophone bars are thinned less, so the ratio $f_2/f_1 = 3/1$. Thus, f_2 is an octave and a fifth above f_1. The 3/1 ratio accounts for the prominence of the sound of a musical fifth in the xylophone's timbre.

Each bar of the marimba, xylophone, and vibraphone is also equipped with a resonating tube placed below it to amplify and draw out the fundamental pitch (at the expense of shortening the bar's vibration time because resonance steals energy from the bar at this frequency). The vibraphone also has an electric motor that rotates paddles within each tube. They look just like rotating dampers in a stove pipe. The paddles cut off the energy supplying the resonator tubes, giving a tremolo effect (periodic amplitude fluctuation plus a small periodic fluctuation in pitch) as they rotate. The speed of rotation can be varied by a control on the motor, and the motor can also be switched off. An interesting additional consequence of the flue arrangement on the vibraphone resonators is that tones last longer, on average, when the paddles rotate than when they are open: less energy is radiated from the bars when the resonators are blocked. Therefore the energy lingers longer on average in the bars when the paddles rotate.

8.7.4 Stiffness of Strings and Inharmonicity

We saw in the discussion of transverse bars that stiffness increases in higher modes, stretching the upper partials of these instruments. The same is true for strings, especially thick strings that increasingly resemble transverse bars the thicker they become. Let's return to the discussion of strings and consider the effects of stiffness on nonideal strings.

While the ideal string vibrates in a series of modes that are perfectly harmonic, actual strings have some internal stiffness, so they are not perfectly elastic. Thus there are actually two restoring forces in a string: tension and stiffness, and the vibrating frequency of each mode in a string is determined by both. According to equation (8.15), tension affects all modes equally. However, stiffness provides proportionally greater restoring force for the higher modes because the higher the mode, the more the string is bent. Therefore higher-numbered modes undergo progressively greater amounts of stiffness. And since frequency of a vibrating string is proportional to stiffness (and tension), an increase in stiffness causes an increase in frequency. Thus the frequencies of the modes of a stiff string spread out in frequency and are no longer exact multiples of the fundamental. The stiffer a string, the less it acts like a string and the more it acts like a bar, according to the taxonomy in table 8.1.

Case Study: The Piano The range of frequencies a piano must reproduce is from about 27 Hz to 4000 Hz, a ratio of more than 1:100. If we used strings of the same tension and mass, and if the highest-pitched string were only 4 in. long, the lowest strings would have to be well over 33 ft long. Clearly, real pianos aren't that enormous. Why? Equation (8.15) suggests that the only parameters affecting the frequency of a vibrating string are length, tension, and mass per unit length. If we want to shorten the bass strings, then we must either decrease their tension or increase their mass per unit length, or some combination of both; or play some other tricks in combination with these.

We could shorten the bass strings if we lowered their tension. But piano strings sound best when they are close to their maximum tension so that they produce a bright and long-lasting tone. So we can make only minor adjustments in tension.

We could shorten the bass strings if we made them thicker. But then they would become more like transverse bars: their overtones would become stretched and they no longer would have strictly harmonic spectra.

Actually, the problem is not so much that bass strings would have inharmonic spectra. By itself, a string with mildly stretched overtones sounds pretty good. In fact, studies have shown that musicians and nonmusicians seem to prefer strings with slightly stretched overtones. The real problem is that the *stretched overtones of bass strings do not line up with the fundamentals of strings tuned to the higher octaves of these bass strings.* Ideally, we'd like the overtones of the bass strings to line up exactly with the fundamentals of the higher-pitched strings, but they don't because of their stiffness.

Piano makers have employed a variety of strategies to work around this problem. For instance, since thinner strings have less stiffness, they use multiple thinner strings struck simultaneously instead of one thick string. They also wrap wire around strings to increase their mass. Since the string inside the wrapping is relatively thin, overtones are not stretched as much in these strings as would be the case with a solid string of the same thickness. But for compact pianos such as spinetts, where the bass strings must be very short, overtone stretching is a serious challenge to tuning the instrument. In fact, harmonic stretching is a problem even for grand pianos with the longest, thinnest strings. This is just a fact of life for piano tuners.

The work-around employed by piano tuners is to tune the higher-pitched strings progressively sharper so that the harmonics of the lower strings more or less line up with the fundamentals of the higher strings. Spinets, which by design must have the shortest, thickest bass strings, require the greatest progressive sharpening to blend away the significant overtone stretching of the bass notes, whereas concert grand pianos require the least because they can have longer strings.

8.7.5 Air Columns

An air column by itself can never be anything more than a Helmholtz resonator, vibrating in sympathy to a sound caused by another source, so it must be coupled to a sound-producing source, which can be anything that vibrates (table 8.5).

Modes of Vibration Vibration of an air column occurs because of longitudinal displacement of air particles. There are two forms of air columns: those open at both ends, and those open at one end only. Additionally, the profile of the pipe may be cylindrical or conical.

Recall that a node is a point where displacement due to vibration is zero, and an antinode is a point where displacement due to vibration is greatest. At the open end of a pipe, there is a *displacement antinode* because the air inside is free to move in and out of the tube. At the closed end of a pipe, there is a *displacement node* because the air can't move longitudinally (the closed end prevents it). The vibration modes of air columns can be found quickly using the same approach we took for strings.

Pipe Open at Both Ends Clearly, air is free to vibrate in and out of the ends of a pipe open at both ends. That means a pipe open at both ends can only support modes that have displacement antinodes at both ends. The first four displacement modes are shown in figure 8.16. The figure indicates how much particle displacement is possible at each position along the length of the pipe. The actual particle motion in an air column is the same as shown for longitudinal bar vibration in figure 8.13.

Table 8.5
Air Column Instruments

Bar	Xylophone, marimba, and vibraphone; some pipe organ ranks; many automobile horns; some enclosed-reed mouth-blown instruments such as the crumhorn
Lips and bar	Woodwinds; jaw harp
Loudspeaker	Ducted-port loudspeaker enclosure
Lips and mouthpiece	Brass instruments
Fipple	Recorder, pennywhistle, most pipe organ ranks
Lips and fipple	Flutes and fifes

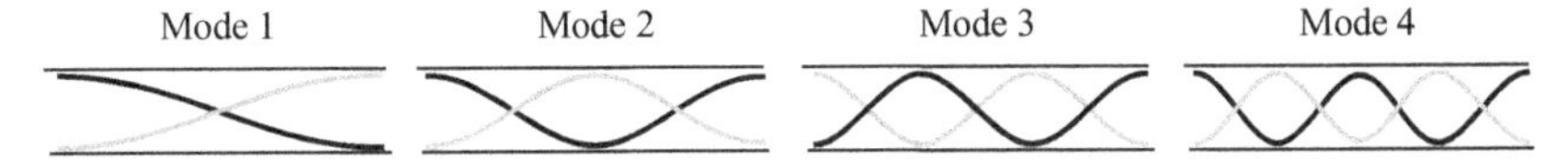

Figure 8.16
Displacement modes of open-ended pipe.

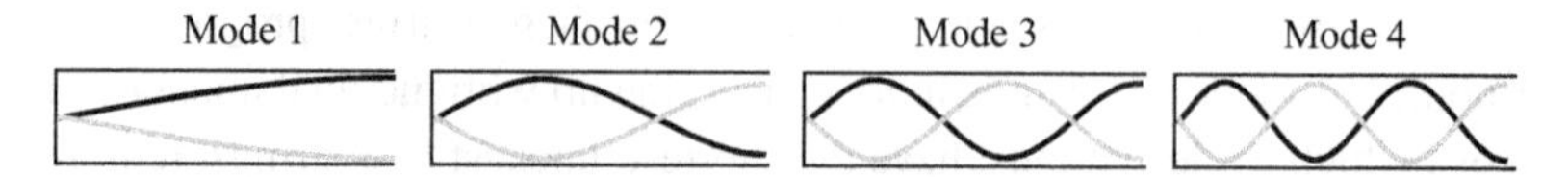

Figure 8.17
Displacement modes of closed-ended pipe.

The wavelength of mode n is $\lambda_n = 2L/n$, where L is the length of the tube. Therefore the frequency of mode n of a pipe open at both ends is

$$f_n = \frac{v_t}{\lambda_n} = \frac{n2v_t}{L}, \qquad n = 1, 2, 3, \ldots. \qquad \textit{Frequency Modes of a Pipe Open at Both Ends} \quad (8.20)$$

Equation (8.20) is a slight simplification because the effective length of the tube is actually a little longer than its physical length. The air in the column recruits air near the end of the tube into its vibration pattern, and an end correction scaling must be applied to obtain a reasonable estimate of the effective length. The end correction depends upon the geometry of the opening (see section 8.3.3).

Pipe Closed at One End A pipe closed at one end can only support modes that have displacement antinodes at the open end and displacement nodes at the closed end. The first four are shown in figure 8.17.

- Mode 1 is one quarter of a sine wave, so $\lambda_1 = 4L$.
- Mode 2 is three quarters of a sine wave, so $\lambda_2 = 4L/3$.
- Mode 3 is five quarters of a sine wave, so $\lambda_3 = 4L/5$.
- Mode 4 is seven quarters of a sine wave, so $\lambda_4 = 4L/7$.

Extracting the pattern, we see that the wavelength

$$\lambda_n = \frac{4L}{n}, \qquad n \text{ odd}.$$

Thus, the closed-ended pipe only exhibits odd harmonics, and it sounds an octave below an open-ended pipe of the same length. The equation for the mode frequencies of the closed-ended pipe is

$$f_n = \frac{nv_t}{4L}, \qquad n \text{ odd}. \qquad \textit{Frequency Modes of a Pipe Closed at One End} \quad (8.21)$$

This is also a slight simplification because of the need for an end correction.

From this we can explain why a clarinet sounds an octave lower than a flute in spite of being approximately the same length: the flute functions as a pipe that is open at both ends, whereas the clarinet is closed at one end.[11] The same fact explains why the spectrum of a flute includes all harmonics, whereas that of the clarinet contains only odd harmonics. Differences in their harmonic spectra also account for what happens when they are overblown. The flute *overblows at the octave:* it sounds eight diatonic steps above standard fingering. The clarinet *overblows at the twelfth:* it sounds twelve diatonic steps (an octave plus a fifth) above standard fingering.

Tube with Conical Bore The bores of flute and clarinet are both approximately cylindrical and are approximately the same length. The oboe, bassoon, and saxophone have approximately conical bores and are all closed at one end.

Since the oboe is about as long as a flute but closed at one end, we might naively predict that the oboe should, like the clarinet, be able to play an octave below the flute. But, in fact, the oboe and flute have about the same bottom pitch. Why?

The simple answer has to do with the physics of the conical bore of the oboe. In cylindrical tubes sound propagates as a virtually plane wave. (The smaller the bore, the more it is like a plane wave, but as the diameter gets large in comparison to the length, the waves start to become more spherical.) If we ignore the very small effect of air absorption of the sound along the tube, the amplitude of the signal is relatively constant along its length.

But because sound spreads out as it moves toward the open end of a cone, we must take into account the effects of attenuation of the signal is it travels toward the open end. Recall from equation (4.36) that intensity I falls off with the square of the distance r, so $I = 1/r^2$. Recall also that amplitude A is proportional to the square root of intensity, so $A = \sqrt{I}$. Therefore amplitude diminishes as $1/r$ along the inside of a cone.

Conical tubes, like cylindrical ones closed at one end, must have a displacement node at the closed end and a displacement antinode at the open end. But the wavelengths that fit must take into account the $1/r$ amplitude scaling (figure 8.18). For a tube with conical bore of length L, the wavelengths that fit are the sinusoids

$$\frac{1}{r}\sin\frac{\pi r n}{L}$$

for $n = 1, 2, 3, \ldots$ because they all have a node at $r = 0$ and an antinode at $r = L$. (To understand the node, think carefully about the value of this expression as r goes to zero; to understand the antinode, think about its value as r goes to L.) Because all n sinusoids fit, the spectrum contains all harmonics. Because the wavelength of the conical bore's fundamental is $2L$, its fundamental pitch is the same as a cylindrical bore of the same length. The silver flute and oboe are approximately the same length; the bottom note of the silver flute is C4, and the oboe's bottom note is a half-step lower, B♭3.

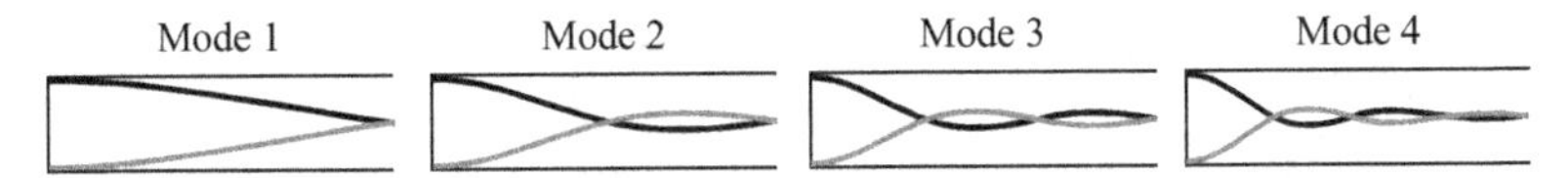

Figure 8.18
Harmonic pressure waves of a conical bore.

8.8 Two-Dimensional Vibrating Elements

There are many musically interesting vibrating surfaces, including stretched membranes and plates.

Stretched membranes such as drums are the two-dimensional equivalent of the stretched string, where the restoring force depends upon tension. Like the ideal string, the ideal membrane is infinitely flexible, infinitely thin in cross-section, and uniformly stretched by a force sufficiently massive not to be affected by the motion of the membrane. The overtone series of stretched strings is harmonic, but stretched membranes have inharmonic spectra. Besides many percussion instruments, instruments with stretched membranes include the resonator for banjos and the Hindustani sarod and esraj.

Plates are the two-dimensional equivalent of the transverse bar, where the restoring force depends upon inner stiffness of the plate material. Whereas stretched membranes must always be fastened at the rim, plates can be clamped at the edge, supported at the edge, supported at the center, or completely free to vibrate. A piano sounding board can be thought of as a plate supported at the edge. A cymbal is a plate supported at the center. Although analytic solutions for arbitrary two-dimensional geometries can certainly be derived, this section focuses on circular shapes.

Both concentric and radial vibration modes are possible for circular vibrating elements. Circular modal geometries are traditionally classified by two numbers, the first indicating the *number of radial nodes,* and the second indicating the *number of concentric nodes* (always including the node at the rim). General membrane vibration modes are shown in order of increasing modal frequency in figure 8.19. This classification applies to circular stretched membranes and also to circular plates clamped or supported at the edge because these systems always have a node at the rim (they are clamped). However, it does *not* apply to circular plates supported at the center or free because these systems have an antinode at the rim.

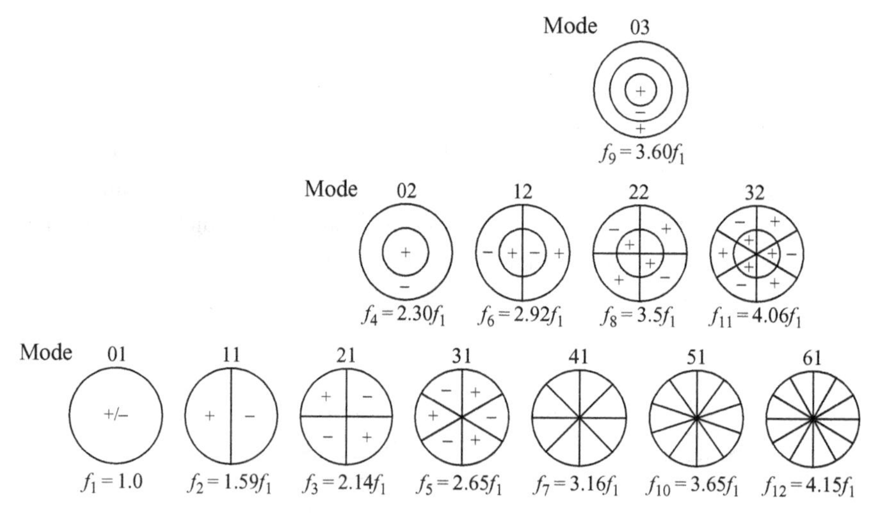

Figure 8.19
Modes of two-dimensional vibration.

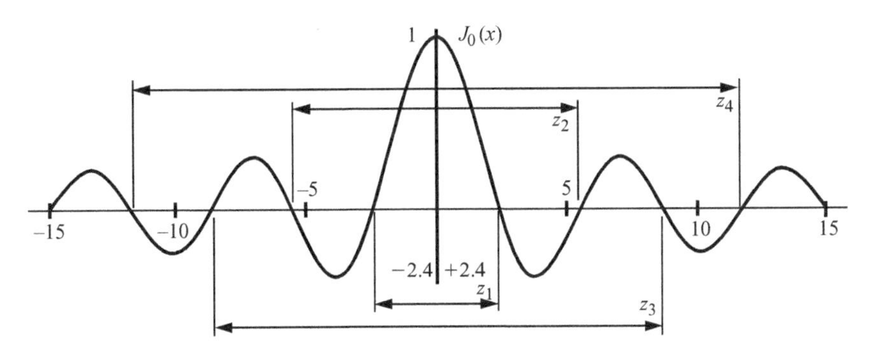

Figure 8.20
Bessel function of the first kind, order 0, $J_0(x)$.

8.8.1 Bessel Functions

It can be demonstrated that the contour of the surface for each mode shown in figure 8.19 is given by a Bessel function of the first kind. There are families of these Bessel functions ("of the first kind," "of the second kind," etc.). Each family is made up of functions of integer orders. Figure 8.20 shows a Bessel function of the first kind, order 0 in the range $-15 < x < 15$. Bessel functions of the first kind are traditionally denoted by the letter J, with a subscript indicating the order. Thus, the function shown in figure 8.20 would be written $y = J_0(x)$. Bessel functions of the first kind resemble damped sinusoids because their peak amplitudes gradually diminish as the index x increases. They have the characteristic shape of a cross-section of a vibrating two-dimensional object such as a drum head. We might imagine it could be a stop-action photograph of a water wave emanating from where a drop of water fell into a pond.

8.8.2 Stretched Circular Membranes

The fundamental frequency f of a vibrating membrane is conventionally given as

$$f = \frac{0.765}{d}\sqrt{\frac{T}{\sigma}} \qquad \textit{Stretched Membrane} \quad (8.22)$$

where σ is the area density, d is the diameter, and T is the tension. Hall (1980) specifies area density for a Mylar tympani head (with 2 mm thickness) as $\sigma = 0.26$ kg/m^2. Tympani drums come in many sizes, and their tightness is frequently adjusted during performance. But assuming a tympani drum with $d \approx 0.6$ m and $T \approx 2 \times 10^3$ N/m, the fundamental would be $f \approx 112$ Hz, around the pitch A2.

Bessel functions of the first kind $J_n(x)$ can be used to model the vibrating modes of a circular stretched membrane. It is easiest to start with mode 01, the fundamental mode, shown in plan in figure 8.19 and in elevation in figure 8.20.

The roots of the Bessel function (the places where it crosses zero) indicate the location of the nodes of the concentric modes. Circular stretched membranes must always have a node where they

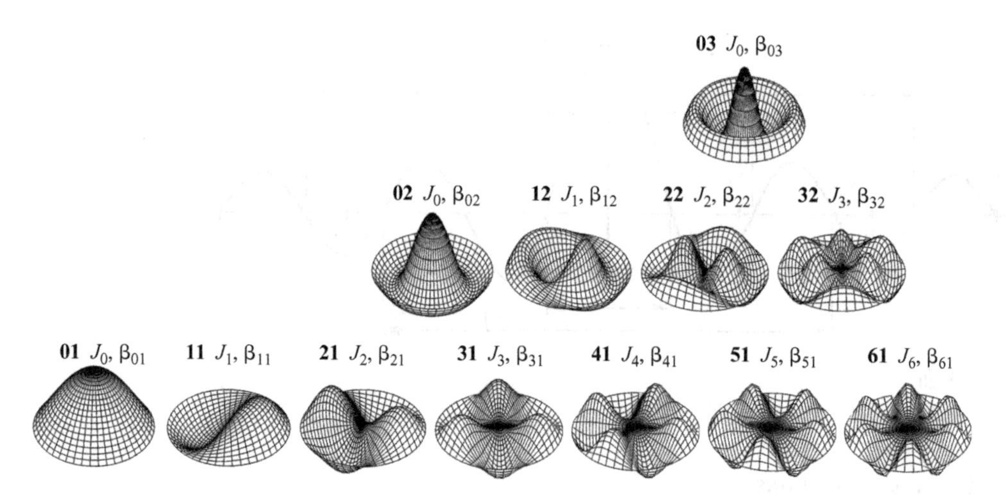

Figure 8.21
Vibration modes of a drum head.

are clamped at the edge. For instance, figure 8.20 shows that the first roots of $J_0(x)$ are $x \approx \pm 2.4$. Thus the shape of mode 01 vibration is defined as $J_0(x)$ over the range of approximately $-2.4 < x < 2.4$, labeled z_1 in figure 8.20. If this section of the Bessel function is spun 360° around its y-axis to create a circular surface, we get the shape for mode 01 shown in figure 8.21.

The location of the next roots of $J_0(x)$ are at $x \approx \pm 5.5$, labeled z_2 in figure 8.20. This section of $J_0(x)$ corresponds to mode 02 in figure 8.21. This mode has two circular nodes, one at the outer edge and the other about halfway toward the center. If the radius of the outer node is 1.0 m, the radius of the inner node would be about $2.4/5.5 = 0.436$ m.

Following this pattern, the shape of mode 03 vibration is defined as $J_0(x)$ over the range of approximately $-8.6 < x < 8.6$, labeled z_3 in figure 8.20.

For circular membranes, the frequencies of the modes are given by

$$f_{mn} = f \cdot \beta_{mn}, \qquad \textit{Drum Head Mode Frequencies} \quad (8.23)$$

where f is the fundamental frequency of the membrane, and β_{mn} is the nth root of the mth-order Bessel function of the first kind. (For convenience, I count the first root of the Bessel functions as $n = 1$.) Figure 8.22 shows $J_m(x)$ for $m = 0, 1, 2, 3$. The function β_{mn} is just the list of all the places where the Bessel functions are zero, sorted by Bessel function order.

Unfortunately, the roots of the Bessel functions are not evenly spaced, and no simple equation is known for finding them. However, we can approximate their values. For instance, we've already observed that $\beta_{01} \approx 2.4$, and $\beta_{02} \approx 5.5$. Thus, by (8.23), if the frequency of $f_{01} = 240$ Hz, the frequency of f_{02} would be about 550 Hz, and so on.

The equation used to plot the vibration pattern of the drum modes shown in figure 8.21 is

$$J_m(rx)\cos(m\phi)\cos(tx), \qquad \textit{Drum Vibration} \quad (8.24)$$

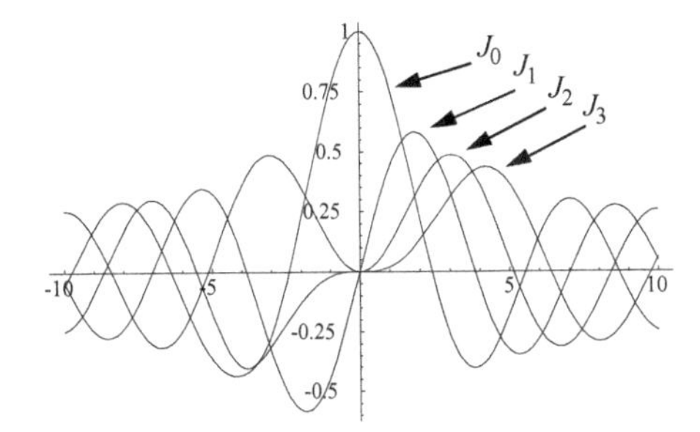

Figure 8.22
Bessel orders 1 to 4.

where m is the number of circular nodes in the mode, r and ϕ are the polar coordinates of the point on the surface of the drum being evaluated, and x is the value of the nth root of the mth-order Bessel function. For example, the surface of mode 01 (figure 8.21) is defined as $m = 0$, $x = 2.4$ (which is the first root of J_0, referred to as β_{01}). We can evaluate any point on the surface of the drum by specifying its location in polar coordinates via radius r, which varies over the unit distance 0 to 1, and ϕ, which varies from 0 to 2π. The parameter t specifies the phase of the drum mode's vibration. If this function is plotted such that t goes gradually from 0 to $2\pi/x$, we see one complete vibration of the drum head for that mode.

Equation (8.24) was used to plot the concentric and radial modes shown in figure 8.21, which correspond to the modes shown in plan in figure 8.19. Mode 01 has only one concentric node at the outer edge where it is clamped. The $0n$ modes (consisting of the set of modes 01, 02, 03, . . .) are strongly excited when energy is injected into the center of the membrane. Mode 01 makes the surface move uniformly up and down and radiates energy into the surrounding air very efficiently because it pushes air directly away from the membrane's entire surface. Consequently, the energy in this mode radiates into the surrounding air very quickly, and the sound dies away rapidly—so rapidly that for most drums one simply hears a thump from this mode after the mallet strikes it. Succeeding $0n$ modes radiate progressively less efficiently than mode 01, so the energy given them by the initial mallet strike is conserved through time. Thus they contribute slightly more to the ringing sound of the drum because they dissipate less quickly.

The $1n$ modes (modes 11, 12, . . .) are strongly excited when the drum is struck between the center and outer edge. Mode 11 radiates sound less efficiently than mode 01 because it merely sloshes the surrounding air laterally back and forth from one side of the membrane to the other as the two halves alternately rise and fall. Because little energy is dissipated, this mode strongly contributes to the sound of the drum through time. Higher $1n$ modes contribute progressively less energy to the overall sound.

Studies show that the modes that most strongly contribute to the ringing tone of a timpani drum include the 11, 21, 31, 41, and 51 modes.

The frequencies of the stretched membrane partials can also be approximated with the formula $f\ .\ 1.2\sqrt{n}$, where f is the fundamental frequency given in equation (8.22) and n is the partial number.

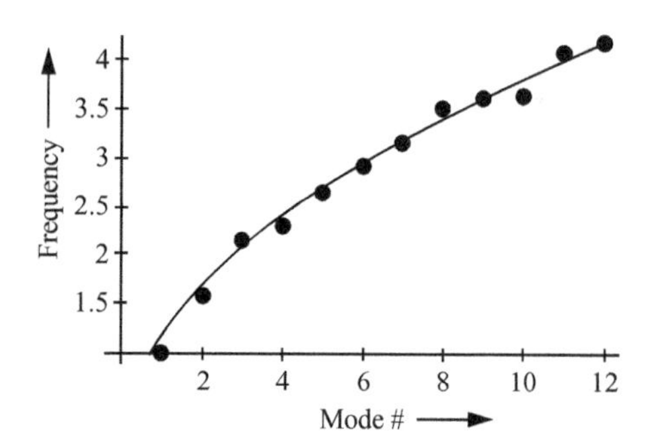

Figure 8.23
Stretched membrane mode frequencies.

Figure 8.23 shows a plot of the frequency coefficients for the first 12 modes in figure 8.19. The solid line behind the coefficients in figure 8.23 is an approximate fitted curve.

Notice that the higher partials of a stretched membrane increasingly crowd together with increasing mode number, in contrast to the transverse bar, where higher partials spread apart with increasing mode number. This accounts for the dense sound of drums in comparison to bar instruments: drum partials tend to stack up closer and closer together as frequency rises.

8.9 Resonance (Continued)

This section continues the discussion of resonance that began with the Helmholtz resonator (in section 8.3.3). The aim here is to create a solid framework in preparation for a more detailed mathematical treatment in volume 2, chapter 6.

Resonance lies at the heart of virtually every kind of musical instrument.

Resonance is the tendency of a system to vibrate sympathetically at a particular frequency in response to energy induced at that frequency.

Resonance requires two elements: a *driving force,* represented as a function of time, $r(t)$, and a *driven vibrating system* such as a spring/mass combination. A system that contains both these elements is called a *driven harmonic oscillator*. The driving force is the input to the vibrating system, and the *forced motion* is the output. This is in contrast to the *free motion* of a vibrating spring (see figure 8.4), which receives no external force after initial excitation.

While the driving force $r(t)$ can be any function, we get a clearer view of resonance by observing periodic inputs, and we get the clearest view from studying sinusoids, such as

$$r(t) = A\cos \omega t, \qquad \textit{Driving Force} \quad (8.25)$$

where t is time, A is the *driving amplitude,* $\omega = 2\pi f$, and f is the *driving frequency*. Of course, more complicated periodic and nonperiodic signals can be used, but here I limit the discussion to sinusoids.

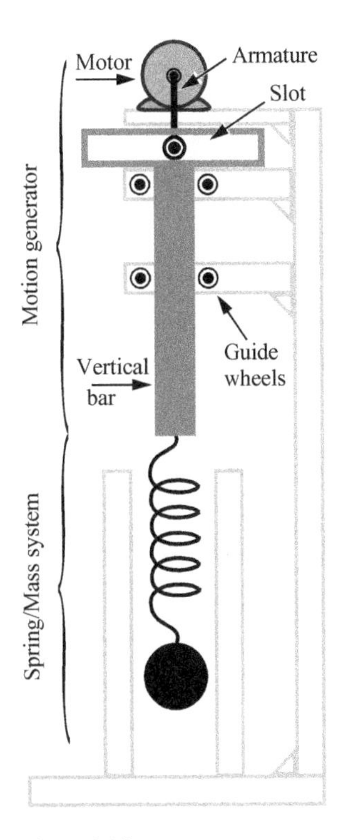

Figure 8.24
Driven harmonic oscillator.

8.9.1 Driven Harmonic Oscillator

Many musical instruments can be broken down into a part that *generates* vibrating energy and a part that *modifies* vibrating energy to create that instrument's particular sound. For example, the breath of a flute player is shaped by the resonance of the flute to produce the flute's characteristic sound. In order to understand the vibrations of such instruments, we can study an equivalent but simpler system consisting of a harmonic oscillator driven by a variable-speed motor (figure 8.24).

What happens when we vibrate a harmonic oscillator? How can we characterize its motion? We want to understand how the natural vibrating frequency of the spring/mass system responds to the frequency of the driving force.

The first thing we need is a driving force that will produce sinusoidal motion, as defined in equation (8.25). In figure 8.24 the driving force is provided by a *motion generator* that consists of an armature attached to a motor shaft. A wheel at the end of the armature is captured in the slot of a horizontal bar that is attached to a vertical bar. Together, the two bars make a T shape. The T-shaped bars can only move vertically between four guide wheels. The motor and guide wheels are mounted on a rigid framework, and the mass is restrained so it can only move up and down.

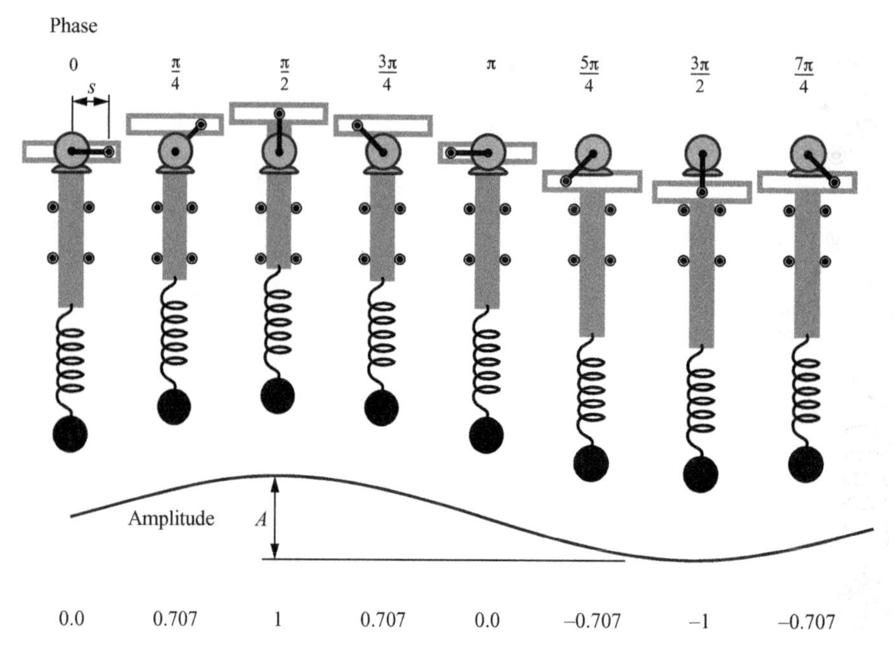

Figure 8.25
Phases of the driven harmonic oscillator.

As the motor turns, and the T bar configuration rises and falls in sinusoidal motion, the generator raises and lowers the mass via the spring.

Figure 8.25 shows successive snapshots of the system for one rotation of the motor shaft. The phase angle of the armature is given for each position, together with its corresponding amplitude. Comparing the displacement of the system to the sinusoidal line below them demonstrates that the motion of the system is indeed sinusoidal and is related directly to the phase of the motor shaft.

The vertical distance traveled by the motion generator is the driving amplitude A in equation (8.25). If the length of the armature is s, then the peak-to-peak amplitude of the generator $A = 2s$. The revolutions per second of the motor (and hence, complete sinusoidal periods of the motion generator) corresponds to the driving frequency f in equation (8.25).

8.9.2 Response Amplitude

The displacement of a spring and mass from moment to moment depends upon the interplay of Hooke's law and Newton's first law of motion (see section 8.4). The spring/mass system has a natural resonant vibrating frequency f_r, but now we must also take into account the fact that it is driven by $r(t)$, the periodic function of the motion generator. When the driving frequency equals the natural vibrating frequency, $f = f_r$, the spring/mass system will respond by vibrating strongly in sympathy with the driving force. When $f \neq f_r$, the response of the spring/mass system will be less strong.

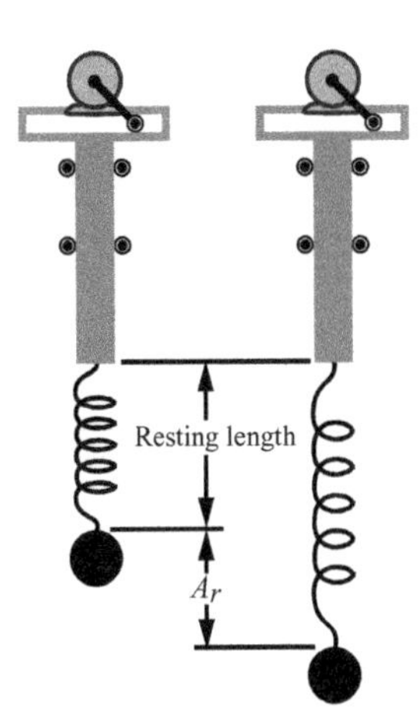

Figure 8.26
Response amplitude.

The response of the system is the amount of movement made by the mass. If the response of the system to the driving force is strong, the mass at the end of the spring will move a great distance, causing the spring to bend. If the response is weak, the spring will bend very little or not at all. So we can characterize the *response amplitude* A_r as the difference between the length of the spring/mass system when it is not being driven (its *resting length*) and its length while it is being driven.

We measure A_r by observing how much the spring is flexed—either compressed or expanded. Thus A_r is the *change in the length of the spring* from its resting length. When A_r is positive, the spring is stretched; when A_r is negative, the spring is compressed (figure 8.26).

To study resonance, we want to compare the magnitude of the response amplitude A_r to the magnitude of the driving amplitude A, moment by moment.

8.9.3 Visualizing Driven Oscillation

Let's set the motion generator to a low frequency. A low frequency is any frequency f that is substantially lower than the natural vibrating frequency f_0 of the harmonic oscillator. We indicate this by requiring $f \ll f_0$. Next, we position the armature so that it is horizontal and facing to the right (the position shown in the leftmost drawing in figure 8.25), and switch it on. As it begins turning counterclockwise, the rising force of the motion generator displaces the spring, which passes the force along to the mass. By Newton's first law of motion, the inertia of the mass applies a counterforce to the change in applied force. The spring, being flexible, stretches to make up the difference between the rising force of the generator and the counterforce of the mass's inertia. As the spring stretches, by Hooke's law, it applies a greater force to the mass, which consequently accelerates upward.

As the armature rotates toward the vertical position, it no longer lifts the spring/mass system so quickly, but the mass continues to rise because of Newton's first law of motion. The spring—squeezed between the generator and the mass—compresses until its counterforce balances

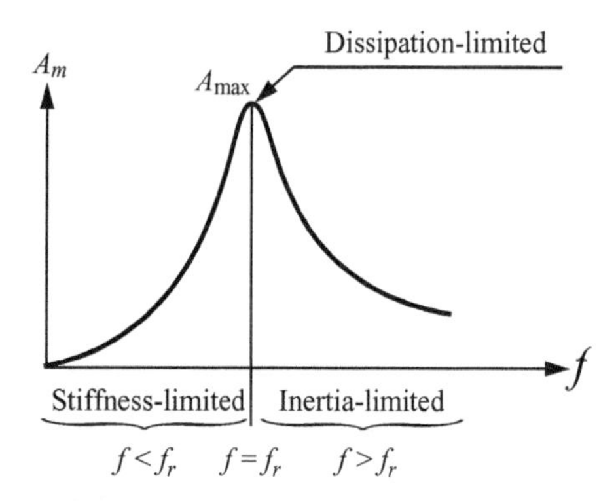

Figure 8.27
Resonant spectrum.

the upward force of the mass. As the armature starts down, the spring is further compressed, increasing the force on the mass until it accelerates downward as well. The rest of the cycle continues in this manner.

8.9.4 Varying the Driving Frequency

As we increase the driving frequency f, the force supplied by the generator to the mass increases, the counterforce of the mass's inertia increases, and the flexion of the spring increases to compensate. Consequently, the magnitude of the response amplitude A_r grows.

We might expect A_r to continue to grow as we increase the driving frequency, but when f (the driving frequency) is equal to f_r (the resonant frequency), A_r stops growing and, for higher frequencies, begins to shrink. As f continues to increase, A_r shrinks even more. Let's define *maximum amplitude* A_{max} as the amplitude at which A_r achieves its greatest value, and the *resonant frequency* as the frequency f_r at which $A_r = A_{max}$. Then, by definition, the maximum response of the spring/mass system to the generator occurs when $f = f_r$ (figure 8.27).

Stiffness-Limited Vibration For very low driving frequencies, where f is near zero, the acceleration applied to the mass by the generator is small, so the inertial counterforce of the mass is also small. Since the force of the spring's stiffness is much greater than the counterforce of the mass's inertia, the displacement of the mass closely tracks the displacement of the spring, which in turn closely tracks the displacement of the driving force. Since the mass, the spring, and the driving force are all moving together at the same speed in the same direction at the same time, they are *in phase*. For frequencies below resonance, the response amplitude A_r is *stiffness-limited* because the spring's stiffness limits the magnitude of A_r.

At low frequencies most of the energy expended by the generator to accelerate the mass is stored in the mass as kinetic energy (and the rest is stored in the spring as flexion). All energy stored in the mass (and the small amount stored in the elastic force) is returned to the generator when the mass is decelerated by the generator. So over time no work is done. (Of course, some energy is dissipated because of friction, which is ignored here.)

Inertia-Limited Vibration For high frequencies, where $f_r \ll f$, the acceleration applied to the mass by the driving force is very large, and so the inertial counterforce of the mass is also very large. The spring is literally caught between these two forces and must flex quite far to span the distance between the accelerating generator and the lagging mass. The mass will barely have begun to accelerate in one direction before the spring starts tugging at it from the other direction. As a consequence the mass moves less and less in either direction as frequency rises above f_r. The response amplitude A_r is *inertia-limited* for frequencies above resonance because the mass's inertia limits the magnitude of A_r.

Most of the energy expended by the generator to accelerate the mass is stored in the spring as flexion (some is stored in the mass). All energy stored in the elastic force (and energy stored in the mass) is returned to the generator when the spring is unflexed (and the mass is decelerated). So over time no work is done.

Dissipation-Limited Vibration Note that the energy stored by the spring and the mass is conserved (see section 4.17). The energy dissipated by the system consists of such nonconservative forces as heat and sound radiation. *The conservative forces maintain the resonance; the nonconservative forces dissipate or radiate the system's energy away.*

Near the resonant frequency, where $f_r \cong f$, the elastic force and inertial force come into balance. The relative positions of the mass, spring, and armature are such that the generator performs positive work on the mass *throughout its cycle,* so energy flows constantly from the generator to the mass through the spring. The phase of the generator leads the mass by one quarter of a cycle. The spring and mass trade energy between each other, never returning it to the generator.

If the spring/mass system continuously absorbs energy from the generator without ever returning any of it, we might expect that A_r would grow without bound. That's true except for one thing. A_r tends to grow without bound at resonance, and the velocity of the mass also tends to grow without bound. But the increased velocity causes energy to be dissipated at a faster rate, radiated as more intense sound and heat. At some value of A_r, the energy being received by the spring/mass system from the generator balances the energy being dissipated by the spring/mass system, and the amplitude reaches its maximum, A_{max}.

The amplitude of the oscillation is *dissipation-limited* when $f = f_r$. This suggests an alternative definition of resonant frequency:

Resonant frequency is the frequency that is most effective at enabling a vibrating system to return to its original energy level by dissipation.

Given the propensity of systems to seek the most efficient way to return to their original energy levels, it seems entirely reasonable that the world should be filled with resonant systems.

If the rate of energy dissipation is small, A_{max} can become large enough to destroy the system because there is nothing to stop the escalating amplitude of the mass as it continues to receive energy.

Figure 8.28
Tacoma Narrows Bridge disaster.

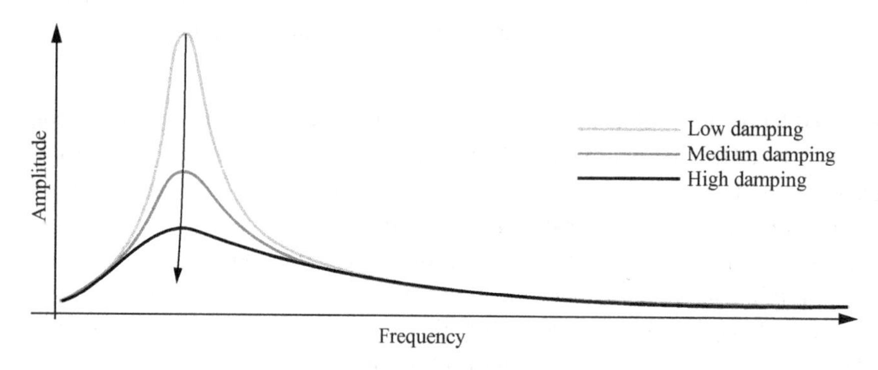

Figure 8.29
Effect of damping on resonance.

The well-documented catastrophic failure of the Tacoma Narrows Bridge is the often-cited case in point for this phenomenon (figure 8.28). (The common explanation that the bridge failed because it resonated with a frequency component of the howling wind is not necessarily incorrect, but in fact it was probably not the simple linear resonance being described here that destroyed the bridge. The exciting force of the wind was itself affected by the vibrational response of the bridge. The result was a recursive nonlinear dynamic system (Lazer and McKenna 1990; McKenna 1999) (see section 8.10.1)).

8.9.5 Damping

What is the effect of various rates of dissipation on resonance? Damping refers to how efficiently energy can be dissipated by a vibrating system.

Suppose we increase the amount of friction the mass undergoes while moving up and down (see figure 8.25). We could do this, for example, by suspending the mass in a liquid of some kind. The viscosity of the fluid resists the vertical vibrating motion of the mass in proportion to the rate at which the mass is drawn through the fluid: the faster its velocity, the greater the drag. The effect of greater damping on a resonant system is to reduce and broaden the resonant curve (figure 8.29).

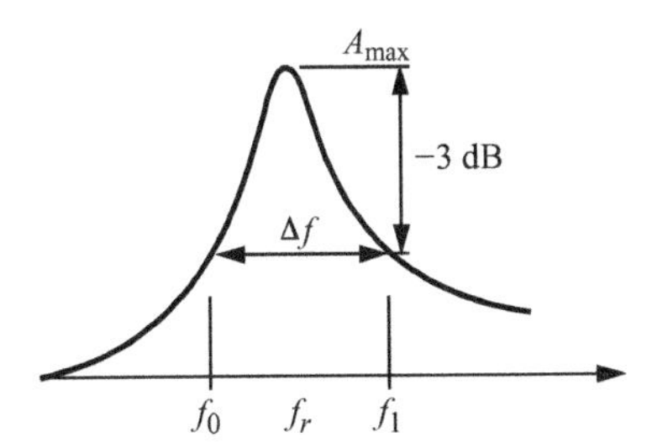

Figure 8.30
Quality factor.

Note that the peak resonant frequency declines slightly as damping increases, as indicated by the curved line drawn through the peaks in figure 8.29.

Different degrees of resonance are required for different purposes in musical instruments. The resonance peak of an organ pipe must be very narrow so that it sounds just one frequency. But the resonance of a piano sounding board should be as broad and flat as possible so that it will respond to all frequencies the same. Similarly, it is important for loudspeakers to have as broad a resonance as possible so as not to overly color the sounds they reproduce. Since damping broadens the resonant peak, pianos and loudspeaker systems often are designed to be highly damped.

8.9.6 Bandwidth and Quality Factor

As shown in figure 8.30, we can characterize the sharpness of a resonance by comparing its height to its girth at some particular distance down from the top of the peak. Starting from A_{max}, the apex of the curve, we drop down a distance of 3 dB.[12] The frequencies where this line intersects the skirts of the curve are f_0 and f_1, and the span of frequencies $\Delta f = f_1 - f_0$ is the bandwidth of the resonator. The ratio of the resonant frequency to bandwidth 3 dB down from peak amplitude is a frequency-independent measure of the steepness of the curve that engineers call *quality factor.* It is defined as

$$Q = \frac{f_r}{\Delta f}. \qquad \textit{Quality Factor} \ (8.26)$$

Q indicates how much more a driven oscillator absorbs power at its resonant frequency than it does at a standard distance from the resonance frequency. In figure 8.29, the most highly damped resonance has the lowest Q.

8.9.7 Phase Delay

For the harmonic oscillator in figure 8.25, if we plot the phase delay between the angular position of the motor arm and the linear position of the mass for various values of Q (figure 8.31), we see that the higher the Q, the more abruptly the system transitions from in-phase to out-of-phase motion. For a high-Q resonator at low f, phase delay remains near zero until f nearly equals f_r, at

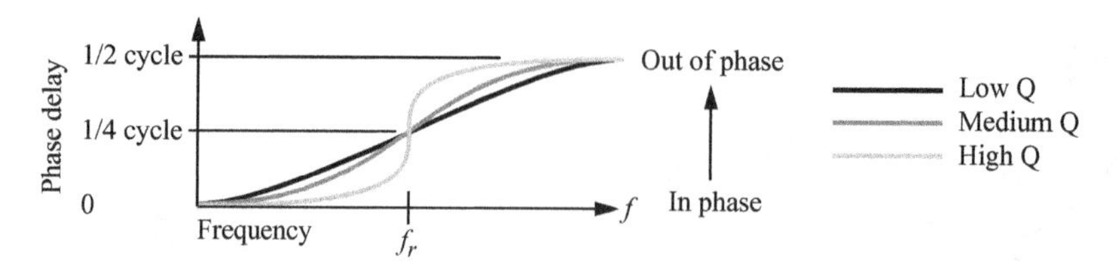

Figure 8.31
Phase delay for various quality factors.

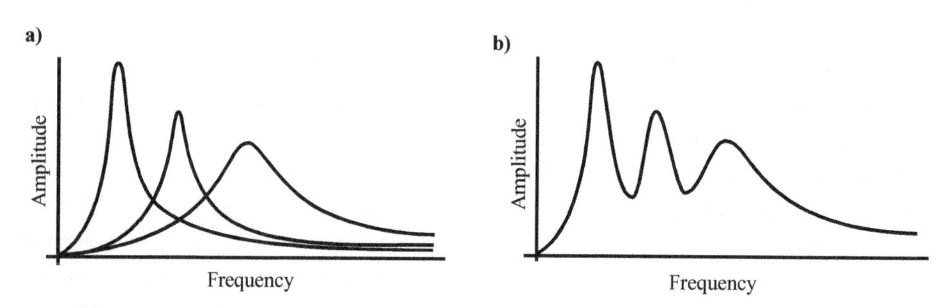

Figure 8.32
Combining resonances.

which point small additional increments in f result in large increases in phase delay. On the other hand, highly damped (low-Q) oscillators build up phase delay gradually.

8.9.8 Resonance with Multiple Degrees of Freedom

These ideas about resonance can easily be extended to more complicated vibrating systems with multiple degrees of freedom. Each vibrating mode simply has its own resonant response, characterized by a resonant frequency f_r and Q (figure 8.32a). The total response of such a system is the combination of these resonant curves (figure 8.32b).

8.10 Transiently Driven Vibrating Systems

When a performer starts to play a note on a sustaining instrument such as a pipe organ, the vibration of the instrument builds up gradually over time during the onset, or attack, phase of the note (figure 8.33). When the performer stops playing, it gradually returns to silence during the decay phase. The attack and decay phases are known collectively as *transients.*

Attack Suppose we set the speed of a driven harmonic oscillator's motor to its resonant frequency f_r, then switch on its power. This would be analogous to blowing into a flute or organ pipe, bowing a string, starting to sing, and, in general, beginning a sustained tone. Even if the

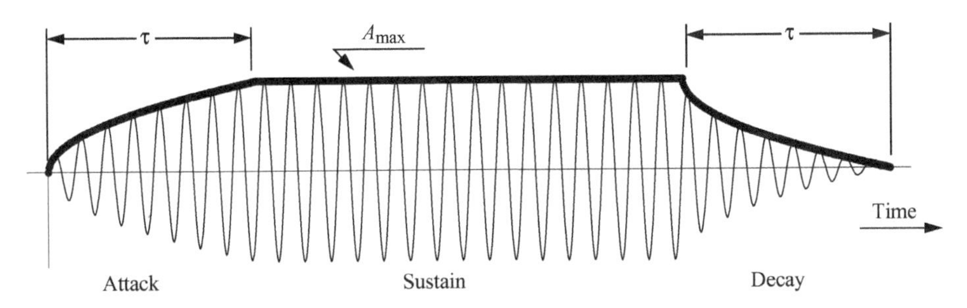

Figure 8.33
Amplitude envelope of a harmonic oscillator.

motor starts turning instantaneously, it still takes time for the response of the system to reach A_{max} because

- The mass/spring system absorbs energy at a constant rate, causing the amplitude of vibration to grow.
- As the amplitude of vibration grows, the system dissipates energy at an increasing rate.

Because the rate of dissipation increases with increasing amplitude, growth in amplitude gradually slows as dissipation approaches equilibrium with the applied force. The higher the Q, the greater is the system's energy storage capacity at frequency f_r, and the longer it takes to reach an equilibrium between the applied force and dissipation.

Steady State When energy is dissipated at the same rate that it is applied, amplitude growth stops, and a *steady state* is achieved. We say the resonator is *ringing* at f_r.

Decay When the applied energy is withdrawn, we enter the decay phase, and the system behaves exactly as described in section 8.4. In a highly damped system (low-Q), vibration ceases quickly because the dissipation rate is high. But a high-Q resonator has little dissipation, so the energy drains away more slowly.

Release A fourth state, release, characterizes the final sound some instruments make if they are stopped from vibrating by dampers. For example, when lifting the key on a harpsichord before the tone has died away, there's a slight buzz as the damper presses down on the string to stop it from vibrating.

8.10.1 Resonance, Recursion, and Xeno's Paradox

Suppose we examined the amplitude A of the decay at regular intervals, and tabulated a list of the results over N sample times. We'd have a sequence of samples $\{A_0, A_1, \ldots, A_{N-1}\}$. We select one of those samples, A_n, where $0 \le n < N - 1$. We can compute the next value in the sequence, A_{n+1}, by multiplying A_n by some factor $0 < d < 1$, corresponding to the rate of dissipation. If the

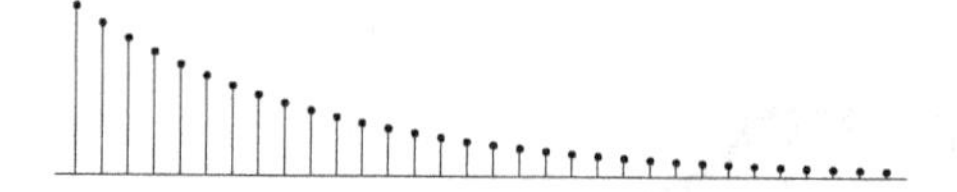

Figure 8.34
Exponential decay.

amplitude of the current cycle is A_n and the dissipation is d, then the amplitude of the next sample will be

$$A_{n+1} = A_n \cdot d. \quad (8.27)$$

So the energy in the vibrating system at each moment depends upon how much energy there was in the previous moment, times a constant factor that determines how much will be dissipated away as sound and heat. Mathematicians call such systems recursive; physicists call them dynamical.

We can evaluate the decay curve by letting $a_n = a_{n+1}$ and then repeating (8.27) forever to generate the rest of the curve. For example, if we set $a_0 = 1$ and $d = 0.9$, we have the sequence {1.0, 0.9, 0.81, 0.729, . . .}. Plotting these points reveals an exponential decay (figure 8.34). This function never reaches zero.

A curious aspect of resonant systems is that, theoretically, they never stop vibrating. This is an acoustic incarnation of Xeno's paradox (see appendix A). Suppose we measure the amount of time it takes for the amplitude of a note to drop to one half of A_{max} and call it $t_{1/2}$, the *halving time*. If $A(t)$ is a measure of the amplitude at time t, and $A(0) = A_{max}$, then we can express halving time as

$$\frac{A(0)}{A(t_{1/2})} = \frac{A_{max}}{2}. \quad (8.28)$$

Since the amplitude drops to 1/2, the vibrational energy in the system drops to one fourth of its original value at $t_{1/2}$, and

$$\frac{A^2(0)}{A^2(t_{1/2})} = \frac{A^2_{max}}{2^2}.$$

If we wait until an additional time interval $t_{1/2}$ has elapsed, the amplitude will be one fourth and energy one eighth of the original. At each subsequent time interval $t_{1/2}$, the amplitude will again be halved and the energy quartered, *but there is still energy present proportional to what was there before*. Thus, unless we wait for eternity, the amplitude never reaches zero (unless it was zero to begin with).

The target value that the amplitude is heading toward (but never reaches) is the *asymptote*. During the attack phase, the asymptote is A_{max}, during the decay phase it is zero.

The Exponential Function and the Time Constant We've seen that the transients of all linear resonant systems—where the rate of energy loss or gain is proportional to the current energy—have a characteristic exponential shape. This includes virtually all musical instruments and sound in reverberant spaces. The exponential function commonly used in musical applications to model this is

$$y = E(t) = A_{max} e^{-t/\tau}, \qquad \textit{Exponential Decay} \quad (8.29)$$

where A_{max} is peak amplitude, time $t \geq 0$, and the *time constant* τ is the characteristic rate of decay. The asymptote of the exponential decay function is zero.

Conventionally, τ is the time it takes for $E(t)$ to decay by $1/e$, that is,

$$\frac{E(\tau)}{E(0)} = \frac{1}{e} \cong \frac{1}{2.78} \cong 0.36\,, \qquad (8.30)$$

corresponding to a drop of $10 \log(1/e) = -4.34$ dB SIL (see equations (5.31) and (5.32)).

The attack envelope is the inverse of (8.29):

$$y = E(t) = A_{max} - A_{max} e^{-t/\tau}. \qquad \textit{Exponential Attack} \quad (8.31)$$

The asymptote of the exponential attack is A_{max}. Figure 8.33 shows an example of exponential attack and decay envelopes.

Solution to the Paradox What's the solution to the paradox of the never-ending exponential envelope?

Often, the force of friction in a vibrating system increases at low amplitude, erasing the little remaining energy in the vibrating system at an accelerated rate and helping to mark the end of its sound. A nonlinear friction function at low velocity also explains why the brakes in a car sometimes start to grab just as it approaches a complete stop. But even if energy is depleted at an accelerated rate, there's theoretically still some there forever.

Perceptually, a decaying sound will become inaudible if it drops below the threshold of hearing or the ambient noise level, whichever is higher. So the empirical solution to this version of Xeno's paradox is to decide on a time after which we consider the amplitude to be insignificant.

T60, Decay Time, and the Meaning of Silence How long does it take for sound in a concert hall to decay into silence? The reverberation time of a hall (see section 7.13.2) is one of the key determinants of its acoustical quality: if reverberation lasts too long, music and speech tend to become blurred, reducing intelligibility. A short reverberation time may improve intelligibility but may make the room sound dead.

A rule of thumb used widely in architectural acoustics is that a sound with initial amplitude A_{max} becomes insignificant after it has decayed by 60 dB SPL. This time, called t_{60}, is a measure of the reverberation time of a room. Some cathedrals have a t_{60} time of about 10 seconds or more; the t_{60} time of concert halls usually lasts a few seconds. The t_{60} time of a bedroom may be a few milliseconds, and the t_{60} time of a good anechoic chamber should be vanishingly close to zero.

Since -60 dB $= 20 \log_{10} 0.001$, t_{60} can be thought of as the time it takes A_{max} to decay by a factor of 1000. That is, if $A(t)$ is the amplitude of a sound at time t and $A(0) = A_{max}$, then

$$\frac{A(t_{60})}{A(0)} : \frac{A_{max}}{1000}.$$

We can relate t_{60} to the halving time $t_{1/2}$ by noting that $A(t_{60})$ corresponds to about ten halvings of amplitude, that is,

$$\frac{A(t_{60})}{A(0)} = \frac{A_{max}}{1000} \approx \frac{A(10 \cdot t_{1/2})}{A(0)},$$

so $t_{60} \approx 10t_{1/2}$. Similarly, we can relate t_{60} to the time constant τ by solving (8.29) for τ: $\ln(1000)\tau \cong 6.91\tau$. Thus, t_{60} is just under seven time constants long. The energy being radiated at time t_{60} is the square of the amplitude, or

$$\left(\frac{A_{max}}{1000}\right)^2 = \frac{(A_{max})^2}{1{,}000{,}000},$$

which corresponds to a drop in sound intensity of 60 dB SIL. If the original intensity is, say, 100 dB, then 100 dB – 60 dB = 40 dB SIL, approximately the same as the threshold of ambient noise in quiet listening environments, which is a workable definition of silence.

8.10.2 Why High-Frequency Components Die Out Faster

Vibrating systems with many degrees of freedom have multiple resonances, and overall the response of the system is a composite of the individual resonances of the modes (see section 8.9.6). It follows that each degree of freedom n has its own damping time τ_n. Characteristically, high-frequency vibration modes have the smallest τ_n.

To see why, consider two identical masses m_1 and m_2 vibrating in simple harmonic motion with identical amplitude A but with frequencies $f_1 < f_2$. By equation (5.27), we know that the energies of the two masses are $E_1 = m_1(2\pi A f_1)^2$ and $E_2 = m_2(2\pi A f_2)^2$, and so $E_1 < E_2$, that is, m_1 has less energy than m_2. Since at any moment of time, the rate at which energy is radiated is proportional to the total amount of energy, m_2 loses energy faster than m_1, and its vibrations are damped out more quickly (Ruiz 1969).

Figure 2.24 shows the evolution through time of the harmonics of a musical instrument tone. Each instrument has a characteristic way in which its partials evolve through time, which can be understood by examining the interplay of the vibration modes, the forces the player exerts upon the instrument, and the coupling of the instrument to the air (see volume 2, chapter 6).

8.11 Summary

We examined the mathematical formulas that determine the sound of conventional musical instruments. We combined Hooke's law with Newton's second law of motion to observe in detail the movement of simple harmonic motion.

A vibrating system has degrees of freedom, which can interact additively. Musical instruments can be organized by the similarity of their mathematical equations. The parameters are dimension, restoring force, and vibrating element.

We considered the mathematics of strings, string modes, and standing and traveling waves. Bars and strings are governed by the same equations, but because bars are stiffer, they are nonharmonic. Young's modulus is used to determine the vibrating properties of strings and bars.

We examined air columns in light of the Helmholtz resonator and developed models for the vibration of pipes open at one end or both ends, which differ from models of conical pipes. Drums are two-dimensional vibrating analogues of strings and bars. Their vibrating modes are characterized by Bessel functions of the first kind.

The discussion of resonance was continued by examining the behavior of a driven harmonic oscillator. The behavior of a resonator changes as a function of driving frequency, resonant frequency, and the amount of damping in the system. The result is characterized in terms of quality factor Q and phase delay. We considered transiently driven vibrating systems, such as when a musical instrument starts and stops a note, and observed a paradox related to Xeno's.

8.12 Suggested Reading

Askill, J. 1979. *The Physics of Musical Sound*. New York: Van Nostrand.

Moravesik, M. J. 1987. *Musical Sound: An Introduction to the Physics of Music*. New York: Paragon.

Pierce, John R. 1983. *The Science of Musical Sound*. San Francisco: Scientific American Books.

Rigden, J. S. 1977. *Physics and the Sound of Music*. New York: Wiley.

Rossing, Thomas D. 1983. *The Science of Sound*. Reading, Mass.: Addison-Wesley.

White, H. E., and D. H. White. 1980. *Physics and Music*. New York: Holt, Rinehart, and Winston.

Wood. A. 1975. *The Physics of Music*. New York: Wiley.

9 Composition and Methodology

[The Analytical Engine's operating mechanism] might act upon other things besides number, were objects found whose mutual fundamental relations could be expressed by those of the abstract science of operations, and which should be also susceptible of adaptations to the action of the operating notation and mechanism of the engine . . . Supposing, for instance, that the fundamental relations of pitched sounds in the science of harmony and of musical composition were susceptible of such expression and adaptations, the engine might compose elaborate and scientific pieces of music of any degree of complexity or extent.
—Ada Lovelace[1]

The best view of musical composition is provided by a study of methodology. So understanding methodology is the first aim of this chapter. The subject of methodology encompasses most human activities, including the arts and sciences. Approaching composition this way has the great advantage of enabling us to relate the arts and sciences, to see their similarities and differences in sharp relief.

Studying the methodology of composition provides a crisp and efficient way to identify and compare the aesthetic aims of particular composers and schools of composition. This is of great benefit because we can then accurately compare and contrast the wide panorama of interests and values that have concerned composers over the ages.

The second aim of this chapter is to study the development of artificial composing systems. Perhaps surprisingly, the foundations of this field are over a thousand years old. The implications of these ideas stir far-reaching and provocative questions about the nature of music and composition.

The third aim of this chapter is to show how compositional principles can be expressed in a computer programming language and to develop a set of tools that can be adapted for readers' own music research. A simple but powerful music programming language called MUSIMAT is presented, and many of the methods discussed are shown in MUSIMAT code. The chapter provides computational strategies for composing music and insights about the nature of composing.

9.1 Guido's Method

Around 1026 the learned Benedictine and famous music theoretician Guido d'Arezzo developed a way to teach his students composition which, in his religious environment, amounted to composing plainchant melodies to accompany sacred texts in Latin. This same Guido invented the

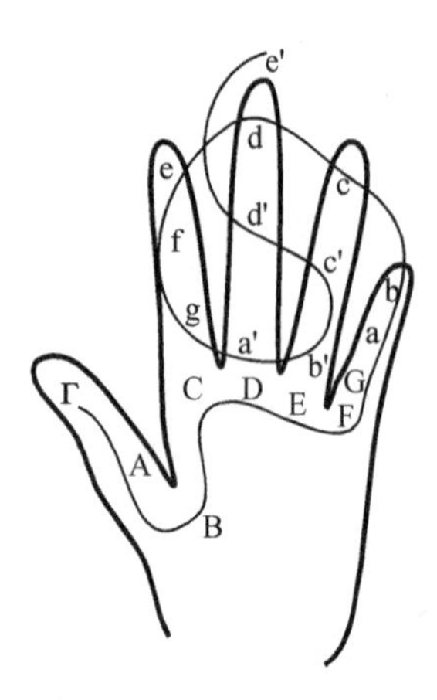

Figure 9.1
Guidonian hand. (Adapted from Apel 1944.)

medieval music theory of hexachords, a group of six diatonic tones with a semitone interval in the middle, e.g., C␣D␣E˅F␣G␣A. He also assigned letter names to the diatonic scale and invented the solmization syllables *ut* (do), re, mi, fa, sol, la that are familiar to music students.

Guido developed a method of memorizing the musical scale and its solmization syllables, historically called the *Guidonian hand* (figure 9.1). By pointing to parts of the hand a choir master can indicate the next note of the melody to be sung. Although it is just a simple way of associating pitches with positions on the hand, it came to epitomize the entire system of the church modes in medieval Europe. It became such a powerful metaphor that conservative music theorists of the late Middle Ages used it to resist the introduction of chromaticism by saying that the new scale degrees were "not in the hand" (Apel 1944).

Guido's combination of theoretical prowess and practical aptitude laid the foundations of *objective composition,* which I define as the use of naturalistic (nonsubjective) processes in composition. Although we associate the development of automated composition with the twentieth century, Guido's work demonstrates that it is a quite ancient practice.

Guido published his composition method for students as part of a treatise for singers titled *Micrologus*. This was an important source for the development of *organum,* the earliest type of polyphonic music in Europe. There is debate as to whether Guido was seriously proposing his method as a means of composing music, or if it was just a didactic aid for teaching composition. But in any event, he managed for the first time in history to objectify a way of composing music into a definite set of rules. Although elementary, his method is the prototype for all objective compositional systems from that day to this. It can be used for thought experiments on the general nature of composition.

Guido's first step was to construct a table of correspondences between the notes of the scale and the vowels contained in the Latin text that is to be set to music. First he laid out the pitches of the double octave, which was the standard compass of vocal music of his time (figure 9.2). Against

Figure 9.2
Vowel/note correspondence.

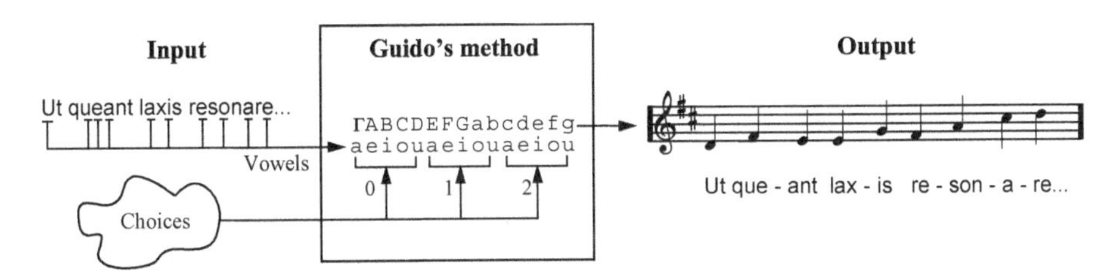

Figure 9.3
Block diagram for Guido's method.

this he placed three iterations of the vowel sequence, *a e i o u*:

Γ	A	B	C	D	E	F	G	a	b	c	d	e	f	g	a′
a	*e*	*i*	*o*	*u*	*a*	*e*	*i*	*o*	*u*	*a*	*e*	*i*	*o*	*u*	*a*

Guido then selected a Latin text and extracting the vowels from each word, set about looking up corresponding pitch values from his table. Since this method supplies three choices for each vowel (and four choices for the vowel *a*), the method has multiple solutions for each text. Following this procedure, he composed a melody for the entire text that changed pitch on every vowel. Figure 9.3 shows a block diagram outlining Guido's method. There are two inputs, the vowels of the Latin text and the choices made by the subjective judgment of the composer that determines from which vowel group to draw the pitch. The output is the resulting plainchant melody. The Latin text shown is the one Guido famously used as the basis of his solmization system, but Guido's original melody is replaced here by one generated by Guido's method.

For a one-vowel text, there are 3 possible one-note "melodies"; for a two-vowel text, 3^2 melodies of two notes; and for an N-vowel text, 3^N melodies. The number of possible melodies grows explosively for longer texts. Guido suggested that anyone who felt his system was too constraining should expand it by adding another line of vowels under the notes with a different starting point, doubling the number of choices. Guido's intention was to ease his students into the deep ocean of unlimited possibilities with small steps by the shore.

But even if choice is constrained by the method, composers still must exercise their subjective faculties to develop a pleasing and musically interesting line. Guido suggested that by selecting only the best excerpts from several attempts, composers could obtain a composition perfectly adapted to the text and meeting the requirements of good compositional practice.

9.2 Methodology and Composition

Methodology is what allows us to construct the wheel, plant crops, solve equations, and write symphonies. Methodology is the DNA of human culture: it carries information that enables societies to function and to persist from generation to generation. The proper study of composition is the study of methodology, and the full appreciation of methodology requires an understanding of algorithm.

9.2.1 Algorithm

Algorithm is the most highly qualified methodology. The word comes from *algorism,* which means to calculate with Arabic numerals.[2] According to Donald Knuth (1973) algorithm is a broader concept, covering any set of rules or sequence of operations for accomplishing a task or solving a problem so long as it demonstrates each of the following five characteristics:

- *Finiteness* The method must not take forever.
- *Definiteness* Each step must have a significance that is commonly understood.
- *Input* The method must have valid materials or information upon which to operate.
- *Output* The method must produce at least one result, generated by applying the method to the inputs.
- *Effectiveness* The method must always produce the same output from the same input; the result must not depend upon unknowns (e.g., a miracle, a coin toss, or the phase of the moon); and there can be no ambiguous outcomes (e.g., dividing by zero is not allowed because the result is undefined).

A method that meets all these requirements is called algorithmic.

According to Knuth, methods also display aesthetic traits, or "goodness." These include efficiency, simplicity, grace, elegance, parsimony (no extraneous steps or rules), and tractability (easily adapted to a variety of circumstances). A method's goodness is also demonstrated by how well it reveals our understanding of the problem being solved.

9.2.2 Euclid's Method

By way of example, consider the problem of finding the greatest common divisor (GCD), which is the greatest number that divides two numbers without remainder. This comes in handy when reducing two numbers to their lowest form, so as to reduce interval ratios to their lowest common denominator. For example, the GCD of 9 and 12 is 3. We just "know" that, but *how* do we know it, and how can we represent this knowledge to someone else? And how can we find the GCD of 91 and 416, which we almost certainly do not "know"? Euclid developed the following method to solve this class of problem for positive integers.

Euclid's Method

1. Given two numbers, m and n both greater than zero, find their remainder after integer division.
2. If the remainder is 0, the answer is n.
3. Otherwise, let $m = n$, and let $n = r$, and start over.

Table 9.1
Euclid's Method, 9 and 12

Step	m	n	r
1	9	12	9
2	12	9	3
3	9	[3]	0

Table 9.2
Euclid's Method, 91 and 416

Step	m	n	r
1	91	416	91
2	416	91	52
3	91	52	39
4	52	39	13
5	39	[13]	0

The results for 9 and 12 are shown in table 9.1, and the results for 91 and 416 in table 9.2.

9.2.3 Is Euclid's Method Algorithmic?

Yes, Euclid's method is algorithmic.

- It is *finite* (it will always eventually reach $r = 0$).
- It is *definite* (for positive integers) because the meaning of division and remaindering for positive integers is unambiguous. (But if we extend the positive integer range to include zero, the method is no longer definite because division by zero is undefined.)
- It has *inputs, m* and *n*, and an *output,* the answer.
- It is *effective* because there is no miraculous, random, or subjective element in Euclid's method; it always gives the correct answer (barring mistakes in arithmetic).

It meets Knuth's aesthetic criteria as well: it is simple (requiring only three steps), elegant (lovely to think about), and parsimonious (it gets straight to the point). It is also tractable because it can easily be adapted (e.g., to a computer).

9.2.4 Is Guido's Method Algorithmic?

No, Guido's method is not algorithmic. Subjective choice is required for Guido's method, so it fails the definiteness criterion. But it meets all other criteria, including Knuth's aesthetic criteria. We can call methods like Guido's *nondeterministic methodologies,* but I prefer a more concise name: *art.* The characteristic feature of art of all kinds is that it *combines objective criteria and methods with choice making*. The difference between art and algorithm is that deterministic methodology (algorithm) always produces the same result from the same inputs, whereas nondeterministic methodology may produce variable results even with the same inputs.

Choice making, such as thinking of a number between 1 and 6, is always subjective. So-called objective choice, such as tossing a six-sided die, is actually just the delegation of subjective choice to an external process. (Have you ever flipped a coin to make a choice and then decided to do the opposite?) A delegated external choice-making entity is an *oracle*. If a method requires consulting an oracle of any kind, it is automatically art, not algorithm. We should also distinguish between choice making and choice accepting. The latter is always subjective.

9.2.5 Why Study Methodology?

In order to create their methods, both Euclid and Guido had to reach inside their own subjectivity, to hold their goals in mind while simultaneously observing their own mental processes long enough to objectify what they discovered into a set of rules. Because this requires considerable mental discipline, I believe that we only develop methods where we care deeply about the aim of the method.

This suggests that the study of methods can reveal our values and hidden assumptions. For example, we observe that Guido's method constrains the music to follow the words, thereby revealing Guido's belief that the purpose of music was to set off the biblical text, much the way a ring sets off a jewel.

The guiding principle of this chapter is that *the analysis of methodology can reveal the aesthetic agenda of its creator*. Thus, by examining the methods of composers, we can understand the inner significance of their music. After building some tools and skills, I discuss some ground-breaking compositional methods for the purpose of examining their underlying values, as a way of helping us to establish our own.

9.3 MUSIMAT: A Simple Programming Language for Music

If we wish to use computers to operate on music, as Ada Lovelace suggested, we must find ways to represent music that both composer and computer can understand. The representation must be intuitive, and yet definite enough to be computable. It must provide expressive control over the musical materials we wish to operate upon.

In order to study methodologies, we must have a completely definite language with which to express them. A programming language is a specialized means of describing rule systems and methods. MUSIMAT is a programming language designed specifically for the subjects presented in this chapter. A tutorial introduction to MUSIMAT is given in appendix B.

It is possible to read this chapter without knowing MUSIMAT. However, I highly recommend spending the time needed to understand it before proceeding. Many of the examples in this chapter are expressed in MUSIMAT, and though I summarize everything in nontechnical language as well, readers won't be able to adapt and use this information without understanding the language in which it is written.

9.4 Program for Guido's Method

With the MUSIMAT programming language we can program a version of Guido's method.

First, we transform Guido's vowel sequences to pitches:

```
PitchList guidoPitches =
   {G3,A3,B3,C4,D4,E4,F4,G4,A4,B4,C5,D5,E5,F5,G5};
```

See section B.2.1 for a description of `PitchList`.

Then we need a source of judgment for which of Guido's three vowel sequences should be chosen. We'll use the integer `Random( )` method to generate random values. Combining these, we obtain the program for Guido's method:

```
PitchList guido(String text) {
   PitchList G;                       //place to put the melody
   Integer k = 0;                     //indexes G
   Integer offset;                    //indexes guidoPitches[ ]

   //evaluate one character of the text at a time
   For (Integer i = 0; i < Length(text); i = i + 1) {
     Character c = text[ i ];         //get a character of the text

     If ( c == 'a' ) { offset = 0; }
     Else If ( c == 'e' ) { offset = 1; }
     Else If ( c == 'i' ) { offset = 2; }
     Else If ( c == 'o' ) { offset = 3; }
     Else If ( c == 'u' ) { offset = 4; }
     Else { offset = -1; }            //the character is not a vowel

     If ( offset != -1 ) {            //if the character is a vowel. . .
          Integer R = Random( 0, 2 ); //returns 0, 1, or 2
          Integer n = ( 5 * R ) + offset;
          G[ k ] = guidoPitches[ n ];
          k = k + 1;
     }
   }
   Return( G ); //return the list of pitches composed
}
```

The program indexes one `Character` at a time of `text`. If `Character c` is a vowel, it calculates `offset` based on which vowel it is. If it is not a vowel, the program sets `offset` to –1 so that the final step is skipped. If it is a vowel, the program chooses a random number 0, 1, or 2, corresponding to the three possible outcomes for each vowel. This is multiplied by 5, corresponding to the number of vowels, and added to `offset` to arrive at the index of the selected element in the list of `guidoPitches`. The selected `Character` from that list is then stored in `PitchList G`. The method is repeated until `text` is exhausted. `PitchList G` then contains the list of pitches composed for this text. As its final action, the `PitchList G` is returned to the calling program.

To invoke the function `guido()`, we need a Latin text. I'll use the first phrase of the text Guido used to name the solfeggio syllables, the medieval hymn *Sanctus Joharines* (St. John). This program fragment prints a list of pitches:

```
Print(guido("Ut queant laxis resonare."));
```

An example result of this method is shown in figure 9.3.

9.5 Other Music Representation Systems

There are a virtually unlimited number of approaches to the representation of music, depending upon one's aims. The aim of MUSIMAT is compactness and expressivity for composition. A short list of some important music representation and programming systems, drawn from the extensive literature on the subject, follows:

- *MIDI* Musical Instrument Digital Interface, a still prevalent standard for encoding and transmitting musical gestures between computers and music synthesizers (Loy 1985) provides a very simple and concrete mapping from musical keyboards, nobs, and sliders to musical sounds. In its original form no specific mapping of sounds was stipulated. One MIDI synthesizer might play a particular note using string tones, whereas another would use bassoons. More recently, General MIDI, a standard set of timbres, was adopted. This standard stipulates a common mapping of timbres that every conforming synthesizer must implement. Scores played on any General MIDI synthesizer realize a similar orchestration (Jungleib 1996). MIDI presents a normative and limiting conception of music (F. R. Moore 1988), but it is very widespread.
- *CHARM* Common Hierarchical Abstract Representation of Music provides a way of looking at music that is useful for musicological analysis (Wiggins, Harris, and Smaill 1989).
- *SCORE* A music printing system developed over the last 30 years by Leland Smith, SCORE can be used for high-quality printing of common music notation, tablature, and other nonstandard musical formats.
- *DARMS* This is an early, overly ambitious (flawed but interesting) music description language, developed by Ray Erickson (1975) and Stephen Bauer-Mengelberg.
- *GUIDO* An extensible text-based score representation language for notation software, composition, analysis, and performance developed by the Salieri Project at the Technical University in Darmstadt (Hoos et al. 1998).
- *DMIX* Developed by Daniel Oppenheim (1996), DMIX combines graphical sound editing, algorithmic composition, computer programming, and real-time interaction and improvisation.
- *Kyma* This is a sound specification language developed by Carla Scaletti (1991). It uses a general specification of sounds as the building blocks of composition. "The structure of a composition in this language is the set of traces left by the compositional process, that is, each composition

contains within it a record of how it was composed. This record serves as one of the many possible analyses of the composition" (Scaletti 1989, 43).

- *Max* A graphical programming language developed by Miller Puckette, Max enables composers to create music-processing systems by connecting processing icons on a screen much the way one would plug Moog or Buchla analog synthesizer modules together.

9.6 Delegating Choice

The agency of compositional choice (see figure 9.3) can be delegated from one person to another or from a person to an objective process such as rolling dice.

9.6.1 Subjective Choice

A composer may delegate choice of musical elements to an assistant or amanuensis. This is a common practice, for example, among famous Hollywood movie composers. Although the head composer may stipulate criteria, the actual composing is done by assistants. By delegating, the head composer loses some control over the result.

Even if a composer writes every note in the score in minute detail, its realization will necessarily include many chance elements introduced by the performer. Conventional rules for classical performance interpretation are just one of the uncertainties affecting the composer's music. Others include who performs it, the venue, the choice of instrumentation and equipment, whether it is broadcast or recorded, other compositions on the program, their order in the program, and so on. The composer's instructions may be ignored altogether. Most of these uncertainties remain even if the realization consists of playing prerecorded music. Of course, there are many forms of music, such as American jazz, where uncertainty predominates because the music is more-or-less improvised.

Even some classically trained twentieth-century composers experimented with delegating additional elements of the composition process to performers. An early piece of this type was Karlheinz Stockhausen's *Klavierstük XI* (1956), where the score for piano solo consists of 19 disjointed fragments of music notation of varying lengths, placed on a large sheet of cardboard with plenty of empty space between them. Stockhausen indicated that the performer should play the fragments in any order "that catches his eye" and should also choose tempo, dynamic level, and type of attack during the performance.

Other composers of that era, including John Cage, Earle Brown, and Pierre Boulez, devised pieces that invite performers to determine some part of the structure of the music they are playing. One can make an analogy between such open compositions and the mobile sculptures of Alexander Calder (figure 9.4). Calder's mobiles are fixed structures, but the parts can move relative to one another. The possible shapes are determined by the artist, but virtually infinite configurations are possible. The compositions of Stockhausen, Cage, Feldman, and others raised some serious

Figure 9.4
Alexander Calder mobile: *Untitled,* 1942 (Cat. A15493).

questions about the future of Western classical music. Potter (1971) writes,

> Many questions arise regarding formal openness: Is a composer abdicating his responsibility when he relinquishes formal control? Is formal determination an (the) essential task of the composer? Should a group of unordered musical segments be considered a single piece? What purpose is served by allowing the performer to order the material he performs? Is a listener aware of the formal openness? Should he be? Can a single performance (live or recorded) be a self-sufficient artistic statement, or is more than one performance necessary to expose the formal variation possible? (120)

Though these are excellent questions that deserve answers, I have a slightly different agenda to pursue here, which will nonetheless lead back to these kinds of questions, but from a broader perspective.

9.6.2 Objective Choice

The time-worn approach to delegating choice to an objective process is to use dice or an urn of numbered balls: a ball is pulled "at random" from the urn, and the color or number on the ball is used to determine the choice. But more fanciful ways have been advanced specifically for composing music by chance. Mauritius Vogt (1719) suggested a method of composing by bending hobnails into various shapes, then casting them on the ground and interpreting the rise and fall of the music by the way the hobnails fell. William Hayes (1751) suggested that the composer spray ink from a brush onto music manuscript paper, then add note stems, staves, barlines, and all the rest according to signs drawn from a pack of cards.

In fact, any objective process can be used whether it is random or not. Employing chance as an agent of choice has a very long history in human culture. For example, wind chimes and aeolian harps harness random natural forces to create pleasing—dare I say musical?—sounds.

Have you ever used a coin toss to decide what to do? This can help get you unstuck if you are truly undecided or really don't care about an outcome. But after the coin was tossed, did you really follow the coin's dictate, or did you back out and choose the outcome yourself after all? Perhaps the coin's choice just didn't feel comfortable? *A chance process is unlikely to make the same quality of choices that a person would.* Actually, this could be good or bad.

On the plus side, chance decisions can help prevent a student of composition from being overwhelmed by the vastness of possible outcomes. Guido may have had this point in mind. Or a composer might look to an objective process to suggest a novel direction to take to get past unconscious biases. The composer Herbert Brün (1970) wrote about using computers to provide a random choice-making element while composing:

> Whereas the human mind, conscious of its conceived purpose, approaches even an artificial system with a selective attitude and so becomes aware of only the preconceived implications of the system, the computers would show the total of the available content. Revealing far more than only the tendencies of the human mind, this nonselective picture of the mind-created system should be of significant importance.

The composer David Cope (1996) reported that overcoming composer's block was one reason he developed his ambitious Experiments in Music Intelligence (EMI) system (see section 9.24). The results of a chance process can provide a welcome new perspective that gets a composer out of a rut.

On the minus side, pure chance has no regard for what a composer thinks or would prefer. Constraining pure chance so that it does what composers want (mostly) occupies a great deal of the effort on automatic composing systems.

9.6.3 The Role of Interest in Music

Music is a delicate balance between what is familiar and what is surprising. And the ultimate source of surprise is chance. But this approach is not without risk. A truly random process such as flipping a coin displays neither skill nor taste at composing because it has no awareness of the music it is being used to create. It does not, in and of itself, learn from its mistakes or make inferences about its experiences. It does not favor particular outcomes, and as a consequence, its results have an undesirable "wandering" quality.

If we want to incorporate chance into composition—and if we care about the interest of our listeners—we must become students of *interest* and look for ways to increase the likelihood that the choices made on our behalf are interesting, because without interest there is no music, only noise. This problem was solved very cleverly in the antique automatic composing system described in the next section.

9.6.4 *Musikalische Würfelspiel*

Whereas Guido's composing method was intended to be driven by human choices, a related technique, *Musikalische Würfelspiel,* which arose during the European classical era, was intended to be driven by a throw of the dice. The music engineering problem solved by this system was how to direct unregarding chance operations to make musically suitable choices and compose interesting music.

In 1757 in Berlin, Johann Philipp Kirnberger published *Der allezeit fertige Polonaisen und Menuetten Komponist,* roughly translated as "The Ever-Ready Composer of Polonaises and

Minuets." Like Guido's, Kirnberger's intent was to provide a simplified means of composing music. In his preface Kirnberger states that the reader "will not have to resort to professional composition." Although using the technique required no musical training, it required considerable compositional skill to create it in the first place. Composers of preeminent stature, including Wolfgang A. Mozart, Joseph Haydn, and C.P.E. Bach, developed *Würfelspiel* techniques (Potter 1971).

Because the aesthetics of the European classical era were so strict, it was possible to construct a simple music-making game for composing minuets, trios, and other incidental works. The method consisted of applying the outcome of throwing dice (or spinning a spinner, or similar actions) to choosing which of several possible musical motives would be selected from tables of precomposed musical figures. A well-formed piece of music in the classical style would result. The reason that chance does not cause the resulting musical composition to wander is because the compositions are prestructured to be musically interesting by the master composers who set up the tables.

Figure 9.5 shows a fragment of *Würfelspiel* minuet trios attributed to Joseph Haydn (O'Beirne 1968). Only the first two phrases of two of the six original minuet variations are shown, enough to give an idea of how the method works. Variations a and b can be played as perfectly acceptable minuet trios. But the composer cleverly arranged for all variations to have the same harmonic plan and close-enough voice leading so that others could create new minuet trios by interleaving measures from any of the variations so long as they are taken in order across the page. We can create a derivative variation, for example, by alternating measures of a and b, $\{a_1, b_2, a_3, b_4, a_5, b_6, a_7, b_8\}$, which sounds like a pleasant minuet trio.

Altogether, there are six variations of 16 bars in the full score, which by the rule of enumeration would mean there are 6^{16} variations. However, because some of the measures are identical in some of the variations, there are actually "only" 940,369,969,152 enumerations. O'Beirne (1968) lays

Figure 9.5
Musical dice extract, attributed to Joseph Haydn.

out the case for Haydn's authorship in an article that publishes all six of these minuet trios in full. He writes that if attribution to Haydn is correct, "One question remains: have we found one new Haydn item? or six? or—in view of the intended permutation possibilities—something more like 1,000,000,000,000 new Haydn trios!"

There was one other ingredient in the typical *Würfelspiel* setup: the measures were not laid out as obviously as in figure 9.5. Instead, the variations were chopped up into one-measure chunks and entered in an indexed table in random order. This served no purpose other than to obscure the underlying mechanism, to make the process seem more "magical" to the user.

Componium Diedrich Nikolaus Winkel (1773–1826), rightly the inventor of the metronome (see section 2.6.2), is credited as the first to construct an automated music composing machine. (Tiggelen 1987; Buchner 1956). In fact, it seems likely that what he did was to adapt elements of Kirnberger's *Würfelspiel* idea to mechanical form. Winkel's Componium is basically a barrel organ, an *orchestrion,* like the ones used to accompany merry-go-rounds. These instruments encode music by pins protruding from the surface of the barrel that key organ pipes or other musical instruments to play as the drum rotates.

Unlike a standard orchestrion, the Componium was equipped with a second barrel (see figure 9.6). The first barrel encodes several variations of short musical works. A few barrels survive, containing works by Mozart, Moscheles, and Spohr. The second barrel, in conjunction with a complicated gearing apparatus, determines which of the variations will be played from measure to measure, providing a large enumerative set of possible compositions.

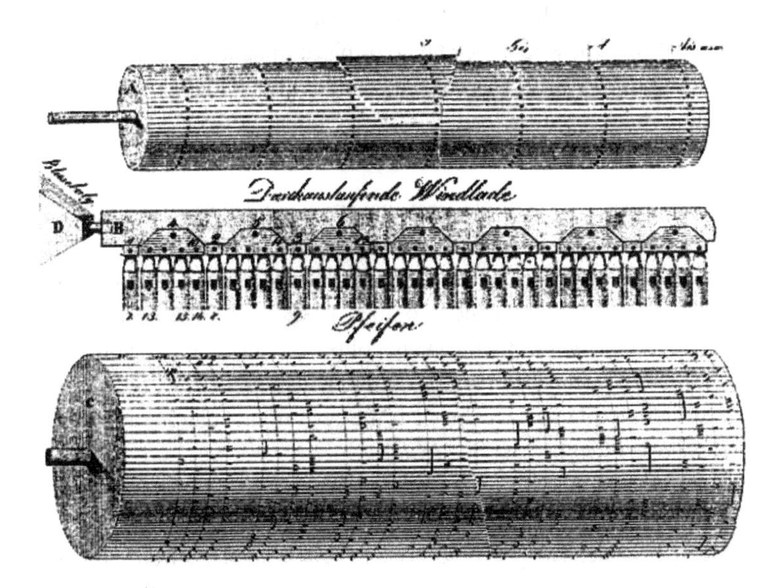

Figure 9.6
Componium of D. N. Winkel. (Buchner 1956.)

Origins of *Würfelspiel* *Würfelspiel* emerged from the era in which probability calculus was pioneered by Blaise Pascal and work on permutations and combinations was done by Jacob Bernoulli. Gerigk (1934) writes,

> This sort of musical game was in the air in the second part of the eighteenth century, though clearly regarded as entertainment only. This is, e.g., expressed by Kirnberger, whom we should also regard as the father of musical literature of this sort in the preface to his composition of this kind (1757). Every game is after all a mirror of the ideas of the times: the rationalistic epoch considers the possibility of mechanical composition.

Many systems of this type were published, including one by Peter Weleker in London in 1775 under the amusing title, *A Tabular System Whereby Any Person without the Least Knowledge of Musick May Compose Ten Thousand Different Minuets in the Most Pleasing and Correct Manner,* which seems to follow Kirnberger's lead (Köchel 1862).

Turning the Tables *Würfelspiel* uses chance as an alternative to personal choice for decisions we do not wish to make or cannot make ourselves. But there are many other reasons to use chance as a source of choice.

The American composer John Cage (1961) was well known for using chance techniques and purposeful silence in his compositions. The way in which he incorporated chance operations in the act of composition invited natural forces to speak directly through his music. Of course, the idea of appreciating the aesthetics of natural forces channeled through the arts did not originate with Cage. We listen to wind chimes and aeolian harps for much the same reasons. Some forms of Japanese painting utilize imperfections in the paper to the same end.

For another example, Santillana and von Dechend (1969), in their landmark work *Hamlet's Mill,* discuss several games from ancient times in which the choice of piece to be moved in chess, for instance, was determined by a throw of the dice. Called "The Game of the Gods" or "Celestial War," these games are documented in texts dating from the fourteenth and fifteenth centuries in India and China.

Chance as Oracle In life, we are affected by natural forces that are beyond our ability to predict and that appear to be utterly random. We observe the effects of our own willful actions and presume by analogy that random natural events can be seen as the "will" of natural forces acting upon us. We externalize our personal will and project it by analogy onto a "cosmic will." If we ourselves deliberately generate chance occurrences such as by throwing coins, we can endow the outcome with prophetic value because the chance occurrences are presumably in alignment with this same natural "cosmic will." Perhaps this idea explains why chance is the basis of oracular methods such as Tarot card readings and the Chinese oracular text, the *I Ching*.[3]

At the root of these oracular methods is the belief that the chance processes on which they are based are synchronized with the "cosmic will": the same force that determines outcomes in our lives determines the chance process used by the oracle. The psychologist Carl Jung coined the word *synchronicity,* which he defined as "meaningful coincidence." This is a purely descriptive term that denotes an association between an objective event and its subjective significance; Jung did not imply a necessary causal association between an objective event and any personal subjective meaning. Of course, many people through the ages have believed that they (or their local soothsayer) knew how to interpret the synchronistic thread between an event and its subjective meaning. But interpreting an oracle requires a way of decoding the message that is supposedly implied by the "cosmic will." As the histories of supplicants at Delphi can attest, this is generally very difficult to do.

However, even if we don't use an oracle in an interpretive way, we can still consider that chance operations incorporated into an art form allow nature to speak to us through that art, and we can appreciate the message aesthetically even if we don't claim to understand it.

9.7 Randomness

Randomness is literally in the eye of the beholder. We can derive randomness from any natural process, such as the flood tides of the Nile, drawing numbered balls from an urn, the motion of wind or waves, the distribution of ink splotches on a page, the motion of atoms near an electrode, or throwing dice.

Heitor Villa-Lobos used the skyline of New York City to create the melody for his composition *New York Skyline*. John Cage composed *Atlas Eclipticalis* using astronomical charts. Charles Dodge used fluctuations of the earth's magnetic field to create a large work of electronic music titled *The Earth's Magnetic Field*.

Are the New York skyline, the contents of astronomical charts, and fluctuations in the magnetic field random processes? Aren't building heights in New York a function of the building codes? Isn't the distribution of stars a function of the laws of gravitation? True, but the central question is epistemological, not physical: *can we determine a formula that exactly characterizes the phenomenon?* If not, it is a random process *to the observer.* Note that this implies there is no randomness without an observer.

9.7.1 What Constitutes Randomness?

The crucial characteristic of useful random processes is that chance events must be independent of each other. By *independent* I mean that even knowing a very large set of outcomes does not help us guess any other outcomes. If the outcomes of a random process are absolutely independent, then the process is infinitely random and aperiodic. Such a sequence constitutes an inexhaustible source of surprise and novelty.

A random process can be viewed as distributed in time or in space. For example, the location of stars in the night sky constitutes a very slowly changing random function of position, one that evolves over millions of years. More typically, we examine the sequential outcomes of a spatially localized random process like a coin toss, which we view as a function of time. This in turn suggests additional qualities of a random process:

- *Uniform distribution* Does the sequence enumerate all possible outcomes? Are all values more or less equally likely, or are some regions favored over others?

- *Uniformity over ranges* What are the permutational characteristics of the sequence? Are there patterns in the data—ranges of numbers that resemble each other in some predictable way? Are subsequent values correlated somehow with previous values?
- *Uniformity over frequency* What is the rate of change of values in the sequence? Do the magnitudes change slowly or quickly? If magnitudes change quickly, the data form a jagged series of abrupt peaks and valleys, corresponding to high frequencies. If magnitudes change slowly, the next number in the sequence won't be very far from previous values, and the data form a smoother, less jagged curve, corresponding to low frequencies.[4]

9.7.2 Pseudorandomness

Computers can only execute methods that are strictly algorithmic, and the effectiveness requirement for algorithms rules out anything that depends upon unknowns; hence computers cannot be a source of true random sequences *by design*. If a computer ever did anything genuinely random, it would have to go in for repairs. Nonetheless, mathematicians have spent a fair amount of effort trying to develop computable sources of randomness. John von Neumann (1963), mathematician and pioneer in computer science, recognized this contradiction and is widely quoted as having said, "Anyone who considers arithmetical methods of producing random digits is, of course, in a state of sin." This is a droll remark, coming as it does from a pioneer of deterministic methods of generating random numbers.

By the principle of independence, we only know that a sequence is perfectly random if it never repeats its choices in whole or in part. But in practice a sequence does not have to be perfectly random to be useful. It need only be "random enough" to surprise us. So randomness is essentially an empirical criterion that we use to characterize processes we can't predict.

Although computers can't generate pure random sequences, there are numerical techniques that allow computers to generate number sequences that are "random enough" for practical use. However, all such computer-generated sequences eventually repeat, so they are *pseudorandom*. I present a simple but effective approach to generating pseudorandom numbers in section 9.7.3, but a brief digression into polynomials is required first.

Polynomials We can express any number N as a *polynomial* of integers in base b, for instance, 123 in base 10, written as

$$123 = (3 \times 10^0) + (2 \times 10^1) + (1 \times 10^2).$$

The ratio of some numbers in some bases produces an infinite *polynomial expansion,* such as

$$\frac{10}{3} = (3 \times 10^0) + (3 \times 10^{-1}) + (3 \times 10^{-2}) + \cdots = 3.33333\ldots.$$

Sometimes a *cyclic polynomial* is produced, such as

$$\frac{13}{7} = 1.\underbrace{857142}\underbrace{857142}\ldots$$

Irrational numbers represented as polynomials in any base produce an infinitely noncyclic sequence of digits. For instance, the irrational number $\pi = 3.1415927\ldots$ shows no apparent pattern in its infinite polynomial expansion. New techniques are available to calculate arbitrary digits of π with good efficiency and without having to know the preceding digits, making this a possible source of random values.[5] We can calculate successive random digits from the fractional values of an irrational number. Cyclic polynomials are quite easy to calculate, and we can generate sequences that are quite long and have good uniformity.

Converting Polynomials to Digit Sequences We can convert any polynomial sequence into a sequence of digits as follows. For some radix base, b, let f be a fraction: $f = a_1 b^{-1} + a_2 b^{-2} + \cdots + a_n b^{-n}$. All the values of a must lie within the radix, that is, they must satisfy $0 \le a_i < b$. (For instance, the decimal system has radix 10, and so values must lie between 0 and 9.) If we multiply the fraction f by b, we will shift the first fractional digit, a_1, out of the fraction and into the units place:

$$bf = a_1 + a_2 b^{-1} + a_3 b^{-2} + \cdots + a_n b^{-n+1}.$$

In this way we have isolated a_1 in the units place. If we repeatedly multiply the result of the previous step by b, we push the next digit out of the polynomial's fractional value into the units place. For example, let $f = 0.2615$, and $b = 10$:

$$0.2615 \cdot 10 = 2.615$$

$$2.615 \cdot 10 = 2\,6.15 \leftarrow$$

$$26.15 \cdot 10 = 26\,1.15 \leftarrow$$

$$261.5 \cdot 10 = 261\,5.0 \leftarrow$$

We can use this technique to extract successive digits to form random number sequences.

9.7.3 Linear Congruential Method

Equation (9.1) shows the linear congruential method for generating random numbers, introduced by D. H. Lehmer in 1948 (Knuth 1973, vol. 2):

$$x_{n+1} = ((a x_n + b))_c, \qquad n \ge 0. \tag{9.1}$$

The notation $((x))_n$ means "x is reduced modulo n." The result is the remainder after integer division of x by n (see appendix A).

Equation (9.1) is a *recurrence relation* because the result of the previous step (x_n) is used to calculate a subsequent step (x_{n+1}). It is *linear* because the $ax + b$ part of the equation describes a straight line that intersects the y-axis at offset b with slope a. *Congruence* is a condition of equivalence between two integers modulo some other integer, and refers here simply to the fact that modulo arithmetic is being used.

For successive computations of x, the output will grow until it reaches the value c. When c is exceeded, the new value of x is effectively reset to $x - c$ by the modulus operation. A new slope will grow from this point, and this process repeats endlessly.

The result can be quite predictable depending upon the values of a, b, c, and x_0. For instance, if $a = b = x_0 = 1$, and $c = \infty$, an ascending straight line at a 45° slope is produced. However, for other values, the numbers generated can appear random.

In practice, the modulus c should be as large as possible in order to produce long random sequences. On a computer, the ultimate limit of c is the arithmetic precision of that machine. For example, if the computer uses 16-bit arithmetic, random numbers generated by this method can have at most a period of $2^{16} = 65{,}536$ values before the pattern repeats.

The quality of randomness within a period varies depending on the values chosen for a, x, and b. Much heavy-duty mathematics has been expended choosing good values (Knuth 1973, vol. 2). For 32-bit arithmetic, Park and Miller (1988) recommend $a = 16{,}807$, $b = 0$, and $c = 2{,}147{,}483{,}647$.

The linear congruential method is appealing because once a good set of the parameters is found, it is very easy to program on a computer. The `LCRandom()` method returns a random number by the linear congruential method each time it is called:

```
//Constants from Park and Miller
Constant Integer a = 16807;  // a, b, c and x are global constant values
Constant Integer b = 0;
Constant Integer c = 2147483647;
Integer x = 1;               // x stores the value produced by
                             // LCRandom between invocations

Integer LCRandom(){
    x = Mod(a * x + b, c);   // update x based on its previous value
    Integer r = x;           // x may be positive or negative
    If (r < 0)               // force the result to be positive
        r = -r;
    Return(r);
}
```

The parameters `a`, `b`, and `c` are constant (time-invariant) system parameters. Parameter `x` is initialized in this example to 1, but it can be initialized to any other integer. The value of `a * x + b` is calculated, the remainder is found modulo `c`, and the result is reassigned to `x`.

While the value of `x` is less than `c`, `x` grows linearly. When the expression `a * x + b` eventually produces a value beyond the range of `c`, then `x` is reduced modulo `c`. The random effect of this method comes from the surprisingly unpredictable sequence of remainders generated by the modulus operation, depending upon careful choice of parameters.

The calculation of `x` ranges over all possible positive and negative integers smaller than the value of `±c`. But it is generally preferable to constrain its choices to a range. To make this conversion easier, we force the result to be a positive integer.

Seeding the Random Number Generator Unlike the natural sources of randomness, `LCRandom()` will always produce the same sequence with the same initial parameters. Different sets of pseudorandom sequences can be generated by varying the initial value of `x`, as with the following function:

```
SeedRandom(Integer s) { x = s; }  // set global variable x to seed s
```

This function allows us to set the initial value of `x`. If we initialize `x` to a parameter such as the current time in seconds from some fixed moment, then we start at a different place in the pseudorandom cycle each time (although, of course, this is finite, too, because the sequence length is necessarily limited).

The linear congruential method is simple and efficient, but it is hardly the best source of random values. Even ignoring the fact that it repeats, its uniformity is not wonderful. Knuth (1973, vol. 2) cautioned, "Random number generators should not be chosen at random." For superior techniques, see Press et al. (1988, 210). However, this method is very simple to implement and has the advantage over natural random processes of providing the same pseudorandom sequence if seeded with the same values.

Random Real Numbers The `LCRandom()` method returns integers between 0 and c. It is straightforward to map its output to any range of `Real` values between an upper bound `U` and a lower bound `L`:

```
Real Random(Real L, Real U) {
    Integer i = LCRandom();        // get a random integer value
    Real r = Real(i);              // convert it to a real value
    r = r/Real(c);                 // scale it to 0.0 <= r < 1.0
    Return(r * (U - L) + L);       // scale it to the range L to U
}
```

First, we use `LCRandom()` to get a random integer. Recall that `LCRandom()` forces the result to be positive. We promote its random integer result to `Real` and store it in `r`. Next, we divide it by `c` so its range is `0.0 <= r < 1.0`. Finally, we scale it by the difference between `U` and `L`, and add `L`, so that the random value is bounded above by `U` and below by `L`. That way we can get a random result from a particular range of values that we can stipulate.

Random Integer Numbers Scaled to an Arbitrary Range We can adapt the `Random()` function to return integers within a specified integer range. When a real value is converted to an integer, we truncate (discard) the fractional part, leaving the integer part. For example,

```
Real x = 3.14159;
Integer i = Integer(x);
Print(i);
```

prints `3`. Truncation is equivalent to the *floor function,* written $\lfloor 3.14159 \rfloor = 3$.

Here is a method to generate integer random values over an integer range.

```
Integer Random(Integer L, Integer U) {
   Real rL = L;                         // convert L to Real
   Real rU = U + 1.0;                   // convert U to Real, add 1.0
   Real x = Random(rL, rU);             // get a real random value
   Return( Integer(x));                 // return it as an integer
}
```

Note that I added 1.0 to the upper real boundary. Truncation of the random result necessitates slightly increasing the top end of the range of choice. For example, in order to choose a value in the integer range 0 to 9, we must generate a random real value x that lies in the range $0.0 \le x < 10.0$. This gives an equal chance of obtaining an integer in the range 0 to 9.

9.8 Chaos and Determinism

Dynamics is a field of classical mechanics that studies how force affects motion of material bodies through time. A system is *dynamical* if its subsequent state depends upon its current and previous states. A flying airplane is an example of a dynamical system. Suppose x_n represents the current position of an airplane, and x_{n+1} represents its next position. Then the relation between these two positions,

$$x_{n+1} = f(x_n), \tag{9.2}$$

is dynamical because its subsequent state (x_{n+1}) is a function f of its previous state (x_n). Equation (9.2) is another example of a recurrence relation because it shows the relation between subsequent values of a function.

A dynamical system may depend upon its current inputs as well as its past outputs. For example, the airplane's position will also depend upon the operation of its controls and the forces of the air.

A system is *deterministic* if every cause has a unique effect. The uniqueness requirement goes from cause to effect but not necessarily from effect to cause. For example, the function $y = x^2$ is deterministic because y can be predicted given x, but one can't necessarily deduce x given y because there may be two choices.

Because the `LCRandom()` method is a deterministic way of generating what appear to be random values, it is a *chaotic system*. The term *chaotic* has been taken by physicists to mean a deterministic system that appears to be random, such that it is impossible to make long-range predictions about its behavior. Although it repeats over large spans of time, a simple system like `LCRandom()` can behave so unpredictably in the short run that it would be very difficult to deduce its rather simple generating structure from its output alone.

A chaotic system is one that we know to be deterministic but that appears to be random. A truly random system is nondeterministic. Therefore, pseudorandom systems are chaotic, not random.

9.8.1 Sensitivity to Initial Conditions

A key characteristic of chaotic systems such as `LCRandom()` is their *sensitivity to initial conditions*. Even the smallest changes to variable `x`, the random seed, can lead over time to totally different behaviors of the system to a point where the differences far overshadow the similarities.

Natural examples of chaotic dynamical systems include the earth's atmosphere and the vibrations of virtually all sources of musical sound, such as the scrape of a bow on the strings or the turbulent flow of air from the player's lips over the fipple of a flute. Small differences in initial conditions can be amplified by such systems to such an extent that any error in measuring the initial conditions can render any long-range forecast of system behavior wildly inaccurate, even if there is no further disturbance to the system. The weather from day to day is never exactly the same. Notes played on a flute, though they may sound alike, are never exactly the same. Our ears gloss over these differences, hearing sound categorically. But if we wish to understand the precise mechanism of a dynamical system so as to accurately predict its behavior over time, the initial conditions must be known exactly.

By using more accurate measurements on such natural systems, we can reduce but not eliminate measurement uncertainty. But only if we measured with infinite precision—an impossible task—would we be able to eliminate all uncertainty, and only then would the initial conditions allow us to obtain utterly predictable behavior from a model of a dynamical system. The implicit Western scientific assumption has been that we can continue to shrink the uncertainty of a dynamical system's outcome by measuring its initial conditions with ever greater precision. Thus, we assume that more nearly perfect predictions could be made by supplying more precise initial conditions.

However, through the work of the mathematician Henri Poincaré (1854–1912), we know that there are systems whose long-term predictions are not improved by increased precision of the initial conditions. While studying the gravitational influences of three bodies upon each other, he discovered that under certain circumstances, even if the initial uncertainties are infinitesimal, the predicted outcomes can be so different that the deterministic prediction is really no better than if the prediction had been made by chance. This is how sensitivity to initial conditions is tied to the appearance of randomness.

To illustrate this point, Edward Lorenz (1972), another pioneer in chaos theory, wrote a paper titled "Predictability: Does the Flap of a Butterfly's Wings in Brazil Set off a Tornado in Texas?"[6] Unlike the debate about the number of angels that can dance on the head of a pin, the answer to Lorenz's question (yes) has dramatic consequences for the limits of epistemology. Many if not most of the basic systems in life, such as the weather, are chaotic dynamical systems, and we are unable to predict the long-range behavior of any such system whose initial conditions we don't know with infinite precision. Alas for the human condition, this explains why we are blind to the future until it is upon us. This is the glass cage that confines our Faustian desires.

9.8.2 Complexity Theory

Complex dynamical systems such as clouds can be seen from a reductionistic perspective as merely disorganized collections of water droplets. However, these systems also have an evident self-organizing flow. For example, the shape of a cloud will grow and transform in a manner that reveals an emergent internal structure. We can summarize this by saying that unconstrained complex dynamical systems have a natural and innate tendency to move toward complexity.

A *complex system* contains elements that are both differentiated (specialized or compartmentalized) and integrated (connected or unified) on all levels of scale. Its complexity comes about through the interaction of internal and external constraints. For example, the internal constraints of a cloud are the molecular forces of the air and water, and the external constraints are the winds that drive it. The internal constraints of the brain are the synaptic connections, and the external constraints are the flow of information from outside events and other minds. The internal constraints of music are the criteria of musical perception and cognition, and the external constraints are the flow of expectation from musician to listener and the return flow of interest from listener to musician.

When a system is not in complexity, it tends toward monotony (saturated integration) or cacophony (disintegration).

What benefit does complexity provide to a system? Why do clouds not make geometric patterns in the sky or devolve into utter randomness? The reason is that when a system moves toward complexity, it is in its most stable, adaptive, and flexible state. When a brain is stuck in linear thinking or lost in confusion, it may not thrive. When music is not in complexity, we stop listening.

These characteristics of self-regulation are cornerstones of healthy responsiveness to life and mental well-being. How appropriate that stability, adaptability, and flexibility are also hallmarks of successful music. How interesting it is that these qualities emerge through the dynamic interplay of differentiation and integration on all levels of scale in a musical work. How natural it seems to think of music as embodying these core principles of stamina and health. Here is the foundation for a music theory that weaves together information theory, chaos theory, complexity theory, cognitive psychology, and nonlinear dynamics in a way that honors music's therapeutic capacities.

9.9 Combinatorics

The discussion now shifts to more practical concerns. If composing is about methodologies of choice, it is worth wondering about the range of choices that various musical systems provide to the composer. These questions are studied by the field of combinatorics.

As the name suggests, *combinatorics* is the study of how sets can be combined in patterns. This includes enumerating all the possible permutations of a set. Some musical questions opened up by combinatorics include the number of orderings of musical motives within a scale, the total number of diatonic scales, and the number of possible melodies of a certain length. In the early twentieth century, composers of the second Viennese school associated with Arnold Schoenberg borrowed ideas from the mathematics of combinatorics to construct a radically different kind of music than had ever been heard before. This brief study of combinatorics leads to an overview of their techniques.

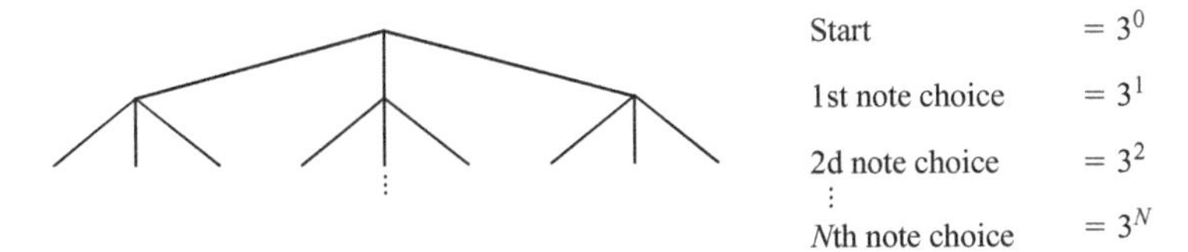

Figure 9.7
Guido's method expressed as a tree of possibilities.

9.9.1 Enumeration

If we examine the possible outcomes of Guido's method, we see that there are three choices at each step. For a one-vowel text there are 3 possible one-note melodies; for a two-vowel text, there are 3^2 melodies of two notes; and for an N-vowel text there are 3^N melodies. Thus the number of possible melodies grows exponentially for longer texts (figure 9.7). This demonstrates the principle of enumeration:

If there are m outcomes of operation 1, and then n outcomes of operation 2, the composite number of outcomes of operation 1 followed by operation 2 is m times n.

For instance, for Guido's method, m and n are both 3. So for step 2, the number of outcomes is $3 \cdot 3 = 9$ (see figure 9.7).

Enumerating the possibilities of something means itemizing all possible outcomes. Counting all such outcomes is to enumerate them. For instance, how many 12-note melodies can be formed from the dodecaphonic scale? By the principle of enumeration, there are 12 possibilities for the first note, and then 12 possibilities for the second note, and then . . . through 12 steps. So the answer is $12 \cdot 12 \cdot 12 \cdot \cdots = 12^{12} \approx 8.9 \times 10^{12}$, which is nearly nine trillion. The set of melodies includes, for instance, the ascending and descending chromatic scales, the first 12 notes of Antonio Carlos Jobim's *One Note Samba* transposed to all 12 pitches, and the first 12 pitches of every part of every symphony, and opera that has ever been written, or could ever be written, in a dodecaphonic scale. There are more 80-note chromatic melodies than there are subatomic particles in the universe (assuming there are 10^{80} or so such particles). I suppose this makes it pretty unlikely that future composers will run out of material to work with!

9.9.2 Permutation

The principle of enumeration answers the question, How many total outcomes are possible? The principle of permutation answers the question, How many *unique orderings* are possible?

For instance, how many ways are there to order the sequence, a, b, c? We find out by swapping the elements around until we run out of unique orderings. Let's use the method where we swap the last two elements, then the previous two elements, and so on (figure 9.8). We can create six permutations this way before the reordering procedure recreates the original ordering. So there can be six permutations of three things. But how could we discover the number of possible permutations without having to reorder and inspect them?

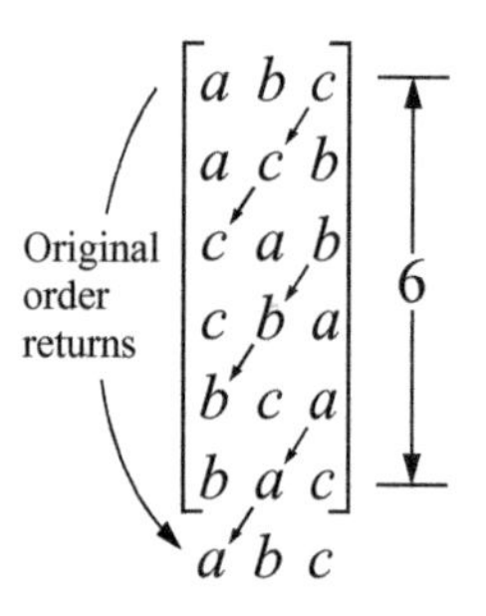

Figure 9.8
Permutation of three objects.

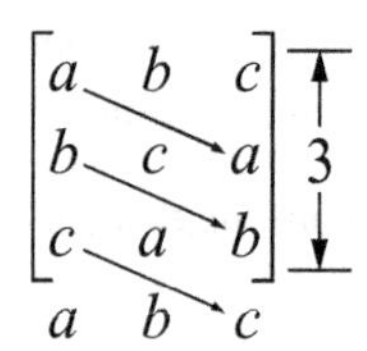

Figure 9.9
Example of circular permutation.

Let's find the solution through another musical example. How many unique 12-tone rows are there in the set of dodecaphonic scales? Recall that when we enumerated all the 12-note melodies, we could pick from all 12 pitches at every step. That led to melodies with repeated notes. But a tone row is defined as a melody of 12 nonrepeating pitches, so we must exclude whatever pitch is chosen from subsequent choices. We can choose from 11 pitches for the second note, 10 for the third, and so on. Otherwise, the process is just like enumeration. Thus the number of unique orderings is $12 \cdot 11 \cdot 10 \cdot 9 \cdot \cdots = 12! \cong 4.7 \times 10^8$, or slightly more than 470 million 12-tone rows in the dodecaphonic system. As one might expect, there are substantially fewer permutations of 12 tones than there are enumerations of them. Thus there are $n!$ permutations of n objects. Going back to the first example, there are $3 \cdot 2 \cdot 1 = 6$ permutations of three objects.

9.9.3 Circular Permutation

A *circular permutation,* or *rotation,* occurs when the element at one end is lopped off and attached to the other end circularly (figure 9.9). There are n circular permutations of n objects. Rotation can be to the left or right by one or more places.

Here is a method that rotates a list by an arbitrary number of places either to the right or left:

```
Rotate(IntegerList Reference f, Integer n, Integer i = 0){
  n = Mod(n, Length(f)); //constrain rotation to length of list
```

```
   Integer x = f[i];        //store f[i] for use after recursion
   If(i < Length(f)-1)      // reached the end?
      Rotate(f, n, i+1);    // no, call Rotate() recursively

   // continue from here when the recursion unwinds
   Integer pos = PosMod(i+n, Length(f)); // index list modulo its length
   f[pos] = x;              // assign value of x saved above
}
```

This example uses recursion to perform its function. It takes three arguments, a list `f`, the number `n` of positions to rotate by, and `i`, the index for where to begin, usually set to zero. If `n` is positive, the list is rotated to the right that many places; if `n` is negative, the list is rotated that many positions to the left. The first step is to constrain `n` modulo the length of the list so that any amount of rotation can be handled.

The declaration `IntegerList Reference f` requires a bit of explanation. We want `Rotate()` to modify the list that is supplied. But functions are ordinarily supplied only with copies of the value of the actual arguments (see appendix B, B.1.22). The word `Reference` in the declaration tells MUSIMAT that it should supply `Rotate()` with the actual variable named when the function is invoked. Thus changes to the list handed to `Rotate()` will persist after the function is finished.

We need to make sure the variable `pos` stays within the range of valid list elements, which naturally suggests the use of `Mod()`, except that `Mod()` can return negative values. But list indexes must be strictly positive. So we use a function called `PosMod()`, which returns only the positive wing of modulo values (see appendix A.6).

Table 9.3 shows left and right rotation by various amounts for a list `L` defined as

```
IntegerList iL = {0, 1, 2, 3, 4, 5};
```

9.9.4 Partitioning

Suppose we want to create a 12-tone row consisting of the 12 pitches partitioned into three motives of six notes, three notes, and three notes each. How many different motives could there be? Clearly,

Table 9.3
Left and Right Rotation

Rotate(iL, −n, 0)							Rotate(iL, n, 0)						
$n = -0$	0	1	2	3	4	5	$n = 0$	0	1	2	3	4	5
$n = -1$	1	2	3	4	5	0	$n = 1$	5	0	1	2	3	4
$n = -2$	2	3	4	5	0	1	$n = 2$	4	5	0	1	2	3
$n = -3$	3	4	5	0	1	2	$n = 3$	3	4	5	0	1	2
$n = -4$	4	5	0	1	2	3	$n = 4$	2	3	4	5	0	1
$n = -5$	5	0	1	2	3	4	$n = 5$	1	2	3	4	5	0
$n = -6$	0	1	2	3	4	5	$n = 6$	0	1	2	3	4	5

the total number of unique tone rows is still 12!. But the number of unique motives should be fewer than 12! because the tone row space is divided into groups.

Here's a possible way of doing it:

1. Assign pitches to the motives. For instance, assign pitches C–F to the six-note motive, pitches F♯– G♯ to the first three-note motive, and pitches A–B to the second three-note motive.
2. Order the pitches in the first motive.
3. Order the pitches in the second motive.
4. Order the pitches in the third motive.

We don't know how many outcomes are possible for step 1 yet, so let's call this the unknown, x, for now.

Steps 2, 3, and 4 are ordering operations. Because ordering operations are permutations, there are 6!, 3!, and 3! orderings in steps 2, 3, and 4, respectively.

Note that steps 1–4 *enumerate* the steps of creating a tone row according to the motivic arrangement. Remembering the rule for enumeration, that means the total number of unique tone rows would be $6!3!3!x$. But since the total number of unique tone rows is 12!, we can equate these two pieces of information, yielding $6!3!3!x = 12!$. Solving for x yields the number of unique motives:

$$x = \frac{12!}{6!3!3!} = 55{,}440 = \binom{12}{6,\ 3,\ 3}. \tag{9.3}$$

Thus 12 pitches can be partitioned into 55,440 motives of three subsets of six, three, and three notes. The rightmost term in (9.3) shows how partitioning is notated. It is read as "the number of ways 12 objects can be partitioned into groups of six, three, and three."

Generalizing from this particular solution, we can express partitioning N objects into p subsets of r_p elements as

$$\binom{N}{r_1, r_2, r_3, \ldots, r_p} = \frac{N!}{r_1!r_2!r_3!\ldots r_p}. \qquad \textit{Partitioning} \tag{9.4}$$

9.9.5 *N* Objects *R* at a Time

Suppose we select seven pitches from the 12 semitones and order them into a seven-note melody. How many such melodies are there? How many ways are there to select seven notes out of 12? We can think of this as a kind of partitioning because ordering the melody partitions it into eight subsets: the first note, the second note, and so forth, up to the seventh note. The eighth subset is the unchosen pitches out of the original 12, which is $12 - 7 = 5$ pitches. So we can use the partitioning formula, (9.5), as follows:

$$\binom{12}{1, 1, 1, \ldots, 5} = \frac{12!}{5!} = 3{,}991{,}680,$$

that is, 3,991,680 melodies of 12 pitches taken seven at a time. Because $5! = (12 - 7)!$ we can express the same thing this way:

$$\frac{12!}{(12-7)!} = 3{,}991{,}680.$$

This is convenient because 12 represents the total number of elements and 7 represents the size of the partition. Abstracting based on this example, in general there are

$$\frac{n!}{(n-r)!} \tag{9.5}$$

permutations of n objects taken r at a time.

9.9.6 Combinations

How many seven-note *scales* are there in the 12 pitches of the dodecaphonic system? This is like taking N *unordered* objects R at a time. It seems reasonable to expect that there will be fewer scales of seven pitches than melodies of seven pitches because melodies can repeat a note whereas scales cannot.

We divide the pitches into two groups: seven chosen pitches, and five unchosen pitches. By (9.4), there must be

$$\frac{12!}{7!(12-7)!},$$

or 943 such scales. This is a rather large number in comparison to the dozen or so of those commonly in use. So, in general, a partitioning of $\binom{n}{r,\, n-r}$ possible outcomes equals $\frac{n!}{r!(n-r)!}$ actual outcomes.

This is used commonly enough to have its own notation and is usually written

$$\frac{n!}{r!(n-r)!} = \binom{n}{r}. \qquad \textit{Taking n Unordered Objects r at a Time} \tag{9.6}$$

9.10 Atonality

Combinatorics can guide us to a deeper understanding of the compositional aims of atonal music. I examine this in some depth because composers of atonal music pioneered the use of explicit compositional methodology to a degree that had not previously been attempted in music. Thus atonal music is a fruitful field of study for compositional methodology.

In the early part of the twentieth century, Arnold Schoenberg, Alban Berg, and Anton Webern, the composers of the so-called second Viennese school, developed a compositional method based on note patterns that contain all 12 pitches (see section 3.16).

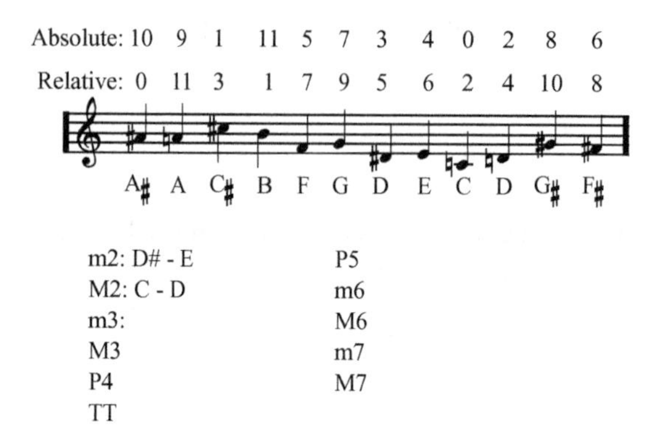

Figure 9.10
Tone row for Schoenberg's *Fantasy for Violin and Piano,* Opus 47.

In Schoenberg's original method, each composition was organized around a particular ordering of the 12 pitch classes of the chromatic scale that he called a *tone row* (see section 2.4). No pitch appears more than once within the row, so none of the 12 pitch classes is favored. Schoenberg's idea was to use this method to remove any vestiges of tonal harmony from his music, hence to compose atonal music.

For example, the row shown in figure 9.10 appears in Schoenberg's *Fantasy for Violin and Piano,* Opus 47. The pitch classes can be numbered in two ways:

- Absolute pitch numbers, indexed by chromatic half steps above C
- Relative pitch numbers indexed by half steps from the first pitch in the row

Relative indexing has some advantages that will become evident later, so I use that from now on.

The basic method is as follows. Each time a new tone is needed in the composition, the composer picks the next pitch class in the row, circling back to the first pitch class when the list is exhausted.

Since the primary aim is to remove tonal references, and other considerations are secondary, the way in which each pitch class is projected into the composition is left up to the composer. The pitch classes can be freely applied to any octave, assigned to any instrument, and given any desired dynamic level, rhythmic value, or performance articulation. Some pitch classes derived from a row might be used to generate a musical line while others might be used to spell a chord, for example.

The tone row and the plan for how it is to be projected into the composition are separate steps taken prior to actual composing. This planning stage is called *precomposition.* The following sections describe some of the theory of sets and sequences upon which atonal music theory is based.

9.10.1 Series

In general, a *set* is an unordered collection of any size. A *series* is a particular ordering of a set. A tone row is a series based on a set of pitch classes (Forte 1973). A tone row may contain all or

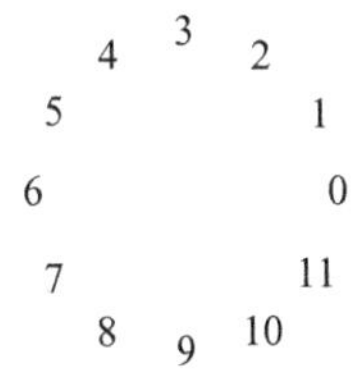

Figure 9.11
Chroma.

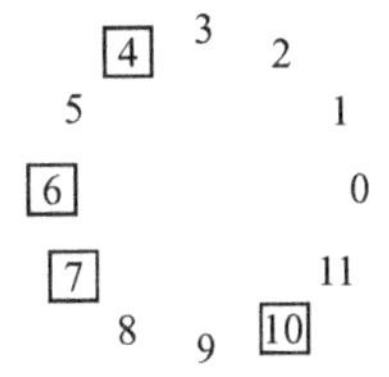

Figure 9.12
The set {4, 6, 7, 10}.

part of the available pitch classes. Since the distinguishing characteristic of each pitch class is its chroma (see figure 6.5), the pitch classes can be characterized circularly (figure 9.11).

There are over 470 million 12-tone rows in the dodecaphonic system (see section 9.9.2). If we add to this all rows of less than 12 tones, there are a great many more. But many of them share characteristics that make them seem related. How can we tell them apart, and how can we characterize their similarities?

For example, consider the set of pitch classes {4, 6, 7, 10}. By octave equivalence as well as by circular permutation, we could relate this set to the sets {6, 7, 10, 4}, {7, 10, 4, 6}, and {10, 4, 6, 7}. These sets are equivalent except for their starting points (figure 9.12). They are *equivalent under circular permutation.* It would be nice to give them a name that reflects their equivalence. We could name the whole collection after just one of them, but which of these permutations should we consider to be the principal one?

Since the distinguishing characteristic of a row is the placement of different-sized intervals, let's arbitrarily make a rule that the *normal form* of a set lists the pitch classes in *ascending numeric order* (corresponding to counterclockwise motion around the circle) in the *intervallically most compact form.* A set is most compact whose interval size between the first and last pitch class is smallest, modulo 12. For example, with the preceding set permutations, and using the notation $((x))_{12}$ to denote x modulo 12, we have

{4, 6, 7, 10}	{6, 7, 10, 4}	{7, 10, 4, 6}	{10, 4, 6, 7}
$((10-4))_{12} = 6$ (boxed)	$((4-6))_{12} = 10$	$((6-7))_{12} = 11$	$((7-10))_{12} = 9$

Because the order $\{4, 6, 7, 10\}$ yields the smallest difference (6) between first and last pitch, this is the normal form for this set. The name for this set of permutations is then [4,6,7,10]. (The set name is written with brackets and commas without spaces.) This is the way to name all sets that are equivalent under circular permutation.

If multiple orderings tie for compactness, we need another rule to break the tie. In this case let's make a rule that the normal form for a set is the one *most compact to the left*. So, for example, for the set $\{0, 3, 6, 7, 9\}$,

	$\{0, 3, 6, 7, 9\}$	$\{3, 6, 7, 9, 0\}$	$\{6, 7, 9, 0, 3\}$	$\{7, 9, 0, 3, 6\}$	$\{9, 0, 3, 6, 7\}$
Tied	$((9-0))_{12} = 9$	$((0-3))_{12} = 9$	$((3-6))_{12} = 9$	$((6-7))_{12} = 11$	$((7-9))_{12} = 10$
Tied	$((7-0))_{12} = 7$	$((9-3))_{12} = 6$	$((0-6))_{12} = 6$		
Most compact		$((7-3))_{12} = 4$	$((9-6))_{12} = 3$		

so we name this set [6,7,9,0,3].

If a set is so regular that there is no tie breaker, then pick the ordering that begins with the lowest number. For example, $\{2, 6, 10\}$ has permutations $\{6, 10, 2\}$ and $\{10, 2, 6\}$. Name this one [2,6,10].

Let's denote the operations required to normalize a set as $N(x)$, where x is the set to be normalized. Then, for example, we can write

$$N(\{0, 3, 6, 7, 9\}) = \{6, 7, 9, 0, 3\}.$$

Set Classes In the previous example, several sets were seen to be related in an algorithmic way (by circular permutation), so we grouped them together under the name of one of the sets that had a particularly elegant form: [2,6,10]. A *set class* is a named group of sets that are equivalent under specific conditions. In the example, the sets

$$\{\{2, 6, 10\}, \{6, 10, 2\}, \{10, 2, 6\}\}$$

form a set class named [2,6,10] of sets that are equivalent under circular permutation. There are many ways in which sets can be related into set classes, but the following relations are particularly useful.

Transposition The sets $\{4, 6, 7, 10\}$ and $\{9, 11, 0, 3\}$ are related by transposition because if we transpose each pitch class in the first set up by five semitones (modulo octave equivalence), it equals the second set. Therefore, these two sets are *equivalent under transposition* (see section 2.5.4). We can define transposition as

$$T_n(x) = ((x + n))_{12}, \qquad \textit{Transposition} \quad (9.7)$$

where x is the pitch class to be transposed, and n is the number of degrees by which to transpose it. To transpose up by 5 we can write

$$T_5(\{4, 6, 7, 10\}) = \{9, 11, 0, 3\}.$$

To transpose down by 4,

$$T_{-4}(\{4, 6, 7, 10\}) = \{0, 2, 3, 6\}\,.$$

There are 12 unique transpositions (including the zeroth) of 12 pitch classes. We collectively name them after the transposition with the smallest initial value, so this set class would be named [0,2,3,6].

Inversion We can create a mirror image of a set by subtracting each pitch class from the number of elements it contains. We can define inversion as

$$I(x) = ((N - x))_N, \qquad \textit{Inversion} \quad (9.8)$$

where N is the number of available pitch classes, in this case, 12. The modulo operation is needed to handle the case where $x = 0$. For example, the sets {4, 6, 7, 10} and {8, 6, 5, 2} are *equivalent under inversion* because

$$((12 - 4))_{12} = 8$$

$$((12 - 6))_{12} = 6$$

$$((12 - 7))_{12} = 5$$

$$((12 - 10))_{12} = 2.$$

So we can write $I(\{4, 6, 7, 10\}) = (8, 6, 5, 2\}$. Because of this relation, we can also classify {8, 6, 5, 2} as a member of the set class [4,6,7,10].

It is easy to visualize the effect of inversion by imagining a line bisecting the circle of pitch classes horizontally (figure 9.13). Pitch classes related by inversion are mirror opposites above and below this line. This figure shows the original form {4, 6, 7, 10} being inverted by reflection across the bisecting line into {8, 6, 5, 2}.

Retrograde Sets that are related by having their members in reversed order are *equivalent under retrogression*. If $R(x)$ denotes the retrograde of a set x, then we can write, for example, $R(\{4, 6, 7, 10\}) = \{10, 7, 6, 4\}$ and also classify {10, 7, 6, 4} as a member of set class [4,6,7,10].

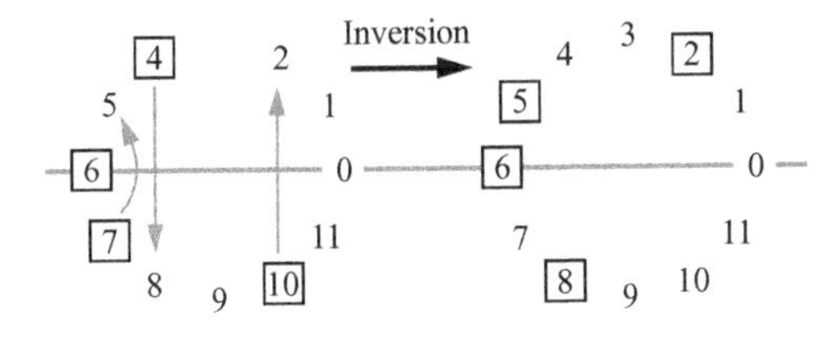

Figure 9.13
Inversion.

Prime Form All the sets that are equivalent under circular permutation, transposition, retrogression, and inversion can usefully be grouped into a single set class because they can all be derived from each other using these operations. But which set should we select as the *primogeniture*—the "mother of all sets"—in its class? The standard convention is to choose the set that

- Is in normal form
- Is most compact to the left
- Is transposed so that its first pitch class starts at zero.

This set is the *prime form* of the set class. All other members of the set class are derived from the prime form.

We find the prime form of {4, 6, 7, 10} as follows. Make its first pitch class start at 0: T_{-4}({4, 6, 7, 10}) = {0, 2, 3, 6}, and make sure the result is in normal form, which it is. We must compare this with its inversion to see which is more compact, so: *I*({4, 6, 7, 10}) = {8, 6, 5, 2}, and its normal form is *N*({8, 6, 5, 2}) = {2, 5, 6, 8}.

Finally, we must transpose it, so T_{-2}({2, 5, 6, 8}) = {0, 3, 4, 6}. Since {0, 2, 3, 6} is more compact to the left than {0, 3, 4, 6}, we name this set complex [0,2,3,6].

Interval Classes So far we have examined just the pitch class content of sets, but a set's interval content is what provides its musical signature. The interval content is the set of interval classes between all pitch classes of the set. For example, the interval classes for the set {4, 6, 7, 10} are

Interval	4–6	4–7	4–10	6–7	6–10	7–10
Distance	2	3	6	1	4	3

This set of intervals, {2, 3, 6, 1, 4, 3}, is the intervallic signature of this set. These interval distances appear in every member of the set class, giving all members of the class the particular sound of the set class. To borrow an example from tonal harmony, the triads sound like triads because they share the same interval distances: major third, minor third, and perfect fifth (see section 3.10.2). Similarly, augmented and diminished triads are distinct to our ears because of their characteristic interval distances.

We can characterize the intervallic profile of a set by making a *histogram* (a simple ordered tally) of the number of intervals it contains. In the preceding example, there are two instances of interval distance 3, and all the rest of the intervals appear only once. If we think of the various interval classes as making up a set of orthogonal dimensions in interval space, we can consider the number of repetitions of each interval as the length of a vector in that interval's dimension. The combination of all these vectors makes a single, unique multidimensional vector characterizing the intervallic content of the set. For example, the interval class vector for the set shown above is:

Interval class	1	2	3	4	5	6
Quantity	1	1	2	1	0	1

so the *interval class vector* is [1,1,2,1,0,1], which is the profile of its unique intervallic content and hence the signature of its unique sound.

Cardinality The number of unique pitch classes in a set is its *cardinality*. The maximum cardinality in the dodecaphonic system is, of course, 12, and there is only one class in this set: the *aggregate set*, containing all 12 pitch classes. The minimum cardinality of a set of intervals is 2. Cardinalities between 2 and 11 have the following Latin names: diad, trichord, tetrachord, pentachord, hexachord, heptachord, octachord, nonachord, decachord, and undecachord.

Complement Relation If a set class contains fewer than 12 pitch classes, the pitch classes that are left out are its *complement set class*. For example, the whole-tone scale has two versions: {0, 2, 4, 6, 8, 10} and {1, 3, 5, 7, 9, 11} that are complement set classes (see section 2.5.7).

9.11 Composing Functions

In mathematics a function is *composable* with another if it can be the other's argument. For instance, if $y = f(x)$, and $z = g(y)$, then $z = g(f(x))$ is the *composition* of g with f. Consider the definitions $f(x) = x + 1$, and $g(x) = x^2$. If $z = g(f(x))$, then $z = x^2 + 2x + 1$. To be composable, the range of f must be a subset of the domain of g.

Following Ada Lovelace's train of thought quoted at the beginning of this chapter, let's use MUSIMAT to create a short excerpt of atonal music using function composition.

9.11.1 Precomposition

The process of composing atonal music is typically divided into two parts.

- Precomposing: assembling the musical materials
- Composing: applying the assembled materials in a design

MUSIMAT already has a number of data types and operations, but a few more are needed:

- To represent pitches as symbols with integer values:

```
Integer C = 0, Cs = Db = 1, D = 2, Ds = Eb = 3 . . ., B = 11;
```

- To represent motives as lists:

```
IntegerLista = {F, F, G, A}; IntegerListb = {F, A, G}; IntegerListc = {F, E};
IntegerList d = {Bb, A, G, F}; IntegerList e = {E, C, D, E, F, F};
```

- To combine motives and concatenate lists:

```
IntegerList y = Join(a, b, a, c, a, d, e);
```

(y is defined as the list of pitches of the tune "Yankee Doodle.")

- To transpose a pitch set:

```
IntegerList transpose(IntegerList L, Integer t) {
    For(Integer i = 0; i < Length(L); i = i + 1)
        L[i] = Mod(L[i] + t, 12);
    Return(L);
}
```

- To invert a pitch set:

```
IntegerList invert(IntegerList L) {
    For(Integer i = 0; i < Length(L); i = i + 1)
        L[i] = Mod(12 - L[i], 12);
    Return(L);
}
```

- To take the retrograde of a set:

```
IntegerList retrograde(IntegerList L) {

    Integer n = Length(L);
    IntegerList R = L;                  // make a new list as long as L
    For(Integer i = 0; i < n; i = i + 1)
      R[i] = L[n - i - 1];
    Return(R);
}
```

9.11.2 The Set Complex

Using these tools, we can create a matrix containing the prime form, inverse, retrograde, and all transpositions of any row, called the *set complex*. The purpose of these transformations is to generate variants that are related to the original intervallic structure of the prime row, to be used as material in developing compositions.

`Matrix` is simply a two-dimensional grid, or list of lists, all of the same length. The individual elements of `Matrix` can be accessed by extending the index operator [...]. The first operand is the matrix, the second is the row position, and the third is the column position. (Whether row or column comes first is arbitrary. The following order is called row/column order.)

M =

	0	1
0	A	B
1	C	D

For example, for this matrix, `M[0][0] == A`, `M[0][1] == B`, `M[1][0] == C`, and `M[1][1] == D`. The following is a method for creating a set complex. It basically copies the prime form to the zeroth row, then copies the inverse form to the zeroth column, then for each other cell in the matrix sums the corresponding row and column value modulo the length of the prime form:

```
Matrix setComplex(IntegerList prime) {
   Matrix M;
   Integer len = Length(prime);
   IntegerList inverted = invert(prime);
   For (Integer i = 0; i < len; i = i + 1) {
        M[0][i] = prime[i];
        M[i][0] = inverted[i];
   }
   For (Integer i = 1; i < len; i = i + 1) {
        For (Integer j = 1; j < len; j = j + 1) {
          M[i][j] = Mod(M[i][0] + M[0][j], len);
        }
   }
   Return(M);
}
```

To demonstrate these tools, table 9.4 shows the set complex for Schoenberg's Opus 23 #5, *Five Piano Pieces*. The prime set {C♯, A, B, G, G♯, F♯, A♯, D, E, D♯, C, F} is shown in numeric form along the top row. Prime rows are read left to right, retrograde rows right to left, inverse rows top to bottom, and retrograde inverse rows bottom to top.

This completes the precomposition phase. Now it's time to look at methods to traverse the rows created with the preceding techniques to generate a composition.

9.12 Traversing and Manipulating Musical Materials

Having arranged the materials from which a composition is to be derived, we now consider methods to traverse these materials in structured ways. Following are a few ways rows can be traversed to structure tonal or atonal melodies, rhythms, dynamics, articulation, instrumentation, or anything else that can be parameterized.

9.12.1 Deterministic Serial Methods

This section demonstrates some methods for iterating through tone rows. They are deterministic because their outcomes do not rely on chance. They are serial because they iterate through lists. Their use is not limited to tone rows but can be extended to arbitrary lists of data.

The basic idea is to supply a list of musical materials to a method that will select and return list elements one at a time in a chosen order.

Cycle This method iterates a sequence either forward or backward. It can either select successive elements or skip through the list. When it reaches the end of the list (either end), it starts over at

Table 9.4
Set Complex for Schoenberg's Opus 23 # 5

Prime

Inverse

0	8	10	6	7	5	9	1	3	2	11	4
4	0	2	10	11	9	1	5	7	6	3	8
2	10	0	8	9	7	11	3	5	4	1	6
6	2	4	0	1	11	3	7	9	8	5	10
5	1	3	11	0	10	2	6	8	7	4	9
7	3	5	1	2	0	4	8	10	9	6	11
3	11	1	9	10	8	0	4	6	5	2	7
11	7	9	5	6	4	8	0	2	1	10	3
9	5	7	3	4	2	6	10	0	11	8	1
10	6	8	4	5	3	7	11	1	0	9	2
1	9	11	7	8	6	10	2	4	3	0	5
8	4	6	2	3	1	5	9	11	10	7	0

Retrograde Inverse

Retrograde

the other end. Its inputs are

- The list to traverse
- The previous position in the list
- Whether to move forward (prime) or backward (retrograde)

Its output is the next element in sequence based on its previous position in the list. As a side effect, it updates its position in the list.

If it traverses the list forward, it returns to the head of the list when it goes past the tail. If it traverses the list in retrograde, it returns to the tail of the list when it goes past the head.

In the following code example, setting `inc` to `1` moves forward one element every time `cycle()` is called, and setting `inc` to `-1` moves backward one element at a time. Setting `inc` to any other value skips through the list by that amount, wrapping around at the ends.

```
Integer cycle(IntegerList L, Integer Reference pos, Integer inc) {
  Integer i = PosMod(pos, Length(L)); // compute current index
  pos = PosMod(pos + inc, Length(L)); // compute index for next time
  Return(L[i]);
}
```

The `pos` argument keeps track of the position in the list. We wish to delegate to `cycle()` the job of managing the list position, so we declare `pos` as a `Reference` argument. Thus, when `cycle()` updates `pos`, the corresponding actual argument is changed. (If `pos` were not a `Reference` variable, any changes `cycle()` made to its value would be lost when it returns (see appendix B, B.1.22).

Here's an example of invoking `cycle()`:

```
IntegerList L = {10, 11, 12};
Integer myPos = 0;
Integer n = 2 * Length(L) -1; // go 1 less than two times through list
For (Integer i = 0; i < n; i = i + 1)
  Print(cycle(L, myPos, 1)); // 1 = forward direction
Print("myPos=", myPos);
```

This program prints `10, 11, 12, 10, 11`. Last, it prints `myPos=2`, proving that `cycle()` is changing the `myPos` parameter.

Palindrome We can iterate a sequence in prime order until the last element in the sequence is reached, then iterate the sequence retrograde until the first element in the sequence is reached, then repeat.

```
Integer palindrome(IntegerList L, Integer Reference pos, Integer
   Reference inc) {
   Integer curPos = pos;
   Integer x = cycle(L, pos, inc);
   If (curPos + inc != pos){
      inc = inc * (-1); // change direction
      pos = curPos;
   }
   Return(x);
}
```

This method calls `cycle()` to do most of its work. Like `cycle()`, this method updates `pos`, but it also must update its increment argument, `inc`, because whenever it hits the end of the list, we want it to reverse the direction of traversal rather than start over. The extra work done by this method is to change the increment and reset the position when either end of the list is reached. Here is an example of invoking `palindrome()`.

```
IntegerList L = {10, 11, 12};
Integer myPos = 0;
Integer myInc = 1; // can be any positive or negative integer
For (Integer i = 0; i < 2 * Length(L); i = i + 1)
     Print(palindrome(L, pos, inc));
```

prints `10, 11, 12, 12, 11, 10`. Note that the end of the list is printed twice. This makes it a so-called even palindrome. It would be an odd palindrome if it were `10, 11, 12, 11, 10`. It is left as an exercise for the reader to adapt `palindrome()` to generate odd palindromes.

Permutation Iterate the supplied sequence in prime order until exhausted, then permute the entire row by `inc` steps and repeat from the beginning.

```
Integer permute(IntegerList L, Integer Reference pos,
            Integer Reference count, Integer inc) {
    Integer curPos = pos;           // save current position
    Integer x = cycle(L, pos, 1);   // update pos and get list value
    count = count + 1;              // increment counter
    If (count == Length(L)){        // have we output L items from list?
        count = 0;                  // reset count
        pos = curPos + inc;         // permute position for next time
    }
    Return(x);
}
```

Here is an example of invoking `permute()`.

```
Integer inc = -1;
Integer pos = 0;
Integer perm = 0;
For (Integer i = 0; i < 3 * Length(L); i++)
     Print(permute(L, pos, perm, inc));
```

prints `10, 11, 12, 11, 12, 10, 12, 10, 11`. Because `inc = -1`, it skips back one place in the row every time. The trigger for it to skip is when it has output as many elements as are in the list.

Transpose The dodecaphonic pitch classes are not tied to any octave. In order to realize music from a tone row, its intervallic content must be translated to actual pitches of the musical scale. One way to do this is to supply a pitch offset that transposes across pitch space (i.e., without limiting it just to the range of pitch classes).

```
Integer transpose(Integer p, Integer off){
    Return(p + off);
}
```

The C major diatonic scale in the fourth piano octave can then be given as follows:

```
IntegerList Cmaj = {C, D, E, F, G, A, B}; // define C major scale
For (i = 0; i < Length(Cmaj); i = i + 1){
    L[i] = transpose(L[i], 4 * 12); // shift all up 4 octaves
Print(Pitch(Cmaj));
```

Table 9.5
interpTendency Example

Row A	0	2	4	6	8	10	12
$f = 0.00$	0	2	4	6	8	10	12
$f = 0.25$	3	4	5	6	7	8	9
$f = 0.50$	6	6	6	6	6	6	6
$f = 0.75$	9	8	7	6	5	4	3
$f = 1.00$	12	10	8	6	4	2	0
Row B	12	10	8	6	4	2	0

This prints `{Cn4, Dn4, En4, Fn4, Gn4, An4, Bn4}.`

Interpolated Tendency Mask We can produce a new row that is a mixture of two other rows. Let's have a variable that varies continuously between 0.0 and 1.0 such that when it is 0.0, the output row is exactly the same as the first row; when it is 0.5, the output is exactly halfway between the first and second; and when it is 1.0, the output is exactly the second row. For example, suppose the first pitches in each row are 3 and 9, and the interpolation parameter is 0.5. Then the expected result would be 6 because 6 lies halfway between the two values. If the interpolation parameter were 0.0, we'd select 3, and if it were 1.0, we'd select 9.

Table 9.5 shows what happens if row A = {0, 2, 4, 6, 8, 10, 12} and row B = {12, 10, 8, 6, 4, 2, 0}, and f is set successively to 0.0, 0.25, 0.5, 0.75, and 1.0. When $f = 0$, we select the prime row, when $f = 1.0$, we select the retrograde row, and in between, we select weighted mixtures.

We use *unit interpolation* to find intermediate values that lie a certain distance between two known points. If u is the upper bound and l is the lower bound and f is a control parameter in the *unit distance* from 0.0 to 1.0, then

$$y = f \cdot (u - l) + l \qquad \textit{Unit Interpolation} \quad (9.9)$$

sets y to a value close to u if $0 \ll f$; it sets y to a value close to l if $f \ll 1$; it sets y to a value exactly halfway between u and l if $f = 0.5$. Here is the function for unit interpolation:

```
Real unitInterp(Real f, Integer l, Integer u){
    Return(f * (u - l) + l);
}
```

This is a `Real` function because `f` must be a `Real` to take on fractional values. When we use it as follows, we convert the `Real` result back to an `Integer` by rounding:

```
Integer interpTendency(
    Real f,                              // factor ranging from 0.0 to 1.0
    List L1, Integer Reference pos1,     // list 1 and its position parameter
    List L2, Integer Reference pos2,     // list 2 and its position parameter
```

```
    Integer inc                    // amount by which to adjust position
) {
    Integer x = cycle(L1, pos1, inc);
    Integer y = cycle(L2, pos2, inc);
    Return(Integer(Round(unitInterp(f, x, y))));
}
```

This function can perform a couple of neat tricks. First, we can have the function return exactly `L1` or `L2` by setting `f = 0.0` or `f = 1.0`, respectively. By setting `f = 0.5`, we get the average of the two rows. By gradually changing the value of `f` from 0.0 to 1.0, we mutate `L1`, transforming it gradually until it becomes `L2`. Also, the lengths of `L1` and `L2` need not be the same. If `L1` has a length of 5 and `L2` a length of 6, it will take $5 \cdot 6$ iterations before the pattern repeats. Both lists use the same increment, but redesigning this to use separate increments would provide for even more possibilities.

Linear Interpolation Linear interpolation allows us to map a range of values so that it covers a proportionately wider or narrower range. Figure 9.14 shows linear interpolation from the range 1–4 on the left being mapped to the range 3–9 on the right. The value 3 on the left corresponds by linear interpolation to 7 on the right. Linear interpolation maintains the linear proportions of the two number lines: 3 is two-thirds of the way from 1 to 4, and 7 is two-thirds of the way from 3 to 9.

Linear interpolation is a slight generalization of unit interpolation, as follows. If x_{max} is the upper bound and x_{min} is the lower bound, and x is a parameter in the range $x_{min} \le x \le x_{max}$, then

$$y = \frac{x - x_{min}}{x_{max} - x_{min}} \cdot (y_{max} - y_{min}) + y_{min} \qquad \textit{Linear Interpolation} \quad (9.10)$$

sets y to a position within the range $y_{min} \le y \le y_{max}$ that is proportional to the position of x within its range. Here's the definition of linear interpolation in MUSIMAT:

```
Real linearInterpolate(
    Real x,                   // value ranging from xMin to xMax
```

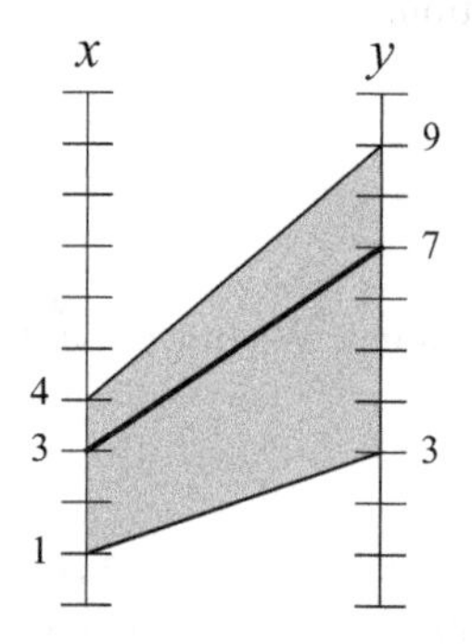

Figure 9.14
Linear interpolation.

```
    Real xMin,                 // minimum range of x
    Real xMax,                 // maximum range of x
    Real yMin,                 // target minimum range
    Real yMax                  // target maximum range
) {
    Real a = (x - xMin) / (xMax - xMin);
    Real b = yMax - yMin;
    Return(a * b + yMin);
}
```

We also can use linear interpolation to map an entire function to a different range. We do so by applying linear interpolation to every point on the function. For example, we can scale a chromatic melody to occupy a wider or narrower tessatura as follows:

```
IntegerList stretch(IntegerList L, Integer yMin, Integer yMax) {
    Integer xMin = Min(L);    // find the list's minimum
    Integer xMax = Max(L);    // find the list's maximum
    For (Integer i = 0; i < Length(L); i = i + 1) {
        L[i] = linearInterpolate(L[i], xMin, xMax, yMin, yMax);
    }
    Return(L);
}
```

For example, invoking `stretch()` with these arguments

```
IntegerList x = stretch(L, 24, 47);
```

will scale the row to cover a two-octave range and offset it upward by one octave. If the input is

```
IntegerList L = {0, 8, 10, 6, 7, 5, 9, 1, 3, 2, 11, 4},
```

then `x` will be `{24, 40, 44, 36, 38, 34, 42, 26, 30, 28, 47, 32}`. It can also be used to compress rows. With the same input, `stretch(L, 0, 5)` will produce `{0, 3, 4, 2, 3, 2, 4, 0, 1, 0, 5, 1}`.

9.12.2 Deterministic Rhythmic Techniques of Joseph Schillinger

Joseph Schillinger, a refugee from Soviet Russia, became a prominent music theorist in New York in the 1930s and counted among his students the famous jazz musicians George Gershwin and Benny Goodman. In his book *The Mathematical Basis of the Arts* (1948) he was highly critical of art theory, writing, "It is time to admit that esthetic theories have failed in the analysis as well as the synthesis of art. These have been unsuccessful both in interpreting the nature of art and in evolving a reliable method of composition."

He was looking to establish a scientific theory of art and to put practical methods into the hands of artists, giving them a mathematician's vision of the nature and extent of their domain. This, he hoped, would help free musicians from the deadening weight of musical tradition, much as Schoenberg hoped atonal composing techniques would do the same.

Schillinger envisioned development of "instruments for the automatic composition of music," including rhythm, melody, harmony, harmonization, counterpoint, and timbre. His name for such instruments was *Musamaton*. He collaborated with Leon Theremin to create a device he dubbed the *Rhythmicon,* which he used for "the composition and automatic performance of rhythmic patterns." He looked toward the use of such devices by anyone, not requiring special training, "suitable for schools, clubs, public amusement places, and homes."

He wrote a large, deeply flawed, two-volume tome, *The Schillinger System of Musical Composition.* Some of his ideas seem banal, others are incomprehensible, and he expressed his musical formalisms using a pseudomathematical notation of his own design, accompanied by often cryptic explanations that usually served to mystify the reader. He criticized the work of famous composers such as Beethoven, rewrote compositions of J. S. Bach to "improve" them, and in general displayed an arrogance that undercut his message (Backus 1961). Nonetheless, for the intrepid, there are interesting ideas in his work, particularly regarding rhythm, an otherwise quite neglected subject in music theory.

He began with the observation that music is a time-based art where continuous time is broken into pulses. Schillinger's idea is that rhythm arises through the "interference" of two sources of pulse. For example, consider two harmonically related pulse generators (figure 9.15). The major generator produces three pulses in the same time as the minor generator produces two. Schillinger called the resultant pattern pulse interference, although this is a confusion because the result is actually the product of the two functions, whereas interference implies addition (see section 7.7).

All the pulse interference patterns that can be produced by the ratio of any two integers form an inversely symmetrical pattern around their midpoint. Transitions in the pulse interference functions represent rhythmic stress points in the resulting rhythmic pattern. Their interpretation depends upon the musical context. For example, the interference pattern shown in figure 9.15 can be interpreted trivially in any of the three ways shown in figure 9.16.

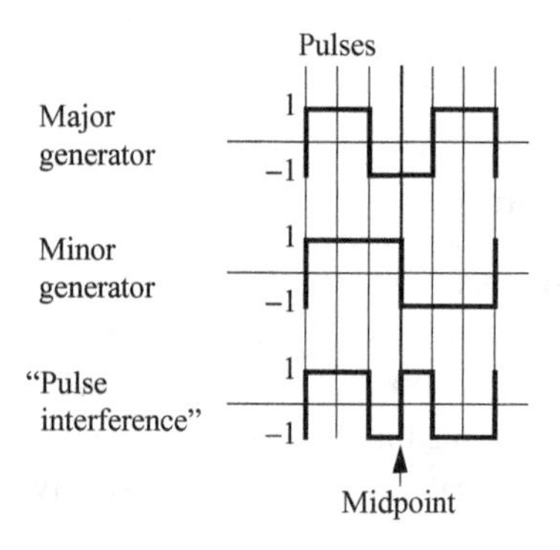

Figure 9.15
Pulse interference.

Figure 9.16
Schillinger's pulse interference patterns.

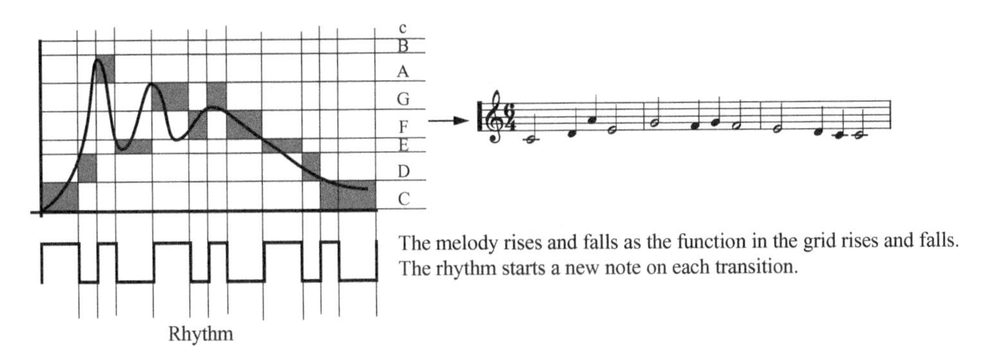

Figure 9.17
Generating a melody with Schillinger's interference patterns.

Such patterns can be applied to many musical contexts. For example, we can create a melody using the pulse interference pattern shown in figure 9.15 as a trigger function to select a pitch from another function representing pitch displacement (figure 9.17). The function shown in the grid is an arbitrary shape that determines the melody; the function labeled Rhythm is an independently generated interference pattern that determines the rhythm. Applying the interference pattern to the melody shape produces the sequence of notes shown to the right in figure 9.17.

The pulse interference pattern is projected across the x-axis, and the diatonic scale is projected across the y-axis. Notes are placed where the transitions in the rhythmic pattern intersect the pitch displacement function. The interference pattern determines the note's duration. The composer John Myhill adapted this technique in his 1965 composition *Scherzo a Tre Voce* for computer-synthesized tape alone (Ames 1967).

9.12.3 Representing Music with Functions

The basis of Schillinger's compositional idea is to map an arbitrary curve to musical notation by quantization (see volume 2, chapter 1). Of course, the process works in reverse as well: the grid in figure 9.17 can be used to generate the corresponding pitch curve and rhythmic function of any piece of notated music. Mathews and Rossler (1968) developed a graphical language for representing scores of computer-generated sounds that uses this approach. They represented music with continuous time functions that were quantized to obtain pitch and discretized to obtain time.

Mathews produced an interesting demonstration of the flexibility of this approach for composing. He began by generating pitch and rhythm functions for two traditional tunes, the English military anthem *The British Grenadiers* and the American tune *When Johnny Comes Marching Home*. Then he created a new melody by performing linear interpolation on the two sets of functions. When the interpolation parameter was set at 0.0, the method produced *The British Grenadiers,* and set at 1.0, it produced *When Johnny Comes Marching Home*. In between, one heard something that sounded like a mutated combination of both. In his example, he varied the interpolation parameter gradually from 0.0 to 1.0, with the result that the synthesized melody first resembled *Grenadiers,* but *Johnny* gradually emerged from the chaos in the middle and took over. Though it is graceless as a musical étude, Mathews's effort nonetheless is a startling demonstration of how malleable music can be under these kinds of transformations.

9.12.4 Nondeterministic Serial Methods

Deterministic methods produce the same result every time they are presented with the same inputs. The methods discussed in this section rely on randomness, so they are nondeterministic methods.

Sampling without Replacement We can generate a randomly selected 12-tone row, for example, by putting 12 balls in an urn, each marked with one of the chromatic pitch classes, and draw them out one at a time without replacement, thereby guaranteeing that no pitch class is chosen more than once.

`Random(0, 11)` returns a random integer between 0 and 11 with equal probability. But it could return the same value multiple times, so we must keep track of which pitch classes have been chosen to ensure that it eventually picks one of each. This function takes one argument, `N`, determining the length of the row.

```
IntegerList randomRow(Integer N) {
    IntegerList L;              // keep track of pitches chosen so far
    IntegerList M;              // used to build up random 12-tone row
    Integer i;

    // set all list elements to zero, which means "unused"
    For (i = 0; i < N; i = i + 1) {L[i] = 0;}

    // build up M, marking off elements in L when they are chosen
    i = 0;
    While (i < N) {
        Integer x = Random(0, N - 1); // returns integer random value
        If (L[x] == 0) {        // hasn't been chosen yet?
            L[x] = 1;           // mark it "used"
            M[i] = x;           // save result
            i = i + 1;          // increment control variable
        }
    }
    Return(M);
}
```

Note that the second loop keeps repeating over and over until `Random()` has finally selected all `N` pitch classes. It then returns the newly created 12-tone row in `M`. Here is an example row created by `randomRow()`:

```
{0, 6, 2, 9, 7, 5, 4, 10, 8, 3, 1, 11};
```

Every pitch class is represented exactly once.

Shuffle We can create a random permutation of a row rather as one would shuffle a deck of cards. If we distinguish between the cards and their position in the deck, shuffling consists of swapping the positions of all cards a pair at a time. First, we need a way to swap the position of two cards in the deck. We can swap the position of two elements in `IntegerList` like this:

```
IntegerList swap(IntegerList L, Integer from, Integer to) {
    Integer x = L[to];          // save target value
    L[to] = L[from];            // swap from → to
    L[from] = x;                // swap to → from
    Return(L);
}
```

To shuffle an entire deck of cards (or row of pitch classes), we visit each position in the list from first to last in order and swap the card at each position with a card at a randomly chosen other position. Because we use `Random()` to choose the position of the other card to swap, the "other" position can be any position in the deck, including the currently selected position; thus we may occasionally swap a card with its own position, leaving it where it was. However, in a subsequent step, that card might be chosen to be swapped elsewhere.

```
IntegerList shuffle(IntegerList L) {
    IntegerList M = randomRow(Length(L)); // elements to swap
    For (Integer i = 0; i < Length(L); i = i + 1) {
        Integer j = M[i];
        L = swap(L, i, j);
    }
    Return(L);
}
```

The first step is to generate a new row with `randomRow()`, which is stored in `IntegerList M`. Successive values of `i` and successive elements of `M` give the indexes of the elements in `L` that are to be swapped. Suppose we have

```
L = {0, 6, 2, 9, 7, 5, 4, 10, 8, 3, 1, 11}; // source row
M = {5, 1, 0, 4, 6, 7, 9, 3, 10, 8, 11, 2}; // row created in shuffle
```

Then each row in table 9.6 shows the intermediate values of `L` as its elements are being swapped. The pattern starts out like this: swap the value in position 0 and the value in position 5; swap the value

Table 9.6
An Example of Shuffling a Set

i =	0	1	2	3	4	5	6	7	8	9	10	11
L =	0	6	2	9	7	5	4	10	8	3	1	11
0	5	6	2	9	7	0	4	10	8	3	1	11
1	5	6	2	9	7	0	4	10	8	3	1	11
2	2	6	5	9	7	0	4	10	8	3	1	11
3	2	6	5	7	9	0	4	10	8	3	1	11
4	2	6	5	7	4	0	9	10	8	3	1	11
5	2	6	5	7	4	10	9	0	8	3	1	11
6	2	6	5	7	4	10	3	0	8	9	1	11
7	2	6	5	0	4	10	3	7	8	9	1	11
8	2	6	5	0	4	10	3	7	1	9	8	11
9	2	6	5	0	4	10	3	7	9	1	8	11
10	2	6	5	0	4	10	3	7	9	1	11	8
11	2	6	8	0	4	10	3	7	9	1	11	5

in position 1 with itself; swap the value in position 2 and the value in position 0; swap the value in position 3 and the value in position 4; and so on. The result is that every element of the input row is swapped randomly with another element, but there's a chance it might be swapped with itself.

Random Tendency Mask We can use a row to specify an upper boundary and another row to specify a lower boundary, and then pick a pitch in this range. We can pick any pitch in the range, either the median pitch or a random pitch or even all pitches, depending upon what we want to use it for. This example returns a random value lying between two rows that act as fences to limit the random range.

```
Integer randTendency(IntegerList L1, Integer Reference pos1,
                    IntegerList L2, Integer Reference pos2, Integer inc) {
    Integer x = cycle(L1, pos1, inc);
    Integer y = cycle(L2, pos2, inc);
    If (x < y)
        Return(Random(x, y));
    Else
        Return(Random(y, x));
}
```

For example, if `L1` and `L2` are as shown in the following table, the values in the middle row are random values chosen from between.

Row L1	0	8	10	6	7	5	9	1	3	2	11	4
Output row	0	1	8	6	7	5	5	3	5	4	9	7
Row L2	4	0	2	10	11	9	1	5	7	6	3	8

9.12.5 Serialism

Schoenberg and his school were amplifying musical trends of their time to deconstruct tonal expectation and key-centeredness in European art music. Rows and their treatment were chosen to defeat the tendency to hear tonal centeredness of any kind. Functional harmony was banished; even the too frequent repetition of a pitch was taboo lest it lend a tonal center to the music.

But it would be a disservice to Schoenberg and his school to imply that their music followed a deconstructionist agenda to the exclusion of all else. They offered the intervallic structure of the row and the organization of set forms as the new ligatures holding their music together. Perle and Lansky (1981) write,

> Perhaps the most important influence of Schoenberg's method is not the 12-note idea in itself, but along with it the individual concepts of permutation, inversional symmetry and complementation, invariance under transformation, aggregate construction, closed systems, properties of adjacency as compositional determinants, transformations of musical surfaces through predefined operations, and so on.

But deconstructionism, once set into motion, rarely stops until it has devoured everything. Some composers of the post–World War II era observed vestiges of other traditional techniques in the music of Schoenberg and Berg. They noted that Schoenberg and Berg treated the 12-tone row as a theme to be developed, a practice that harked back to the classical technique of theme and variations—*thematicism*. They idolized the work of Schoenberg's pupil Anton Webern because he eschewed thematic development, building up compact, jewel-like compositions from as few as three notes. For example, in his *Concerto for Nine Instruments* written in 1934, all pitches are derived from the simple motive B-B♭-D (prime form) and its retrograde, inversion, and retrograde-inversion. His systematic treatment of pitch, rhythm, dynamics, and articulation was taken by these younger composers as a model for a new form of music.

The composer Olivier Messiaen in France extended the 12-tone pitch-ordering technique of Schoenberg's school to all other parameters of music, although he was working with modal pitch structures, not 12-tone rows (Messiaen 1942; Drew 1954/1955). Inspired by Webern and Messiaen, other composers, including Pierre Boulez in France and Milton Babbitt in the United States, adapted Messiaen's ideas back to atonal practices, and *totally organized music,* or *serialism,* was born. According to Stuckenschmidt (1969), "Serial techniques are essentially a systematic transference of Schoenberg's 12-tone technique to elements of musical sound other than pitch."

This idea interlocked with two others. Just as the tones in a 12-tone row were decoupled in significance from each other, the serialist composers decoupled all parameters of the musical note from each other. Pitch, register, tone color, and dynamic level became independent. Just as all tones were used in a 12-tone row, the serialist composers employed the entire available range of every other musical parameter—high to low, loud to soft, fast to slow, bright to dull—without preference.

The dodecaphonists observed the tonal equivalence of the equal-tempered scale and sought to construct a new musical aesthetic that reflected this equality. To do so, they developed a 12-tone method that deconstructed tonal expectation and key-centeredness. The notion of tonal equivalence was extended by the serialists to project a uniform proportionality between all musical parameters and all combinations of musical parameters.

Stuckenschmidt (1969), who witnessed the premiers of the European serialist composers in the 1950s, wrote, "The impression made by all these works, even on a listener who had read the commentaries beforehand, was one of chaos" (214).

The composer György Ligeti (1965) wrote, "Now that hierarchical connections have been destroyed, regular metrical pulsations dispensed with, and durations, degrees of loudness, and timbres have been turned over to the tender mercies of serial distribution, it becomes increasingly difficult to achieve contrast" (16). These compositions often projected a static quality, a musical equivalent of alphabet soup (see section 9.15 for why these effects occur). Ligeti (1965) summed it up: "Serial music is doomed to the same fate as all previous sorts of music; at birth it already harbored the seeds of its own dissolution" (14).

9.13 Stochastic Techniques

With every musical parameter now serially ordered, there was even less familiar structure for listeners to rely upon to orient themselves in the music. The composer Iannis Xenakis (1955) criticized serialism as follows:

> Linear polyphony destroys itself by its very complexity; what one hears is in reality nothing but a mass of notes in various registers. The enormous complexity prevents the audience from following the intertwining of the lines and has as its macroscopic effect an irrational and fortuitous dispersion of sounds over the whole extent of the sonic spectrum. There is consequently a contradiction between the polyphonic linear system and the heard result, which is surface or mass.

Echoing the same sentiment, the composer Gottfried M. Koenig (1970) wrote, "The trouble taken by the composer with series and their permutations has been in vain; in the end it is the statistical distribution that determines the composition."

Believing that the listener experiences only the statistical aspects of serial music, these composers reasoned that a better approach would be to compose directly using probabilistic instead of serial techniques. Xenakis (1955) writes,

> This contradiction inherent in [serial] polyphony will disappear [and] what will count will be the statistical mean of isolated states and of transformations of sonic components at a given moment. The macroscopic effect can then be controlled by the mean of the movements of elements which we select. The result is the

introduction of the notion of probability, which implies, in this particular case, combinatory calculus. Here in a few words, is the possible escape route from the "linear category" in musical thought.

Xenakis (1971) was reacting against serialism and also aligning himself with a worldview then developing in the physics of quantum mechanics: "It is a matter here of a philosophic and aesthetic concept ruled by the laws of probability and by the mathematical functions that formulate that theory, of a coherent concept in a new region of coherence." Xenakis's attempt to align music aesthetics with a natural theory is not a new enterprise, of course, but dates back at least to the early Renaissance music theorist Gioseffo Zarlino, who championed the view (as did others) that music imitates nature (see section 9.17.5).

While some of Xenakis's examples in his book *Formalized Music* describe methods for organizing music for traditional instruments, elsewhere in this work he presents a more abstract kind of sound organization. He asserts, "All sound is an integration of grains, of elementary sonic particles, of sonic quanta." Xenakis was influenced by the seminal work of Dennis Gabor, who in 1947 observed an isomorphism between the Fourier series and a quantum analysis of sound (see volume 2, chapters 9 and 10).

Given the burden of computation required by a statistical approach to composition, it is not surprising that composers like Koenig and Xenakis turned to computers to help compose musical works. Xenakis (1971) enthused, "With the aid of electronic computers the composer becomes a sort of pilot: he presses the buttons, introduces coordinates, and supervises the controls of a cosmic vessel sailing in the space of sound, across sonic constellations and galaxies that he could formerly glimpse only as a distant dream." These composers believed that statistical composing systems using computers would allow them to shift their attention from the surface of the music to its inner structure.

9.14 Probability

Suppose a player with eyes closed strikes a piano key at random. What is the chance that the struck key will be middle C? A standard piano has 88 keys, so to a first approximation, we'd expect the possibility to be 1 out of 88. But because the white keys are larger than the black keys, all outcomes are not equally likely. To study this more closely, let's define some terms.

- *Sample space* The set of possible outcomes.
- *Event* The outcome of a random process, such as a roll of the dice.
- *Probability* The relative liklihood of an event, usually expressed as a real number in the range $0 < p < 1$.
- *Probability distribution* A function, graph, or listing of the probabilities of the sample space that shows how probability is distributed among the possible events.
- *Uniform distribution* If all events in a sample space are equally likely, the resulting distribution is said to be uniform.

- *Discrete distribution* A distribution is discrete if the events in the sample space can be individually distinguished. Tossing coins or dice or picking a note on a keyboard are examples of discrete distributions.
- *Continuous distribution* A distribution is continuous if the events in the sample space cannot be individually distinguished. Temperature and frequency are examples of continuous distributions.
- *Random variable* Let s be the sample space consisting of both sides of a coin, which can be represented as the set {Heads, Tails}. When a coin is flipped, outcome R must be one of Heads or Tails. In order to construct the probability distribution, we set a random variable x in turn to each possible outcome of the sample space s and determine the probability that x is equal to outcome R, as follows:

$$f(x) \equiv P(x = R) = \begin{cases} .5, & x = \text{Heads} \\ .5, & x = \text{Tails} \end{cases}$$

which is read as "The probability distribution function f of random variable x is defined as the probability that x equals outcome R, which is .5 if x is heads and .5 if x is tails." The random variable indexes the probability distribution function in order to determine the value of the function at that index.

We can use these terms to classify chance operations for further study. For example, tossing a coin has a sample space consisting of two outcomes, Heads or Tails, and the probability is 1/2 for either Heads or Tails if the coin is true, so its discrete probability distribution is uniform. Tossing a single die has six possible outcomes; if the die is true, each outcome has a probability of 1/6, so its discrete distribution is also uniform.

9.14.1 Discrete Distribution

The sample space of one die has $d = 6$ outcomes. Suppose we roll a white die d_w and a black die d_b. If we distinguish the event $\{d_w = 1, d_b = 2\}$ from the event $\{d_w = 2, d_b = 1\}$ and tally up the combination of all possible outcomes, we find that the sample space is the product: $d_w \cdot d_b = (6 \cdot 6) = 36$. The states are enumerated in table 9.7. Each number in the table grid is the

Table 9.7
Sample Space: Sum of Two Dice

	Black Die					
White Die	1	2	3	4	5	6
1	2	3	4	5	6	7
2	3	4	5	6	7	8
3	4	5	6	7	8	9
4	5	6	7	8	9	10
5	6	7	8	9	10	11
6	7	8	9	10	11	12

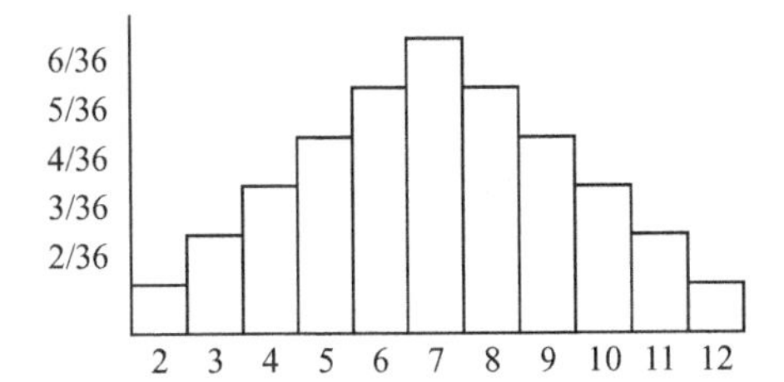

Figure 9.18
Probability for the sum of dice.

sum of the two dice. Note that only one dice combination sums to 2, one sums to 12, and six combinations yield 7. We'd rightly expect that the more combinations sum to the same value, the more probable those outcomes will be. So we'd expect a roll of two dice to be most likely to sum to 7 and least likely to sum to either 2 or 12. The corresponding probability distribution for the sum of the dice is shown in figure 9.18.

Interestingly, if we roll two dice and tally them separately, the probability distribution of all faces is uniform. But if we *sum* two dice, some combinations are more likely because some combinations are more numerous than others, as shown in figure 9.18.

A fundamental insight of probability theory is that if a random variable x has distribution $f(x)$ and a random variable y has distribution $f(y)$, then the distribution of the sum of the two random variables $f(x + y)$ is the *convolution* of $f(x)$ and $f(y)$ (F. R. Moore, 1990). (To understand the mathematical reason for this, see volume 2, chapter 4.) Figure 9.18 shows the convolution of two uniform distributions.

9.14.2 Continuous Distribution

Suppose a violinist with eyes closed stops the G string (which is pitched a fourth below middle C) somewhere along its length. What is the chance that the violinist stops the string at exactly middle C, 261.626 Hz? Because the string is continuous, there are in fact an infinite number of frequency gradations along its length, just as there are an infinite number of points along its length.[7] So the likelihood that the violinist will stop the string at any particular pitch is infinitesimal. How do we study continuous distributions if every event is infinitely improbable? We finesse this problem by assigning probabilities to subsets of the sample space, effectively breaking the continuous space into discrete regions. We ask questions like, What is the probability that the violinist stops the string within a half step of middle C? A positive probability can be assigned to such an event.

This example shows that probability only operates on discrete sample spaces, and if we must operate on a continuous variable such as frequency or temperature, we must first break the continuum into a discrete sample space. If we take this region size to the infinitesimal limit, we are in effect operating on a discrete sample space of infinitesimal dimensions. But then we are back to the situation where the probability of each infinitesimal outcome is infinitely small.

9.14.3 Uniform Distribution

Let's return to the example of striking piano keys at random. Assume (incorrectly) that the outcomes are all equally likely and that the probability of actually striking a key is 1. Then the probability of striking a *particular* key (such as middle C) is the probability of striking *any* one key divided by the number of keys, or 1/88.

If the events in a sample space are all equally likely, we can define the uniform probability distribution function $f(x)$ as

$$f(x) \equiv P(x = R) = \frac{1}{s}, \tag{9.11}$$

where R is a particular outcome (e.g., the struck key is middle C), s is the number of events in the sample space, x is the random variable, and $P(x = R)$ is the probability that x is R.

The number of keys s on an organ keyboard is 60, so striking middle C in a random attempt is somewhat more likely on this instrument. Along the same lines, the chance of striking any key in the middle octave of the piano is 12/88. The probability of striking any pitch class C is 8/88 because there are eight C keys on the standard piano.

9.14.4 Nonuniform Distributions

It's time to face up to the fact that more area on a piano keyboard is covered by white keys than black, so the likelihood of striking a black key at random is less than striking a white one. The ratio of the area occupied by all the white keys k_w to the total keyboard area k_a expresses the probability of striking a white key:

$$p(w) = \frac{k_w}{k_a}, \tag{9.12}$$

where $p(w)$ is the probability of striking a white key. There are $n = 2$ kinds of keys. If $p(w) \neq 1/n$, the probability distribution is not uniform. If $p(w) > 1/n$, striking a white key is more probable.

By inspecting a piano keyboard, we can estimate that the ratio of white key area to total key area is $p_w \approx 3/4$. By this analysis, the odds are that a white key would be randomly selected about 75 percent of the time and a black key the remainder of the time (figure 9.19). This plot is a *probability distribution function* because it expresses how probability is distributed over the sample space.

Let s be the sample space of all white and black piano keys, which can be represented as the set {White, Black}. The outcome R must be one of White or Black. We construct the probability distribution function by setting a random variable x in turn to each possible outcome of the sample space s and determining the probability that x is equal to outcome R. Unlike in the coin example, this distribution is not uniform:

$$f(x) \equiv P(x = R) = \begin{cases} .75, & x = \text{White} \\ .25, & x = \text{Black} \end{cases}$$

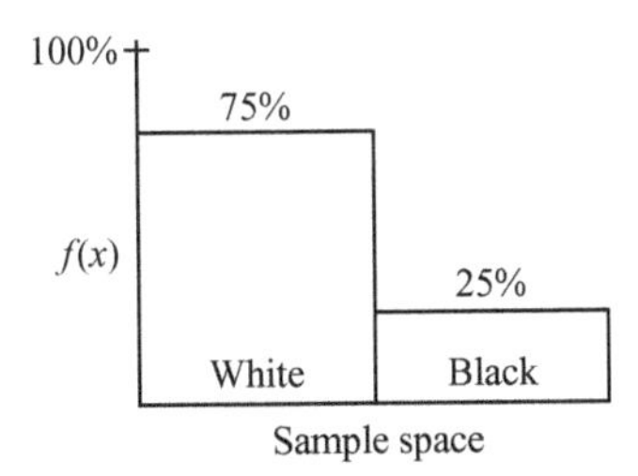

Figure 9.19
Piano key probability distribution.

In general, if the sample space $s = \{x_0, x_1, \ldots, x_n\}$, the probability that some event R is equal to a particular x in s is the function

$$f(x) \equiv P(x = R) \tag{9.13}$$

for any x. This is read as, "The probability that a random event R will result in an outcome x is defined by the function f." For example, $f(\text{Black}) \equiv P(R = \text{Black}) = 25$ percent, and $f(\text{White}) \equiv P(R = \text{White}) = 75$ percent.

9.14.5 Generating Outcomes from Probability Distributions

Probability distributions allow us to analyze random systems like dice and coins, but we can also use them to synthesize random numbers that are distributed in probability according to our choosing. We can use such systems to drive compositional processes to automatically generate music according to rules that we supply.

Say, for instance, we wish to use a random system to create a melody so that it favors lower pitches in the scale. Let's limit the sample space to one octave of the chromatic scale. We can represent this as a probability inequality:

$$f(x) \equiv P(R = \text{C}) > P(R = \text{C\#}) > P(R = \text{D}) > \cdots > P(R = \text{B}).$$

To be specific, suppose we want to create a probability distribution function that is 12 times more likely to pick C than B, 11 times more likely to pick C♯ than B, 10 times more likely to pick D than B, and so on. The probability distribution function would look like the one in figure 9.20.

We know what we want, but how do we get it? So far, the only things we have to work with are a random number generator, `Random()` (see appendix B, B.1.27) and a probability distribution function (figure 9.21).

9.14.6 Cumulative Distribution Function

Let's rotate each of the weights in figure 9.21 and then concatenate them. Their sum is 78, so we divide the length of each weight by 78 so that the weights sum to a length of 1.0 (figure 9.21). We have effectively divided up the x-axis in the unit interval into 12 areas that are proportional to the weights in the

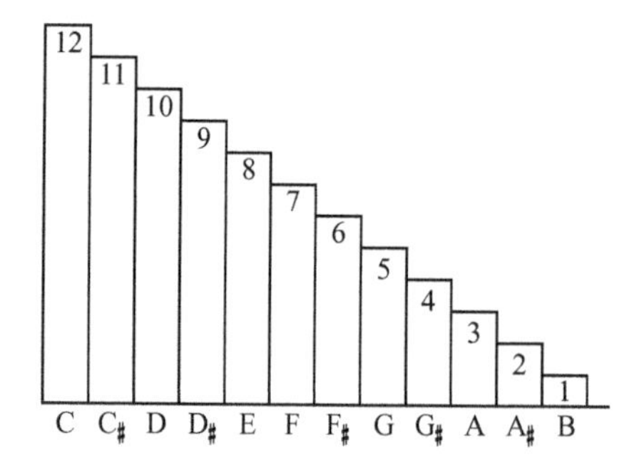

Figure 9.20
Chromatic probability distribution.

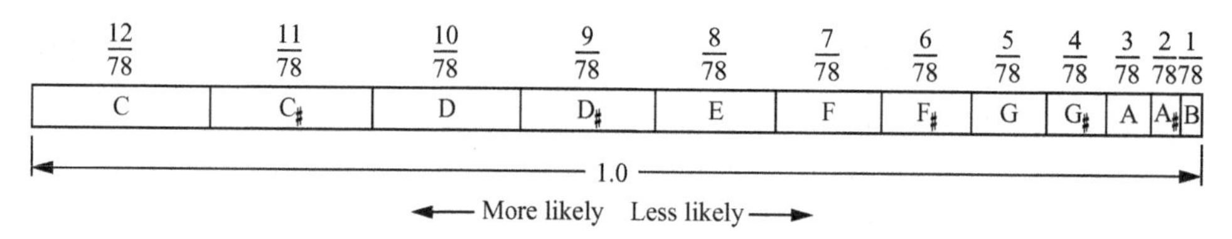

Figure 9.21
Chromatic probability distribution.

original distribution. Now we pick a random number in the unit interval with the `Random()` function, see which interval the number would fall in, and then determine the chosen pitch. The probability that a particular interval will be chosen is proportional to the extent of its footprint on the *x*-axis.

How can we represent this formally so that a computer can do this? First, the statement

```
RealList f = {12.0, 11.0, 10.0, 9.0, 8.0, 7.0, 6.0, 5.0, 4.0, 3.0, 2.0, 1.0};
```

defines the weights for each pitch, lowest to highest, left to right. Note that the type of list `f` is `RealList`.

Next, we normalize the weights so that they sum to 1.0 (see appendix A, A.3):

$$\sum_{n} f(n) = 1.0.$$

Normalizing is done in two steps:

1. Find the sum of all weights:

```
Real sum(RealList L){
  Real s = 0.0;
  For (Integer i = 0; i < Length(L); i = i + 1){
      s = s + L[i];
  }
```

```
    Return(s);
}
```

Given the definition of `RealList f` above, `Print(sum(f))` prints `78`.

2. Divide each weight by `sum(f)` so that the sum of the weights equals 1.0:

```
RealList normalize(RealList L, Real s){
    For (Integer i = 0; i < Length(L); i = i + 1){
        L[i] = L[i]/s;
    }
    Return(L);
}
```

Given the definition of `RealList f` above, the statements:

```
RealList r = normalize(f, sum(f));
Print realToRational (r)); // realToRational is a built-in function
```

prints `{12/78, 11/78, 10/78, 9/78, 8/78, 7/78, 6/78, 5/78, 4/78, 3/78, 2/78, 1/78}`.

After these two steps, `r` will look like figure 9.20 except that all values are scaled down by 78. (The built-in `realToRational()` function is described in appendix B, B.2.2.)

Next, we create a function such that each step along the x-axis accumulates all the weights to its left with its own weight (figure 9.22). The first column has a height of 12/78, the second of 12/78 + 11/78, the next of 12/78 + 11/78 + 10/78, and so on. This function is called a *cumulative distribution function*.

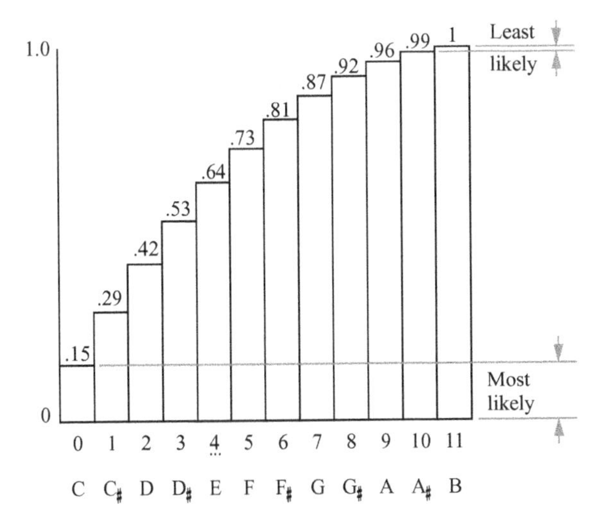

Figure 9.22
Cumulative distribution function.

If we index the *y*-axis of figure 9.22 with a random value in the unit interval, the corresponding *x*-axis value will be one of the 12 pitches of the scale. Furthermore, the choice will more likely fall on a step that occupies a wider footprint on the *y*-axis, corresponding in this case to the lower pitches of the scale, just as we wanted. We can create the cumulative distribution function in figure 9.22 as follows:

```
RealList accumulate(RealList L){
    For(Integer i = 1; i < Length(L); i = i + 1){
          L[i] = L[i] + L[i - 1];
    }
    Return(L);
}
```

Starting with the second element in the list (indexed as 1), we replace this element with its original value plus the value of the previous element. As we proceed through the list, each list element will be equal to itself plus all previous elements. Given the preparation of the `RealList r` performed above, `Print(accumulate(r));` prints `{0.15, 0.29, 0.42, 0.54, 0.64, 0.73, 0.81, 0.87, 0.92, 0.96, 0.99, 1.0}`.

We have prepared the cumulative distribution function, and now we can access it with a random value to select a pitch. Pick a number in the unit interval to be the next note of the melody:

```
Real R = Random();
```

`R` will fall within the range of one of the 12 steps in figure 9.22 because both `Random()` and the cumulative distribution function exactly span the unit interval, 0 to 1. For example, if `R` equals 0.1, then by inspection of figure 9.22, we can see that `R` lies within the first step, which covers the interval [0, 0.15], so the pitch that this value of `R` selects is C.

To automate this, we start at the top end of the cumulative distribution function and work down. As we go, we compare the value of `R` to the current step size. We've gone one step too far when the value of `R` exceeds the step size, so we return the previous step as the answer, and stop.

```
Integer getIndex(IntegerList L, Real R){
    Integer i;
    For (i = Length(L) - 1; i >= 0; i = i - 1){
         If (R > L[i]){
              Return(i + 1);
         }
    }
}
```

We can invoke `getIndex()` as follows:

```
Real R = Random();
```

Figure 9.23
(Boring) musical example of weighted random values.

```
Integer p = getIndex(f, R); // where f was defined previously
Print(p);
```

If `R` is 0.1, then `p` prints `0`. Now let's bring all the pieces together. Here is a program that creates a melody of 25 pitches favoring pitches that are at the low end of the chromatic scale:

```
RealList f = {12.0, 11.0, 10.0, 9.0, 8.0, 7.0, 6.0, 5.0, 4.0, 3.0, 2.0, 1.0};
StringList n = {"C", "C#", "D", "D#", "E", "F", "F#",
          "G", "G#", "A", "A#", "B", "c"};
f = normalize(f, sum(f)); // replace f with its normalized form
f = accumulate(f);        // calculate cumulative distribution function

StringList s;             // a place to put the result
For (Integer i = 0; i < 25; i = i + 1){
   Integer p = getIndex(f, Random());
   s[i] = n[p];
}
Print(s);                 // print the melody
```

Running this program will generate something like figure 9.23, depending upon the values produced by `Random()`. As we see, lower pitches are favored in approximately the proportions we specified. The longer the sample melody, the more likely the pitch choices would conform on average to the distribution function.

Unfortunately, this melody is dreadfully dull, but it strictly obeys our requirements. This goes to show that one only gets back from an approach like this exactly what one specifies. A more graceful melody might rise to its climax gradually, then fall at the end. The following example accomplishes this by selecting among a set of probability distributions at different points of the melody.

```
Integer N = 13; //each list specifies 13 pitches

RealList a = { //force choice to be pitch C
  1, 0, 0, 0, 0, 0, 0, 0, 0, 0, 0, 0, 0
};
RealList b = { //force C#, D, D#, E, or F
  0, 1, 1, 1, 1, 1, 0, 0, 0, 0, 0, 0, 0
};
```

```
RealList c = {  //force F#, G, G# A, A#, or B
    0, 0, 0, 0, 0, 0, 1, 1, 1, 1, 1, 1, 0
};

RealList d = {  // force pitch c an octave above
   0, 0, 0, 0, 0, 0, 0, 0, 0, 0, 0, 0, 1
};

// indicate what percentage of the score is completed
Real progress(Integer p, Integer L){
    Return Real(p) / Real(L);
    // L is the total number of notes and p is the current note
}
randomMelody(RealList a, RealList b, RealList c, RealList d){
   Integer K = 25;              // we'll play 25 notes
   Integer highPoint = Integer(K * 2.0/3.0);
   normalize(a, sum(a)); normalize(b, sum(b));
   normalize(c, sum(c)); normalize(d, sum(c));
   StringList s;                // a place to put result
   For (Integer i = 0; i < K; i++){
       RealList f;
       If (i == 0 Or i == K - 1) // force first and last notes to
            f = a;              // be pitch C
       Else If (progress(i, K) < 0.30)// less than 30% of the way?
            f = b;              // force lower hexachord
       Else If (progress(i, K) < 0.60)// between 30% and 60%?
            f = c;              // force upper hexachord
       Else If (i == highPoint)// force high point to be high c
            f = d;
       Else If (progress(i, K) < 0.80) // between 60% and 80%?
            f = c;              // force upper hexachord
       Else                     // otherwise force lower hexachord
            f = b;
            f = normalize(f, sum(f)); // replace f with its normalized form
            accumulate(f);
            Integer p = getIndex(f, Random());
            s[i] = n[p]; // n is StringList defined in previous example
    }
    Print(s);
}
```

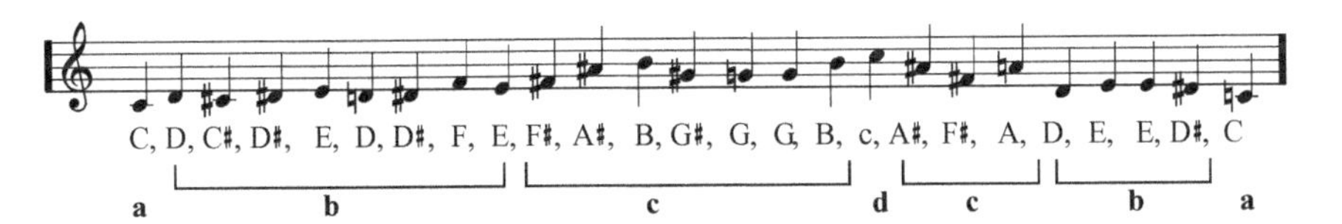

Figure 9.24
(Less boring) musical example of weighted random values.

Running this program will generate something like figure 9.24, depending upon the values produced by `Random()`. The distributions responsible for each section are shown in the figure.

The musical example in figure 9.24 is certainly an improvement, but I doubt it would win any prizes. Certainly a composer of a melody takes its *whole shape* into consideration during writing, but successive weighted random selections are *completely independent of the past and future.* Many composers have used techniques like this to obtain freedom from predictable musical contexts. But we must have a way to correlate past and future choices to the present before random choice techniques are of use in those musical styles that manipulate listener expectation. The next section lays the foundations for a mathematics of expectation.

9.15 Information Theory and the Mathematics of Expectation

Information is a property of a message that is transmitted from a sender to a receiver via a signaling system (see section 6.1). We know intuitively what information is, but when we look more deeply, it has some unusual characteristics. For example, it is possible to quantify the amount of information in a message.

Suppose you receive a letter from a friend announcing her engagement. The letter only contains information if you don't already know that she's engaged, that is, if you were *uncertain* about the contents of the letter. For example, if a friend had told you the news by phone before the letter arrived, the letter carries no information; in fact, it is *redundant.* We see from this example that information and redundancy have a curious relation to uncertainty.

Shannon and Weaver (1949) developed *information theory* to study the quantitative aspects of information. They were not concerned with the qualitative meaning or value of information but strictly focused on how much information was communicated by different kinds of messages. Consider the problem, for instance, of music dictation. Traditionally, music students are taught music dictation by a professor who plays music that the students must learn how to write down. Suppose we are students who are required to take a class that the syllabus calls Music Dictation from Hell 101. Our sharpened #2 pencils are at the ready, poised over blank music paper.

On the first day, the professor says, "Class, I am going to play one note, middle C, over and over again for the next hour. Be sure to write them all down correctly." He then goes to the piano and plays C, C, C, C, C, C, ... Any time someone coughs or there is a loud disturbance, we can't hear the piano, so we write C, C, ?, C, C, ?, C, ... No matter, we know that the missing notes are C.

On the second day, only the students who need this class to graduate show up. The professor says, "Class, I am going to play the C major scale for the next hour. Be sure to write them all down correctly." So we write C, D, E, F, G, A, B, C, D, E, ... Occasionally, we nod off, missing a note or two, so we write C, D, ?, ?, G, ... No matter, we know exactly what the missing notes are. As before, the message is almost totally redundant, but the redundancy allows us to recover from any transmission errors.

On the third day, a miserable handful of students straggles into the room. "Class, I am going to play each of the 88 keys on the piano in random order, but you'll be glad to know that I won't repeat any key until I've played every one of them. Be sure to write them all down correctly." We scribble furiously: C3, G♯4, B♭5, F7, G♮1, ... It seems impossibly difficult at first. Any disturbance in the room means we've irretrievably lost that note because we can't predict what it will be. But we discover that it becomes easier as we go along because we know the professor won't repeat any note until he's played all the others. By the time he has played 78 notes, if we miss a note it's not too bad because there are only ten possible notes left that he could play. And when he has played 87 notes, the eighty-eighth note is a certainty; we don't even need to hear it to write it correctly. The information content of each subsequent note declines while its redundancy increases because each new note played narrows the choices of what notes can be played subsequently.

On the fourth day, you and I are the only two students desperate enough to show up. The professor says, "Class, I am going to play each of the 88 keys on the piano in random order, and I may repeat keys any time I like. Be sure to write them all down correctly." The only source of information about what note will be played next is the note itself. Information in each note is very high, redundancy is very low. But we hear patterns occasionally as we go along. We write, "Repeated B♭5 four times in a row" or "Played melody of *Moonlight Sonata*" as a shorthand. These shorthands allow us to recover information and squeeze out redundancy in what we write, because otherwise we'd have to enter all the notes or write out the melody of the *Moonlight Sonata*. (This kind of information recovery, by the way, is similar to one stage in the process used by MP3 encoders to compress musical sound.)

On the fifth day, the professor does not come, but there's a note on the piano that says, "Go down to the beach and write down every note you hear in the ocean's waves. Be sure to write them all down correctly." We go to the beach. Overwhelmed by the multitude of frequencies in each splash of the waves, we cannot write down anything.

Throughout the week, the professor played patterns that went from great certainty to great uncertainty. We became aware that the amount of information carried by what the professor actually played was a function of our uncertainty about what *could* be played. The more we knew about what was coming, the less information was conveyed by what was communicated. Every constraint the professor imposed on his freedom of choice resulted in a decrease of information in the music itself. We observed the value of redundancy to help prevent information loss when noise disrupts the communications channel. We also learned that we have emotional reactions to different degrees of information and redundancy.

9.15.1 Entropy and Redundancy

Shannon and Weaver (1949) formalized their ideas about information using the concept of entropy, which they adapted to their purposes from the physical sciences. In chemistry, *entropy* is a measure of the ways in which the energy of a molecular system is distributed among the motions of its particles, its *thermodynamic probability*. In information theory, *entropy* is a measure of the ways in which the information of a signaling system is distributed among its communications. A highly entropical microparticle distributes its energy widely among its possible motions. A highly entropical signal requires a large number of independent facts in order to fully communicate it. In terms of the Music Dictation from Hell example, days 1 and 2 were low-entropy days and the rest were high-entropy days.

9.15.2 Surprisal

On day 1 of Music Dictation from Hell, the probability that the next note would be the same as the preceding note was 1.0, because there was no unexpectedness or *surprisal* about what note the professor would play. As the probability of an event decreases from 1.0 toward 0, the surprisal goes from zero to infinity. But what is the exact trajectory of this relation?

Recall day 4 of Music Dictation from Hell. As the professor plays notes at random over the entire range of the keyboard, suppose you and your friends devise a game to pass the time, betting on which key the professor will play next. You wager that the next note will be below the midpoint of the keyboard.[8] The probability is 44/88 = 1/2 that you will be right. If the professor's next note is as you predicted, you are pleasantly surprised, and your friends mark down 1; otherwise they mark down 0. Since there are only two possible outcomes, the amount of surprisal requires one binary digit, called one *bit,* to represent (see volume 2, chapter 1).

Suppose you take a bigger risk and wager that the next note will be in the bottom quarter of the keyboard. The probability is 22/88 = 1/4. Since your risk has doubled, you'd be twice as surprised in the event you guessed correctly. You'd need two bits to represent the amount of surprisal. With two bits, you can represent a magnitude of 4.

Wagering that the next note is in the bottom eighth has probability 11/88 = 1/8, requiring three bits to represent the amount of surprisal because you can represent a magnitude of 8 with three bits. Probability of 1/16 requires four bits of surprisal; probability of 1/32 requires five bits. These examples can be expressed as follows:

$$p = \frac{1}{2^s}, \qquad \textit{Probability and Surprisal} \quad (9.14)$$

where p is probability and s is surprisal. Equation (9.14) finds probability given surprisal. To find surprisal given probability, we solve (9.14) for s:

$$s = \log_2 \frac{1}{p} = -\log_2 p = -\frac{\ln p}{\ln 2}, \qquad (9.15)$$

where $\ln x$ is the natural logarithm to the base e.

For example, the probability p of predicting the next individual key the professor plays is 1/88, and its corresponding surpisal is 6.46. One advantage of surprisal is that where probabilities multiply, surprisals merely add. For example, the probability of guessing two individual keys in succession is $1/88^2 = 1/7744$, but the surprisal is merely $6.46 + 6.46 = 12.92$.

We can extend (9.15) to represent the surprisal for every key. Let each key be labeled x_i, $i = 1, 2, \ldots, M$, where $M = 88$ is the number of keys. Let the probability that the ith key is pressed be P_i. Then the surprisal of the ith key's being played can be defined as

$$s_i = -\log_2 P_i. \qquad \textit{Surprisal} \quad (9.16)$$

The negation in (9.16) reminds us that the surprisal of an event *increases* as its probability *decreases*.

Suppose the professor plays a total of N notes. If the ith key is played N_i times, then the *average surprisal* of all pitches in the melody would be

$$H(X) = \sum_{i=1}^{M} \frac{N_i}{N} s_i, \qquad \textit{Average Surprisal} \quad (9.17)$$

where X represents all possible keys on the piano keyboard.

As the total number of notes N increases to infinity, the ratio N_i/N tends to P_i. By combining this with the definition for s_i given in (9.16), we have

$$H(X) = -K \sum_{i=1}^{M} P_i \log_2 P_i. \qquad \textit{Uncertainty} \quad (9.18)$$

$H(X)$ is a measure of the *uncertainty* of the system, and K is a positive constant of proportionality. By suitable adjustment of K, we may choose any base for the logarithm. Use of base 2 logarithms is fairly standard, but in general Shannon and Weaver defined the information in a system X as

$$I(X) = -K \sum_{i=1}^{M} P_i \ln P_i. \qquad \textit{Information (Entropy)} \quad (9.19)$$

They noted a striking resemblance of this equation to the equation relating thermodynamic probability to entropy:

$$H = -k \sum_{i=1}^{M} W_i \ln W_i, \qquad \textit{Thermodynamic Probability (Entropy)} \quad (9.20)$$

where W_i is the thermodynamic probability of each state, k is Boltzmann's constant, equal to 1.3807×10^{-23} JK^{-1}, and H is the resultant entropy. They then related entropy to information by the simple expedient of the ratio k/K.[9]

When $\log_2 x$ is used, the unit of entropy is called a *bit* (though this definition is more flexible than the bits in a computer memory). When $\ln x$ is used, the unit is called a *nat*. For $\log_{10} x$ the unit is called a *hartley*.[10]

Summarizing equations (9.19) and (9.20), Shannon (1948) makes the following points:

- Entropy H of a communications channel will be zero "if and only if all the P_i but one are zero, this one having the value unity. Thus, only when we are certain of the outcome does H vanish. Otherwise H is positive." This case corresponds to day 1 of Music Dictation from Hell.

Only absolute certainty banishes entropy absolutely.

- "For a given n, H is a maximum and equal to $\log n$ when all the P_i are equal (i.e., $1/n$). This is also intuitively the most uncertain situation." This case corresponds to day 4 of Music Dictation from Hell.

The most uncertain situation has the maximum entropy.

- "Any change towards the equalization of the probabilities $P_1, P_2, \ldots, P_n$ increases H." Conversely, any change that makes probabilities less equal reduces H. For example, on day 3 of Music Dictation from Hell, H was gradually reduced as notes that were played were removed from the pool of possible notes.

9.15.3 Department of Redundancy Department

We can use the definition of maximum entropy to show the relation of entropy to redundancy. Redundancy relates the actual entropy $H(X)$ to its theoretical maximum, $\log N$, as follows:

$$R(X) = 1 - \frac{H(X)}{\log N}. \qquad \textit{Redundancy} \quad (9.21)$$

Because redundancy is normalized for the length of the communication, it is actually more useful than entropy as a way to compare sequences.

Information theory presents us with the somewhat counterintuitive outcome that the greatest amount of information is associated with the greatest degree of uncertainty. But information is not the same thing as knowledge.

Information relates to the breadth of what could be communicated. Knowledge is a distillation of the regularity and order arising from a communication.

9.16 Music, Information and Expectation

Ordinarily, music, like most systems, contains some entropy and some redundancy. In Music Dictation from Hell, we saw that the extremes of entropy and redundancy kill our interest. If the degree of redundancy is too high (as on days 1 and 2), the music is too predictable, and the listener eventually gets bored and stops listening. If the degree of entropy is too high (as on days 4 and 5), the music is too unpredictable, and the listener eventually gets frustrated and stops listening. In between is where music happens: when entropy and redundancy sustain a fluid, dynamic balance, there is enough regularity to orient the listener in the music but also enough novelty to preserve interest. This suggests that, in general,

Composing is about the manipulation of interest, affect, and attention.

This shouldn't be too surprising: after all, the human neocortex is a very refined organ of expectation. A fundamental job of the neocortex is anticipating what may happen next. One of the ways we entertain ourselves is by exercising this faculty in play.

Susan Langer (1953) characterized music as a kind of emotional algebra: "Music conveys general forms of feelings, related to specific ones as algebraic expressions are related to arithmetic [expressions]."

Leonard Meyer (1956) proposed an "affect theory of music," writing "Emotion or affect is aroused when a tendency to respond is arrested or inhibited. . . . What a musical stimulus or a series of stimuli indicates . . . [is] not extramusical concepts and objects but other musical events *which are about to happen*. . . . Embodied musical meaning is, in short, a *product of expectation*" [italics added]. Meyer has precisely defined musical meaning, and it bears repeating:

Expectation is a prediction based on current and past experiences. Musical meaning is a function of expectation.

Aristoxenus said much the same when he wrote,

> Musical cognition implies the simultaneous recognition of a permanent and a changeable element . . . for the apprehension of music depends upon those two faculties, sense perception and memory; for we must perceive the sound that is present, and remember that which is past. In no other way can we follow the phenomenon of music.[11]

If audition and memory are the engines that drive expectation in music, expectation itself is the beginning and end of music. Freyd (1987) developed what she calls "representational momentum" to characterize expectation of movement: "The perceptual system is geared to perceive transitions in real time" (428). In other words, the brain constantly anticipates the future. This must be so; how else would we catch a baseball, drive a car, or comprehend the rise and fall of a melody? Freyd (1993) writes, "Just as time is a dimension in the external world, inseparable from the other physical dimensions, so might time be a dimension in the represented world [in the mind]" (105).

Many traditional compositional practices are aimed at securing and maintaining the listener's interest through expectation. Consider the following musical motive:

If I then play

you become aware that I am sequencing a motive rising by diatonic steps, and you may expect I will repeat it. If I meet your expectation by extending the sequence

representational momentum increases and entropy decreases. I risk losing your attention because you now recognize the pattern, and since there is hardly any new information in it, you may start to lose interest. If instead of the previous motive, I play as a final motive

I have frustrated your representational momentum by shifting your attention from the horizontal melodic sequence to the vertical harmonic resolution. Surprise renews interest. There is also the satisfaction of arriving at a complete musical thought by cadencing.

9.16.1 The Golden Mean

Here is the entire phrase just described:

Notice that the sequence's momentum is broken about two thirds of the way through by the cadence. It is very common for musical patterns to veer off in a new direction near ratios of the golden mean. This proportionality appears in musical structures of all kinds and all levels of compositional scope, ranging from motivic fragments to cycles of works. For example, the boundary between the exposition and development section in many Mozart sonatas begins in the vicinity of (and sometimes even exactly on) the measure that divides the movement by the golden mean (Putz 1995; Kay 1996). Similar arrangements appear in the works of Beethoven, Webern, and many other composers (Novden 1964). Did these composers intentionally structure their music to have these proportions? We know that Mozart was fascinated by mathematics, but there's little hard evidence one way or the other. On the other hand, in his work *Music for Strings, Celeste and Percussion,* Béla Bartók used the golden mean so accurately, so often, and at so many structural levels simultaneously that it is easy to assume he did so intentionally (Lowman 1971).

But proportional analysis of music only goes so far. Music more resembles objects shaped by natural forces than objects shaped by axiom. For example, Putz (1995) found that large structures in Mozart's sonatas came statistically much closer to the golden mean than smaller structures. He attributed this—correctly I believe—to the tendency of natural proportional structures, such as segment sizes of shells and branching patterns in plants, to become increasingly approximate at the extremes of scale.

Another reason for the limited success of proportional analysis of music is that a high degree of strict proportionality on many levels of scale is highly redundant, and this is inconsistent with the compositional exploitation of expectation and surprise. Structural predictability can only be useful to a composer up to a point because music is designed to gain and maintain interest, and this requires a certain degree of structural ambiguity. Consequently, materials may be ordered, combined, disordered, and recombined in a manner that defies easy analysis. Meyer (1956) writes,

> Weak, ambiguous shapes may perform a valuable and vital function . . . for the lack of distinct and tangible shapes and of well-articulated modes of progression is capable of arousing powerful desires for, and expectations of, clarification and improvement. . . . some of the greatest music is great precisely because the composer has not feared to let his music tremble on the brink of chaos, thus inspiring the listener's awe, apprehension and anxiety, and, at the same time, exciting his emotions and his intellect.

Information theory and its relation to expectation and surprise show up even at metalevels of the composition process. Wherever there is a belief, there is an opportunity for its deconstruction, with all the same consequences for expectation and surprise. For example, it seems John Cage's primary aim was not to maintain an audience's interest. Rather, he wanted to allow natural processes to manifest directly in his music, in part, I suppose, because *this would deconstruct compositional methodology based on interest and expectation*. Where Schoenberg and his school sought to erase the expectation of tonal harmony, Cage and others sought to erase the expectation of expectation itself. (Note that this still requires a sense of expectation.) Thus, deconstructionism can be seen as the play of information and expectation in the realm of belief systems.

9.17 Form in Unpredictability

Music is like a field, bordered on one side by order and regularity and on the other by surprise and irregularity, and the most effective musical domains lie in the middle ground between these borders. Redundant elements communicate a *sense of order* that is embodied, for example, in the regularities between the various parts of a musical composition. *Taste* is reflected in the entropical elements, and *style* is revealed in the pattern of trade-offs made by the composer between order and taste. If we appreciate the sense of order, taste, and style in music, we appreciate the *intelligence* that informs the composer's work.

The efforts in this chapter to generate compositions by rule have so far shown no particular musical intelligence. Because all values chosen by the `Random()` function are strictly independent, the music created directly from it is unsatisfying; it lacks the glue—redundancy—that binds music together. But there are mathematical forms, called *fractals,* that reveal a deep inner structure, very similar to the complex inner structures of music, that combine varying degrees of predictability and unpredictability in one contour.

9.17.1 Self-Similarity

Consider the Weierstrass function shown in figure 9.25. Like ocean waves, it is shaped from point to point with a balance of predictability and unpredictability. This balance extends across different levels of magnification: the shape of the smaller parts resembles the shape of the larger parts, and vice versa, demonstrating *self-similarity* at various scales. For example, the contours inside the two boxed sections of the curve in figure 9.25 are similar. This calls to mind the proverb "The more things change, the more they remain the same." A structure is self-similar if, when magnified, its structure remains similar to the original scale. But what defines similarity in this case?

Let's examine how energy is distributed in the partials of the Weierstrass function. Figure 9.26a shows the power spectrum of this function on a linear scale (see volume 2, chapter 3). A power spectrum is basically a means to observe where there is energy in a signal. Most of the energy in this signal is near 0 Hz, and energy drops off quickly with increasing frequency, but it's a little hard to see what's really happening. A clearer picture emerges from figure 9.26b, which depicts the same spectrum, but with log frequency and log amplitude shown on the x- and y-axes, respectively. Viewing the power spectrum as a log-log plot reveals the essential detail of the spectral plot. The lines through the peaks of both plots show the ratio of $1/f$, where f is frequency. The tips of the

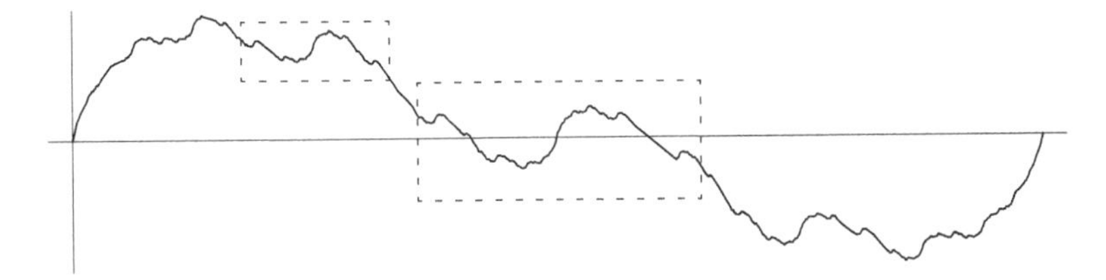

Figure 9.25
Weierstrass function.

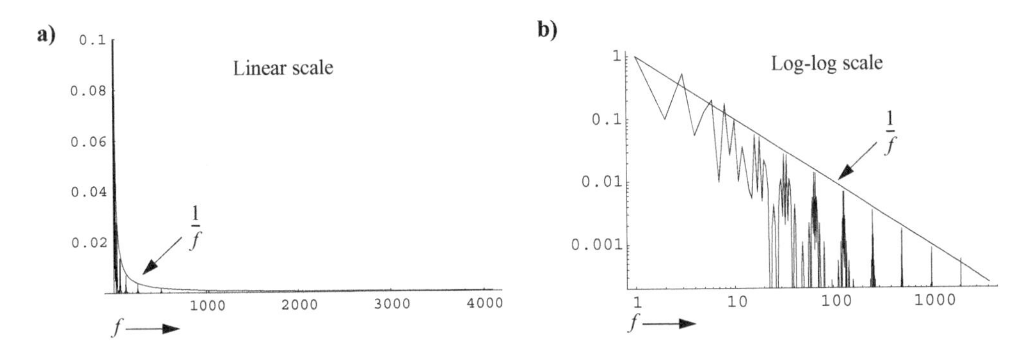

Figure 9.26
Weierstrass function power spectrum.

spectral components seem to track this $1/f$ line. We say that the Weierstrass function has a *spectral tendency* of $1/f$, meaning that the intensity of frequency nf has $1/n$ the intensity of frequency f. This corresponds to a roll-off of high-frequency energy at the rate of about −3 dBSIL per octave. So how does this characterize similarity? And what does this form of similarity have to do with music?

9.17.2 Fractal Geometry

The ordinary materials of Euclidian geometry, such as lines, surfaces, and volumes, are organized by their *dimension,* which can be intuitively defined as the number of numbers needed to uniquely locate a point in space. To locate a point some distance along a line or curve, one number suffices, so lines and curves are one-dimensional. One number also serves to measure the distance from a point on the circumference of a circle, or to measure along the edge of an object.

For a point on a plane, two numbers are required, so a plane is two-dimensional. We can organize the two numbers in a variety of ways. Typically, we establish an orthogonal coordinate system with linear dimensions and express points in Cartesian coordinates such as $[x, y]$. But we could consider other non-Euclidian two-dimensional "spaces," such as telephone numbers that consist of a three-digit exchange number followed by a four-digit line number. We could also consider the nonlinear two-dimensional surface of a Möbius strip (figure 9.27) or a deformed surface such as a balloon. Three numbers are required to describe a point in 3-D space, and so forth.

The characteristic size of objects in Euclidian spaces changes in a regular way as the extent of their linear dimensions change. For example, if a line is doubled in length, its characteristic size also doubles. Doubling the length of a square's side multiplies its area by 4. Doubling the length of a cube's side multiplies its volume by 8. Abstracting from these examples, if D is dimension

Figure 9.27
M. C. Escher, *Möbius Strip II,* 1963.

and L is a scaling coefficient, then the characteristic size s of an object is given by $s = L^D$. Solving for D, we have

$$D = \frac{\ln s}{\ln L}. \qquad \textit{Dimension} \ (9.22)$$

Euclidian geometry covers the cases where $D = \{1, 2, 3, \ldots\}$, $D \in I$. However, there are structures that look like curves, such as the one in figure 9.25, but that don't behave like curves because position along the curve can't be described as a one-dimensional offset from some other point. Such shapes do not yield integer values for D and do not obey the scaling rule for Euclidian geometries. These shapes are not mere pathological[12] curiosities. They reflect the structures of coastlines, the branching of plants and blood vessels in the lungs, the annual flood tides of the Nile, and many other natural phenomena, including music. To accommodate such geometrical anomalies, mathematicians have had to devise more nuanced definitions of dimension, allowing for *fractional dimensions*. Objects with fractional dimension were nicknamed fractals by Benoit Mandelbrot (1977).

The Koch Snowflake A simple fractal example is the Koch snowflake. To generate this shape, begin with a triangle, such as an equilateral triangle with sides of length 1. Then, for each side, divide the length by 3, and build another triangle with its base upon every middle segment and its apex pointing outward. Last, discard the base segment, leaving only the sides. The first four approximations are shown in figure 9.28.

The shape becomes ever more detailed, and in the limit as the number of iterations goes to infinity, the distance between any two points along the curve becomes infinite, even though the area bounded by the curve remains finite. Therefore, in the limit, it is impossible to determine a length along the boundary. The structure shows similarities at all levels of magnification, so it is self-similar, which makes sense, considering how it is constructed.

The Koch snowflake and the Weierstrass function are examples of *deterministic fractals* because they are defined by an algorithm or mathematical formula. As shown in figure 9.28, the *regularities* of deterministic fractals are self-similar. There are also nondeterministic fractals, or *random fractals,* that more closely resemble the natural shapes of coastlines, mountain ranges, and natural musical signals. The *irregularities* of random fractals are statistically self-similar. Although deterministic fractals are infinitely self-similar, the self-similarity of natural fractal shapes tends to break down at very large and very small scales.

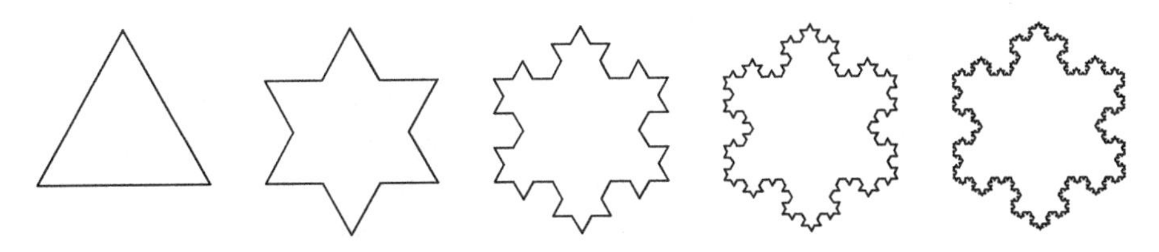

Figure 9.28
Koch snowflake.

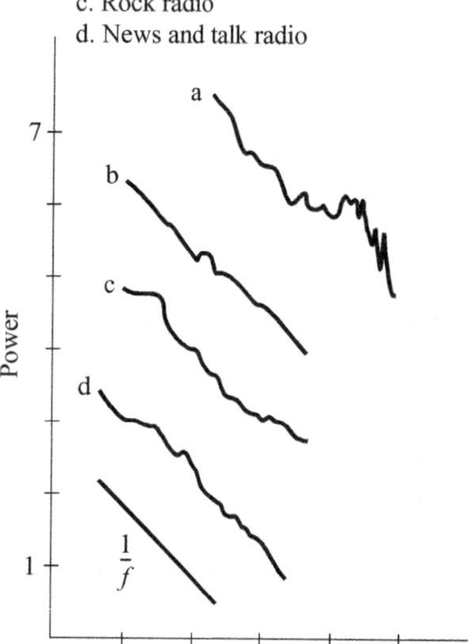

Figure 9.29
$1/f$ spectra. (Voss and Clarke 1975; 1978.)

9.17.3 Self-Similarity in Music

Richard Voss and John Clarke, when they were graduate students at the University of California in Berkeley, observed that a great deal of music, when examined over a long enough time span, appeared to have spectral tendency of $1/f^{v}$, $1 < v < 2$. They observed this by connecting a spectrum analyzer to the output of an AM radio. The frequency components revealed self-similar musical structure, especially for frequencies below 1 Hz. Since frequencies below about 50 Hz correspond to rhythmic and structural elements in music, they reasoned that the compositional structure of music—sections, phrases, motives, and note durations—exhibit a $1/f$ spectral tendency, revealing an even balance between entropy and redundancy.

Figure 9.29 shows some of their results. The Scott Joplin piano rags were averaged over an entire recording, perhaps an hour of music. Voss and Clarke (1974; 1978) attributed the high variation in this curve between 1 and 10 Hz to strongly characteristic rhythmic elements in Joplin's music. The rock music station recorded over a 24-hour period shows a spectral bump at about an hour's duration, perhaps corresponding to station breaks. Wondering how universally this result would hold, Voss and Clarke repeated the experiment with recorded music from a wide variety of musical ages, locations, and styles. All their subjects showed $1/f$ spectral tendencies, especially at very low frequencies. They believed they'd found experimental evidence that music favors this spectral tendency universally. Although the effect may not be universal, approximate self-similarity of musical structures has been widely demonstrated.

9.17.4 Generating Scaling Signals

The work of Voss and Clarke has evoked a great deal of interest and some controversy. Musicologists have surveyed a great deal of music for fractal elements, and composers have experimented with fractal designs. To do either requires a way to observe and generate fractals. Following are some techniques to generate fractal signals.

Generating Deterministic Fractal Signals Although it appears to be random, the generating equation for the Weierstrass function is strictly deterministic, like the Koch snowflake. In fact, it is just a variation on Fourier synthesis, summing a number of sinusoids at various harmonics (see volume 2, chapter 9). Unlike Fourier synthesis, the harmonics and their corresponding amplitudes are in an exponential rather than a linear sequence:

$$w(t) = \sum_{k=0}^{N} r^{kH} \sin(\pi r^{-k} t) . \qquad \textit{Weierstrass Function} \ (9.23)$$

Figure 9.25 shows the Weierstrass function for $r = 0.5$, $H = 1.0$, and $N = 32$. If we could hear the waveform in figure 9.26 it would sound like a rich pipe organ tone.

In equation (9.24) the parameter r, called the *lacunarity,* controls the texture of the spectrum. It can usefully vary over the range $0 < r \le 1$. H is called the Hurst exponent, or more intuitively, the self-similarity parameter or long-range correlation parameter. It has the range $0 < H \le 1$ and controls the spectral tendency because it determines the amplitudes of the harmonic sequence. H is related to the fractional dimension $D = 2 - H$ (Falconer 1990). As H goes to 0, high frequencies in the spectrum become stronger until, when $H = 0$, the spectrum no longer drops off in amplitude with higher frequencies. The Weierstrass function varies in dimensionality between 1-D and 2-D as H varies. Near $H = 0$ the curve is so dense that it seems to fill up the whole plane and so has dimensionality near 2-D.

Brownian Noise and the Random Walk We can relate the independent values of a uniform random number generator in such a way that they show interdependence and correlation across time and so achieve self-similarity. As a model of this process, consider the *random walk* of a drunk person who repeatedly stands up and stumbles off in an independent random direction, falls down, and starts off again and again. Clearly, where the drunk was a moment ago determines the possible places he will fall next, so there is a sense of history, albeit a quixotic one, to the process. If U_s is a uniform real random sample and x_n is the current point, then Brownian noise is defined as

$$x_n = U_s + x_{n-1} . \qquad \textit{Brownian Noise} \ (9.24)$$

Because subsequent points depend upon current and previous points, this is a recursive process.

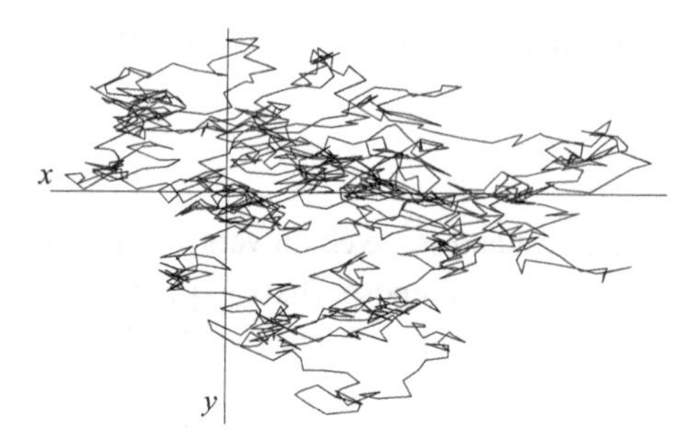

Figure 9.30
Brownian motion.

Brownian motion was first identified by Jan Ingenhousz in 1785, but it was named for Robert Brown, who rediscovered it in 1827 while watching the dance of pollen grains in a drop of water under a microscope. Albert Einstein identified this in 1905 as the effect of molecules of water, excited by heat, striking the pollen grains. Brownian motion describes the movement of microparticles in liquids and gases. Their movement is subject to Newton's first law of motion, so their inertia would make them want to travel in a straight line, but they can move only so far on average (the *mean free path*) without bumping into other microparticles, which sends them off in new directions. (Calculus alert!) A function is integrated by adding each subsequent point on the function to its previous point. Brownian motion can be viewed as the integral of uniform random noise. Figure 9.30 shows an example of Brownian motion in two dimensions.

Because this movement depends not on an absolute position but rather on its previous relative position, the range of x is theoretically without bounds. For example, if U_s happens to favor positive outcomes in the long run, x_n could grow toward positive infinity. Because computers have limited precision, an adjustment must usually be made to keep the random walk within computable limits. Here is a simple Brownian number generator (F. R. Moore 1990):

```
Real brownian(Real x, Real w, Real B){
    Real R;
    Do {
        R = x + Random( -w, w );
    } While ( R > B Or R < -B );
    Return R;
}
```

Parameter `x` is either the initial value of the random walk or the value last calculated by `brownian()`. Parameter `w` is called the window parameter because it determines the maximum amount by which the value of `x` can change at one time. Parameter `B` is the bounds, limiting the Brownian motion to within its range. This method departs from strict Brownian motion by retrying the random choice until the new value lies within this range.

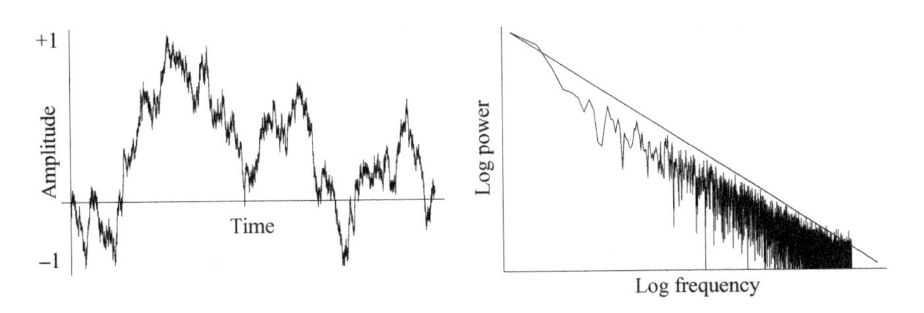

Figure 9.31
Brownian noise and its power spectrum.

We call the `brownian()` method each time we want a new Brownian number, passing it either an initial value or the value of its previous output. For example, the following code generated the function shown in figure 9.30.

```
Real x = 0.0;
Real y = 0.0;
For (Integer i = 0; i < 1000; i = i + 1) {
      x = brownian(x, 0.5, 0.5);
      y = brownian(y, 0.5, 0.5);
      PlotPoint(x, y);    // plot a point on a graph at location [x, y]
}
```

A Brownian noise signal and its power spectrum on a log-log plot are shown in figure 9.31. The straight line in the figure traces the contour of $1/f^2$ for reference.

Fractional Brownian Motion The preceding Brownian number generator produces a high degree of local similarity because subsequent points are constrained to remain relatively close to previous points. But because the random increment at each step is independent, Brownian motion typically only shows self-similarity in a region of its spectrum, so its fractal quality degenerates with scaling.

Fractional Brownian motion (fBm) is like Brownian motion, but the increments are no longer independent. Instead, just as low-frequency ocean waves extend their influence over many cycles of higher-frequency waves, in fBm, local rapidly fluctuating values are influenced by broader, slower-moving values extending proportionately over the entire spectrum. As fBm is magnified, it retains its statistically self-similar shape, and so it is fractal regardless of magnification.

Think of it this way. If we had an ideal tape recorder that accurately recorded all frequencies, and we gradually increased the speed of a tape recording of Brownian noise, the character of the noise would change (from a relatively low-frequency "whoosh" to a higher-frequency "whish"). But a recording of fBm noise will sound the same regardless of playback speed. All speeds sound the same because both the signal and the spectrum are self-similar at all levels of scale. A number of methods can be used to generate fBm noises.

Randomized Weierstrass Method One way to generate fBm noise is to add a random phase term to the Weierstrass function:

$$w(t) = \sum_{k=0}^{N} r^{kH} \sin(\pi r^{-k} t + \Phi(x)), \tag{9.25}$$

where $\Phi(x) = \pi U_s r^{kH} x$. In the function Φ, the parameter x allows us to set the strength of the effect. The strength of phase randomization is scaled as frequency rises so that the overall spectrum remains approximately $1/f$, depending upon the choice of parameters.

Voss's Method Martin Gardner (1978) reported a fractal noise generator attributed to Voss. A set of random variables x_k are summed on each sample n, and the result is output. The random variables are updated at different rates. If $((n))_{2^k} = 0$, then the kth variable is assigned a new random number U_s. The index k ranges from 0 to $N-1$. So x_0 is randomized every sample, x_1 is randomized every other sample, x_2 is randomized every fourth sample, and so on, until finally x_{N-1} is only randomized every 2^{N-1} samples. We can express the formula as follows:

$$f(n) = \sum_{k=0}^{N-1} \{(((n))_{2^k} = 0), (x_k \leftarrow U_s \text{ else } x_k)\} \tag{9.26}$$

where U_s ia source of random numbers. We can code this method as follows:

```
Real VossFracRand( Integer n, RealList L ) {
    Real sum = 0.0;
    Integer N = Length( L );
    For(Integer k = 0; k < N; k = k + 1) {
        If (Mod(n, Pow(2, k)) == 0) {
            L[k] = Random(-1.0, 1.0);
        }
        sum = sum + L[ k ];
    }
    Return(sum);
}
```

The following creates and prints a list of 128 fractal noise samples over four octaves:

```
RealList L = {Random(), Random(), Random(), Random()};
RealList R;
For (Integer n = 0; n < 128; n = n + 1) {
    R[n] = VossFracRand(n, L));
}
Print(R);
```

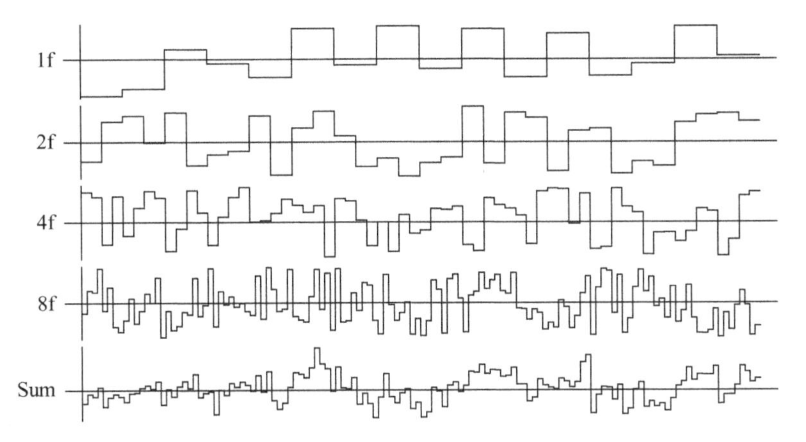

Figure 9.32
Voss's fractal generator.

Figure 9.32 shows how this noise is constructed by this method. Each function changes at a rate twice as fast as the previous function, and the functions are summed. Random values in the rapidly changing functions have only local influence, whereas values in the slowly changing functions extend their influence over many samples of the summed result, giving the result a fractal characteristic.

Spectral Filtering Method We can generate noise with an arbitrary spectral tendency by scaling the power spectrum of uniform noise. In fact, completely arbitrary noise functions can be obtained this way, fractal and otherwise. The method is to compute the Fourier transform of a noise signal, scale its power spectrum as we like, then retransform with the inverse Fourier transform (see volume 2, chapter 4).

9.17.5 Composing with 1/*f* Noise

In their experiments Voss and Clarke (1978) showed that a $1/f$ spectral characteristic was widespread in the structure of music. They conjectured that compositions created with a $1/f$ spectral characteristic would sound the most like music. To test this hypothesis, they synthesized melodies of three types using a computer: the first type made tone and rhythmic selections with a uniform $1/f^0$ noise generator, the second type used an fBm $1/f^1$ noise generator, and the last used a Brownian $1/f^2$ noise generator. For each generator type, they created melodies of two octave compass, using pentatonic, diatonic, and chromatic scales. They only conducted informal listening tests, but they reported that the consensus of listeners was that the fractally generated examples sounded the most like music.

Mandelbrot's (1977) reaction to the work of Voss and Clarke was to note that "[music] teachers insist that every piece of music [should] be 'composed' down into the shortest meaningful subdivisions. The result is bound to be scaling!" (375). Though the work of Voss and Clarke has drawn widespread interest and seems self-evident, it has been subjected to some skeptical analysis by, among others, the musicologist Nigel Nettheim (1992), who sought to evaluate and confirm their results.

Nettheim complained, for example, that analyzing long swaths of music broadcast over a radio would combine spectral contributions from many composers and ages, including announcer's messages, commercials, and other extraneous nonmusical material. For his own observations, he limited the analysis window to the duration of individual musical works and observed greater diversity of spectral tendency for different kinds of music. He also found that the fractal dimension was often closer to 2 (Brownian) than 1 (fractal). Nettheim's results were extended by Boon and Decroly (1995). Neither Nettheim nor Boon and Decroly refuted the basic premise of Voss and Clarke that there is an approximate self-similar structure to the power spectrum of music at low frequencies, but they showed that there is greater spectral variation, and pointed the way to more rigorous application of the technique in musicology.

Plato said, "For when there are no words (accompanying music) it is very difficult to recognize the meaning of the harmony and rhythm, or to see that any worthy object is imitated by them."[13]

By "any worthy object," Plato meant any natural object. To Voss, the appearance of fractal structure in music bolstered the theory that art imitates nature. This idea has been championed in virtually every age from the ancient Greeks to the present. But few natural processes seem to be inherently musical. So the question arises, If art imitates nature, exactly what is being imitated? Voss's answer is that musical signals, like so many other biological and natural signals, reveal a self-similar character.

9.18 Monte Carlo Methods

Lejaren Hiller and Leonard Isaacson (1959) are generally regarded as the first to seriously study composition of music with computers. They used the Illiac computer at the University of Illinois to create an experimental composition entitled *Illiac Suite for String Quartet* in 1957. As with Xenakis's work, chance techniques play a large role in this work, though for quite different purposes. Hiller and Isaacson (1959) write,

> The process of musical composition can be characterized as involving a series of choices of musical elements from an essentially limitless variety of musical raw materials. Therefore, because the act of composing can be thought of as the extraction of order out of a chaotic multitude of available possibilities, it can be studied at least semi-quantitatively by applying certain mathematical operations deriving from probability theory and certain general principles of analysis incorporated in a new theory of communication called information theory. It becomes possible, as a consequence, to apply computers to the study of those aspects of the process of composition which can be formalized in these terms.

Hiller and Isaacson wanted to use computers to model the composing process itself unlike Xenakis who saw them merely as an aid to human composers. So Hiller and Isaacson's investigation was conducted in the then-novel field of cybernetics. Their approach was to reduce the rules of various compositional styles—ranging from rudimentary species counterpoint to free atonality—into a set of numeric determinants that could be incorporated into programs running on the Illiac computer.

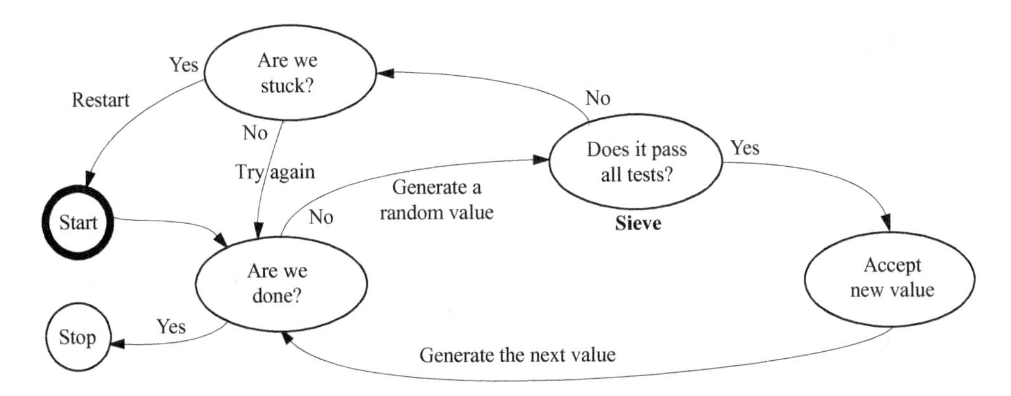

Figure 9.33
Random sieve method.

Any technique that uses probability to study complex systems can be called a Monte Carlo method. These methods were so named because of the similarity of probabilistic simulations to games of chance and because Monte Carlo, the capital of Monaco, was famous for gambling. These techniques are now used in many of the physical sciences. Hiller and Isaacson pioneered their use in music. Two notable methods they used are the random sieve method and Markov chains.

9.18.1 Random Sieve Method

With this method, choices made by a random number generator are accepted or rejected depending upon whether they obey certain rules, rather as a sieve strains out some objects and allows others to pass. One version of this method is outlined in figure 9.33. We begin at the Start state, and if we are not done, we generate a random value that we subject to tests. If it passes all tests, we accept the new value, and if we are not done, we go on to the next choice. If it does not pass all tests, we check to see how many times in a row we've failed to pass the tests. If we've failed so often that we believe we are stuck, we abort the process and restart. If we're not stuck, we try again with a different random choice.

For instance, *Experiment One* and *Experiment Two* (as movements of the *Illiac Suite* were called) were based on the rules of species counterpoint that were formalized by Fux (1725) in his work *Gradus ad Parnassum*. Fux's method is still widely taught in counterpoint classes today.

Hiller and Isaacson expressed Fux's rules in numerical terms that could be represented in a computer program. If a random choice would construct a harmony that violates the encoded rules of counterpoint, for example, movement by parallel or direct unisons, fourths, fifths, or octaves (figure 9.34), then their system would discard the choice and try again until no rules were violated. The successful choices were then appended to the end of the musical composition being generated, and the process was repeated until the composition was of the desired length. They conducted numerous tests of this kind at each step of the composing process.

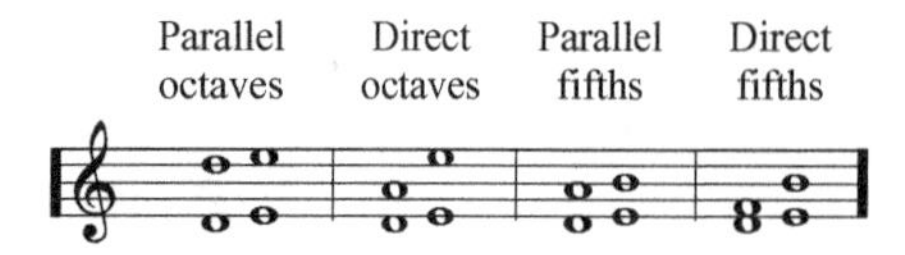

Figure 9.34
Parallel motion.

They found that for complex rule sets the situation can sometimes arise where there is no choice that doesn't break some rule. As a trivial example, suppose one rule establishes a range that the melody must lie within, but another requires it to skip outside of this range. The program would never finish running because no solution exists. The approach taken in figure 9.33 allows the program to restart if the number of unsuccessful trials exceeds some predefined threshold.

9.18.2 Backtracking

A finer-grained recovery technique is to *backtrack* if forward movement seems impossible. To do so, the current choice that appears to be stuck is aborted and the previous choice is repealed as well, forcing a new choice for the previous state. Then progress is attempted from there. If this still doesn't work, the next previous choice is repealed, and so forth. Stanley Gill (1963) was apparently the first to demonstrate the use of backtracking for composition in a 1963 piece composed for the BBC in the style of Arnold Schoenberg.

Gill avoided stalemates between conflicting rules by prioritizing them. When evaluating the suitability of a particular choice, his program calculated a score of demerits based on how many rules that choice would violate and how important the violated rules were. The choice with the lowest demerits was accepted unless no choice produced a score low enough, in which case the program would backtrack. Gill's program extended a small number (eight) of competitive versions of a composition in progress. At each step, one would be extended by a certain length (one beat), then evaluated for its goodness. Versions that were unfruitful were eventually abandoned automatically by his method.

Prioritizing the rules allowed Gill to adjust the rate of composition. If the criteria for extending a sequence were too severe, the program would make no progress; if they were too lenient, it would rapidly produce a composition of poor quality. He scaled the demerit score at each step by an adjustable coefficient that allowed him to mediate the rate of composition. The adjustable coefficient was itself determined by a negative feedback process so that the rate of composition remained relatively steady.

9.18.3 Searching

The random sieve method generates music much as one might try to find one's way through a maze: the rules are like the walls of the maze, and the random number generator is how one chooses a new direction to try. Backtracking is a strategy for recovering from dead ends. In any event, what these methods are doing is searching for solutions—looking for a way through the maze.

In general, two search strategies can be used, depending on the purpose (Ames 1983).

Comparative Search We may be in the maze purposely in order to map it. In that case, we want to systematically enumerate every possible solution from every possible entry point to every possible exit. We can then *compare* the goodness (in Knuth's sense of the term) of all possible solutions and arrive at the optimal one. In this case, we'd use a deterministic method of choosing a new direction at each step to be sure we traversed the entire maze from every possible direction of attack. We'd use backtracking to get out of dead ends.

Constrained Search We might simply want to exit the maze as quickly as possible without having to compare all possible solutions. This is a good approach if the time to find a solution is limited, if any solution will do, or if we believe good-enough solutions are plentiful. We may be forced to use this approach if the maze is so extensive that comparative search is not feasible. Composing music and playing chess can be thought of as very extensive mazes indeed, so this technique is often used in these cases. We'd use a random method of choosing a new direction at each step and employ backtracking to get out of dead ends. Gill's method of scaling the demerits of each choice is rather like adjusting the height of the barriers of the maze, allowing us to jump over low hurdles to speed progress (possibly to the detriment of the quality of the solution).

Bach Chorale Harmonization with Constrained Search One of the gold standards for modeling composition with computers is to replicate or create new works in the style of J. S. Bach's 389 chorale harmonizations.[14] The chorales were originally simple unaccompanied melodies that Bach arranged in a homophonic chordal style to be sung by church choirs. Because the style is so definite and regular, and because virtually every composition student is required to study them, these chorales have become a kind of standardized "laboratory rat" for such tests.

Kemal Ebcioglu (1986; 1988) used constrained search with prioritized rules and backtracking to model composing two-part species counterpoint, and he later used these techniques in his impressive program for harmonizing the chorale melodies of J. S. Bach. He programmed a computer with general rules about harmonic part writing based on the theories of Schenker (1935) and added specific information about Bach's chorales using a logic programming language. He created new chorale harmonizations that emulated Bach's style very closely. In fact, some of Bach's chorale harmonizations emerged verbatim from his system.

9.19 Markov Chains

Even if we were able to identify all the rules that characterize a particular musical style (and that's a big "if"), there is still a great deal of difference between music that breaks no rules and music that shows taste. Certainly a critical element of a composer's aural sensibility is a sensitivity to musical context, but none of the methods discussed so far take the surrounding music into account to determine subsequent choices.

Markov chain techniques are sensitive to their immediately preceding context, so they can create contextually appropriate outcomes. Markov chains use recently chosen states to

Figure 9.35
Chorus from *Oh Susanna* by Stephen Foster.

influence the probability of subsequent choices. Another advantage of Markov chains is that the rules driving the process can be readily discovered from existing compositions. Thus, it is possible to use Markov chains to compose music that is like other music. Harry Olson (1952) used them to construct musical examples that resembled the works of the composer Stephen Foster, and Hiller and Isaacson (1959) used them to compose a movement of the *Illiac Suite*. The technique is widely used.

9.19.1 Markov Chain Orders

Markov chains are ordered by how much recent history is taken into account when determining the next state. Following Olson's lead, let's analyze a Stephen Foster song, *Oh Susanna,* using various orders of Markov process. By focusing just on the chorus of the tune, we can keep the analysis from becoming too long-winded. Figure 9.35 shows the chorus, which has 25 notes (not counting rests), labeled R_0 to R_{24}.

9.19.2 Zeroth-Order Markov Process

Since the weighted choice technique (see section 9.14.4) takes no account of any previous states, it is defined as the zeroth-order Markov process, H_0. Even simple weighted choice is useful for matching the static event frequency of data drawn from the real world.

We create the probability density function for *Oh Susanna* by counting how many times each pitch is visited as a ratio of the total number of notes:

C	D	E	F	G	A	B
4/25	5/25	5/25	2/25	5/25	4/25	0

The counts are expressed as a fraction of the total number of notes. A table like this of event occurrences is called a histogram.

Feeding the *Oh Susanna* probability density function into the weighted choice technique would generate a new melody with pitches in roughly the same proportions as *Oh Susanna,* but the new melody would probably have little if any of the musical character of the original.

9.19.3 First-Order Markov Process

Since music unfolds in time, the context of each note consists of the note or notes that precede it. If we want to incorporate context into our analysis, we must study how notes succeed each other in the melody. For each note, let's tabulate the note that follows it. We can distill from this information what the probability of the next note will be, given the current note.

Markov Analysis We create a first-order Markov analysis by the following steps:

1. Catalog the note transitions. We pair each note in the melody with the note that follows it. If we let the first note (F) be the current note, then the second note (also F) is the next note. So the first transition in the melody is F → F. If we now make note 2 (F) be the current note, then the next note is note 3 (A). So the second transition is F → A. The third transition is A → A, and so on.

The *transition table* (table 9.8) tabulates this information. Each cell stands for a transition from a particular current note to a particular next note. The row indexes the current note, and the column indexes the next note. Thus, the first transition, F → F, is indicated by a 1 in row F, column F. The second transition, F → A, is indicated by a 2 in row F, column A. The third transition, A → A, is indicated by a 3 in row A, column A, and so forth.

2. Tally up the number of transitions in each cell (table 9.9). What we end up with is essentially a set of zeroth-order Markov histograms in the rows. When we go to generate a melody based

Table 9.8
Markov Order 1 Transitions for *Oh Susanna*

	Next						
Current	C	D	E	F	G	A	B
C	0	9, 11, 19	0	0	0	0	0
D	10, 24	23	12, 20	0	0	0	0
E	8, 18	22	21	0	13	0	0
F	0	0	0	1	0	2	0
G	0	0	7, 17	0	6, 14	15	0
A	0	0	0	0	5, 16	3, 4	0
B	0	0	0	0	0	0	0

Table 9.9
Markov Order 1 Tallies for *Oh Susanna*

	Next						
Current	C	D	E	F	G	A	B
C	0	3	0	0	0	0	0
D	2	1	2	0	0	0	0
E	2	1	1	0	1	0	0
F	0	0	0	1	0	1	0
G	0	0	2	0	2	1	0
A	0	0	0	0	2	2	0
B	0	0	0	0	0	0	0

on this analysis, we select a particular histogram row depending upon which note is the current note.

3. Convert the rows into cumulative distribution functions. First we normalize each row. We want to adjust each histogram so that the sum of its probabilities equals 1. (If any row sums to 0, we set all elements of that row to 0.) This is shown in table 9.10.

4. Transform each column into a cumulative distribution function by summing each cell with all cells in the row to its right (table 9.11). The table is finally in a cumulative distribution format we can use to synthesize a first-order Markov melody. It determines subsequent notes based on how probable the transition is in the original melody. The method of traversing this function is the same as that described in section 9.14.6.

Markov Synthesis When using table 9.11 to generate a melody, we pick a starting note at random from the sample space, {C, D, E, F, G, A} (pitch B is ignored because nothing transitions to or from it). Let's make F the current note. Table 9.11 shows that there is a 50/50 chance that the

Table 9.10
Normalized Markov Order 1 for *Oh Susanna*

	Next						
Current	C	D	E	F	G	A	B
C	0	3/3	0	0	0	0	0
D	2/5	1/5	2/5	0	0	0	0
E	2/5	1/5	1/5	0	1/5	0	0
F	0	0	0	1/2	0	1/2	0
G	0	0	2/5	0	2/5	1/5	0
A	0	0	0	0	2/4	2/4	0
B	0	0	0	0	0	0	0

Table 9.11
Markov Order 1 Distribution Function for *Oh Susanna*

	Next						
Current	C	D	E	F	G	A	B
C	0	1	1	1	1	1	1
D	2/5	3/5	1	1	1	1	1
E	2/5	3/5	4/5	4/5	1	1	1
F	0	0	0	1/2	1/2	1	1
G	0	0	2/5	2/5	4/5	1	1
A	0	0	0	0	2/4	1	1
B	0	0	0	0	0	0	0

next note will be F or A. (This may be easier to follow by reference to table 9.10.) Suppose A is chosen; it is now the current note. Then there is a 50/50 chance that the next note will be G or A. Suppose G is chosen; it is now the current note. Now it is twice as likely that E or G will be the next note than that A will be. We proceed like this until we have enough notes. Figure 9.36 is an example generated automatically from this data set with starting pitch F. Only the pitches were synthesized; the rhythms were copied from the original to aid comparison. This method carries a hint of the musical character of the original into the synthesized melody.

A first-order Markov process asks, Given the immediately preceding state x_{n-1}, what is the likelihood that the current state R is x_n? This is written using *conditional probability* notation,

$$q = P\left(R = x_n \middle| x_{n-1}\right),$$

which is read as "Given the condition that x_{n-1} is the preceding state, let q be the probability that state R equals x_n."

Directed Graph Another way to represent the first-order Markov transition information we have developed is to show it as a *directed graph,* which illustrates the flow of possibilities from state to state. States are represented by circles, and transitions from state to state are represented as arcs (lines with arrows). The directed graph of the chorus for *Oh Susanna* is shown in figure 9.37.

Figure 9.36
Oh Susanna chorus synthesized by first-order Markov process.

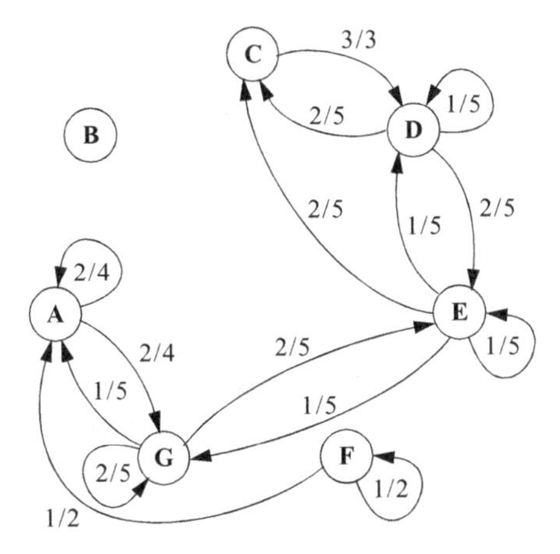

Figure 9.37
Directed graph of *Oh Susanna,* first-order Markov analysis.

The diatonic pitches of the scale are shown in circles. The arcs are labeled with their transition probabilities.

When synthesizing a melody, notice that once we leave pitch F, we can never return to it, because no pitch besides F ever transitions to F. Pitch B is unreachable. Markov synthesis is free to cycle among the remaining pitches. Because it contains cycles, it is a *directed cyclic graph* (DCG). If there were no cycles in the graph, it would be a *directed acyclic graph* (DAG).

9.19.4 Second-Order Markov Process

Second-order Markov analysis basically asks, Given two events in sequence, what is the probability of the next event? We express the probability as

$$q = P\left(R = x_n \big|_{x_{n-1},\, x_{n-2}} \right),$$

which is read as "Let q be the probability that R equals x_n, given that x_{n-1} and x_{n-2} precede it in sequence." We could represent the first few second-order transitions for *Oh Susanna* like this:

$$F:F \to A\,,\; F:A \to A\,,\; A:A \to A,\; A:A \to G\,,\; A:G \to G\,, G:G \to E, \ldots$$

How many possible second-order transitions are there for the diatonic scale? First-order Markov analysis involves two notes (current and next) and so has $7^2 = 49$ orderings. Second-order Markov analysis involves three notes (previous, current, and next), and by the rule of enumeration, there are $7^3 = 343$ possible orderings. We still want to represent the transitions as a two-dimensional matrix so that, as before, each row represents a zeroth-order Markov density function that determines the probability of the next note. We can manage this by marking the rows as the *pair* of previous and current pitches, and the columns as the next pitch. For the diatonic scale, this requires 49 rows and 7 columns, still a pretty big table, but to save room we can leave out any rows that have no transitions.

The analysis is shown in table 9.12. To conserve space, the transition event order and the normalized probability distributions are shown in the same table. For example, the listing for the first transition, $\mathrm{F{:}F} \to \mathrm{A}$, reads 2 (1.00), which means the target pitch A is the second note in the melody (counting from 0), and the probability of this transition is 1.00. Sometimes more than one note shares the same transition. For example, $\mathrm{E{:}C} \to \mathrm{D}$ is shared by notes 9 and 19.

Figure 9.38 shows an example second-order melody synthesized from table 9.12. The melody length and rhythms are the same as the original to facilitate comparison, although they could also be synthesized from a Markov analysis. Note the direct quotation of the original in the first six notes. Because it takes more of the preceding music into account when choosing the next note, melodies created from higher-order Markov synthesis carry over more of the exact phrasing of the original melody.

If we start the Markov synthesis on other than the F:F transition, we enter the analysis matrix at a different position, and different patterns are synthesized. Table 9.13 shows a few example note sequences generated from beginning table 9.12 at different initial transitions.

Table 9.12
Second-Order Markov Analysis of *Oh Susanna*

	Next						
Current	C	D	E	F	G	A	B
D:C	0	11 (1.00)	0	0	0	0	0
E:C	0	9, 19 (1.00)	0	0	0	0	0
C:D	10 (0.33)	0	12, 20 (0.67)	0	0	0	0
D:D	24 (1.00)	0	0	0	0	0	0
E:D	0	23 (1.00)	0	0	0	0	0
D:E	0	0	21 (0.50)	0	13 (0.50)	0	0
E:E	0	22 (1.00)	0	0	0	0	0
G:E	8, 18 (1.00)	0	0	0	0	0	0
F:F	0	0	0	0	0	2 (1.00)	0
E:G	0	0	0	0	14 (1.00)	0	0
G:G	0	0	7 (0.50)	0	0	15 (0.50)	0
A:G	0	0	17 (0.50)	0	6 (0.50)	0	0
F:A	0	0	0	0	0	3 (1.00)	0
G:A	0	0	0	0	16 (1.00)	0	0
A:A	0	0	0	0	5 (0.50)	4 (0.50)	0

Table 9.13
Other Second-Order Markov Note Sequences from *Oh Susanna*

D:C	D	C	D	E	G	G	E	C	D	E	G	G	A	G	G	E	C	D	C	D	E	E	D	D	C
E:C	E	C	D	E	E	D	D	C	D	C	D	C	D	E	G	G	E	C	D	E	G	G	E	C	D
C:D	C	D	C	D	C	D	E	G	G	E	C	D	C	D	E	E	D	D	C	D	E	G	G	A	G
D:D	D	D	C	D	C	D	E	G	G	E	C	D	C	D	E	E	D	D	C	D	E	G	G	A	G
E:D	E	D	D	C	D	C	D	E	E	D	D	C	D	E	E	D	D	C	D	E	E	D	D	C	D
D:E	D	E	E	D	D	C	D	E	E	D	D	C	D	E	G	G	E	C	D	E	G	G	E	C	D

Figure 9.38
Oh Susanna chorus synthesized by second-order Markov process.

9.19.5 Third-Order Markov Process

Third-order Markov transitions require three notes of context. The first few transitions are

F : F : A → A, F : A : A → A , A : A : A → G, A : A : G → G , ...

The analysis is shown in table 9.14. As in table 9.12, the transition event order and the normalized probability distributions are shown in the same table to conserve space.

Figure 9.39 shows an example third-order melody synthesized from table 9.14. Again, the melody length and rhythms are the same as the original to facilitate comparison, although they could

Figure 9.39
Oh Susanna chorus synthesized by third-order Markov process.

Table 9.14
Third-Order Markov Analysis of *Oh Susanna*

	Next						
Current	C	D	E	F	G	A	B
C:D:C	0	11 (1.00)	0	0	0	0	0
G:E:C	0	9, 19 (1.00)	0	0	0	0	0
D:C:D	0	0	12 (1.00)	0	0	0	0
E:C:D	10 (0.50)	0	20 (0.50)	0	0	0	0
E:D:D	24 (1.00)	0	0	0	0	0	0
E:E:D	0	23 (1.00)	0	0	0	0	0
C:D:E	0	0	21 (0.50)	0	13 (0.50)	0	0
D:E:E	0	22 (1.00)	0	0	0	0	0
G:G:E	8 (1.00)	0	0	0	0	0	0
A:G:E	18 (1.00)	0	0	0	0	0	0
D:E:G	0	0	0	0	14 (1.00)	0	0
E:G:G	0	0	0	0	0	15 (1.00)	0
A:G:G	0	0	7 (1.00)	0	0	0	0
G:A:G	0	0	17 (1.00)	0	0	0	0
A:A:G	0	0	0	0	6 (1.00)	0	0
F:F:A	0	0	0	0	0	3 (1.00)	0
G:G:A	0	0	0	0	16 (1.00)	0	0
F:A:A	0	0	0	0	0	4 (1.00)	0
A:A:A	0	0	0	0	5 (1.00)	0	0

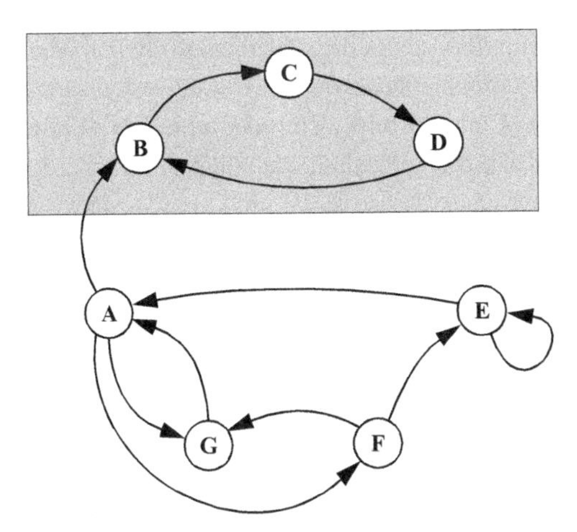

Figure 9.40
Degenerate cycle.

also be synthesized from a Markov analysis. The transition probabilities are now so constrained that a major chunk of the original melody is quoted (motive A in the figure). Only the last measure (motive B) is different. To see how this came about, note that the Markov synthesis simply repeated the melodic fragment C. So B is really just part of C, which is part of A. What happened?

Basically, we hit a cycle in our analysis where a state returns back on itself (see section 9.19.3). Cycles that can't be escaped once they are entered are *degenerate*. The group {C, D, B} in figure 9.40 is a degenerate cycle. The other states are cyclic but not degenerate. This becomes an increasing problem with higher-order Markov synthesis.

9.19.6 *N*th-Order Markov Process

The general form of an *N*th-order Markov process can be expressed as $X = M_N(x)$, where M is a Markov analysis function of order N, x is the sequence to be analyzed, and X is the set of probability distribution functions that result from the analysis. *N*th-order Markov synthesis can be expressed as $y = M_N^{-1}(X)$ where y is the melody synthesized by the process.

As the order increases, we're more likely to get significant chunks of the original in the synthesized melody. At a sufficiently high order, depending upon the material, we will get the entire original. This happens for our example melody with fourth-order Markov synthesis. Thus, although arbitrary-order Markov processing is theoretically possible, for most realistic applications, analysis beyond about the fourth order may not be particularly meaningful.

9.20 Causality and Composition

On hearing Lejaren Hiller's *Illiac Suite,* John Pierce (1983) reported that it "sounds pleasant, but it wanders, and so does the listener's attention." Markov techniques are strictly reactive to the immediately preceding events and do not lend themselves to following an overall plan.

It is worth pausing for a moment to look at our assumptions about the role of causality in music. A system is *causal* if it references only current and past input and past output. Causal systems may not reference future input or current or future output. That is to say, a causal system can't know the future. (These ideas are formalized in the discussion of the canonical filter in volume 2, chapter 5.) Certainly, listening to music we've not heard before is a causal process: we can't know what we'll hear until we hear it, and we can't know our reaction until we have it.

It is easy to assume that because listening is causal that composing must somehow be, too. Some forms of composition, such as improvisation, are primarily causal, and many of the techniques discussed in previous sections, especially Markov chains, give the impression that composing starts with the first note and proceeds to the last in a direct sequence. This is hardly ever the case in practice.

If a role of the composer is to manipulate expectation, then the composer must be of two minds, one part imagining what the listener's expectations will be in time, and the other part keeping a "timeless" plan of the composition in mind. We may think of such a plan as a static design, like an architectural drawing of a building. But any such plan is itself the result of the composer's pursuing an underlying goal: the aim, motive, or reason why the composer is writing the music. The act of reducing one's vision of a composition to a finished score is a teleological process, a process that works backward from the composer's goal.

Composer Herbert Bielawa and Paul Craner developed a teleological process to automatically compose chorale harmonizations.[15] Early in his harmony theory teaching career, Bielawa had students tally the types of chordal root movements in the Euro-classic music they were studying. No matter what the piece was, as long as it was Euro-classic, they would come up with very similar graphs. He eventually boiled it down to a rule: good progressions are up a second, down a third, and either way a fourth or fifth (dominant to tonic).[16] The method Bielawa developed to embody this rule was in essence a simple first-order Markov process, but with a twist. To overcome the aimlessness of Markov chains and solve some tricky problems with cadencing, his program composed backward, beginning with the final cadence.

Ordinarily, one would want "good" root movements (up a fourth or down a fifth) to be selected most often and down a third less often. And although "bad" root movements were rare, they did happen occasionally in real music, so these also had small but nonzero probability. However, to compose the music backward, Bielawa had to flip the probabilities so that all the "good" root movements had to be temporarily "bad" ones, and vice versa. Ultimately, retrograding the generated composition automatically made good progressions out of the bad ones.

9.21 Learning

Hiller and Isaacson's experiments were in the spirit of research efforts to embody expert knowledge about real-world problems in computer programs, known as *Artificial Intelligence* (AI).

The classical AI approach attempts to reduce the subject knowledge domain to its essential rules, much as Fux did for counterpoint. Based on what we've seen of Monte Carlo techniques, at least the following difficulties can be identified with using rule-building systems to model intelligence:

- Determining appropriate rules can be difficult (or impossible), even for experts, if the knowledge is not available to consciousness. Rule-building AI techniques are difficult to apply to subjective elements such as taste, preference, and style.
- Unlike people, rule systems cannot of themeslves adapt to new information or incorporate new rules. That is, they cannot learn of their own accord but require the introspection and programming skill of trained experts.
- The more rules there are, the harder it can be to add new rules without breaking or distorting the logical structures already encoded. As we saw with the random sieve method (section 9.18.1), it is easy to introduce rules that contradict each other. The system becomes more fragile as rules are added.
- True expertise means knowing the rules of a discipline as well as the exceptions that prove the rules. So there must be rules about when the regular rules apply and when they don't. This calls for metarules that enable or disable other rules in certain contexts. This leads to hierarchies of rule systems. As a system of rules grows more complex, it becomes progressively harder for it to change or adapt to new information and novel circumstances. It becomes more brittle as the depth of hierarchy increases.

Capturing real-life expertise by compiling lists of rules tends to create rule systems that are brittle and fragile. In contrast, human knowledge remains relatively flexible in the face of novel insights and developments. It does not seem likely that human learning happens by piling up lists of rules. If that were true, then the more we know, the longer it would take us to react to circumstances, assuming some finite time to evaluate each rule. Rule-based classical AI is not a very probable model of human cognition.

Whatever musical knowledge is, it certainly seems to arise from experience and is thus learned. We appear to learn music by using cognitive strategies that are built into our brains. We apply these cognitive strategies to our experience of music, and somehow the result is knowledge of music. From this knowledge arises affinity for certain forms of music, and musical taste arises. What are these cognitive strategies? What is learning, and how can we model it?

9.21.1 A Self-Learning Grammar

Teuvo Kohonen (1989) has described "a self-learning grammar, the rules for which are automatically and systematically constructed on the basis of exemplary material." The method, which he calls dynamically expanding context (DEC), is like Markov analysis, but instead of fixed-order analysis it uses an order of analysis that grows automatically as necessary to resolve conflicts in the rules. Thus general rules are gradually replaced with specific ones, mastering the maximal degree of complexity with the minimal amount of exemplary materials. This exhibits a form of learning because the rules evolve with increased experience.

DEC is a form of *unsupervised learning* because no a priori knowledge of music is embedded in the DEC method. However, to fully exploit the method, it is necessary to carefully formulate the exemplary material. Like the Markov process, DEC can be driven to synthesize compositions.

The method is best illustrated by considering an example that Kohonen provides. Consider a melody as a sequence of musical elements to which letters have been assigned:

ABCDEFG . . . IKFH . . . LEFJ . . . .

As with first-order Markov analysis, we start by examining the transitions. We eventually notice that there is a three-way conflict for which symbol may follow F: it may be G, H, or J. Using Markov techniques, we would assign probabilities to the outcomes based on their frequency. But Kohonen's approach is to resolve this conflict by enlarging the context. We take the symbol in front of F for additional context (like dynamically jumping to a second-order Markov analysis for just this rule). But there is still a two-way conflict because the successor to E:F could be G or J. Adding a second symbol before F fully disambiguates the three cases. While third-order analysis is required for F, it is overkill for other symbols, such as H, which is fully defined by the second-order rule $K:F \rightarrow H$, and for C, which is fully defined by a first-order rule $B \rightarrow C$. We wish to avoid overspecifying the production rules because—as we saw with higher-order Markov processes—too much context means the rules are too specific and rigid. DEC thus dynamically expands rules only to the extent required to resolve conflicts.

DEC Analysis Kohonen's method is to iteratively scan the training data starting with low-order rules and apply progressively higher-order rules to problem cases until all conflicts are resolved. When a conflict is observed, the existing rules are marked invalid, and new rules are substituted that contain more context. Iteration over the input continues until no further changes to the rules are necessary.

For example, consider rule construction for F. Its first appearance is $F \rightarrow G$. Because it is not already in memory, we create an entry for it as follows:

Rule no.	Left part	Right part	Valid
1	F	G	true

Next we find $F \rightarrow H$ in the input and observe that it conflicts with rule 1. This requires two actions: first invalidate rule 1, then (because it is not already in memory) insert a second-order rule for $K:F \rightarrow H$. Memory now looks like this:

1	F	G	false
2	K:F	H	true

Finally, we find $F \rightarrow J$ in the input, and observing its conflict with first-order rule 1, we enter it as a second-order rule:

3	E:F	J	true

Having exhausted the input, we iterate again from the beginning. When we come to F, we search memory and discover the invalid rule $F \rightarrow G$. We now expand its context by one order, creating a new rule $E : F \to G$. However, we now observe conflict between $E : F \to G$ and $E : F \to J$. We must invalidate rule 3 and enter a new third-order rule 4, as follows:

3	E:F	J	false
4	D:E:F	G	true

As we continue to scan the input for F, we'll eventually discover the invalid rule 3, which we evaluate at a higher order and enter as a new rule 5:

5	L:E:F	J	true

Further iterations over the input do not cause any changes to the rules, so we are done, and we observe that rules 2, 4, and 5 remain valid.

DEC Synthesis Suppose so far we have generated the sequence CDEF. To extend the sequence with a valid next symbol, we first search through memory for first-order rules $F \to ?$. We find rule 1, $F \to G$, which is invalid. Finding no other valid first-order rules, we try second-order rules $E : F \to ?$ and find rule 3, $E : F \to J$, which is also invalid. Having exhausted second-order rules, we look for third-order rules $D : E : F \to ?$ and finally find rule 4, $D : E : F \to G$, which is valid, so the next new symbol we generate is G.

Like Markov synthesis, the output is made up of subsequences of the original material, so that the flavor of the original is preserved, but not its ordering. DEC synthesis, like Markov synthesis, contains a random element, but unlike in Markov synthesis, the occurrence of successive notes does not follow their probabilities in the input. DEC synthesis proceeds as though we always used the highest-order Markov analysis available for each rule. Kohonen suggests that if the results generated this way are too normative, more variance in the productions can be achieved by using lower-order rules, ignoring their validity.

9.21.2 The Nature of Learning

One could say that Markov techniques and Kohonen's DEC technique "learn" to recognize the features of the materials they are given. But they are unable to generalize from what they know to what they do not. If we study a corpus of music, say, the fugues of J. S. Bach, we not only learn the individual works, but as an automatic by-product our cognitive apparatus also distills out a sense of what a fugue is, so that if we later hear a fugue by Mozart, we recognize its form; we don't have to be retrained. The Markov and DEC techniques, like all methods considered so far, fail to generalize at all.

There are other important characteristics of natural learning that are also missing, such as *pattern completion,* for example, our ability to identify major or minor harmonies from a fragment of melody. If I show you a letter that is partially occluded, you are still able to recognize it (figure 9.41a). If it is too occluded to narrow it down to one letter, you can still easily identify the possibilities (figure 9.41b). In my college music appreciation classes, the professor would often test our knowledge of the musical repertoire we were studying by playing a randomly selected excerpt of music by "dropping the needle" (a phrase referring to the days of vinyl records). Even if we had

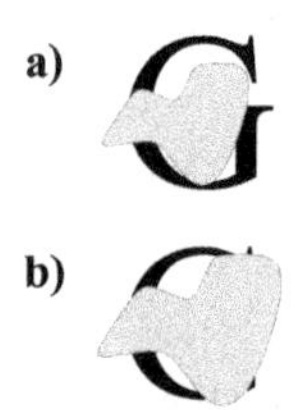

Figure 9.41
Occlusion.

listened to a piece only a few times or the excerpt lasted no more than a second or two, we would instantly be able to identify it. It's an incredible skill our brains have, if you think about it.

Another difference between natural cognition and the kinds of machine cognition described so far is that people can apply *multiple simultaneous constraints,* but standard computers act sequentially. When improvising music, a multitude of constraints operate simultaneously, guiding the musician's choices in the moment.

Our ability to handle multiple simultaneous constraints allows us to mediate the influence of syntax on semantics, and vice versa (McClelland, Rumelhart, and Hinton 1986). Consider the sentence "I saw the Grand Canyon flying to New York." We see that syntax constrains the assignment of meaning but does not determine it. We understand through the interplay of multiple sources of knowledge. Such structures of knowledge have been variously called frames (Minsky 1974), schemata (Bobrow and Norman 1975), and scripts (Schank and Abelson 1976). But rather than being static objects in memory, scripts appear to interact with each other to capture meaning in novel situations. How do we do this? And can machines do it, too?

9.22 Music and Connectionism

We have all learned skills, such as playing a musical instrument, juggling, or riding a bike, that we can do without understanding how we do them. We usually learn and teach these skills by example, not by rule. How do we learn to improvise music? How do we develop a personal musical style? How do we learn to distinguish the characteristic musical swagger of Beethoven's music from Schubert's? We know what we know, but we don't necessarily know how we know it. Since we are clearly able to learn these things, an obvious place to look for solutions is the brain.

9.22.1 Neural Models of Cognition

Neurobiology has shown that the brain can be modeled as a massively interconnected set of neurons operating in parallel. Cognitive psychologists and computer scientists have studied the properties of brain models using artificial neural networks. These models store knowledge in the *connection strengths* between simple processing units, much as our brains store knowledge in the connections between neurons. Because many neurons are acting concurrently in parallel, these models are called

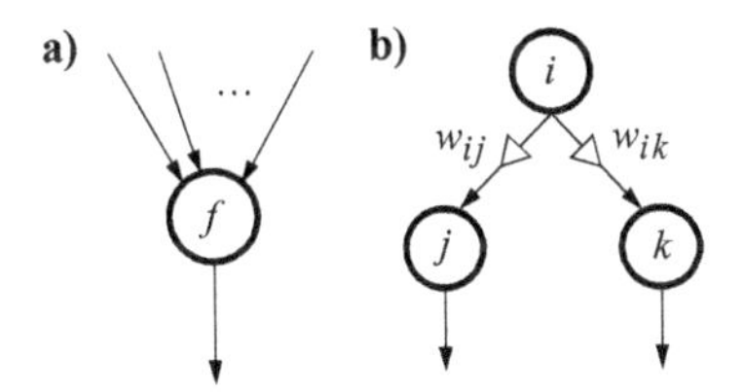

Figure 9.42
Networks.

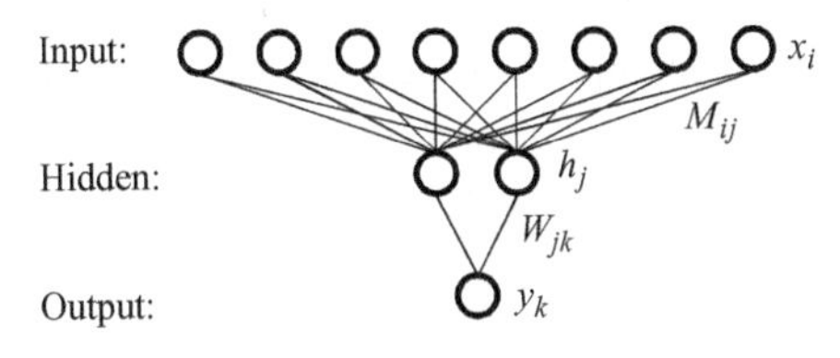

Figure 9.43
Simple feed-forward network.

parallel distributed processing (PDP) or *connectionist* models of cognition. Because the connection strengths between neurons is quantifiable, knowledge in a network is represented in a quantifiable way. Whereas knowledge in rule-based systems tends to be brittle, knowledge in networks can change and adapt as new knowledge is acquired and old knowledge is forgotten.

9.22.2 Artificial Neural Networks

An artificial neural network is simply an interconnected set of simple computational units (figure 9.42a). Usually, the processing performed by all units is the same. Each unit receives inputs from other units and produces a single output, which can be connected to one or more other units. Each connection between units has a unique strength that can be adjusted, so the influence of the units upon each other can vary. In the network shown in figure 43b, the connection strength between units i and j is called w_{ij}, and the connection strength between units i and k is w_{ik}. A strong positive output from i would tend to inhibit k if w_{ik} is negative and to excite k if the weight is positive. If the weight w_{ij} is zero, then the driving unit i has no influence on the driven unit j.

A simple feed-forward network having three layers is shown in figure 9.43. The output of each unit is fanned in to the input of each unit in the next layer. Each unit in the input layer x_i is connected through weights M_{ij} to hidden units h_j, which connect to the output unit y_k through weights W_{jk}. The hidden layer is so named because its values are not directly observable from outside the network. (There are always weights on the lines connecting units, but conventionally they are not explicitly drawn so as to keep down the clutter in the interconnection diagrams.)

The processing performed within the individual units can be as simple as just summing all inputs to produce the output. More typically, the units will use the sum of their inputs to index a nonlinear function of some kind. The function result is then output from the unit. We can express the input to each unit h_j from the input row of the network x_i as follows:

$$h_j = f\left(\sum_{i=1}^{N_i} w_{ij} x_i\right), \tag{9.27}$$

where w_{ij} is the weight from the ith to the jth unit, N_i is the number of input units, and f is some nonlinear function (Dolson 1989). Equation (9.28) also describes the connection from the hidden units to the output.

The nonlinearity of the function in each unit is the key to giving neural networks the ability to make decisions. Without this feature, the output of a unit would simply be proportional to its input. A nonlinear function allows *quantitative* changes in the input to result in *qualitative* changes in the output, such as turning a unit on or off. This capability allows neural networks to translate from subsymbolic activation levels to symbolic knowledge. The two most common choices for nonlinear functions are the hard-limiting *signum function:*

$$\operatorname{sgn}(x) \equiv \begin{cases} x < 0, & -1 \\ x = 0, & 0 \\ x > 0, & 1 \end{cases} \tag{9.28}$$

and the soft-limiting logistic function (figure 9.44):

$$f(x) = \frac{1}{1 + e^{-x}}. \qquad \textit{Logistic Function} \tag{9.29}$$

The logistic function played a role in the development of modern neural network theory because it was a component of the proof of an important neural learning technique, *back propagation* (Rumelhart, Hinton, and Williams 1986). In practice, it is just one of many possible "squashing functions" that map real values into a bounded interval.

There are many ways of connecting units. If there are only feed-forward connections (figure 9.45a), there are no loops in the network, and the computation of the output is fairly straightforward. If there are *feedback* connections (figure 9.45b), computation of the output can get complicated

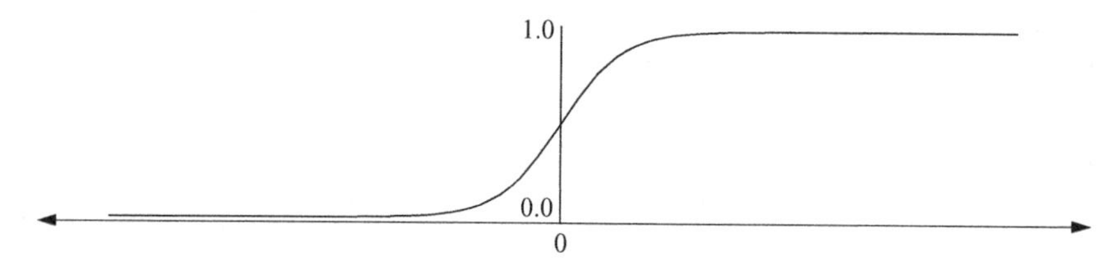

Figure 9.44
Logistic function.

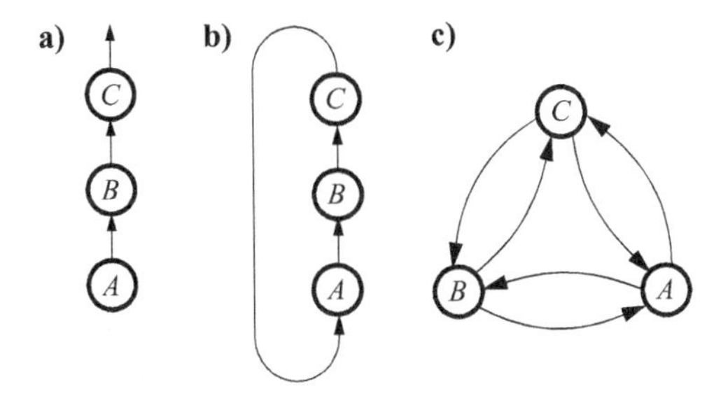

Figure 9.45
Network topologies.

because the output of unit *A* could depend upon unit *B*, which depends upon *C*, which in turn depends upon *A*, and so on. Feedback networks may be partly interconnected (figure 9.45b) or fully interconnected (figure 9.45c) so that all outputs go to all inputs.

Feedback networks can go into oscillation unless they are carefully designed. For every recurrent network, there is a corresponding feed-forward network (Minsky and Papert 1969), so I focus on feed-forward networks here.

Finally, there is the question of assigning weights to the connections. This is where things get interesting. It is possible to assign weights directly to configure a network to perform a particular calculation if we know what the weights should be. More often, we don't know what weights to assign, but we would like the network to discover them. Some networks allow a *supervised learning* method to be employed that automatically adjusts the weights in the network until a training pattern applied to the network's inputs produces the desired value on the output. These networks can learn to produce a desired outcome from a pattern that is applied to their inputs. We only have to show such networks what to do, not how to do it. "Here we have a mechanism whereby we do not actually have to know how to write the program in order to get the system to do it" (Rumelhart, Hinton, and Williams 1986).

As with musical taste and related subjects, we know what we like without necessarily knowing why we like it. If we can show a network examples of good and bad taste, then we can train it to share our taste in music. Once these associations are learned, we can use the network synthetically, to mimic our aesthetic judgments, for example, as a component in a composing program, or analytically, to understand the structure of our aesthetic choices by studying the network's solution.

So far, this is not much different than Markov and DEC techniques, which can also mimic. But what a network can do that Markov and DEC techniques cannot is spontaneously generalize from experience. If we show a trained network an input pattern that it has not previously encountered, it will make an educated guess based on the examples it has seen so far. Thus, knowledge in a suitably trained PDP network can retain a degree of flexibility and adaptability to the unknown.

Pattern completion is another form of generalization that networks can perform. If I play you a few notes from the middle of a familiar tune, you can generally pick up the tune and sing the rest of it. Pattern completion is crucial to our experience of music because this is how we perceive regularity and novelty. We can even model creativity as generalization if we think of it as providing novel responses to novel conditions.

9.22.3 Computing Taste, a Neural Evaluator of Intervals

As a simple example, let's teach a network to appreciate our taste in musical intervals. Of course, for a simple task like this we could simply write out a table, such as table 3.5, stipulating which intervals we find consonant and dissonant. But suppose we know what we like without knowing why. We provide a trainable network with example intervals and provide additional input that gives approval or disapproval based on our preferences, which the network will learn.

Once the network is trained, we can inspect its interconnection strengths to deduce what it knows about our preferences, thus aiding our ability to capture hard-to-explain knowledge. However, the interpretation of trained networks is usually nontrivial. Network analysis may be straightforward for simple networks and simple problems. But for more complicated tasks, the network's solution will tend to be *distributed* throughout the network, carried in the overall pattern of activity, rather than being *localized* in any particular unit, so the network as a whole must be analyzed for these cases (Rumelhart, Hinton, and Williams 1986). Also, the network may not necessarily find the optimal solution.

The following example uses an effective training method called *back propagation of error,* which is available for feed-forward networks like the one shown in figure 9.43. Remarkably, even fairly trivial-looking networks like that one can learn and retain multiple independent facts, just as humans can.

We must specify the significance of the network's inputs and output, which range numerically from 0.0 to 1.0. There are many possibilities to choose from, but the best network designs show a clear relation between the problem at hand and the network topology, and use no more units than necessary. For this example, we use 13 input units: one for each degree of the chromatic scale plus the octave, plus one extra to indicate whether we judge the interval consonant or dissonant that is used during training. When an input unit's activation is 1.0, that degree of the scale is sounding. The consonance/dissonance judgment can be represented as a single output unit. Consonance is associated with 1.0, and dissonance with 0.0. For the purposes of this experiment, let's say that we're not aware that the perfect and imperfect intervals are consonant and the rest are dissonant (see table 3.5). Instead, we train the network with examples of judgment and "discover" this.

Having specified the input and output units, we must decide about the hidden units. Three layers are generally sufficient to compute any function of interest with this method, so although multiple hidden layers can be used, they are not necessary. At present, there is no straightforward method to decide how many hidden units to use. It is generally best to choose the smallest number that is effective, both to simplify the calculations and to make the result as general as possible. So we choose two hidden units for this example.

Overall, we have 13 input units, two hidden units, and one output unit, for a total of 16 units connected by 28 weighted lines. Next we must train the network.

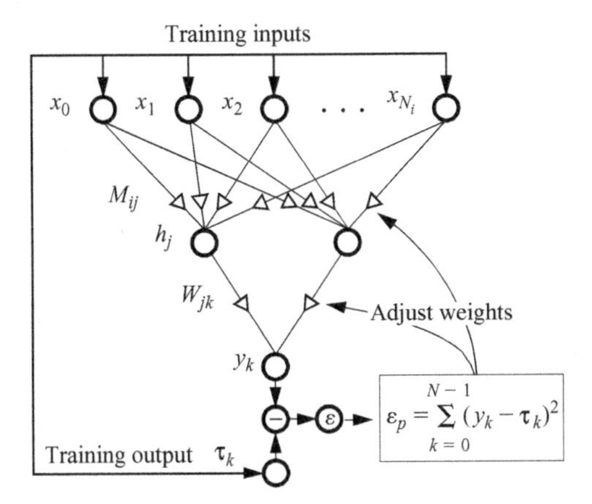

Figure 9.46
Back propagation.

Back Propagation of Error To start, we set all the weights to random values, rather as one shuffles a deck of cards before starting a game. If we then apply a pattern corresponding to some musical interval to the network inputs x_i, the output y_k can be computed directly. Each hidden unit h_j receives the sum of all input units times their respective weights M_{ij}. Each hidden unit uses this weighted sum as an index into the logistic function to produce its respective output. The output unit y_k similarly receives the sum of all hidden units times their respective weights W_{jk} (figure 9.46).

Because we randomized the weights, it is highly unlikely that the output of the network will initially agree with our judgment of an interval's consonance. Let us call the judgment we'd prefer the network to learn the *target,* τ_k. The network's error in judgment is the difference between the network's actual output and the target: $\varepsilon = y_k - \tau_k$. If the output unit's activation level is less than the target level, the error can be reduced by connecting the output unit more strongly to hidden units that are producing a positive value. If the output unit's activation level is greater than the target level, the output unit needs to be connected more strongly to hidden units that are producing a negative value. We can adjust the weights connecting the output unit to the hidden units, but this will not fix the problem by itself because the hidden units themselves also contribute to the error.

Adjusting the hidden units is more challenging because we don't have explicit target values for them—only the output units have targets. Nonetheless, the same basic strategy can be employed. We determine the proportion of error each hidden unit is responsible for and adjust weights connecting them to the input units to minimize this error. The general strategy is to propagate the error backward through the network, adjusting the weights as we go in proportion to their responsibility for the error in judgment at the output, which is how this learning technique came to be called back propagation of error.[17]

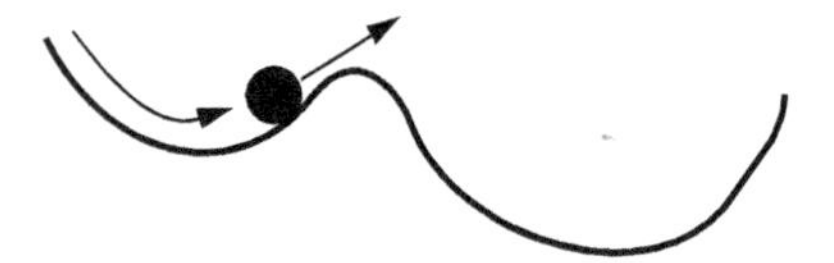

Figure 9.47
Local minimum.

We can obtain the total error on a set of k output units by summing the squares of the individual errors. Squaring the error eliminates the problem of negative errors canceling positive ones. For some input pattern p, the mean squared error is

$$\varepsilon_p = \sum_{k=0}^{N-1} (y_k - \tau_k)^2 .$$

Every time we apply pattern p to the network, we compute ε_p, then make small incremental changes to the weights to reduce ε_p. We continue the process as long as each step continues to reduce ε_p. Eventually, for a well-designed network, ε_p will become quite small. When it has reached a predetermined threshold, we stop the training.

There are things that can prevent ε_p from becoming as small as desired. The network may not be able to find a solution if it has fewer degrees of freedom than the problem space. There may not be enough hidden units, or the problem may not be suitable for the type of network chosen.

A subtler difficulty can arise where the back propagation technique finds an answer but fails to find the optimal answer. To visualize this, imagine letting a marble roll down the side of a basin with a shallow region and a deeper region beyond, separated by a ridge (figure 9.47). The marble might be captured by the upper region of the basin if it does not have enough momentum to ride up over the ridge into the deeper region.

Making small incremental adjustments to the weights of the network is akin to the marble's rolling down the basin. Once the network learns a suboptimal solution, it is unlikely to find a more optimal one because we'd have to allow the error to grow for the network to make it up over the ridge, but we typically stop training if the error starts growing. The shallow basin in this example is called a *local minimum*. This problem can be serious but is rarely fatal. The suboptimal solution the network finds still may be optimal enough. Or we can simply try again with different random weights, which would be like placing the marble in a different part of the basin.

Training the Network to Recognize Consonance of Intervals Because there are 13 inputs to this network, there are a total of $2^{13} = 8192$ possible interval patterns we could present to the network. While we could train the network on all possible patterns, we should not need to do so. Neural networks not only learn by example but also generalize from a limited set of examples, so the solution they find to a subset of the total pattern space can remain valid for patterns the network was not trained on. The network accomplishes this by automatically discovering statistical regularities in the patterns on which it is trained.

Table 9.15
Weights for Interval Consonance Learning Test, Hidden Units

M	C	C♯	D	E♭	E	F	G♭	G	G♯	A	B♭	B	C
h_0	–0.69	2.12	2.22	–0.55	–0.84	–0.55	2.22	–0.86	–0.87	–1.03	2.23	2.05	–0.40
h_1	1.14	–2.98	–2.90	1.21	1.07	1.14	–2.90	1.09	1.01	0.86	–2.90	–3.07	1.38

In this example, we train the network on the 13 chromatic diads (intervals of two tones) in table 3.5. We assume that the perfect and imperfect intervals are consonant, corresponding to an output of 1.0 and the rest are dissonant, corresponding to an output of 0.0. We use two hidden units. When the network has learned our consonance judgments for these intervals, we'll "surprise" it with more complex chords to see how well it can generalize. If the network has done its job, its judgments about these complex chords should be reasonable, even though it has never "heard" them.

The network was trained to recognize the 13 interval training patterns until the normalized mean squared error of all patterns was below 0.001. This required about 14,300 adjustment cycles, using only a few seconds of real time on my notebook computer. At that point, training was stopped. The weights between the inputs and hidden units were as shown in table 9.15.

The rows show the weights connecting the input units to the first hidden unit, h_0, and the second hidden unit, h_1. The weights between the hidden units and the output unit were as follows:

W	0	1
y_0	–6.5	7.5

So, the connection strengths from the first and second hidden units to the output unit were –6.5 and 7.5, respectively.

How the Network Learned the Training Patterns Let's apply a couple of training patterns to get an idea of how the network solved the problem.

Units C and c the octave above are activated. For the octave interval, the network output was $y_k = 0.99$ for a target of $\tau_k = 1.0$, representing consonance (table 9.17). Referring back to equation (9.27), table 9.16 shows how the network computed the output from the training pattern. The outputs of the two hidden units are 0.25 and 0.93 for this example. The calculation for the output unit is shown in table 9.17. The octave produces a strong reading of consonance ($y_0 = 0.99 \cong 1.0$) on the output.

The results of applying the tritone interval to the input units (units C and G♭ are activated) are shown in table 9.18. The outputs of the two hidden units are 0.82 and 0.15 for this example. The calculation for the output unit is shown in table 9.19. The tritone produces a strong reading of dissonance ($y_0 = 0.01 \cong 0.0$) on the output.

Analysis of the Network's Solution Recall that the weights between consonant intervals and the hidden unit h_0 are negative, and those between dissonant intervals and h_0 are positive (see table 9.15). By contrast, the weights between consonant intervals and the hidden unit h_1 were positive, and those between dissonant intervals and h_1 were negative. This means that consonant

Table 9.16
Training the Network on the Octave

Degree	C	C♯	D	E♭	E	F	G♭	G	G♯	A	B♭	B	C
Input	1.00	0.00	0.00	0.00	0.00	0.00	0.00	0.00	0.00	0.00	0.00	0.00	1.00
1st Hidden Unit													
Weight	–0.69	2.12	2.22	–0.55	–0.84	–0.55	2.22	–0.86	–0.87	–1.03	2.23	2.05	–0.40
Product	–0.69	0	0	0	0	0	0	0	0	0	0	0	–0.40
Sum	–1.08												
h_0	0.25												
2d Hidden Unit													
Weight	1.14	–2.98	–2.90	1.21	1.07	1.14	–2.90	1.09	1.01	0.86	–2.90	–3.07	1.38
Product	1.14	0	0	0	0	0	0	0	0	0	0	0	1.38
Sum	2.52												
h_1	0.93												

Note: Using the sum as the index into the logistic function produces the result.

Table 9.17
Hidden Units for the Octave

Unit	h_0	h_1
Input	0.25	0.93
Weight	–6.46	7.54
Product	–1.63	6.98
Sum	5.35	
y_0	0.99	

Table 9.18
Training the Network on the Tritone

Degree	C	C♯	D	E♭	E	F	G♭	G	G♯	A	B♭	B	C
Input	1.00	0.00	0.00	0.00	0.00	0.00	1.00	0.00	0.00	0.00	0.00	0.00	0.00
1st Hidden Unit													
Weight	–0.69	2.12	2.22	–0.55	–0.84	–0.55	2.22	–0.86	–0.87	–1.03	2.23	2.05	–0.40
Product	–0.69	0	0	0	0	0	2.22	0	0	0	0	0	0
Sum	1.54												
h_0	0.82												
2d Hidden Unit													
Weight	1.14	–2.98	–2.90	1.21	1.07	1.14	–2.90	1.09	1.01	0.86	–2.90	–3.07	1.38
Product	1.14	0	0	0	0	0	–2.90	0	0	0	0	0	0
Sum	–1.75												
h_1	0.15												

Note: Using the sum as the index into the logistic function produces the result.

Table 9.19
Hidden Units for the Tritone

Unit	h_0	h_1
Input	0.82	0.15
Weight	–6.46	7.54
Product	–5.31	1.11
Sum	–4.20	
y_0	0.01	

intervals make the sum of h_0 more negative and the sum of h_1 more positive. Also, the weights feeding the output unit negate activation from h_0 but not from h_1. From the shape of the logistic function, if the sum of the activation from the hidden units is greater than 0.0, the output will be turned on, and if it is less than 0.0, the output will be turned off. So the weights on the hidden units have been trained to make the hidden units sum to a positive value for consonance and a negative value for dissonance. Unit h_0 is turned on strongly for dissonance, and h_1 is turned on strongly for consonance. This is just what we wanted, and we didn't have to program the network to find the solution; it figured it out by itself.

Testing the Network—Can It Generalize? Let's see how well the network generalizes to other intervals and chords. Although we only trained it on diads, the network provides encouragingly good-quality guesses about the consonance of some more complex chords (see table 9.20). The diminished seventh chord is arguably the only bad guess, but maybe it's not really so bad after all. That chord is considered dissonant because of its tritone, but it can also be viewed as three minor thirds stacked up, and the interval of a minor third is considered consonant.

So this worked pretty well. But remember that the network looks for statistical regularity, and all our training examples and test examples have the pitch C in them as the lower tone of the interval. How does the network handle intervals and chords starting on another degree of the scale? Let's test the fifths between F and C and E and B (table 9.21). Both F–C and E–B should be consonant. The fact that they are not suggests that the network has relied on the scale degree rather than the interval to determine consonance.

Like all learners, networks tend to search for regularity. But the most regular solutions are not necessarily the best. For example, a child might incorrectly rely on the regularity of English verbs and say "I swimmed today" instead of "I swam today." The network appears to have stumbled for the same reason. It appears to have associated consonance and scale position instead of consonance and interval size because our limited training set failed to provide examples that would have violated this assumption. This network has not discovered all the underlying relations that account for our consonance judgments, and so it can't generalize correctly in all cases.

We can improve the ability of a network to generalize by increasing the ratio of training examples to hidden units. The greater this ratio, the more the network is forced to generalize. For the preceding examples, the ratio is 13/2 = 6.5. Reducing the number of hidden units to one is not

Table 9.20
Network Consonance Guesses for Complex Chords

	Pattern															
Chord	C	C♯	D	E♭	E	F	G♭	G	G♯	A	B♭	B	C′	Output	Analysis	Quality
Major triad	1.0	0	0	0	1.0	0	0	1.0	0	0	0	0	0	0.99	Strongly consonant	Good
Minor triad	1.0	0	0	1.0	0	0	0	1.0	0	0	0	0	0	0.99	Strongly consonant	Good
7th	1.0	0	0	0	1.0	0	0	1.0	0	0	1.0	0	0	0.83	Fairly consonant despite dissonant major 7th	Good
Dim. triad	1.0	0	0	1.0	0	0	1.0	0	0	0	0	0	0	0.13	Fairly dissonant despite consonant minor 3d	Good
Dim. 7th	1.0	0	0	1.0	0	0	1.0	0	0	1.0	0	0	0	0.77	Should not be consonant	Poor
Cluster	1.0	1.0	1.0	1.0	1.0	1.0	1.0	1.0	1.0	1.0	1.0	1.0	1.0	0.01	Highly dissonant	Good

Table 9.21
Network Performance Starting on Other Degrees

	Pattern															
Chord	C	C♯	D	E♭	E	F	G♭	G	G♯	A	B♭	B	C′	Output	Analysis	Quality
F–C	0	0	0	0	0	1.0	0	0	0	0	0	0	1.0	0.99	Strongly consonant	Good
E–B	0	0	0	0	1.0	0	0	0	0	0	0	1.0	0	0.17	Should be consonant	Poor

an option because the network would no longer be able to learn. But it would be appropriate to expand the training set to include all the rest of the diad intervals. If we expand the training set to include every diad on every possible scale degree, we have 80 training patterns, 44 consonant and 36 dissonant. This is still a small fraction of the 8196 total intervals. In practice, the minimum number of hidden units that can solve this set of training patterns appears to be four, for a training ratio of 80/4 = 20. With these adjustments, the network correctly handles all the judgments.

9.22.4 Generalization as Creativity: Composing with Networks

To use a network to understand musical structure in time, we must have a neural representation of time. Let's say we wanted the network to learn melodies. One approach would be to have as many network inputs as there are notes in the longest melody. Or the network input could be a fixed-size time window that slides over a region of the melody. In either case, this kind of windowing approach represents time as position and converts the problem of learning music into learning spatial patterns.

For example, we could train a network such that when one measure is played, the network produces the next measure in sequence. Or we could train a network to generate the next note in sequence by supplying it with some number of previous notes for context. This would require a feedback arrangement in the network design so that previous outputs could influence subsequent choices. The windowing and context methods could be combined so that the feedback units provide context for whole musical phrases. This could be used to study the motivic structure of melodies, for example.

Peter Todd (1989) describes a process whereby a network was trained using the feedback context method to learn a set of melodies. His approach used the back propagation method but also included a set of feedback units that stored context information about the notes played most recently (figure 9.48). Once trained, it could play back the melodies when keyed to do so by a set of plan network inputs that acted like the buttons on the front of a juke box to select the desired melody.

First, he trained the network to play several melodies correctly. He then experimented with setting the plan inputs to untrained values so as to force the network to generalize from the melodies it was trained to reproduce and thereby to compose new melodies. In this way, Todd used generalization as a model of creativity.

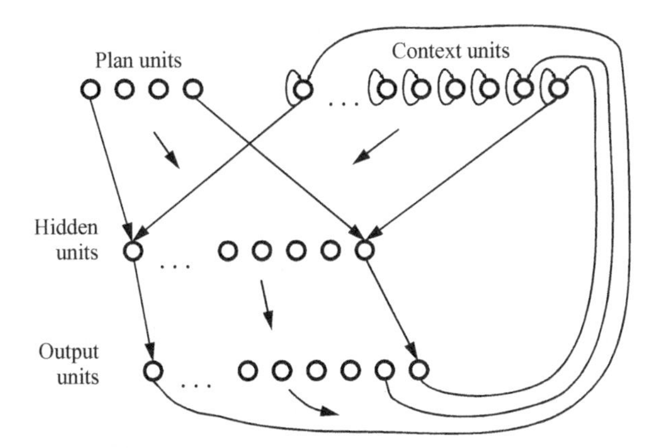

Figure 9.48
Feedback context method.

Todd used very simple folk melodies as training examples, but he could easily have used anything else, including examples composed by a stochastic process or some rule-based approach. Although Todd's network learned only the surface of the melodies, it would be straightforward to extend it to a hierarchical set of networks such that a low-level network responsible for the note-by-note process interacts with higher-level networks responsible for an overall compositional plan.

9.22.5 Bach Chorale Harmonization with Connectionism

A common criticism of connectionist research is that neural network techniques seem to work well on relatively simple proof-of-concept problems but do not scale well to realistic-sized problems traditionally studied in Artificial Intelligence such as playing chess and composing music. The challenge of composing realistic music with neural nets was taken up by Hild, Feulner and Menzel (1991), who developed HARMONET, a program to harmonize chorale melodies in the style of J. S. Bach.[18] Their aim was not only to demonstrate parity with more traditional AI techniques but to exploit the potential for net-based solutions to go, as it were, beyond the rules and penetrate more deeply into the core of a composer's style.

In fact, their approach turned out to be a hybrid of symbolic expert system for some parts of the problem and neural networks for other parts. In particular, they did a fair amount of manual parsing of the chorales to structure the data to create their training set. Then they trained a network with this set to create a "harmonic skeleton" of several chorales. The chorale melody then provided the soprano line, and the harmonic skeleton provided a bass line. Then they had to synthesize the alto and tenor lines, which they did using a standard AI "generate and test" approach. Last, they added passing eighth-note figures characteristic of Bach's style using another network. All networks used a standard back propagation architecture with context units to remember recent events, similar to Peter Todd's approach.

Because Hild and his colleagues don't just use networks throughout, it's not clear that this is the breakthrough realistic-sized problem for connectionist research in music. Nonetheless, they stated that an audience of music professionals had determined HARMONET's output to be "on the level of an improvising organist," and indeed printed scores of their harmonizations seem quite good.

9.22.6 Genetic Programming

We have considered compositional processes over the last thousand years of human history. Time and again, we see a trade-off between generating music and critiquing music. We see it at all levels of the process, from the smallest local detail of a private act of composition to the most public pronouncements of music critics.

In every age, composers put forward their ideas in the context of culture, and critics evaluate them in the same context. Successful works, ideas, and methods survive; unsuccessful ones are scrapped and forgotten. Both composition and criticism adapt to cultural changes. Successful adaptation may mean reproduction (in the sense that children are reproduced from their parents), crossover (swapping elements between successful adaptations the way parents pass their characteristics along to their children), mutation (where novel elements are introduced), permutation, and other reordering processes. What composers do in subsequent days and in subsequent epochs can generally be seen as an evolution from antecedents. Thus, composition can be likened to a natural selection process.

Any process that we can identify we can also model, and a useful computational model for this view of composition is provided by *genetic programming* (Koza 1992). This technique adapts some of the principles of biological natural selection to allow programs to evolve spontaneously.

Suppose we start with a collection of primitive functions to generate and modify basic musical data (such as algorithms to generate and transform a tone row). These are supplied to a genetic programming system, which creates a population of programs that invoke these primitive functions in various random ways. The genetic programming system then executes the population of programs, and their results are evaluated for how well they succeed. This critique is provided by yet another function we must supply that determines how well the programs perform their task, that is, their fitness. Because they were generated randomly, most of the programs probably won't perform very well, but we take those that perform best for subsequent development and discard the rest.

A new set of programs is created from those that survived the previous round by reproduction, crossover, mutation, permutation, and so on. These are tested as before, and the process repeats until some criterion of fitness is achieved.

The good news is, this approach, like artificial neural networks, avoids the requirement of knowing what the solution should be in advance. The bad news is that the solutions may not be optimal; and for realistic-sized problems, solutions may not be scrutable (see especially Todd and Werner 1998).

9.22.7 Summary of Connectionism

A promised advantage of artificial neural networks is that the composer need not invent rules to express preferences. Such preferences are an emergent property of the network. The fact that no music theory is implied in the structure of a network is a benefit because it allows any theory embodied in a model to arise. The ability of a network to generalize from examples provides the composer with ways to go beyond the model in a musically reasonable way.

Such networks can be used to study the psychophysics of sound, the perception of timbre, pitch, and rhythm, tonal analysis, musical instrument fingering, sound synthesis, automatic music classification, recognition directly from the waveform, emotion in music, musical phrasing and interpretation, automatic music manuscript transcription, and many other areas (Todd and Loy 1989).

But both conventional AI and connectionist approaches seem to run out of power when scaled up to the size of problems we'd like them to be able to solve. Perhaps hybrid systems, such as HARMONET, that combine conventional AI techniques with artificial neural networks will eventually succeed where the two approaches separately have faltered. Or perhaps we've simply not found the right model for intelligence yet.

9.23 Representing Musical Knowledge

The terms *arrival* and *departure* are often used in musical analysis because they capture something true about our experience of music. These terms suggest a sense of time and place, and that the music conducts us along a pathway structured by the composition.

Directed graphs embody this sense of place and transition, and we observed the usefulness of directed graphs to characterize the unfolding of a musical theme (see section 9.18.3). Petri's (1979) general net theory extends the directed graph to characterize causal systems of arbitrary morphology and abstraction. Antoni and Haus (1982) adapted them to represent musical structure and knowledge. Haus and Sametti (1991) describe a software tool, ScoreSynth, for analyzing and synthesizing musical scores using Petri nets.

9.23.1 Petri Nets

Petri nets look like directed graphs but with additional elements. As with directed graphs, states are represented as circles, and transitions between states are represented by the movement of tokens along arcs connecting states (see figure 9.37 for an example of a directed graph). But with Petri nets, multiple tokens flow through the net simultaneously. Transitions in the network state can trigger other actions, such as causing transitions to occur in subnets, nested hierarchically. The flow of time can be made explicit in Petri nets. They can handle deterministic and nondeterministic (stochastic) operations. Representation of music structure with Petri nets is compact and expressive. The elements of a Petri net can refer to musical objects such as notes, phrases, motives, sections, and the like, or they can refer to nonmusical objects that manage and control the compositional process.

The basic Petri net elements are places, transitions, and arcs (figure 9.49). Places and transitions are connected by arcs. Two numbers may appear inside a place. The upper number (n) indicates how many tokens it currently contains; the lower (N) indicates the maximum number of tokens it may contain. Transitions control the movement of tokens in the network. Places and transitions can also contain subnets.

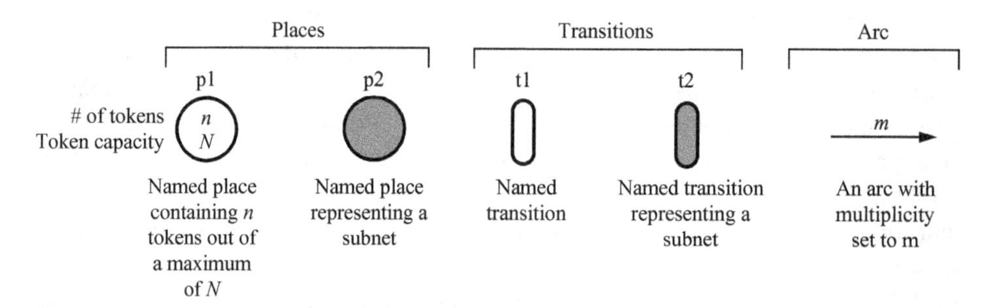

Figure 9.49
Petri net icons.

The Firing Rule The execution of a net is determined by the firing of its transitions. The basic rule for firing is as follows (Haus and Sametti 1991):

A transition may fire if each one of the input places, i.e., places which are connected with oriented arcs to the transition, has at least one token. The transition firing has two effects: to decrement the marking of each input place by one token and to increment the marking of each output place, i.e., a place which is connected with an oriented arc from the transition, by one token.

After the starting of a net, firings follow one another until there are no more transitions which may fire. At the end of transition firings, the execution of the net stops (6).

For example, figure 9.50 shows an elementary sequence. Initially, the input place p1 contains a single token, and the output place contains none. The firing rule indicates that t1 can fire. It decrements the token count in p1 by 1 and increments the output place by 1. This basic firing rule is extended by the following additional rules.

Capacity Each place in a network can be assigned a maximum capacity of tokens, represented as the lower of the two numbers indicated inside places. (If no number is indicated, 1 is assumed.) "Transitions cannot fire if the marking of one output place, at least, will exceed its capacity after transition firing" (Haus and Sametti 1991, 6).

Figure 9.51 shows two examples. In the first case, p2 is full, so t1 cannot fire. The second case represents a conflict, because only one transition can fire. The network determines whether t1 or t2 will fire nondeterministically (stochastically).

Multiplicity If a numerical label is affixed to an arc, called the multiplicity value, then the firing rule must be modified: "A transition may fire if each one of the input places has at least as many tokens as the numerical label on the arc [multiplicity value] connecting the place to the transition."[19]

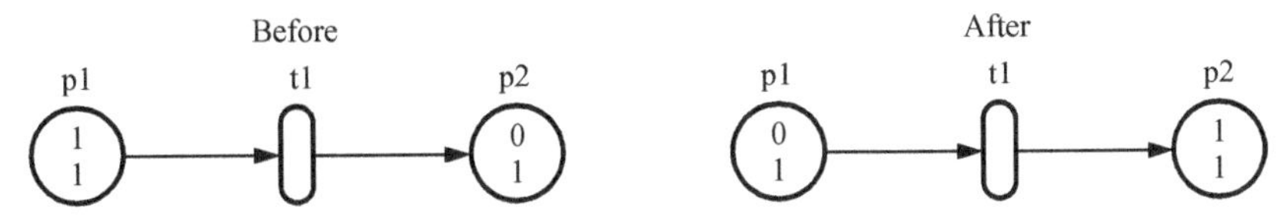

Figure 9.50
Petri net sequence.

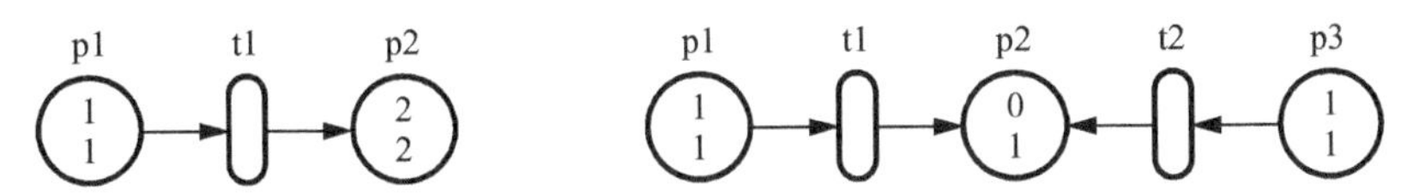

Figure 9.51
Effect of capacity on firing.

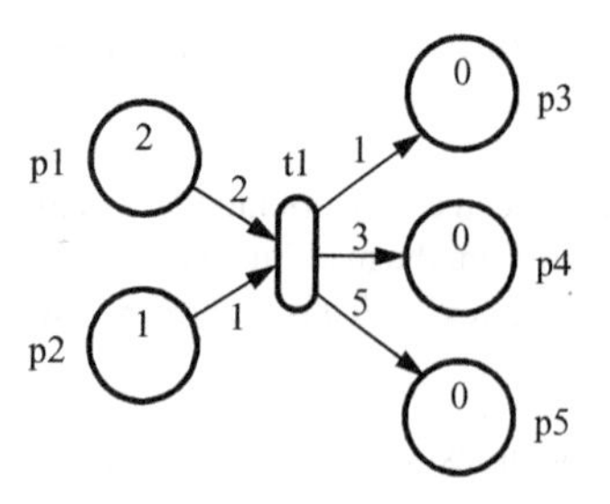

Figure 9.52
Example of firing with multiplicity.

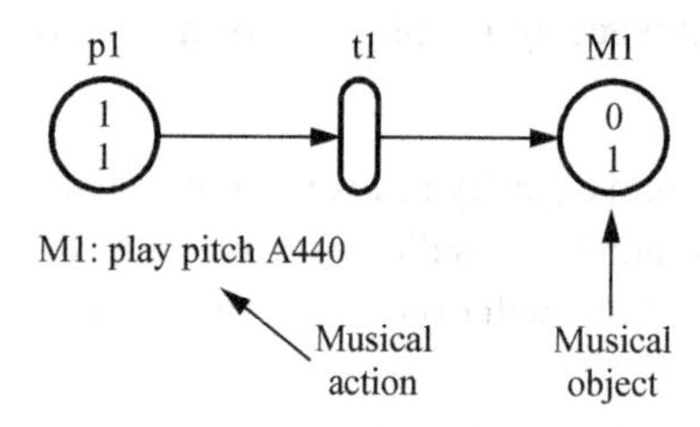

Figure 9.53
Musical objects.

Firing now decrements the number of tokens of each input place by the multiplicity value on the arc connecting the place to the transition, and increments the number of tokens of each output place by the multiplicity value on the arc connecting the transition to the place. Figure 9.52 shows a network with multiplicity ready to fire, and the results after firing.

Firing happens in three stages:

1. The network determines that t1 can fire because each of the input places has at least as many tokens as the corresponding multiplicity value on the arc connecting it to the transition.
2. Upon firing, the transition t1 subtracts the number of tokens from the input places specified by the multiplicity values on the arcs connecting the places to the transition.
3. The transition then adds as many new tokens in the output places as indicated by the multiplicity values on the arcs connecting the transition to the output places. (Multiplicity does not conserve the total number of tokens.)

Musical Objects, Musical Actions We can associate places with any musical significance we like. Places can represent individual notes, phrases, dynamics, motives, and so on. We associate musical meaning to places by affixing labels to them that refer to defined musical objects. In figure 9.53 place M1 has been designated to be a musical object because its name starts with M.

When tokens flow into or out of a musical object, the associated musical action is triggered. If the musical action is a note, it is played; if it is a phrase, the phrase is played; if it is a subnet, the

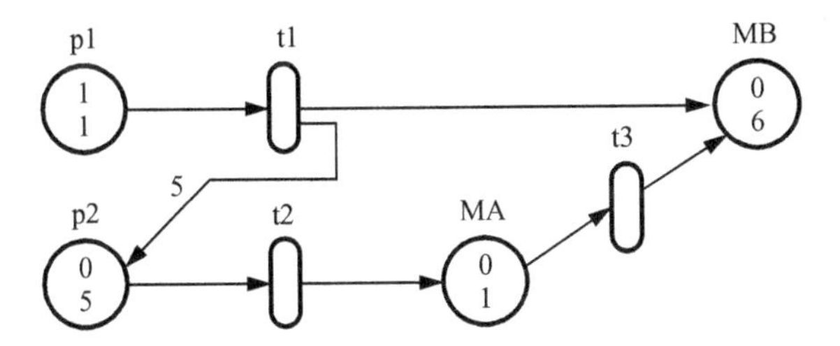

Figure 9.54
Timed net example.

subnet is entered. In figure 9.53 musical object M1 is about to trigger its associated musical action. When t1 fires, M1 will play pitch A440.

Timed Firing But how long will the note in figure 9.53 last? Petri nets have no inherent notion of time or sequence. Many transitions can be qualified to fire at the same time, but the implementation of Petri nets gives no direct control over their order of firing. Also, the duration of firing is assumed to be instantaneous. We must add time structure to the network. Haus and Sametti (1991) took the approach of associating time with musical objects. "When a token is put into a place with an associated MO [musical object] the token cannot be considered for the firing of transitions connected to the place until the associated MO has ended" (8).

To illustrate, suppose that musical object MA is defined to last 1 second (figure 9.54). Every token put into MA is not disposable for 1 second. We also define MB to last 6 seconds per token. Tokens put into places that are not defined as musical objects are immediately disposable. For example, if MA has one token, then t2 is prevented from firing because the capacity of MA is 1 token. Similarly, t3 can't fire because MA does not present its token to t3 until its duration of 1 second has elapsed.

The sequence of firings for figure 9.54 is as follows.

1. t1 fires, subtracting one token from p1 and adding one token to MB and five tokens to p2.
2. MB triggers an instance of its associated musical action, which will last 6 seconds.
3. t2 fires, subtracting one token from p2 and adding one token to MA, which is now at capacity.
4. MA triggers an instance of its associated musical action, which will last 1 second.

These steps transpire instantaneously because they are triggered by places p1 and p2, not musical objects. MB has not reached its capacity, and up to five more instances of MB can be triggered. However, no tokens are available from t1 because p1 is exhausted. MA has reached its capacity, so t2 and t3 are prevented from firing. MA's token will fire t3 in 1 second, so the network must wait until MA's token is available. At this point, we can represent musical actions in progress at time 0 (figure 9.55).

5. When 1 second has elapsed, t3 fires, passing the token from MA to MB. Another instance of MB is triggered.
6. MA is now empty, so t2 fires, passing a token from p2 to MA. Another instance of MA is triggered.

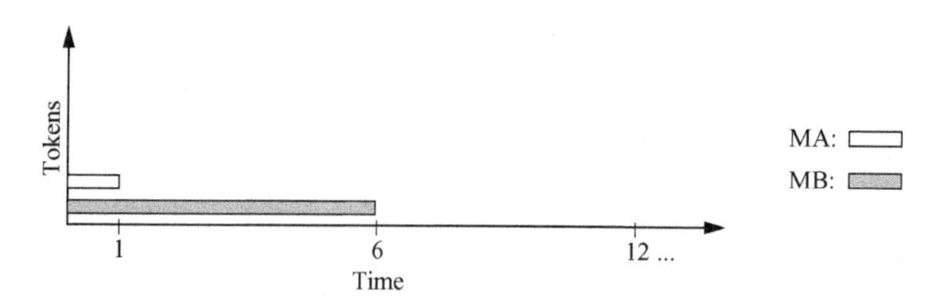

Figure 9.55
Musical actions in progress at time 0.

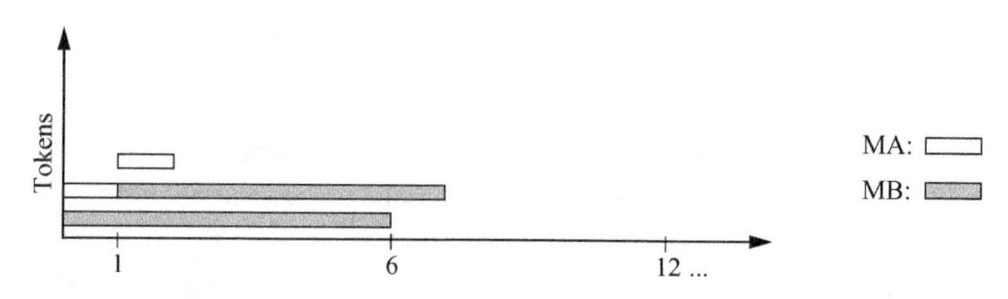

Figure 9.56
Musical actions in progress at time 1.

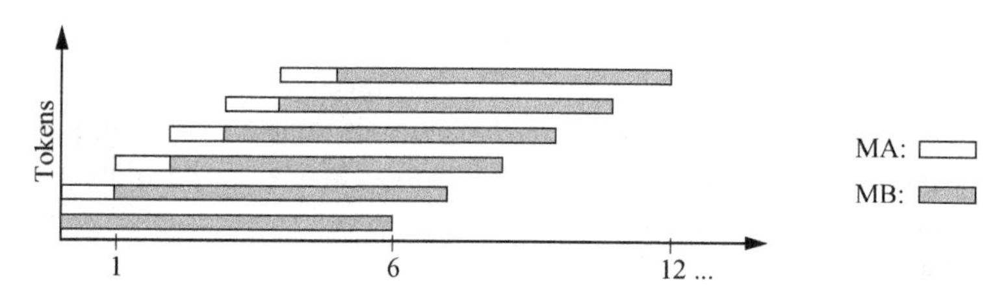

Figure 9.57
Complete result of network execution.

Musical actions in progress at time 1 are shown in figure 9.56.

From this point on, MA and MB will be triggered only by the remaining tokens in p2, consumed in 1 second intervals by MA, then passed to MB. The complete result of network execution is shown in figure 9.57.

Refinement Morphisms Petri nets can be developed through a process of refinement, where a place or transformation can act as a placeholder to be given a more detailed description later. "Refinements can define very complex PN [Petri net] models by means of simple PNs and hierarchical structures, i.e., allowing models to be designed by either a top-down or a bottom-up approach" (Haus and Sametti 1991, 10).

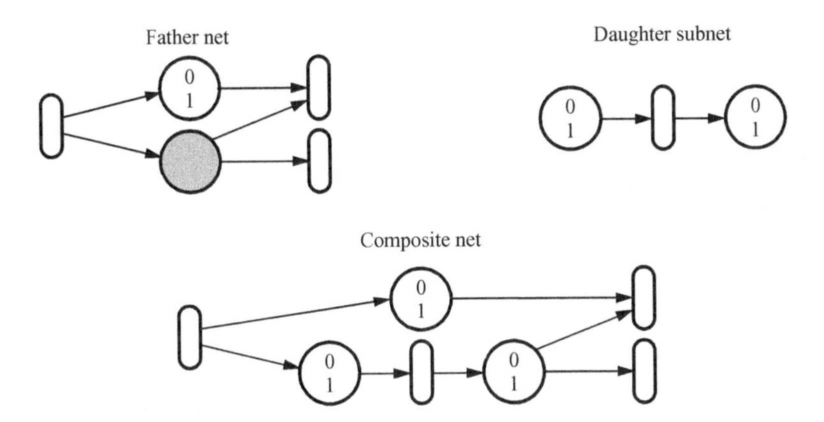

Figure 9.58
Petri net with subnet.

The placeholder node is called the father and the subnet it refers to is the daughter. To associate a daughter subnet with a father place, the subnet must have an input place and an output place. Input arcs to the father place are input arcs to the input place of the daughter net; output arcs from the father place are output arcs from the output place of the daughter net. Transitions can be refined in the same way. An example is shown in figure 9.58.

Building Blocks Some of the basic Petri net building blocks for music are as follows:

- *Sequence (monophony).* The musical object associated with each place is triggered as the token flows along. This can be used to form a melody out of its constituent phrases or a movement out of its sections.

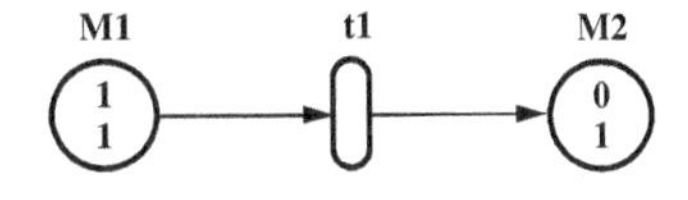

- *Parallel (polyphony).* A single place triggers musical objects M1 and M2 to execute concurrently. This can be used to form a chord or to synchronize polyphonic counterpoint.

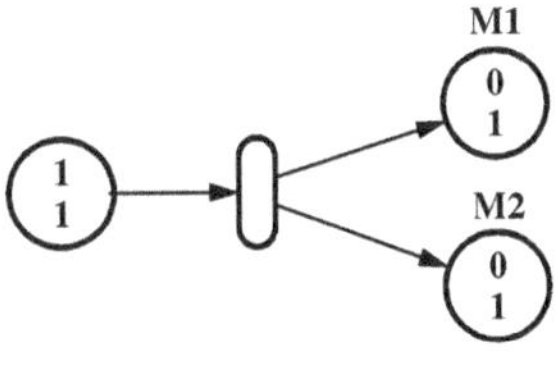

- *Choice.* From a single place, one of two paths can be taken. Since there are two paths but only one token, the net must (stochastically) choose one path.

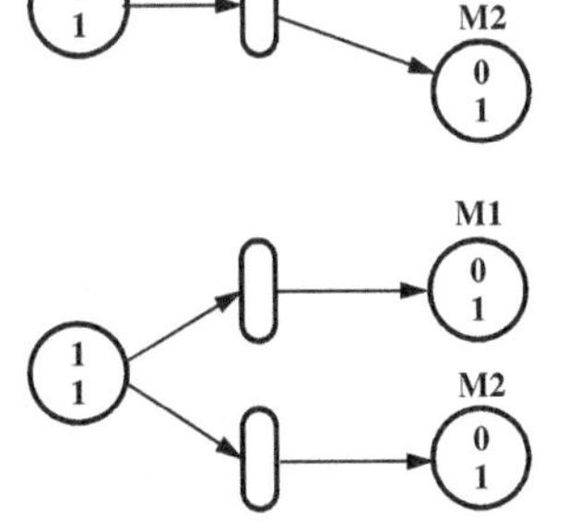

- *Join*. Two input places trigger M1, so two instances of M1 are created and run concurrently.

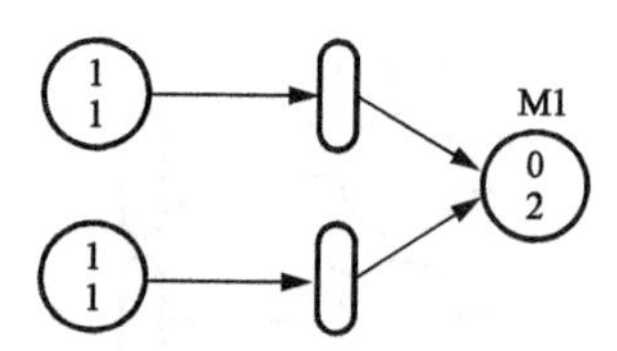

- *Fusion*. Both of the two input places are needed to trigger M1. One instance of M1 is triggered.

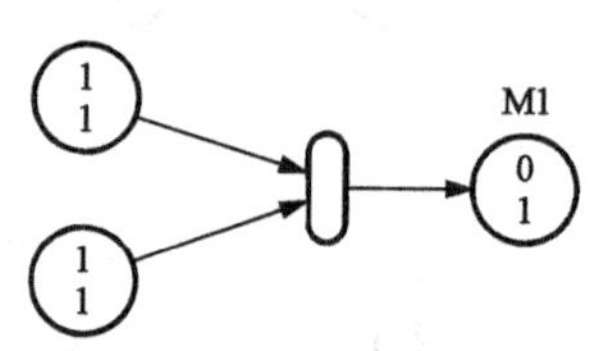

- *Iteration*. The Start place is provided with as many tokens as the required iterations. There must be room for them at the End. In this example, two instances of M1 execute sequentially.

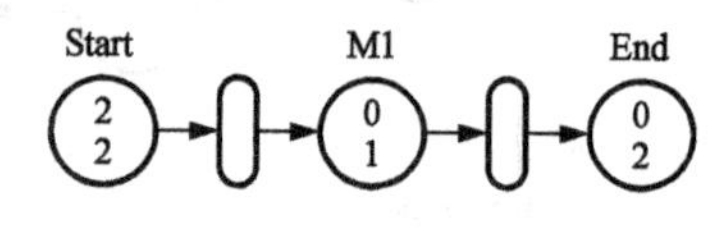

Canon Perpetuus All of these structures can be joined to create more complicated networks. For example, Antoni and Haus (1982) provide a sample analysis of J. S. Bach's "Canon Perpetuus" from his *Musical Offering* for flute, violin, and continuo bass.[20] A brief overview of the flute part will be illustrative. The pitches of the flute part in the score can be grouped into musical motives as follows:

Name	Section
F_1	bars 1–2
F_2	bars 3–10 and the following three notes
F_3	last note of bar 11 to first note of bar 13
F_4	the rest of bar 13 to bar 14
F_5	bars 15–17 and the following note
F_{end}	last note of the flute part

Using these named sections, the flute part can be analyzed as follows:

$$\{F_1, F_2, F_3, F_4, F_5, i(F_1, F_2), rt(F_3), F_1, F_2, F_3, F_4, F_{end}\}, \quad (9.30)$$

where *i*() indicates inversion and *rt*() indicates transposed retrograde. This sequence structure is represented by the Petri net shown in figure 9.59. Place Start contains a token, as does place Ping. In the beginning only the transition leading from Start can fire. When the token reaches F_4, T_1, can fire but T_3 cannot. So the token visits F_5 and goes down the right-hand arm in the figure, eventually reaching T_2. Both F_1 and Pong receive a token, but only the transition from F_1 can fire. So the token visits F_1 through F_4 again. Now T_1 cannot fire but T_3 can. So the token visits F_{end}, then Stop.

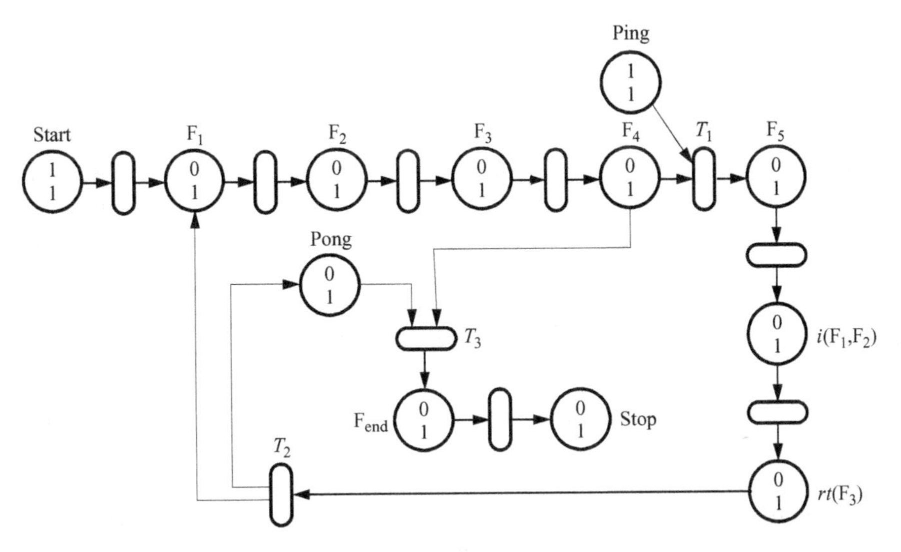

Figure 9.59
Petri net model of flute part from "Canon Perpetuus."

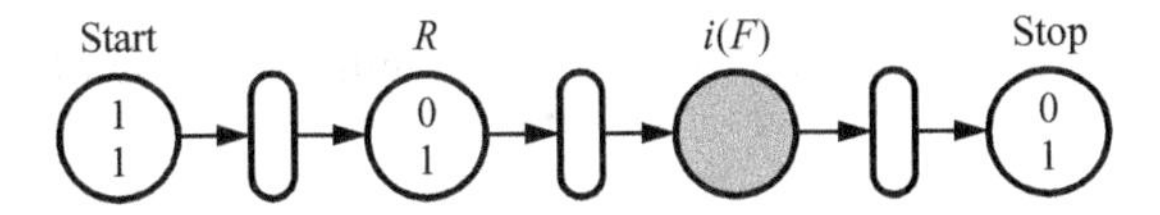

Figure 9.60
Petri net model of violin part from "Canon Perpetuus."

We observe that the violin part of "Canon Perpetuus" is the inversion of the flute part played against the flute with a two-bar delay. Using F to represent the sequence shown in (9.30) except for the final sequence $\{F_4, F_{end}\}$, we can write the violin part as

$$R, i(F), \tag{9.31}$$

where R is the two bars of rest. The Petri net for this sequence is shown in figure 9.60.

Petri nets model the behavior of discrete dynamical systems such as musical scores in a direct and intuitive way. They handle many concepts that are vital to the musical process, including sequence, concurrency, conflict, and resolution. The resulting network descriptions naturally facilitate deeper understanding of the underlying musical system. Petri net representations of music can remove a great deal of redundancy in a musical score, revealing the essential structure of the work. They have been used to analyze music structures of significant size (Haus and Rodriguez 1993). They provide a pragmatic method for hierarchical representation of musical knowledge (Roads 1984). Petri nets can also be constructed from scratch to synthesize musical scores, or nets that are the result of analysis can be modified for subsequent synthesis of related musical works. However, Petri nets can become explosively large when they are used to describe realistically complex systems.

9.23.2 Predicate Transition Nets

High-level Petri nets, also called Predicate/Transition (PrT) nets, have been developed to overcome the problem where Petri nets become unmanageably large when used to model realistically complex problems (Genrich and Lautenbach 1981). The general idea is to attach additional information to the elements of the network to increase their descriptive power.

For example, in Petri nets, tokens are simple featureless counting devices. In PrT nets, tokens have quantity and quality. In fact, they can have multiple quantities and qualities. Algebraic and logical expressions (predicates) can be added to places, transitions, and edges to describe network state and firing. The expressions are evaluated based on the available types and quantities of the tokens flowing through the system.

Consider a single place, called `PianoImprov`, which models the musical resources of the piano part of a musical improvisation. `PianoImprov` contains a collection of tokens representing the pitches the pianist can play. The content of a place is called its *marking*. Suppose that `PianoImprov` is marked by two C4 pitches, three E4 pitches, and two G4 pitches. Because it holds pitches, we say place `PianoImprov` is of type *Pitches*. This means `PianoImprov` can only contain elements of type *Pitches* (figure 9.61).

When an instance of a place is created, it is given an initial marking. The initial marking of `PianoImprov` can be expressed as

$$M_0(\texttt{PianoImprov}) = (2 \cdot \text{A4}) + (3 \cdot \text{B4}) + (2 \cdot \text{G4}) .$$

The markings of a network will vary as the tokens are consumed and produced across the net during operation.

An arc from a place to a transition carries tokens consumed by the transition, and an arc from a transition to a place carries tokens consumed by the place. The label on an arc indicates the number and kind of tokens that can be consumed or produced. We can specify that any quantity of any type of token can travel an arc, or we can restrict the arc to certain quantities and types of tokens. The arc x in figure 9.62 is defined to be of type *Pitches,* and indicates that any number of tokens

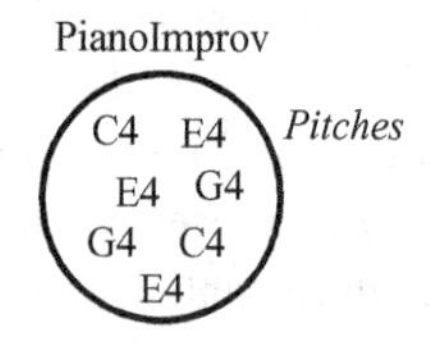

Figure 9.61
PianoImprov 1.

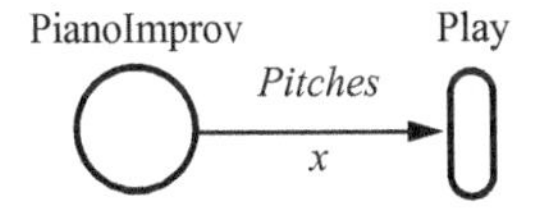

Pitches: {C4, E4, G4}

x: *Pitches*

M_0 (PianoImprov) = 2 · C4 + 3 · E4 + 2 · G4

Figure 9.62
PianoImprov 2.

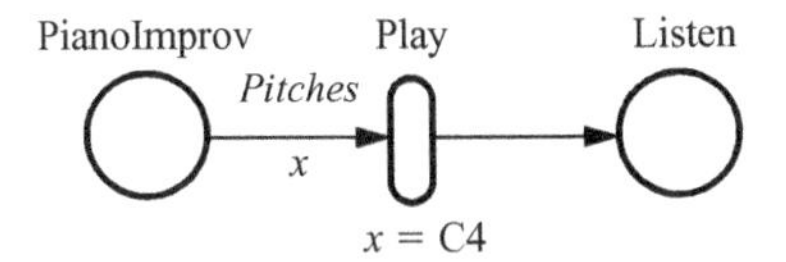

Pitches: {C4, E4, G4}

x: *Pitches*

M_0 (PianoImprov) = 2 · C4 + 3 · E4 + 2 · G4

M_0 (Listen) = ∅

Figure 9.63
PianoImprov 3.

of type *Pitches* can be produced by `PianoImprov` and consumed by transition `Play` in a single transaction. By this rule, one pitch, any combination of pitches, or all pitches can be played by the piano at once.

In figure 9.63 `PianoImprov` is ready to play any of its pitches for `Listen`. `Listen` begins in an empty state. However, by the rule stated beneath the `Play` transition, `Play` can only fire if the received token is the pitch C4, so `Listen` can only hear that pitch.

The network shown in figure 9.63 can be defined as follows:

P = {`PianoImprov, Listen`}

T = {`Play`}

F = {(`PianoImprov, Play`), (`Play, Listen`)},

where P is the set of places, T is the set of transitions, and F defines the arcs that connect places and transitions. Because arcs link places and transitions, F is a list of place/transition pairs.

The combination of typed elements with capacities and predicates makes PrT nets more expressive and representationally compact than Petri nets. Pope (1986) gives an example of the use of PrT nets in a musical context and shows how PrT nets can be abstracted to become the template for deriving other related networks. In this way, PrT networks begin to take on some of the characteristics of object-oriented computer programming languages, such as type inheritance and abstraction, but with the advantage of built-in facilities for modeling the behavior of discrete dynamical systems such as musical scores (Pope 1991). In fact, graphical simulation techniques and high-level computer language design are beginning to converge to the point that practical tools for modeling and emulation of discrete dynamical systems are now commonly available.[21]

What this approach lacks is a built-in mechanism for learning, abstraction, pattern completion, and spontaneous generalization provided by the connectionist approach. Since both PrT nets and connectionist frameworks model dynamical systems, perhaps a hybrid approach combining elements of both would prove sufficiently expressive for problems of realistic scale.

9.24 Next-Generation *Musikalische Würfelspiel*

The tables used to generate *Musikalische Würfelspiel* compositions were each predetermined by a master composer, so a composition in the style of that composer is guaranteed if the method is followed. The composer David Cope (2001) has developed a set of programs he calls Experiments in Musical Intelligence (EMI) that has a similar aim: EMI produces original works in the style of a particular composer by recombining atomized musical quotations derived from that composer's works. But whereas the composers of *Musikalische Würfelspiel* had to compose their own atomized musical tables, Cope's EMI system generates the musical tables that are the basis of the new works to be composed by analyzing the target composer's musical corpus under the direction of a trained operator.

EMI performs its analysis using techniques drawn from natural language processing, augmented transition networks, and other techniques drawn from AI to synthesize new compositions. Like Markov and connectionist approaches, EMI *recomposes* the music of the target composer. The result is highly original (though not always very artful) music with the stylistic signature—in both its surface and deep structure—of the identified composer.

Music expresses its own essential nature much the way that organisms are expressions of their genes. If we can identify the genetic basis (so to speak) of a composer's style in a sufficiently formal way, we should be able to use it to create original compositions in that style. This issue speaks to the desire Schillinger first expressed that theories of art should be generative, not merely analytical (see section 9.11.2).

Cope's aesthetic premise is that new music in the target composer's style can be created through recombinant techniques. EMI is an analysis/synthesis system that creates a database of musical elements by analyzing a composer's works and then interpolates among them in various ways to realize new works. In outline, the method is as follows:

1. The user must select and encode a corpus of musical works from the target composer into a format that EMI can digest. To facilitate pattern recognition, Cope suggests, the selected works

should be relatively homogeneous, similar in overall structure, range, and orchestration. For example, Cope (1999) used Mozart's middle symphonies (numbers 6–31) as an analysis set for a new symphony in Mozart's style. He used a similar approach for a new piano concerto. Both works have been recorded and are available commercially. This is the same initial step that must be employed by any system that learns from a corpus of examples. Clearly, the operator's selections have dramatic impact on EMI's subsequent steps.

2. EMI performs a lexical analysis based generally on Noam Chomsky's theories of the structure of natural languages, and a hierarchical temporal and harmonic analysis of the works based on the ideas of Heinrich Schenker (1935).

3. EMI identifies what Cope calls signatures, which are unique characteristics of the composer's style, using pattern recognition techniques adapted from natural language processing. The analysis ostensibly contains no a priori notion as to the signatures to be found, so the technique can presumably be applied to music of any style. But in fact there is great latitude in this step for the EMI operator to refine the process of signature selection based on the operator's prior experience with the composer's style. Cope has reported, for example, that in the case of the Mozartian symphony and piano concerto, he took great pains to tune EMI's analysis parameters to greatest advantage.

4. Driven by this analysis, EMI then breaks the musical corpus into its fundamental components, which are now ready to be recombined.

5. EMI uses augmented transition networks driven by a random process, and makes refinements by pattern matching based on extracted stylistic features, to recombine the music into an original that preserves the composer's signature style. The creative aspect of EMI reflects many of the techniques described in this chapter. It can provide variation by interposing similar but distinct elements from the analysis. The recombination can take place on several levels of musical scope because the hierarchical analysis provides compositional rules for thematic, middle-ground, and large-scale structures.

6. The original work is then formatted for representation in common music notation to be played by traditional instruments or converted to a format such as MIDI so it can be synthesized.

In the hands of a skilled operator, EMI can produce a believable facsimile of a composer's style from a carefully selected corpus of the target composer's works. Cope has also published examples in the style of J. S. Bach, Frédéric Chopin, and Scott Joplin, among others.

Cope's EMI system is perhaps the most advanced automated composition system extant today and therefore can serve as a good target for analysis and criticism. In the following sections, I discuss some of the important questions raised by his work.

9.24.1 Is EMI Experimental?

Insofar as Cope defines his system as "Experiments in Musical Intelligence," it is fair to ask if Cope's system is truely an experimental method.

There is actually a seventh step in the EMI process, not an official part of Cope's system but obviously of crucial importance: editing. The operator selects among the generated compositions for those good enough to be played in public. Editing is a necessary step because there would be little audience for every one of the possible compositions such a system can create, just as there would have been little audience for what Beethoven threw in his trash basket.

Although the quality of EMI's compositions and their entertainment value is crucial, these are not their most important credentials: the selected compositions, especially the ones created by the author of the method, become official specimens, proof of the effectiveness of the method that created them. But there is a contradiction between hand-selecting the most convincing examples and the requirements of a true experimental method.

The experimental method in science is about testing suspected explanations regarding one's observations. To get at the truth, one conducts experiments that must be carefully constructed to avoid hidden biases and *confounding factors,* undetectable things that might influence the results and lead the research astray. The most common confounding factor is pure chance, where through luck one happens upon a population of specimens that erroneously validates or invalidates the hypothesis.

Because of the danger of confounding factors in experimental design, the scientific method requires that conclusions must not be based merely on anecdotal evidence such as a single instance or a very limited group of specimens or subjects. Regardless of the hypothesis Cope is expressing, and though the published examples of his EMI work are superlative, his specimens constitute a very limited group and therefore must be considered only as anecdotal evidence of EMI's effectiveness.

But EMI should not be singled out here; this criticism can also be directed at every method discussed in this chapter. Hiller and Isaacson (1959), the other composers to use the word *experiment* in the name of their composing system, were the first to face this problem. In creating their *Illiac Suite,* they similarly had to generate official specimens that proved the effectiveness their methodology. In their book *Experimental Music,* they claim that they used no preferential criterion to select example outputs from their program for inclusion in the *Illiac Suite* so as not to color the results. Although their approach helps, it does not prevent their single published composition (the *Illiac Suite*) from being anecdotal evidence.

Use of the terms *experiment* and *experimental* in the arts is difficult because the terms are so freighted with scientific meaning. However, there is a constellation of related words, such as *experience* and *experiential,* that share the same root and that capture the importance of personal observation to both the arts and the sciences. If we think of *experiment* as meaning *qualified experience,* I think we get closer to how artists think about being experimental in their art. Artistic experiment is about considering novel or unusual combinations of elements for the purpose of increasing the horizons of surprisal. I spent a number of years working at the Center for Music Experiment at the University of California, San Diego, and this pretty well characterizes what went on there.

9.24.2 Is EMI Intelligent?

When I heard EMI's Mozartian symphony and piano concerto for the first time, I definitely had the experience of listening to Mozart, or perhaps a good Mozart imitator. To have captured Mozart's style so aptly is such a stunning technical and aesthetic accomplishment that it warrants asking, Is EMI intelligent?

Turing (1950) suggested that if we can't distinguish between an intelligent person's choices and a computer's choices, then it is reasonable to say that the machine is behaving intelligently. Insofar as Cope defines his system as "Experiments in Musical Intelligence," it is fair to ask if Cope's system passes a musical equivalent of Turing's test.

9.24.3 Aural Sensibility

Hiller and Isaacson (1959) attempted to directly encode rules of composition into their system, but they realized that there were limits to what they could accomplish just with the use of rules. Composing is about more than following rules. They wrote, "The composer is traditionally thought of as guided in his choices not only by certain technical rules but also by his 'aural sensibility,' while the computer would be dependent entirely upon a rationalization and codification of this 'aural sensibility.'"

Other systems, such as DEC and neural networks, allow a composer's "aural sensibility" to emerge from experience by inference and generalization, emulating human learning. However, EMI, arguably more successful to date than these other approaches, actually relies much less than they do on reasoning and inference. Instead, it takes a brute force approach. By analyzing a large corpus of works, EMI's analysis phase attempts to provide its synthesis phase with a rich set of options for every choice faced by the target composer. EMI's analysis database is essentially a compendium of answsers to the question, What would Mozart have done in this situation?

This is similar to the approach presumably taken by IBM Corporation's chess-playing program Deep Blue, which managed to defeat chess master Garry Kasparov in 1997. It seems that a large catalog of chess moves was created by analyzing many games of chess masters. During a game, the program would move pieces based on context, selecting from among the moves that were made in similar situations by the masters the program was emulating.

This approach is mostly about modeling the choices already figured out by masters and actually requires little learning or reasoning about music or chess. What is needed is a really big and really fast database and a sophisticated search capability. And although human composers and chess players also learn by example, we are not wired to perform by exhaustive search.

Does this disparity in method disqualify EMI or Deep Blue from being called intelligent? Can a system be intelligent only if its methods are fashioned after our own? Turing urged us not to focus on the process but on the result. He was less concerned about implicit use of reasoning and learning than explicit behavior: Does a computer *seem* intelligent? If so, then it is! Just as we can admire a beautiful sunset without worrying about how it was created, we should certainly be able to enjoy a composition that pleases us without caring who, how, or what created it. I suspect Turing would urge us to quit worrying about intelligence and to relax and enjoy the music.

9.24.4 A Musical Intelligence Test

Cope (1999) comments in the liner notes of his album *Virtual Mozart* that he avoided analyzing Mozart's symphonies beyond #31 "because they are so well known that derivations would have been recognizable." But he goes on to say, "The resultant work does show influences of these later symphonies, however."

It makes sense that Mozart's later symphonies were informed by his earlier ones. But has Cope's EMI system managed to do the same thing? Has EMI identified and developed some elements of Mozart's earlier symphonies into a more mature style? If so, that would seem to be evidence that Mozart's aural sensibility lies within his music where EMI can access it, and that EMI is able to find it, extract it, mature it, and use it as the basis of new compositions. Given what we know of EMI's process, this seems implausable. But according to Turing, we are to consider the system's behavior, not its inner workings, when deciding about intelligence. And so, like a good jury, let us follow the judge's orders, at least for now.

Cope's observation that EMI evidently developed some elements of Mozart's mature style is only anecdotal and subject to interpretation. How could we prove or disprove that EMI (or any other system) can act to develop a more mature style from a less mature one?

Turing's test methodology allows the experimenter to ask any question or pose any problem that would help prove or disprove the intelligence of the system under test. But it's hard to ask questions of a musical score. It is easier to claim that a chess program that beats a master is intelligent because there is a clear criterion: winning. The arts are more ambiguous.

But music can be analyzed. For example, suppose we conduct a test of intelligence on EMI such as the following. We begin with two contrasting premises:

- If a composing automaton is driven by a random process, we should be able to identify the "wandering" quality in its output that was noted by Pierce (1983).
- If a composer's works are informed by his or her prior works and the related works of others, then the same should be true of an artificial composer's works.

This suggests an experiment, as follows. Let E_1 stand for the Mozartian symphony created by EMI for the album *Virtual Mozart*. Let the way EMI composed E_1 be the function f of the set of Mozart symphonies #6 through #31 as follows: $E_1 = f(M_6, M_7, M_8, \ldots, M_{31})$. Now suppose we compose a set of N additional Mozartian symphonies using the same EMI technique:

$$E_2 = f(M_7, M_8, \ldots, M_{31}, E_1)$$

$$E_3 = f(M_8, \ldots, M_{31}, E_1, E_2)$$

$$\vdots$$

$$E_N = f(E_1, E_2, \ldots, E_{N-1}).$$

In other words, each new EMI Mozartian symphony replaces the earliest Mozart symphony in the database until all of Mozart's symphonies are eventually replaced by new EMI Mozartian symphonies.

As EMI symphonies progressively replace Mozart's in the database used to derive subsequent symphonies, how does EMI's musical style evolve? Does the progression of symphonies:

- Present a coherent, self-consistent set of works, as do Mozart's?
- Develop recognizable signatures of Mozart's mature style?
- Suggest how Mozart might have developed with his symphonies had he lived longer?

If any answer is yes, that would be strong evidence that EMI has successfully encoded Mozart's aural sensibility.

Does the progression of symphonies drift off in some other direction that is not Mozartian? Do subsequent EMI compositions become less musically interesting as the database is progressively left to its own devices? If either answer is yes, then perhaps EMI has not encoded Mozart's aural sensibility.

Of course there is also the possibility that EMI remains stylistically stagnant, continuing to churn out endless minor variations on Mozart's symphonies 6–31.

Cope reportedly conducted an experiment like this based on three works by Igor Stravinsky (Holmes 1997). From time to time, he mixed in the work of another contemporary of Stravinsky's to model the way human composers are influenced by their peers. Cope reported that over the course of time EMI developed the style of a mid-twentieth-century Russian-American composer. This effort of Cope's seems a lot closer to the true meaning of *experiment*. There is a hypothesis, a method, and most important, repeatability. Others could conduct this experiment and the results could be subject to peer review, all important aspects of the scientific method. It would be particularly interesting to know if stylistic stagnation resulted if the targeted style were not mixed with others. This might open up an understanding of the interaction of personal creativity and social forces.

9.24.5 Taste, Goodness, and Design

If we stick to Turing's behaviorist approach to measuring intelligence, then I think this member of the jury would have to find EMI guilty of intelligence. But I believe we also have an obligation to reflect upon the model whereby EMI creates its music and compare that to the human process as best we can. And when we do, I believe the jury is still out.

Consider the fact that EMI requires a database of preexisting works. When examined from a functional perspective, its intelligence, like that of every other composing system discussed in this chapter, is derivative of the music it emulates, derivative of the musical experience of its operator in fine-tuning its analysis, and derivative of the knowledge of EMI's creator. How could it be otherwise?

In contrast, Mozart had no corpus of examples of his personal style to draw upon when, as a young child, he began to compose in his highly recognizable signature style. Of course he was profoundly influenced by his teachers and by the music around him, but the origination of his personal style was seemingly guided primarily by his superlative taste, which had no external referent because, manifestly, no one else ever did manage to compose like he did. The same could be said of Beethoven, Schoenberg, and Stravinsky, for example. After hearing the child genius Mozart perform his own compositions, the great composer Joseph Haydn is known to have remarked to Mozart's father, Leopold Mozart, "Before God, and as an honest man, I tell you that your son is the greatest composer known to me either in person or by name. He has taste and, what is more, the most profound knowledge of composition."[22]

And of course, Mozart's ultimate signature is the music he went on to create throughout his career. While the signature of his musical style remained relatively continuous throughout his life, the compositional visions he brought to life seemed discontinuously to spring full-blown, like Athena from the head of Zeus. The art of his mature works seemed to have no antecedent. This characteristic of an artist's work is one of the indicators that predicts enduring fame.

Haydn's taste, Knuth's goodness (see section 9.2.1), and Hiller and Isaacson's aural sensibility are all key aspects of *design*. For example, there are an infinite number of ways to find the greatest common divisor of two integers (including guessing), but we choose Euclid's method (see section 9.2.2) because its design appeals to us—it shows goodness. Design is a key underpinning of mathematics as well: "Mathematics are the result of mysterious powers which no one understands and in which the unconscious recognition of beauty must play an important part. Out of an infinity of designs a mathematician chooses one pattern for beauty's sake and pulls it down to earth" (Morse 1959).

At bottom, whether designing music or mathematics, we reach into ourselves and extract that which most agrees with our natures and the problem that we pose to be solved. This is the fundamental process of art. All else is imitation.

9.25 Calculating Beauty

Hermann Helmholtz (1863) wrote,

> To furnish a satisfactory foundation for the elementary rules of musical composition . . . we tread on new ground, which is no longer subject to physical laws alone. . . . Hence it follows—*that the system of Scales, Modes, and Harmonic Tissues does not rest solely upon inalterable natural laws, but is also, at least partly, the result of aesthetical principles, which have already changed, and will still further change, with the progressive development of humanity.* (250–251; italics in original)

From ancient times, we have sought a rational explanation of nature through scientific enquiry. Since art was considered an imitation of nature, science also studied art. The Pythagoreans were perhaps the first to identify a connection between aesthetics and mathematics: beauty was found to reside in certain mathematical divisions of a vibrating string, and not in others. The distinguishing characteristic seemed to be the harmonious—beautiful—proportions of the division. Thus the Pythagoreans discovered what they believed was a way to study beauty objectively, to quantify beauty by simple integer ratios. Thus beauty could be found in the proportions of a string or in the proportions of the whole universe. This discovery powered aesthetic research and debate for centuries.

The Pythagorean observation of beauty in ratios came to be studied under the name *eurythmics*, the study of harmony and proportion. As a theoretical device, it could be used to create and analyze all forms of art: dance, architecture, music, sculpture, painting, and so on. From antiquity and through the Middle Ages, the sciences of subjective and objective nature thrived together, united as the branches of the *quadrivium*. "Mathematical science . . . has these divisions: arithmetic, music, geometry, astronomy. Arithmetic is the discipline of absolute numerable quantity. Music is the discipline which treats of numbers in their relation to those things which are found in sound."[23]

However, the quadrivium fell apart in the Renaissance. Natural scientists and mathematicians became increasingly uninterested in the arts because—despite the Pythagorean premise—no theory had successfully provided a rational link between aesthetics and proportion. The science of aesthetics fell by the wayside and was deemed unscientific (James 1993). The gulf between the arts and sciences has continued to this day. Some artists rail against reductionistic explanations of creativity. Some scientists question whether aesthetic experimentation can be scientific. But methodological analysis reveals that the disciplines of art and science are cut from the same cloth. Comparing Euclid's method with Guido's method, we saw that they are distinguished only by the role of subjective choice—of nondeterminism—in art. Art is not science, but their methods are more alike than different.

When applied to music, methodological criticism goes quickly to the core of the artist's intention, allowing us to apprehend the deeper significance of their art. Guido d'Arezzo's method shows his concern that music should be subordinate to the sacred Latin text. Schoenberg's combinatoric methods show a desire to deconstruct conventional harmonic expectation. Schillinger's methods followed from his desire to develop generative theories of art. Xenakis' statistical methods reflect his view of the quantum nature of sound. Cage's chance methods serve his aim to deconstruct the expectation of expectation.

Methodological criticism, information theory, psychoacoustics, complexity theory, and other approaches discussed in this chapter are making important contributions to theoretical aesthetics that finally allow the dialogue about the nature of art to move beyond its fixation with Pythagorean proportionality. Perhaps the truest proportions in music are those that relate expectation, interest, entropy, and redundancy; perhaps the truest study of music structure requires understanding the nonlinearities of our perceptual and nervous systems as well as the self-organizing principles of nature.

A Appendix

The wondrous potency of music, which moves the world and compels the spirit, captured in the net of numbers.
—Walter Burkert, *Lore and Science in Ancient Pythagoreanism*

A.1 Exponents

If p and q are any real numbers, a and b are positive real numbers, m and n are positive integers,

$$a^p a^q = a^{p+q} \qquad \frac{a^p}{a^q} = a^{p-q} \qquad (a^p)^q = a^{pq} \qquad (a^m)^{1/n} = a^{m/n}$$

$$a^{-p} = \frac{1}{a^p} \qquad \left(\frac{a}{b}\right)^{1/n} = \frac{a^{1/n}}{b^{1/n}} \qquad a^0 = 1 \quad \text{if} \quad a \neq 0 \qquad (ab)^p = a^p b^p$$

A.2 Logarithms

If $a^p = x$ where a is not 0 or 1, then p is called the logarithm of x to the base a. If x, y, and a are real numbers, then by this definition and the rules for exponents, we can write

$$\log_a xy = \log_a x + \log_a y$$

$$\log_a (x/y) = \log_a x - \log_a y$$

$$\log_a x^y = y \log_a x$$

The irrational number e is called the natural base of the logarithms, and $\log_e x$ is also written as $\ln x$. When written without specifying a base, log implies base e, that is, $\ln x = \log x = \log_e x$. The value of e is irrational; its first few digits are 2.718281828. . . .

To change the base of a logarithm, use the formula $\log_a x = \log_b x / \log_b a$.

A.3 Series and Summations

A *series* is any summation of a repeating pattern of terms. An example of an arithmetic series is $2+4+6+8+\cdots$. Each subsequent term is computed by adding or subtracting a constant amount to the immediately preceding term.

A simple geometric series might be $2+4+8+16+\cdots$. Each subsequent term is computed by multiplying or dividing the immediately preceding term by a constant amount.

Mathematicians have developed a useful shorthand for representing series, *sigma notation*. For example, the equation

$$s = \sum_{n=1}^{4} (5 \cdot n)$$

is shorthand for the equivalent expression

$$s = (5 \cdot 1) + (5 \cdot 2) + (5 \cdot 3) + (5 \cdot 4).$$

We can use it to form the sum of arithmetic and geometric series. The symbol Σ, the Greek character sigma, is used by mathematicians to represent the sum of a sequence of terms. The expression to the right of the sigma, $(5 \cdot n)$, is the *summand*. The numbers below and above the sigma are the *limits of summation*, and the variable n is the *index*.

This example,

$$s(t) = \sum_{n=1}^{4} 5nt, \tag{A.1}$$

can be written equivalently as

$$s(t) = (5 \cdot 1t) + (5 \cdot 2t) + (5 \cdot 3t) + (5 \cdot 4t). \tag{A.2}$$

This is the *expansion* of (A.1). Every point t of the function s is described by the entire summation. The examples above are finite series because the sequences of terms are finite. In the case of an infinite sequence of terms,

$$x_1, x_2, \ldots, x_n, \ldots$$

the corresponding infinite series is

$$x_1 + x_2 + \cdots + x_n + \cdots = \sum_{n=1}^{\infty} x_n.$$

The nth term x_n of a series is the *general term*. An infinite series is *convergent* if its value tends toward a finite sum, otherwise it is *divergent*.

A.4 About Trigonometry

Trigonometry is the study of *trigons,* which are triangles, especially right triangles, that are inscribed within a circle (figure A.1).

The ratio of the diameter to the circumference of a circle is the irrational number $\pi = 3.14\ldots$. Because the radius is half the length of the diameter, the circumference is 2π times the radius.

Angles are commonly measured in degrees, and the angle corresponding to a complete rotation is 360°. There are 2π radians or 360° in a circle (see section 5.2.2). An angle can be measured either clockwise or counterclockwise from a starting point. Conventionally, positive angles are measured counterclockwise from the positive horizontal x-axis of the circle, and negative angles are measured clockwise. Thus, for example, if we picture a circle on a blackboard, an angle of 0° conventionally points to the right along the positive horizontal axis; 90° points straight up; –90° points straight down; and 270° = –90°.

A.4.1 Sine Relation

Suppose we constructed a triangle like the one in figure A.1 so that the length of c (which is both the hypotenuse of the triangle and the radius of the circle) is fixed, but sides a and b are elastic: they can grow and shrink. Also suppose that point p is able to move around the circle, and that point q is constrained to follow it such that the angle $0qp$ is always a right angle. Last, the inner apex of the triangle is always at the center of the circle. These rules basically mean that we are limited to right triangles inscribed in a circle with the triangle's base resting on the horizontal axis. As the angle θ increases and point p moves counterclockwise around the circle, the triangle changes shape in a characteristic way (figure A.2). If we study the way in which the ratio of b/c changes as the angle θ changes, we observe that this relation corresponds to sinusoidal motion.

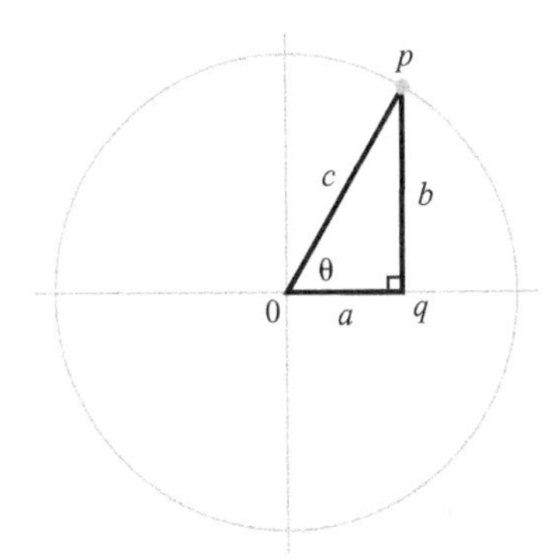

Figure A.1
Right triangle inscribed in a circle.

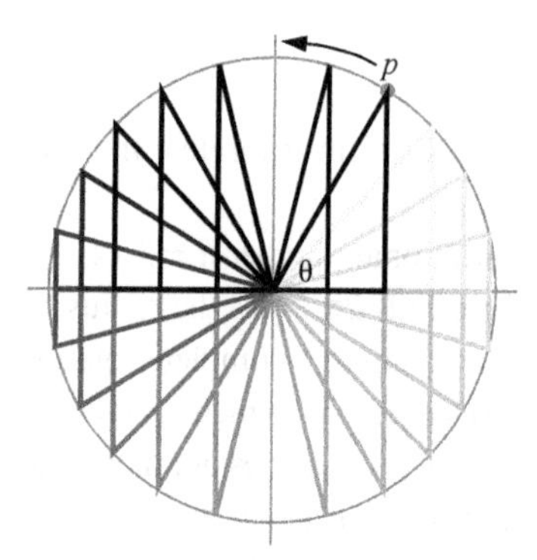

Figure A.2
Family of triangles inscribed in a circle.

That is, the radius c, its angle to the horizontal axis θ, and the ratio b/c are connected to each other by the *sine relation:*

$$\sin\theta = \frac{b}{c}. \qquad \textit{Sine Relation} \quad \text{(A.3)}$$

Consider, for example, when $\theta = 0$. Then c lies along the positive horizontal axis, $a = c$ in length, and $b = 0$, since the "triangle" has no height. Hence sin $0 = 0/c = 0$. When $\theta = 90°$, c lies along the positive vertical axis, $b = c$ and $a = 0$. Hence sin $90° = 1/1 = 1$. See section 5.4 for more details.

A.4.2 Cosine Relation

Consider the relation between sides a and c in figure A.1. The angle θ and the ratio a/c are connected to each other by the *cosine relation:*

$$\cos\theta = \frac{a}{c}, \qquad \textit{Cosine Relation} \quad \text{(A.4)}$$

which is similar to the sine relation except that its values are shifted by 90°, that is, $\cos\theta = \sin(\theta + 90°)$. This makes sense because the cosine involves the ratio a/c instead of b/c, and side a is orthogonal (that is, at a 90° angle) to side b; hence it precedes the sine wave by 90°.

A.4.3 Tangent Relation

Last, consider the ratio b/a, the ratio of the two elastic sides of the triangle in figure A.1. The angle θ and the ratio b/a are connected to each other by the *tangent relation:*

$$\tan\theta = \frac{b}{a}. \qquad \textit{Tangent Relation} \quad \text{(A.5)}$$

When $\theta = 45°$, the triangle is an isosceles right triangle and $a = b$, so tan $45° = 1$. When $\theta = 0°$, tan $\theta = 0/a = 0$. But when $\theta = 90°$, $\tan\theta = b/0 = \infty$.

A.4.4 Relating Tangent, Sine, and Cosine

We can relate these definitions to each other as follows:

$$\tan\theta = \frac{b}{a}, \qquad \sin\theta = \frac{b}{c}, \qquad \cos\theta = \frac{a}{c}, \qquad \frac{b}{a} = \frac{b}{c} \div \frac{a}{c}.$$

$$\therefore \ \tan\theta = \frac{\sin\theta}{\cos\theta}. \qquad \textit{Relation of Tangent, Sine, and Cosine} \ \text{(A.6)}$$

A.4.5 Reciprocal Trigonometric Functions

We form the reciprocals for sine, cosine, and tangent by reversing the order of their ratios. Each of these reciprocals has its own name:

$$\cot\theta = \frac{1}{\tan\theta} = \frac{a}{b}, \qquad \textit{Cotangent} \ \text{(A.7)}$$

$$\sec\theta = \frac{1}{\cos\theta} = \frac{c}{a}, \qquad \textit{Secant} \ \text{(A.8)}$$

$$\csc\theta = \frac{1}{\sin\theta} = \frac{c}{b}. \qquad \textit{Cosecant} \ \text{(A.9)}$$

A.4.6 Inverse Trigonometric Relations

The trigonometric functions determine the angle of the hypotenuse θ from the sides a, b, and c. But what if we know the angle θ and want to use it to find the proportions of the triangle?

The inverse trigonometric functions determine the ratio of the sides from the angle of the hypotenuse against the positive horizontal axis. For instance, the inverse of $\sin\theta$ is arcsine x, also written asin x or $\sin^{-1} x$, where x is a ratio of two sides. The cosine and tangent functions are similarly named, for example, $\arctan x = \text{atan}\, x = \tan^{-1} x$.

But how do we define these functions? At first we might think, just inscribe a triangle in a circle with angle θ, measure its sides, then find their ratio. But there is a problem: because we are measuring angles on a circle, there are actually many angles—infinitely many at multiples of 360°—that correspond to any particular proportion of sides. For example, if $x = b/a = 1/1$, then the triangle is an isosceles right triangle and $\text{atan}\, x = 45°$, but it is also true that $\text{atan}\, x = 45° \pm (k \cdot 360°)$, where k is an integer. So the inverse trigonometric functions are ambiguous.

But, in general, all we usually want is the angle when $k = 0$. So we define a range of *principal values* that covers just these angles. The principal values of the arctangent, arccosine, and arcsine are

$$-90° < \tan^{-1} x < 90°$$

$$0 \leq \cos^{-1} x \leq 180°$$

$$-90° \leq \sin^{-1} x \leq 90°.$$

A.5 Xeno's Paradox

Xeno of Elea reasoned that if one travels distance *d* from point A to point B, one must certainly travel half the distance *d*/2 to point *B* before traveling the whole distance *d*. And from that point one must again travel half the remaining distance *d*/4, and so on. Continuing in this way, he reasoned, one would never reach point B because one must pass through an infinite number of points and that is impossible in a finite time.

Essentially, his argument is that if space and time are composed of indivisible points and moments, these must have some magnitude, and we are faced with the contradiction of a magnitude that cannot be divided. If space and time are divisible ad infinitum, we are faced with the contradiction that an infinite number of points and moments can be added up to make a merely finite sum. Xeno's point is that since multiplicity and motion contain these contradictions, they cannot be real. Therefore, as his teacher Parmenides said, there is only one Being, with no multiplicity, excluding all motion and change.

This is a perfectly fine outcome if you are satisfied with it. If you are not, then a modern way out of this problem is to consider space and time not as a densely packed infinity of points or moments, but as sparsely packed such that no point is next to any other point. Thus, between any two points or moments there is always a third regardless of scale. The advantages of this approach are twofold. First, the nondenumerable infinity of real numbers (and likewise of points in space and of events in time) is therefore much larger than the mere denumerable infinity of integers that Xeno envisioned. Further, the sum of an infinite series of real numbers *can* have a finite sum. This latter point is the clincher.

A.6 Modulo Arithmetic and Congruence

If it's 1:30, and your friend says she'll meet you in 45 minutes, what time will it be when she joins you? If you answered 2:15, then you used modulo arithmetic to obtain the answer. Since there are 60 minutes in the hour, time-based calculations must keep the number of minutes in that range.

We could formalize the example this way. Using "minute arithmetic," we could write $75 \equiv ((15))_{60}$, or $75 \equiv 15 \bmod 60$, expressed as "75 is *congruent* to 15, modulo 60."

In general, if the difference between two integers *r* and *b* can be divided without remainder by another number *m*, then *r* and *b* are congruent modulo *m*. This is written as

$$r \equiv ((b))_m \quad \text{if} \quad (b - r)/m \text{ is an integer.} \tag{A.10}$$

In the example, the quotient of (75 – 15)/60 is an integer, so 75 and 15 are congruent modulo 60.

A common use of modulo arithmetic is to obtain the remainder of integer division. The value *b* is the base and *r* is the remainder. The FORTRAN programming language provides a way to obtain the remainder of two numbers with the function `mod(b, m)`; the C and C++ programming languages define it with a binary infix operator %. MUSIMAT, the programming language invented for this book (see appendix B), defines the remaindering operation as follows:

```
Integer Mod(Integer b, Integer m){
    While(b >= m){b = b - m;}
    While(b <= -m){b = b + m;}
    Return(b);
}
```

Note that `Mod()` can operate on and return negative values. For example, $((-1))_{10} = -1$ and $((11))_{-10} = 1$. In general, the return value will be

$$-m < n < m \quad \text{for} \quad ((n))_m .$$

There are times when it would be convenient to force the remainder r to be a positive modulus number even if the base b is negative. For example, in MUSIMAT, the index of a `List` must be a positive integer. So MUSIMAT has a version of `Mod()` that returns only the positive wing of modulo values:

```
Integer PosMod(Integer b, Integer m){
   While (b >= m) {b = b - m;}
   While (b < 0) {b = b + m;}
   Return(b);
}
```

For example, `Print(Mod(-13, 10));` prints `-3`, whereas `Print(PosMod(-13, 10));` prints 7 (see figure A.3).

Both `Mod()` and `PosMod()` preserve the position within the modulus interval, but `PosMod()` also requires the value to be positive. If we have a `List` of ten elements numbered 0 to 9, we can provide a base of any positive or negative value `b` to `PosMod(b, 10)` and it will coerce the remainder to lie within the valid range of the `List`.

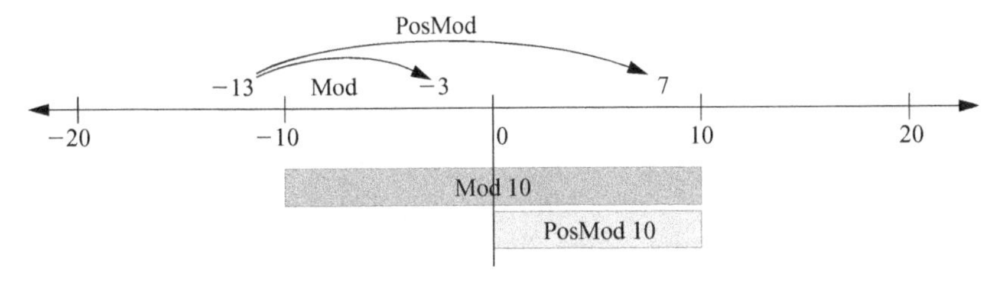

Figure A.3
Signed and unsigned modulus operation.

A.7 Whence 0.161 in Sabine's Equation?

Wallace Sabine derived the constant 0.161 in his equation for reverberation time in a room and verified it both experimentally and theoretically (see equation 7.31). His derivation provides a fundamental view of statistical room acoustics.

In physics the *mean free path* is the average distance a particle can move in a gas without a collision. Sabine (1921) adapted the term to mean the average distance a wave front can travel in a room before being reflected by a wall. He demonstrated that an approximate value for the acoustic mean free path is

$$\text{MFP} \cong \frac{4V}{S} = d, \qquad \textit{Acoustic Mean Free Path} \quad \text{(A.11)}$$

where V is volume and S is surface area of the room. This approximation assumes that the sound field in the room is *diffuse,* that is, the energy density is uniformly distributed throughout the room (*homogeneous*) and is traveling with equal intensity in every direction (*isotropic*). This ideal condition can only be approximated in real rooms because different areas of a hall will have different amounts of absorption, depending upon the hall's geometry and absorptive properties.

The average number of reflections per second $\overline{f_R}$ is the speed of sound c divided by the MFP:

$$\overline{f_R} = \frac{c}{d} = \frac{Sc}{4V}.$$

The average time between reflections is the inverse:

$$\Delta t = \frac{d}{c} = \frac{4V}{Sc}.$$

Although air absorbs some sound, much more sound is absorbed by walls during reflections. So a room with smaller Δt will have a shorter reverberation time than a room with greater Δt.

Intuitively, when the sound source is first turned on in a room, it pumps acoustic energy into the room, and though the walls suck energy out, they don't remove it all. The energy remaining in the room increases its total energy, and we hear an exponential buildup of sound level over time. When the rate of energy entering the room equals the rate of energy leaving the room, equilibrium is achieved, and the energy density plateaus.

Energy density in a reverberant room can be likened to the mean water level in a leaky water tank: the level in the tank is proportional to the input rate of flow and inversely proportional to the output rate of flow. If the input rate of flow goes to zero, water will drain out and the rate of water loss will be proportional to the remaining water level. Similarly, in a room, the rate of energy loss is proportional to the remaining energy density.

Let the energy density in a room at time t be denoted by the function $W(t)$. (Calculus alert!) By definition, the energy rate of change is dW/dt. We can express the observation that the rate of energy change is proportional to the remaining energy by writing

$$\frac{dW}{dt} = kW, \tag{A.12}$$

where k is a constant of proportionality that can be determined for particular rooms. The constant k specifies the steepness of the slope. For example, a highly absorptive room such as an anechoic chamber would have a relatively large value of k, whereas a room with reflective stone walls would have a small value because its reverberation decays much more slowly. Equation (A.12) is a first-order differential equation. The requisite mathematical equipment to solve it is presented in volume 2, chapter 6.

Let's assume a trial solution of the form

$$W(t) = e^{-kt}. \tag{A.13}$$

If we substitute this definition of the function W back into (A.12), we get (by the power rule) the general solution

$$\frac{dW}{dt} = kW = \frac{d}{dt}e^{-kt} = ke^{-kt}. \tag{A.14}$$

When we switch off the input power to measure the reverberation time, let's say the total energy density in the room is initially $W(0) = W_0$. Also set $k = 1/\tau$, where τ is the time constant of the exponential curve. Then we can write the particular solution as

$$W(t) = W_0 e^{-t/\tau}, \tag{A.15}$$

which says that the power W in the room is a decaying exponential function of time t with time constant τ.

The *reverberation time* T_R corresponds to the length of time it takes for the sound to become inaudible, defined as the time required for the sound to decay by 60 dB SIL, that is, to a millionth of its original intensity. We want to know the time T_R required for energy to drop by a factor of 10^{-6} in a particular room, that is, we want a solution to the equation

$$10^{-6} W_0 = W_0 e^{-(T_R/\tau)}.$$

Solving for T_R, we obtain

$$T_R = \log(10^6)\tau, \quad \text{or} \quad T_R = 13.8\tau.$$

Recall the definition of the average time between reflections Δt from (A.11). Setting $\tau = \Delta t$, we obtain

$$T_R = 13.8\frac{4V}{Sc}.$$

Using a value of $c = 342$ m/s for the speed of sound, and combining constants, we have Sabine's equation:

$$T_R = 0.161\frac{V}{S}. \tag{A.16}$$

In the literature, the constant ranges from 0.160 to 0.164, depending upon the speed of sound used.

So is this the right constant? Is this the right equation for the job? Given the number of alternative formulations in the literature over the last hundred years, it would seem that Sabine's remarkable achievement is not without flaws. Even if we correct for air absorption, his formula tends to estimate reverberation times that vary widely from actual results. Also, its statistical nature provides no way to adjust it for environments that are not ideally diffuse.

Another problem is that as average absorption $\bar{\alpha}$ approaches 1, Sabine's equation does not predict that reverberation time goes to 0, even though in an anechoic chamber it effectively does. Eyring (1933) proposed a reverberation formula in which the absorption coefficient is calculated according to $\alpha_E = -\ln(1-\bar{\alpha})$, so that (A.16) becomes

$$T_R = 0.161\frac{V}{-S\ln(1-\bar{\alpha})}, \tag{A.17}$$

which gives a reverberation time of 0 when $\bar{\alpha} = 1$, and reduces to Sabine's formula when $\bar{\alpha} \ll 1$. Many other refinements and alternatives are now available. After a century, reverberation time is still the subject of active research.

A.8 Excerpts from Pope John XXII's Bull Regarding Church Music

The competent authority of the Fathers has decreed that, in singing the offices of divine praise through which we express the homage due to God, we must be careful to avoid doing violence to the words, but must sing with modesty and gravity, melodies of a calm and peaceful character. . . . But certain exponents of a new school, who think only of the laws of measured time, are composing new melodies of their own creation with a new system of notes, and these they prefer to the ancient, traditional music. . . . By some, their melodies are broken up by hocheti or robbed of their virility by discanti, tripla, motectus, with a dangerous element produced by certain parts sung on texts in the vernacular. . . . The mere number of the notes, in these compositions, conceal from us the plainchant melody, with its simple, well-regulated rises and falls which indicate the character of the Mode. These musicians . . . intoxicate the ear without satisfying it, they dramatize the text with gestures and, instead of promoting devotion, they prevent it by creating a sensuous and innocent atmosphere. . . . We are prepared to take effective action to prohibit, cast out, and banish such things from the Church of God. Therefore, . . . We prohibit absolutely, for the future that anyone should do such things, or others of like nature, during the Divine Office or during the Holy Sacrifice of the Mass. . . . However, We do not intend to forbid the occasional use . . . of certain consonant intervals superposed upon the simple ecclesiastical chant, provided these harmonies are in the spirit and character of the melodies themselves, as, for instance, the consonance of the octave, the fifth, the fourth, and others of this nature; but always on condition that the melodies themselves remain intact in the pure integrity of their form, and that no innovation take place against true musical discipline. . . . Made and promulgated at Avignon in the Ninth Year of Our Pontificate (1324–1325). *Corpus juis canonici*. (Hayburn 1979)

A.9 Greek Alphabet

Besides being the alphabet of a major modern civilization, the Greek alphabet (table A.1) is useful not only for the study of mathematics but also for students being rushed for fraternities. It may also come in handy when eating alphabet soup in Greece.

Table A.1
Greek Alphabet

Alpha	Α	α	Iota	Ι	ι	Rho	Ρ	ρ
Beta	Β	β	Kappa	Κ	κ	Sigma	Σ	σ
Gamma	Γ	γ	Lambda	Λ	λ	Tau	Τ	τ
Delta	Δ	δ	Mu	Μ	μ	Upsilon	Υ	υ
Epsilon	Ε	ε	Nu	Ν	ν	Phi	Φ	ϕ
Zeta	Ζ	ζ	Xi	Ξ	ξ	Chi	Χ	χ
Eta	Η	η	Omicron	Ο	ο	Psi	Ψ	ψ
Theta	Θ	θ	Pi	Π	π	Omega	Ω	ω

B Appendix

Mathematics is the music of reason.
—James J. Sylvester

B.1 MUSIMAT

Why did I invent a new programming language when there are so many excellent ones already available? The problem is that most programming languages are more general-purpose than is required for the relatively specialized purposes of this material, and a proper introduction to such a general-purpose language would lead the discussion too far afield. I decided it would be of more service to readers to specialize the language so that its features would match the examples in this book as closely as possible. That way, the focus would remain on the subject being coded rather than on the language being used to code it.

Nevertheless, MUSIMAT is similar to other procedural languages such as C or C++ (see Stroustrup 1991), so if you already know one of these, it should be easy to pick up MUSIMAT. If you don't know any programming language, learning one should be easier after you learn MUSIMAT.[1] I present everything you'll need to know about MUSIMAT in the following sections.

B.1.1 Basic Elements

Virtually all programming languages, including MUSIMAT, share the following characteristics:

- *Flow control* Specifying the order in which the steps are to be taken.
- *Data types* Naming the kinds of objects to be operated on and describing their behaviors. Types of numbers, such as *integer* and *real* are common basic data types.
- *Variables* Names of places to hold data of various types.
- *Operators* A set of actions that can be performed on data. Operations like "add", "assign", and "select" perform well-defined operations on the data.

- *Conditional evaluation* Making decisions based on circumstances and taking appropriate action.
- *Iteration* If an algorithm is to be applied repeatedly to data, for instance, the way Euclid's method does, then we need a way to express this.
- *Recursion* If a future output depends upon a current or previous output as well as possibly the current inputs, we say that the relationship is recursive.
- *Data structures* It is sometimes necessary to group data into collections, such as sets, lists, arrays, and matrices. The types of these data structures can be homogeneous (all alike) or heterogeneous (a mixed bag).
- *Named methods* When we've developed a set of instructions that does something useful, we want to be able to give it a name, like "Euclid's method" or "Guido's method." Since programming languages developed out of the mathematics of functions, we use functional notation to represent the operation of methods.

B.1.2 Statements and Expressions

Most methods read, "Do this, then do that." Each "do this" step is a *statement*. Sequences of statements are read left to right, then down the page. The elements of each statement, called *expressions,* determine what the statement is about. In many programming languages (including MUSIMAT), semicolons (;) separate statements.

B.1.3 Data Types

To begin, we need only two types of numbers, `Integer`, which is a positive or negative whole number, and `Real`, which is an approximate real number. To keep things simple, let's assume for our purposes that we have virtually unlimited precision for computations. As we go along, I introduce additional data types as needed.

B.1.4 Constants

A constant is any number whose value does not change. The number 3 is a constant `Integer`. The number 3.14159 . . . is a constant `Real`.

MUSIMAT also predefines two constants, `True` and `False`, and gives them integer numeric values of 1 and 0, respectively.

B.1.5 Variables

Variables are named places to store data. Names are indicated by one or more upper- or lower-case letters, like `Q`, `n`, or `fred`. Alphabetic case is significant, so `fred` denotes a different variable than does `Fred`. Numbers can also be used in variable names (for example, `Fred33`), but the first letter of a variable name may not be a number.

Since they physically embody data, variables occupy space and time. Variables flow into existence when they are defined, and generally hold their value until the end of the program unless additional steps are taken to change their value or to restrict their existence to a certain region of the program.

B.1.6 Reserved Words

For the language to be unambiguous, we must reserve the meaning of certain words and symbols. Reserved words are distinguished by an initial capital letter and are shown using `a special font`. Reserved words include `If`, `While`, `Do`, `For`, `Repeat`, `Else`, `Halt`, `Real`, `Integer`, `Return`. Some other symbols are also reserved. These symbols can't be used for anything but their designated meaning.

B.1.7 Lists

We can group sets of variables to keep track of their relations. An `IntegerList` represents a collection of `Integer` expressions, for example,

```
IntegerList iL = {1, 1+1, 3, 5-1};
```

defines a list `iL` containing the integers 1 through 4.

A `RealList` represents a collection of `Real` expressions:

```
RealList rL = {1.1, 2.2, 3.3, 4.4};
```

We can obtain the length of a list of any type. For example,

```
Integer n = Length(rL);
Print(n);
```

prints `4`.

B.1.8 Operators and Operands

The symbols + and – are *operators*, and the data they act upon are *operands*. Most operators take two operands, and the operator lies between the operands, for example, `a + b`, and `c / d`. Such operators are called *binary infix*, meaning that the operator lies between two operands. In its binary infix form, the symbol – means subtraction, for example, `a - b`. The *unary prefix* – operator comes before the expression it negates, for example, `-3`.

Multiplication in mathematics is typically expressed by the concatenation of variables, so for instance *at* means the product of variables *a* and *t*. But this can be ambiguous, because *at* could also refer to the single word "at". To avoid ambiguity, the infix operator `*` indicates multiplication, so the product of `m` times `n` is written `m * n`.

B.1.9 Assignment

We can assign the value of an expression to a variable using the assignment operator =. For example,

```
lhs = rhs;
```

assigns the expression `rhs` to `lhs`. The object on the right-hand side of the = sign (i.e., `rhs`) can be any expression. The object on the left-hand side (i.e., `lhs`) must be a variable name, with one exception. For example, the statement

```
s = 3 + 5;
```

sets the value of variable `s` to 8.

The left-hand side of an assignment can also indicate that a certain element of a list is to receive the value on the right-hand side. For example,

```
IntegerList iL = {0, 1, 2, 3};
iL[2] = 10;
```

replaces the third element with `10`, causing the list to become

```
{0, 1, 10, 3}
```

Note that lists are indexed starting at 0. So writing

```
iL[0] = 55;
```

causes the list to be

```
{55, 1, 10, 3}
```

B.1.10 Relations

Relational operators compare numeric values. For example, in the expression `x < y`, if `y` is greater than `x`, the value of the expression is `True`; otherwise the value of the expression is `False`. Other relational operations include > for greater than, <= for less than or equal, and >= for greater than or equal.

Because = has already been given the meaning of assignment, we must choose another way to express equality, which we do by putting two equals signs together: ==. For example, the expression `x == y` is `True` if `x` and `y` have the same value.

Inequality is expressed by `!=`, so for example, `x != y` is `True` if `x` and `y` have different values.

B.1.11 Logical Operations

Logical operators compare truth values. For example, the expression `x And y` is true if and only if `x == True` *and* `y == True`. The expression `x Or y` is true if either `x == True` *or* `y == True`.

B.1.12 Operator Precedence and Associativity

In the expression `a * x + b * y`, what is the order in which the operations are carried out? By the standard rules of mathematics, we should first form the products `a*x` and `b*y`, then sum the result. So multiplication has higher *precedence* than addition. The natural precedence of operations can be overridden by the use of parentheses. For example, `a * (x + b) * y` forces the summation to occur before the multiplications.

In the expression $a + b + c$, we first add a to b, then add the result to c, so the *associativity* of addition is left to right. We could express left-to-right associativity explicitly like this: `(((a) + b) + c)`.

The rules of precedence and associativity in programming languages can be complicated, but the programming examples in this section use the following simplified rules.

Expressions are evaluated from left to right, except

- Multiplications and divisions are performed before additions and subtractions.
- All arithmetic operations (`+, -, *, /`) are performed before logical operations (`And, Or, ==, <, >`).
- Parentheses override the above precedence rules.

For details, see section B.3.

B.1.13 Type Promotion and Type Coercion

What if the values in an expression are not of the same type? For example, since both operands in the expression `2/3` are integers, the quotient will be an integer. The quotient of `4.5/2.25` will be a real number because both operands are reals. But what is the quotient of `2/2.25`? Our options are to coerce the numerator to be a `Real` and then perform real division, or coerce the denominator to be an `Integer` and then perform integer division. Which shall it be?

Since the set of all reals includes the set of all integers, it makes sense to *promote* the integer `2` to the corresponding real value `2.0` and then perform real division. MUSIMAT automatically converts `2/2.25` into `2.0/2.25` and then performs real division. In general, integer values are automatically promoted to reals wherever they occur in an expression with reals.

If automatic type promotion is not desired, the type of an expression can be *coerced* by directly indicating its type. Consider the expression:

```
10/Integer(3.33)
```

First, the real value `3.33` is truncated to the integer value `3`, then because both numerator and denominator are now integers, integer division is performed. Beware of things being done for you automatically by computers! You still must pay attention to head off unintended consequences. Consider:

```
26/Integer(2.5)
```

equals 13, but

```
Integer(26/2.5)
```

equals 10.

B.1.14 Accessing List Elements

We can access an element of a list using the index operator `[]`. Suppose we have the following declarations:

```
Integer w = 1, x = 2, y = 3, z = 4;
IntegerList iL = {w, x, y, z};
```

Then the statement

```
Integer c = iL[0];
```

assigns c the same value as w (which is 1). The statement

```
c = iL[3];
```

assigns c the same value as z (which is 4).

The first element on a `List` is indexed by 0, and if a `List` has N elements, the last one is indexed by $N-1$.

B.1.15 Functions

Functional notation in mathematics allows us to encapsulate and name arithmetic expressions. For example, if we have defined the function $f(a, b) = a + b$, then f stands for $a + b$. The value or values in parentheses after a function name, called *arguments,* supply the function with inputs. Functions also typically return a result. For example, using this definition of f, $7 = f(3, 4)$.

Programming languages typically come with a set of predefined functions for the most common necessities, and they also allow new functions to be created. For example, in MUSIMAT all operators also have a functional representation, so writing

```
Real x = Divide(11.0, 4.0);
```

is the same as writing

```
Real x = 11.0/4.0;
```

In this case, x is set to the quotient, 2.75. Real division is performed because both numerator and denominator are reals. If we want to perform integer division, both numerator and denominator must be integers. We could write

```
Integer x = Divide(11, 4);
```

or equivalently,

```
Integer x = 11/4;
```

In either case, x is set to the quotient, 2, and the remainder is discarded. To get the remainder after integer division we can write

```
Integer x = Mod( 11, 4 ); // remainder of integer division
```

The variable x is set to 3, the remainder of 11/4. For positive integers m and n, Mod(m, n) lies between 0 and $n - 1$. Why is this function called Mod instead of, say, Remainder? See appendix A, A.6. The equivalent operator form for remaindering also looks a little strange:

```
Integer x = 11 % 4;
```

The % sign does not have its usual meaning of "percent" in MUSIMAT. Instead, it means "remainder of integer division." Mod and % can only be applied to integer operands.

Some useful built-in functions are not associated with operators. Exponentiation is performed by the function Pow(). These three statements,

```
Real base = 10.0;
Real exp = 2.0;
Real x = Pow(base, exp);
```

are equivalent to writing $x = 10.0^{2.0}$, and the result stored in x is 100.0. Going the other way: Log10(x) is equivalent to $\log_{10} x$, and

```
Real y = Log10(100.0);
```

sets y to 2.

Another built-in function, Print(), allows us to observe the value of a variable or expression. When executed, the statements

```
Real x = 11.0/4.0;
Print(x);
```

display the value of x, or 2.75. The way in which the value is displayed varies with the type of the expression and the type of computer. If the computer is a person, for example, he or she might say "two point seven five." If it is an electromechanical computer, it might show the value on a display screen.[2]

When the predefined function Halt() is executed, the method in progress stops at that step in the program. The argument to Halt(), if any, can be used to indicate the answer or result obtained by the program up to that point.

One final built-in function is Random(), which returns a real number in the range of 0.0 to 1.0 chosen at random.

B.1.16 Conditional Statements

A mathematical notation for determining the sign of a number is

$$y = \begin{cases} x < 0, & a \\ x \geq 0, & b \end{cases}$$

which sets *y* to *a* if *x* is negative; otherwise it sets *y* to *b*. Such relational expressions are called *predicates*. MUSIMAT accomplishes the same thing like this:

```
If (x < 0)
  y = a;
Else
  y = b;
```

In this example, `y` receives the value of `a` if `x` is less than 0; otherwise `y` receives the value of `b`. The `Else` part of this construction is optional. So for example,

```
If (a < b)
    Print(a);
```

prints `a` only if it is less than `b`. `If` and `Else` can be combined to allow chains of predicates:

```
If (x < 0)           // is x negative?
  y = a;
Else If (x == 0)     // it's not negative, but is it zero?
  y = b;
Else                 // neither negative nor zero, x must be positive
  y = c;
```

B.1.17 Compound Statements

Suppose we need to do more than one thing depending on the value of a predicate. If we need to execute multiple statements that depend upon a common predicate, we can group them together into a list of statements. For example, `{m = n; n = r;}` is a list of statements, also called a *compound statement*. Consider steps 2 and 3 of Euclid's method (see section 9.2.2), which can be expressed

```
If (r==0)
  Halt(n);
Else {
  m = n;
  n = r;
}
```

If `r` is not equal to `0`, first `m` is assigned the value of `n`, and then `n` is assigned the value of `r`. We express this in MUSIMAT by making these two steps into a compound statement.

Any legal statement can appear within a compound statement, including other compound statements. This means we can nest compound statements inside each other.

B.1.18 Iteration

We must be able to repeat a statement or statements multiple times. For example, Euclid's method returns to step 1 from step 3, depending upon the value of variable *r* (see section 9.2.2). In

MUSIMAT, the `Repeat` statement causes a statement or compound statement to repeat interminably. This allows us to implement Euclid's method as follows:

```
Repeat{
  r = Mod(m, n);     //remainder of m divided by n
  If (r == 0)  {
    Halt(n);         // halt, and give answer n
  } Else  {
    m = n;
    n = r;
  }
}
```

This code shows an example of nested compound statement lists. The bare syntax of this example is

```
Repeat {... If (...) {...} Else {...}}
```

and the compound statements following `If` and `Else` are nested inside the compound statement following `Repeat`. We can nest compound statements as deeply as we desire.

Since it never stops by itself, the only way to terminate a `Repeat` statement is with a `Halt` statement.[3] It's a crude but effective technique; however, there are more elegant ways to decide how many times to repeat a block of statements. The `Do-While` statement allows us to specify a *termination condition* that is evaluated after the body has been executed. Here is an example that prints the random value assigned to `x` and repeats for as long as `x` is less than `0.9`.

```
Real x;
Do {
  x = Random();  // choose a random value between 0.0 and 1.0
  Print(x);
} While (x < 0.9);
```

Because `Random()` returns a uniform random value in the range 0.0 to 1.0, its value will be less than 0.9 on average 90 percent of the time. It is possible, though unlikely, that this statement would print its value only once, and it is also possible that it could print dozens, even hundreds, of times before halting, depending upon the particular sequence of random numbers returned by `Random()`.

The `For` statement also implements a way of repeating a statement or compound statement a number of times, but it allows us to directly manage the value of one or more variables each time the statements are executed and to use them to determine when to stop. This example prints the integers between 0 and 9:

```
Integer i;
For (i = 0; i < 10; i = i + 1)
  Print(i);
```

The variable `i` is called the *control variable*. The example first sets `i` to `0`, then tests if `i < 10`. Since `0 < 10`, the `Print()` statement is executed. Next, the `For` statement executes the statement `i = i + 1`, which adds 1 to the value of `i`. So now `i` equals 1. Again, the `For` loop tests if `i < 10`, and since `1 < 10`, it executes `Print()` again. It again adds 1 to the value of `i`. So now `i` equals `2`. This process continues until `i == 10`, whereupon the `For` loop terminates because then `i < 10` is `False`.

The `For` statement is a little twisty, so let's take a more careful look at its operation. In general, we can name the parts of the `For` statement as follows:

```
For (initialization; test; change)
   statement
```

where *`statement`* can be a single statement (terminated by a semicolon) or a compound statement (enclosed with curly braces). The `For` statement first executes the *`initialization`* code, then evaluates the boolean expression `test`. If the value of *`test`* is `False`, the `For` statement terminates. If the value of *`test`* is `True`, the *`statement`* is executed, then the `change` expression is executed, and finally the `test` is evaluated again. If the value of *`test`* is `False`, the `For` statement terminates. If the value of *`test`* is `True`, the cycle repeats again and again until the value of *`test`* is `False`.

As a convenience, it is possible to define and set the value of the *`initialization`* variable in one step, so the preceding example could have been written

```
For (Integer i = 0; i < 10; i = i + 1)
   Print(i);
```

B.1.19 User-Defined Functions

MUSIMAT, like most programming languages, allows users to define their own functions. Take Euclid's method, for example. To define it, we must state how the input variables `m` and `n` receive their inputs, and determine what happens to the result when the method halts. We can define a function named `euclid()` in MUSIMAT as follows:

```
Integer euclid (Integer m, Integer n){
   Repeat {
      Integer r = Mod(m, n);
      If (r == 0)
         Return(n);
      Else  {
         m = n;
         n = r;
      }
   }
}
```

The function is declared to be of type `Integer` because it will return an `Integer` result.

Note that `Return(n)` has been substituted for the `Halt(n)` function shown previously. Instead of halting execution altogether, the `Return(n)` statement only exits the current function, carrying with it the value of its argument back to the context that invoked it. The program can then continue executing from there, if there are statements following its invocation. Here's an example of invoking the `euclid()` function:

```
Integer x = euclid(91, 416);
Print(x);
```

which will print 13. If we had used `Halt()` in `euclid()`, we'd never reach the `Print` statement because the computer would stop.

Here's another way to compute the same thing:

```
Print(euclid(91, 416));
```

This way we can eliminate the "middleman" variable `x`, which only existed to carry the value from the `euclid()` function to the `Print()` function. In this example, the call to the `euclid()` function is nested within the `Print()` function. MUSIMAT invokes the nested function first, and the value that `euclid()` returns is supplied automatically as an argument to the enclosing function, `Print()`. Functions can be nested to an arbitrary extent. The most deeply nested function is always called first.

B.1.20 Invoking Functions

We had two situations where the function `euclid()` was followed by a list of arguments, once where it was defined, and another where it was invoked. The arguments associated with the definition of `euclid()` are called its *formal arguments*. They are `Integer m` and `Integer n`. The values associated with its invocation (integers `91` and `416`) are called its *actual arguments*. A function will have only one set of formal arguments that appear where the function is defined. It will have as many sets of actual arguments as there are invocations of the function in a program.

When a function is invoked, three things happen:

1. The *values* of the actual arguments are *copied* to the corresponding formal arguments.

If an actual argument is a constant, its value is simply copied to the corresponding formal argument. Example: `Print(3)` copies `3` to the formal argument for `Print()`.

If an actual argument is a variable, its value is copied to the corresponding formal argument. Example: `Integer a = 3; Print(a)` copies the value of `a` (which is 3) to the formal argument for `Print()`.

If an actual argument is another function, that function is evaluated first, and its return value replaces the function. Example: For the statement `Print(euclid(91, 416))`, first `euclid(91, 416)` is evaluated, and the result (which is 13) is substituted in its place. So the statement becomes `Print(13)`. Finally, the 13 is copied to the corresponding formal argument of the `Print()` function.

2. The body of the function is executed using the values copied to the formal arguments in the first step.
3. The return value of the function is substituted for the function call in the enclosing program.

B.1.21 Scope of Variables

A function's formal arguments are said to have *local scope* because they flow into existence when the function begins to execute and cease to exist when the function is finished. It is also possible to declare other variables within the body of a function. For example, this function defines a *local variable* named `sum`:

```
Integer add(Integer a, Integer b) {//return the sum of a plus b
   Integer sum = a + b;
   Return(sum);
}
```

Like the formal arguments `a` and `b`, the scope of `sum` is local to the function `add()`. They disappear when the function exits. The only thing that persists is the expression in the `Return` statement, which is passed back to the caller of the function.

Local variables can also be declared within compound statements. For example,

```
If (x > 10 And y < 10){
      Integer sum = x + y;
      Print(sum);
}
```

These variables disappear when the compound statement is exited.

Variables declared outside the scope of any function are called *global variables*. They are accessible from the point they are declared until the end of the program. They are said to have *global scope*.

B.1.22 Pass by Value vs. Pass by Reference

Global variables can be accessed directly within functions. For example, this function returns the difference of global variables `x` and `y`.

```
Integer x = 2;                        // x is a global variable
Integer y = 3;                        // y is a global variable
Integer subxy() {Return (x - y);}
```

Referencing global variables directly inside a function is not a recommended practice because it ties the function to particular individual variables, limiting its usefulness.

The reason people are tempted to reference global variables directly inside functions is that ordinarily all that returns from a function is the expression in its `Return()` statement. Sometimes, it's nice to allow a function to have additional side effects. That way functions can affect more than one thing at a time in the program. But there's a better way to accomplish side effects: we can use arguments to pass in a reference to a variable from outside.

As described in the preceding section, ordinarily only the *value* is copied from an actual argument to its corresponding formal argument. But declaring a formal argument to be of type `Reference` causes MUSIMAT to let the function directly manipulate a variable supplied as an actual argument. The function doesn't get the value of the variable, it gets *the variable itself.* When a function changes a `Reference` formal argument, it changes the variable supplied as the actual argument.

We can use `Reference` arguments to allow functions to have multiple effects on the variables in a program. For example, let's declare a function that takes two `Reference` arguments and adds 10 to each of their values.

```
add10(Reference a; Reference b){
  a = a + 10;
  b = b + 10;
}
```

Now let's declare two global variables with initial values:

```
Integer x = 2;
Integer y = 3;
```

Now let's use them as actual arguments to the function and then print their values:

```
add10(x, y);
Print(x);
Print(y);
```

This prints `12` and `13` because the function changed the values of both global variables. This is a very handy trick.

Here are the rules to remember:

- An ordinary (non-`Reference`) formal argument provides its function with a *copy* of its actual argument. Changing the value of an ordinary (non-`Reference`) formal argument inside the function *does not change anything outside the function,* that is, such arguments have *local scope.* The actual arguments are said to be *passed by value* to the formal arguments.
- A `Reference` formal argument provides its function with direct access to the variable named as its actual argument. The actual argument must be a variable. Modifying the value of a `Reference` argument inside a function changes the referenced variable outside the function. Thus, the scope of a `Reference` formal argument is the same as the scope of its actual argument. The actual arguments are said to be *passed by reference* to formal arguments when they are declared to be of type `Reference`.

B.1.23 Type Conversion

We can explicitly convert `Integer` expressions to `Real`, and vice versa. For example:

```
Real a = 10.0/3.0;
Print(a);
```

prints `3.333 . . .`, and

```
Integer b = Integer(a); // convert a to Integer
Print(b);
```

prints `3`.

When assigning `a` to `b`, the `Real` value `a` is converted to an `Integer` by truncating (discarding) the fractional part of `a` (that is, by discarding 0.333…), and the integer residue (3) is assigned to `b`. If we then write

```
Real c =  Real(b);
```

the integer value of `b` (which is 3) is converted to the equivalent `Real` value (3.0), which is stored in `Real` variable `c`.

Converting from `Real` to `Integer`, we have some choices. For example, if

```
Real a = 10.0/3.0;    // Real variable a is set to 3.333 . . .
```

then

```
Real d = Floor(a); // d is set to 3.0
```

sets `d` to `3.0`. The built-in `Floor()` function returns the largest integer less than its `Real` argument. The statement

```
Real x = Ceiling(a);
```

sets `x` to `4` because the built-in `Ceiling()` function returns the smallest integer greater than its argument.

We can round a `Real` to the nearest whole number as follows:

```
Real r = Floor(a + 0.5); // round c to the nearest whole number
```

If `a = 2.4`, then `Floor(a + 0.5)` returns `2.0`. But if `a = 2.5`, `Floor(a + 0.5)` returns `3.0`. `Floor(a + 0.5)` returns `2.0` for any value `a` in the range `2.0` to `2.499...` and returns `3.0` for any value `a` in the range `2.5` to `2.999...`. But we don't have to do rounding ourselves, MUSIMAT has a built-in function:

```
Print(Round(2.49999)); // prints 2.0
```

B.1.24 Recursion

Recursion means referring back to a value we've calculated previously. Consider the factorial operation where 5! means $5 \cdot 4 \cdot 3 \cdot 2 \cdot 1$. We could use a `For` statement to calculate factorials. This function calculates factorials using iteration:

```
Integer factorial(Integer x){
  Integer n = 1;
  For (Integer i = x; i > 1; i = i - 1)
    n = n * i;
  Return(n);
}
```

The statement

```
Print(factorial(5));
```

prints `120`.
We start with `n = 1` and `i = 5`. The `For` loop takes the previous value of `n`, multiplies it by the current value of `i` and reassigns the value to `n`. It then decrements `i` and performs the operation repeatedly so long as `i > 1`.

B.1.25 Recursive Factorial

Here is a more direct approach to computing factorials using recursion. Since $x! = x \cdot (x-1)!$, we can write

```
Integer factorial(Integer x){
  If (x == 1)
    Return(1);
  Else
    Return(x * factorial(x - 1));
}
```

This method has two states. If `x == 1`, we return 1 since 1! is equal to 1. Otherwise, we return `x` multiplied by the factorial of `x - 1`. Consider the statement

```
Print(factorial(5));
```

When the factorial function is called, `x` is assigned the value `5`. Because `5` is not equal to 1, the factorial function evaluates the `Else` statement and calls `factorial(4)`. Because `4` is not equal to 1, the factorial function evaluates the `Else` statement and calls `factorial(3)`, and so on. Eventually, we reach `factorial(1)`, which returns 1, which is multiplied by `2`, then by `3`, then by `4`, and finally by `5`. The top-level `factorial()` function returns the product, `120`, to the `Print()` routine.

B.1.26 Fibonacci Numbers

In the sequence

1, 1, 2, 3, 5, 8, 13, 21, 34, 55, 89, 144, 377, 610, 987, 1597, 2584, . . .

each subsequent term is the sum of its two immediately preceding values. For example, 8 = 5 + 3. This series, invented by Leonardo Pisano Fibonacci (1170–1250), is the solution to a problem he posed in his book *Liber Abaci:* "A certain man put a pair of rabbits in a place surrounded on all sides by a wall. How many pairs of rabbits can be produced from that pair in a year if it is supposed that every month each pair begets a new pair which from the second month on becomes productive?"

Here is an iterative method of computing the Fibonacci sequence:

```
Integer iterFib(Integer n) {
   Integer fn1 = 1;
   Integer fn2 = 1;
   Integer result = 1;
   For (Integer i = 2; i < n; i = i + 1) {
      result = fn1 + fn2;
      fn2 = fn1;
      fn1 = result;
   }
   Return (result);
}
```

Executing this `For` loop,

```
For (Integer i = 1; i < 10; i++)
      Print(iterFib(i));
```

prints the sequence 1, 1, 2, 3, 5, 8, 13, 21, 34. Here is a method that accomplishes the same calculation using recursion:

```
Integer recurFib(Integer n) {
   If (n == 1 Or n == 2)
      Return(1);
   Else
      Return (recurFib(n - 1) + recurFib(n - 2));
}
```

The recursive technique has crisper expressive power than the iterative approach because we see the inner structure of the sequence directly in the method of its construction. However, it is computationally much more expensive, especially for large `n`, because we must call the `recurFib()` method twice at each step, whereas `iterFib()` performs only one addition and minor data shuffling at each step. Here is an example where Knuth's "goodness" criterion depends upon context. If efficiency is paramount, the iterative approach is preferred; recursion is preferable if expressive crispness is most important.

The Fibonacci sequence becomes relevant musically when we examine the ratios of subsequent terms:

$$\frac{1}{1}, \frac{2}{1}, \frac{3}{2}, \frac{5}{3}, \frac{8}{5}, \frac{13}{8}, \ldots$$

The corresponding sequence of quotients is

1, 200, 1.500, 1.670, 1.600, 1.625, 1.619, 1.617, 1.618, . . .

Thus we see that the ratio of adjacent Fibonacci numbers converges rapidly to the value of the *golden mean,* $\Phi = 1.618\ldots$. The Greek letter phi, Φ, is commonly used to stand for the golden mean. This number appears in a wide range of natural designs, including the arrangements of petals in flowers, seed clusters, and pine cones. Studied at least since Euclid wrote his *Elements,* the golden mean has appeared consciously and unconsciously as a central design element in countless musical works (see section 9.16.1).

B.1.27 Other Built-in Functions

MUSIMAT includes standard mathematical functions such as `Sqrt(x) =` $\sqrt{x}$. There are trigonometric functions such as `Sin(x)`, `Cos(x)`, and `Tan(x)`. Arguments to trigonometric functions are in real radian values. Speaking of radian measure, here's an interesting way to compute π to the machine precision of your computer:

```
Constant Real Pi = Atan(1.0) * 4.0; // arctangent of 1 times 4 equals Pi
```

The function `Abs(x)` returns the absolute value of its argument. It works for either `Real` or `Integer` expressions. For instance, both of the following statements will print `True`:

```
If (Abs(-5) == Abs(5)) Print(True); Else Print(False); // Integer Abs( )
If (Abs(-5.0) == Abs(5.0)) Print(True); Else Print(False); // Real Abs( )
```

With no arguments, the built-in function `Random()` returns a random value between 0.0 and 1.0, but if `Random()` is given arguments specifying `Real` lower and upper bounds, it returns a `Real` random value between those boundaries. For example,

```
Real x = Random(0.0, 11.0);
```

returns a random `Real` value in the range $0.0 \le x < 11.0$. Note the range is from 0.0 to almost 11.0.

If `Random()` is given arguments specifying `Integer` lower and upper bounds, it returns an `Integer` random value between those boundaries. For example,

```
Integer x = Random(0, 11);
```

returns a random `Integer` value in the range $0 \leq x \leq 11$. Note the range is inclusive from 0 to 11.

B.1.28 Comments

It is always helpful to readers if programmers insert comments into their programs. In MUSIMAT, any text beginning with two slashes // out to the end of the line is commentary. For example:

```
x = a + b; // this text is commentary
```

Sometimes it's useful to be able to put a comment anywhere, even in the middle of an expression. All commentary between `/*` and `*/` is ignored.

```
x = y /* this commentary is ignored by MUSIMAT */ + z;
```

When the expression is evaluated, all commentary is ignored, so the resulting expression is `x = y + z;`. Commentary between `/*` and `*/` can extend over multiple lines of text, as necessary.

B.1.29 Representing Text

In order to print text, we use a data type called `Character`, which consists of the letters of the Roman alphabet, digits from 0 to 9, and some nonprinting characters like tab, white space, and punctuation. Characters are written in single quotes: `'a'`, `'B'`, `'c'`, and so on. Punctuation marks include `' '` (blank), `','` (comma), `';'` (semicolon), and `'.'` (period). We can spell words and sentences by making lists of characters, for example `{'G', 'u', 'i', 'd', 'o'}`, but this would be excessively tedious. A shortcut for lists of characters is another data type called `String`. For example,

```
String c = "Ut queant laxis resonare";
```

This string is equivalent to, and much simpler than, assembling a list of characters.

Computers operate with binary numbers, not alphabetic letters. So we must associate each character we want to display with a unique binary number. The computer operates only on the binary numeric values; the display screen connected to the computer knows how to convert binary numeric values to the corresponding characters for display.

We need a table listing the association between particular binary values and the corresponding printed characters. This table is called a *character set*. When a key is pressed on a computer keyboard, the keyboard looks up the corresponding binary number in the character set and sends it to the computer. The computer forwards the number to the display screen, which also uses the character set to determine which character to display. Only the keyboard and the screen use the character set; the computer just stores the corresponding binary numbers.

International standard ISO-10646 defines a Universal Character Set, commonly called *Unicode*. To keep things simple, MUSIMAT uses a common subset of Unicode called ASCII (see section B.2). The built-in `Character()` function takes an ASCII character code as its argument and returns the corresponding printable `Character`.

```
Print(Character(65));
```

prints the character `'A'`. The `Integer()` function can take a printable `Character` as its argument and return the corresponding ASCII character code. For example:

```
Print(Integer('A'));
```

prints `65`.

B.2 Music Datatypes in MUSIMAT

This section describes the design of music data types available in MUSIMAT for representing pitch, rhythm, duration, frequency, and loudness.

B.2.1 Pitch

We would ideally like to have a uniform way to represent all pitch systems discussed in chapter 3. It would be convenient if we could do arithmetic on pitches, for example, to find the size of an interval by subtracting two pitches, to calculate the frequency of a pitch, or to get the pitch of a frequency.

Solving the simplest problem first, I designed a data type for the equal-tempered scale using the piano keyboard. This can be generalized to other scales. The gamut of a standard piano keyboard is 88 keys, indexed from 0 to 87, lowest to highest. We start by associating each key number with a name. The lowest pitch on standard pianos is `A0`, corresponding to key 0, and the highest pitch is `C8`, corresponding to key 87. Interval size in degrees is the difference between key indexes. For example, `C4` is key 48 and `F4` is key 53, so the interval `C4` – `F4` corresponds to five semitones, which is the diatonic interval of a fourth.

MUSIMAT comes with a built-in data type called `Pitch`. By default, it assumes 12 degrees per octave, but the degrees can correspond to any frequencies, so for example, it can be used directly to create any dodecaphonic scale. It also can be adjusted to handle scales with other than 12 degrees per octave.

By default, the `Pitch` data type emulates common musical notation conventions regarding scale degrees, interval sizes, and transposition. For example, the pitch `As4` (pitch class A♯ in the fourth piano octave) is defined as

```
Pitch As4 = Pitch(9,1,4);
```

The first number, 9, represents the diatonic degree as the number of semitones above C. Diatonic pitch A is the ninth semitone above C (see figure B.1). The second number, 1, indicates the accidental. In this case, the A is sharped (raised by a semitone). The chromatic scale degree is obtained by adding the diatonic scale degree, 9, and the accidental, 1, which for A♯ yields 10 (see figure B.1). The third number, 4, indicates the octave on the standard piano keyboard.

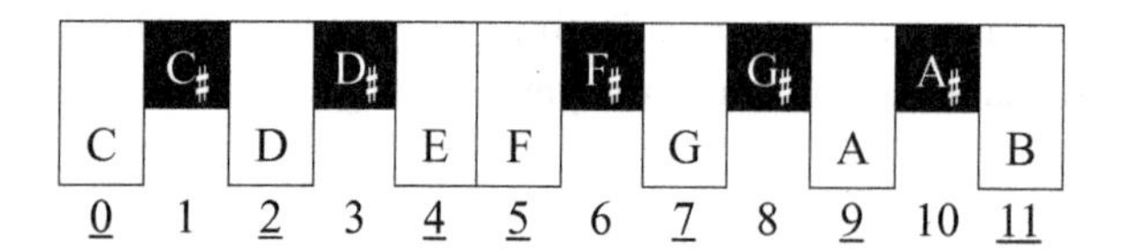

Figure B.1
Diatonic degrees expressed as chromatic pitch classes.

The chromatic degrees from `A0` to `C8` are predefined in MUSIMAT in both flats and sharps. Since `As4` and `Bb4` represent the same chromatic degree, the statement

```
Print(Bb4 == As4);
```

prints `True`. In general, `Pitch` is defined by the triple (*pitch-class, accidental, octave*), where *pitch-class* is an integer from 0 to *N*, and *N* is the number of degrees in an octave.

In defining the pitch A♯4, the triple `(9, 1, 4)` is assigned to the variable `As4`. Variable `As4` contains these three values as one compound entity. This compound value can be passed from one `Pitch` variable to the next. For example, the statements

```
Pitch x = As4; // assign As4 to x
Print(x==As4);
```

print `True`. Arithmetic can be performed on pitches to sharp or flat them. For example, `Print(A4 + 1)` prints `As4`, and `Print(A4 - 3)` prints `Gb4`. Similarly, `Print(A4 * 3)` prints `C12`, and `Print(A4/3)` prints `E1`.

Each element of a `Pitch` can be accessed using these built-in functions:

`PitchClass(Pitch p)`	Returns the diatonic pitch class. For example, if `p` is `As4`, 9 is returned (see figure B.1).
`Accidental(Pitch p)`	Returns the accidental as an integer, where `0` is natural, negative values are increasingly flat, and positive values are increasingly sharp. For example, if `p` is `As4`, 1 is returned.
`Octave(Pitch p)`	Returns the octave on the piano keyboard. For example, if `p` is `As4`, 4 is returned.

These elements can be used to determine the piano key index corresponding to a particular pitch:

```
Integer key(Pitch p) {
  Integer pc = PitchClass(p);          // from 0 .. 11
  Integer acc = Accidental(p);         // -1=flat, 0=natural, 1=sharp
  Integer oct = Octave(p);             // from 0 .. 8
  Return((pc + acc) + 12 * (oct - 1) + 3);// combine
}
```

A way to think about the expression in the `Return()` statement is as follows. Say we want to find the piano key index for `A0`. We know it's the bottom note on the piano, so it should return an index value of `0`. The triple of `A0` is `(9, 0, 0)`. The expression in the `Return()` statement equals `0` for this triple. Similarly, the triple of `A4` is `(9, 0, 4)`, and its corresponding key index is `48`.

Equal-Tempered Frequency Pitch provides a representation of scale degrees and does not denote frequency. We can convert to frequency using any scale system we like, beginning with the equal-tempered scale. We can compute the equal-tempered frequency of a `Pitch`, assuming a reference such as `A4` equals 440 Hz, by adapting equation (3.3), $f_{k,v} = f_R \cdot 2^{v+k/12}$, to compute hertz values from chromatic scale degrees:

```
Real pitchToHz(Pitch p){
  Real R = 440.0;                              // reference frequency
  Real key = PitchClass(p) + Accidental(p);  // get key index
  Real oct = Octave(p);                        // get octave
  Return(R * Pow(2.0, (oct - 4) + (key - 9) / 12.0));
                                               // return frequency
}
```

A way to think about the expression in the `Return()` statement is as follows. The reference pitch is 440 Hz, corresponding to `A4`. So we want the value returned from this function to equal 440.0 when `p` is `A4`. The triple for `A4` is `(9, 0, 4)`, so when `pitchToHz()` is called with `A4`, we want to evaluate $f_R \cdot 2^0$, which can be achieved by subtracting 9 from the pitch and 4 from the octave. Then, executing

```
Print(pitchToHz(A4));
```

prints `440.0`, and substituting any other pitch, regardless of how it is spelled, will produce its proper hertz value. For example, `A0` is 27.5 Hz, `C4` is 261.63 Hz, and `C8` is 4186.01 Hz.

What if we have a frequency x in hertz and want to find its corresponding pitch? The problem is that x may lie in between the pitches of the scale because x can be any frequency. One approach is to compare x to each semitone on the keyboard from lowest frequency to highest, and to stop when the keyboard frequency exceeds x. Then the key one semitone below is the closest corresponding pitch on the keyboard.

```
Pitch hzToPitch(Real x) {              // find pitch closest to x Hz
  For(Integer k = 9+1; k < 88+9; k = k + 1) {// test from As0 to C8
    Pitch p = Pitch(k);                // get pitch of k
    Real f = pitchToHz(p);             // get frequency of p
    if (f > x)                         // have we passed our target?
      Return(p - 1);                   // return previous pitch
  }
```

```
  // If we get here, the Hz value of x is beyond the end of the keyboard
  Return(C8);                          // out of range, clip at C8
}
```

This code returns `A0` if `x` is lower than or equal to `A0`, and it returns `C8` if `x` is greater than or equal to `C8`.

Lists of Pitches We can collect pitches into lists:

```
PitchList shave(C5, G4, G4, Ab4, G4, B4, C5); // shave and a haircut, 2 bits
```

We can do arithmetic on all the pitches in a list. To transpose this pitch list up a whole step,

```
Print( shave = shave + 2 );
```

adds two degrees to every pitch in `shave`, and prints `{D5, A4, A4, As4, A4, Cs5, D5}`. To transpose by an octave,

```
Print( shave = shave * 2 );
```

multiplies every pitch in the list by 2 and prints `{D6, A5, A5, As5, A5, Cs6, D6}`.

Pythagorean Chromatic Scale We can compute the frequency of a `Pitch` in Pythagorean chromatic tuning, assuming a reference such as `A4` equals 440 Hz. We start by computing the frequency of Pythagorean middle C from the reference frequency, using equation (3.11). We define the reference frequencies in MUSIMAT as follows:

```
Real R = 440.0;
Real cPi4 = R * 16.0/27.0; // Pythagorean middle C, 260.74 Hz
```

Next, referring to figure 3.7, we tabulate the ratios of the Pythagorean chromatic scale in MUSIMAT using a `RealList`:

```
RealList pythagoreanChromatic(
  1.0/1.0, 256.0/243.0, 9.0/8.0, 32.0/27.0,
  81.0/64.0, 4.0/3.0, 1024.0/729.0, 3.0/2.0,
  128.0/81.0, 27.0/16.0, 16.0/9.0, 243.0/128.0
  );
```

Last, we define a variation of the `pitchToHz()` function. This version has the same name but takes three arguments instead of one.[4] When supplied with a certain `Pitch p`, it returns the frequency corresponding to its Pythagorean intonation as a `Real` value in hertz.

```
Real pitchToHz(
        Pitch p,                       // pitch
        Real refC,                     // reference frequency
```

```
        RealList scale              // ratios of scale degrees
  ) {
  Integer key = PitchClass(p) + Accidental(p); // get key from pitch
  Real oct = Octave(p);             // get octave from pitch
  Return(refC * scale[key] * Pow(2.0, (oct - 4)));// compute frequency
}
```

The `Return()` statement calculates the frequency of the key from the reference frequency times the ratio for that degree, then adjusts it for the proper octave. Calling

```
Print("A4=",  PitchToHz(A4 , cPi4, pythagoreanChromatic));
```

prints `A4=440.0`, and

```
Print("C4=",  PitchToHz(C4 , cPi4, pyhagoreanChromatic));
```

prints `C4=260.74`, as expected.

Natural Chromatic Scale To create the natural chromatic scale, all we need now is to establish the frequency reference for natural chromatic middle C and tabulate the ratios of the scale.

```
Real  R = 440.0;
Real  cNat4 = R * 3.0/5.0; //264.00 Hz

RealList naturalChromatic(
  1.0/1.0, 16.0/15.0, 9.0/8.0, 6.0/5.0,
  5.0/4.0, 4.0/3.0, 64.0/45.0, 3.0/2.0,
  8.0/5.0, 5.0/3.0, 16.0/9.0, 15.0/8.0
);
```

Then

```
Print("A4=",  PitchToHz(A4 , cNat4, naturalChromatic));
```

prints `A4=440.0`, and

```
Print("C4=",  PitchToHz(C4 , cNat4, naturalChromatic));
```

prints `C4=264.00`.

Sruti Scale As a final example, we adapt `Pitch` to handle nondodecaphonic scales by demonstrating the sruti scale (see figure 3.25). There are 22 degrees in this scale. We start by defining the ratios of the sruti scale:

```
RealList srutiScale(
  1.0/1.0, 256.0/243.0, 16.0/15.0, 10.0/9.0, 9.0/8.0, 32.0/27.0, 6.0/5.0,
  5.0/4.0, 81.0/64.0, 4.0/3.0, 27.0/20.0,45.0/32.0, 729.0/512.0,
```

```
    3.0/2.0, 128.0/81.0, 8.0/5.0, 5.0/3.0, 27.0/16.0, 16.0/9.0, 9.0/5.0,
    15.0/8.0, 243.0/128.0
);
```

We want to preserve the reference A440 Hz and use it to find the frequency of the lowest degree of the scale, as we've done for Pythagorean and natural scales. But which of the 22 degrees should correspond to A440? The sruti scale contains both the Pythagorean major sixth (27/16) and the natural major sixth (5/3). Let's choose the simpler 5/3 ratio at degree 17 to correspond to A440. Then the lowest degree of the sruti scale has the same frequency as the natural chromatic middle C, 264.0 Hz.

```
Real   R = 440.0;
Real   srutiRef = R * 3.0/5.0; // 264.00 Hz
```

Next, we must inform `Pitch` of how many degrees there are per octave, which we can do by finding the length of the list of ratios:

```
SetDegrees(Length(srutiScale)); // set number of degrees in scale
```

The built-in `SetDegrees( )` function adjusts the internal calculations of `Pitch` to the specified number of degrees in the scale. To keep things simple, the degrees of the sruti scale are indicated only by their degree numbers, rather than by trying to extend the Western pitch-naming system. Then the frequencies of particular sruti degrees are computed as follows:

```
For (Integer i = 0; i < Length(srutiScale); i = i + 1) {
    Pitch x( i, 0, 4 ); // pitch, accidental, octave
    Real f = pitchToHz(x, srutiRef, sruti);
    Print(f);
}
```

which prints the frequencies of the sruti scale from middle C as follows:

1	2	3	4	5	6	7	8	9	10	11
264.00	278.12	281.60	293.33	297.00	312.89	316.80	330.00	334.13	352.00	356.40
12	13	14	15	16	17	18	19	20	21	22
371.25	375.89	396.00	417.19	422.40	440.00	445.50	469.33	475.20	495.00	501.19

Other scales, such as Partch's scale and the quarter-tone scale, can be constructed in the same manner. The Bohlen-Pierce scale can also be constructed this way because the `SetDegrees()` function only specifies the number of degrees in the scale and makes no assumptions about octave equivalence.

B.2.2 Rhythm

Duration in common music notation is expressed as a fraction of a whole note. For example a whole note equals four quarter notes:

𝅝 = ♩ + ♩ + ♩ + ♩

We could write this mathematically as follows:

$$\frac{1}{1} = \frac{1}{4} + \frac{1}{4} + \frac{1}{4} + \frac{1}{4},$$

which suggests using rational fractions to represent rhythmic durations. A rational fraction is a ratio of integers. MUSIMAT comes with a built-in data type called `Rhythm`, which emulates common musical notation conventions regarding rhythm. For example, the quarter note is defined as

```
Rhythm Q = Rhythm( 1, 4 );
```

The first number is the numerator of the rational fraction, the second is the denominator. Note that we can't write `Rhythm(1/4)` because the integer quotient of 1/4 is 0 with a remainder of 1; integer division is performed if both the numerator and denominator are integers, which won't work here. Specifying the numerator and denominator separately avoids this problem and has some other numerical advantages as well. Executing `Print(Q);` prints `(1, 4)`. Internally, `Rhythm()` keeps the integer numerator and denominator values separately.

Rhythmic duration can also be given as a real expression. `Print(Rhythm(0.5));` prints `(1, 2)`. How does `Rhythm()` convert this real expression into a ratio of integers? It does so by calling the following function internally:

```
realToRational(Real f, Integer Reference num, Integer Reference den) {
  Constant Integer iterations = 3000000;
  Constant Real limit = 0.000000000001;
  num = den = 1; // start off with ratio of 1/1
  For (Integer i = 0; i < iterations; i = i + 1) {
    If (RealAbs( Real(num) / Real(den) - f ) < limit)
      Return; // we have reached the limit
    Else {
      Real x = RealAbs(Real(num+1) / Real(den) - f);
      Real y = RealAbs(Real(num) / Real(den+1) - f);
      If (x < y)
        num = num + 1;
      Else
        den = den + 1;
```

```
    }
  }
  Return; // if we get here, we've not converged on the limit
} // RealAbs() is just a version of Abs() that uses Real arithmetic
```

Function `realToRational()` takes a `Real` value `f` and attempts to find a rational fraction `num/den` that is as close as possible to it. It starts by setting `num = den = 1` and asking whether `num/den` is already close enough to `f`. If so, it returns. Otherwise, it asks whether `(num+1) / den` is closer than `num / (den+1)`. If so, it increments `num`; otherwise it increments `den` and repeats the process. Because `num` and `den` are `Reference` arguments, any changes to these variables within `realToRational()` are reflected in the value of the actual arguments supplied to it.

This method can be used to find rational approximations to most any real value. For example,

```
Real Pi = 3.14159265;
Print(Rhythm(Pi));
```

prints `(1953857, 621932)`. Note that `1953857/621932 = 3.14159265`, which is pretty close to the value of π. This method is limited by the precision of the computer hardware. The precision of a rational approximation depends upon the value of the built-in variables `iterations` and `limit`. For example, with the values shown in the preceding code, it took `realToRational()` 2,575,787 trials to come up with its best approximation of π, requiring about 1 second on my computer. The `iteration` and `limit` parameters can be set to whatever values produce the optimal performance/accuracy cost/benefit ratio. Barring obscure rhythms (nothing, say, beyond triplet eights), `iterations = 240` and `limit = 1.0/480` should be satisfactory.

Although the details go beyond the scope of this book,[5] here is a sketch of how `Rhythm()` uses `realToRational()`:

```
Rhythm(Real x) {
  Integer num, den; // internal parameters for Rhythm
  realToRational( x, num, den ); // convert x to num / den rational fraction
  // . . .
}
```

When called with a `Real` argument, `Rhythm()` calls `realToRational()` to set its internal integer rational fraction values.

MUSIMAT provides built-in definitions for standard binary divisions of a whole note:

```
Constant Rhythm W = Rhythm(1.0/1.0);
Constant Rhythm H = Rhythm(1.0/2.0);
Constant Rhythm Q = Rhythm(1.0/4.0);
Constant Rhythm E = Rhythm(1.0/8.0);
Constant Rhythm S = Rhythm(1.0/16.0);
```

It is easy to define ternary divisions as well. For example, a triplet eighth is `Rhythm(1.0/12.0)` because there are $3 \cdot 4 = 12$ triplet eighths per whole note. By the same reasoning, a quintuplet eighth is `Rhythm(1.0/20.0)`.

We can express compound rhythms by addition. For example, a dotted half note is

```
Real Hd = Rhythm(1.0/2.0 + 1.0/4.0); // dotted half
```

Equivalently,

```
Real Hd = Rhythm(3.0 / 4.0); // also a dotted half
```

We can also do arithmetic directly with rhythms. For example, `Print(E+S)` prints `(3, 16)`. Also, `Print(W - S)` prints `(15,16)`, `Print(Q * S)` prints `(1,64)`, and `Print(Q / S)` prints `(4,1)`. The last value corresponds to a duration of four whole notes.

We can extract the numerator and denominator from `Rhythm()`:

```
Integer num, den;
Rhythm(E+S, num, den); // assigns rational fraction for E+S to num and den
```

Used this way, `Rhythm()` calculates the rational fraction of its first argument and sets `num` and `den` by reference to the result. For the preceding example, `num` is set to `3` and `den` is set to `16`. We can leverage this capability to obtain the duration of a rhythm as a real number:

```
Real realRhythm(Rhythm x) {
  Integer num, den;
  Rhythm(x, num, den); // find rational fraction for x and set num and den
  Return(Real(num)/Real(den)); // convert num and den to reals and divide
}
```

Then, for example, executing `Print(realRhythm(E + S));` prints `0.1875`.

As with pitches, we can make lists of rhythms.

```
RhythmList R = {Q, E, E, E, S, S, Q};
Print(R);
```

prints `{(1,4), (1,8), (1,8), (1,8), (1,16), (1,16), (1,4)}`.

B.2.3 Tempo

In common music notation, tempo is expressed using Mälzel's metronome markings (see section 2.6.2). For example, ♩= 60MM indicates that the beat or pulse of the music is associated with quarter notes and that there are 60 beats per minute. Thus at ♩= 60MM each quarter note lasts 1 second, and at ♩= 120MM each quarter note lasts 0.5 second. Thus tempo scales the durations of rhythms.

We can emulate this by calculating a tempo factor based on Mälzel's metronome markings. Rhythms are then multiplied by this coefficient to determine their actual duration. First, we need a function that calculates the tempo factor:

```
Real mm(Real beats, Real perMinute) {
  Return(1.0 / (4.0 * beats) * 60.0 / perMinute);
}
```

The `beats` argument is the rhythmic value that gets the beat, and the `perMinute` argument is the number of beats per minute. For example,

```
Real tempoScale = mm( Q, 60.0 ); // 60 quarternotes per minute
```

sets `tempoScale` to `1.0`, and

```
Real tempoScale = mm(Q, 120.0); // 120 quarternotes per minute
```

sets `tempoScale` to `0.5`. Scaling a list of rhythms with `tempoScale` adjusts them to the prevailing tempo. Start with a rhythm list.

```
RhythmList T = {Q, E, E, E, S, S, Q};
Print(T);
```

prints `{(1,4), (1,8), (1,8), (1,8), (1,16), (1,16), (1,4)}`. Now scale it.

```
RhythmList S = T * tempoScale; // tempoScale == 0.5
Print(S);
```

prints `{(1,8), (1,16), (1,16), (1,16), (1,32), (1,32), (1,8)}`.

Though this explicit approach to managing tempo works fine, in fact `Rhythm()` has this calculation conveniently built in. It works in conjunction with a built-in function named `SetTempo()` that implicitly scales all rhythmic durations by the specified tempo factor. So, for example, given the preceding definition of `RhythmList T`,

```
SetTempo(mm(Q, 90)); // set tempo to 90 quarternotes per minute
Print(T);
```

prints `{(1,6), (1,12), (1,12), (1,12), (1,24), (1,24), (1,6)}`. All rhythmic values are scaled implicitly by `Rhythm()`.

B.2.4 Loudness

Loudness is expressed in common music notation using performance indications such as *fortissimo* or *piano* (see section 2.7). But the performed intensity depends upon the acoustical power of the instrument and the interpretation of the performer. A better approach for the purpose here would be to define loudness in objective terms using decibels (see section 5.5.1).

Since microphones and loudspeakers measure and reproduce pressure waves, it is common to use dB SPL in audio work (see equation 5.32). It is also conventional in audio to take the loudest value that can be reproduced without distortion as a reference intensity of 0 dB (see section 4.24.2). Since measured intensities will be less intense than the reference, then by the definition of the decibel they will be expressed as negative decibel levels. We can write, for example, –6 dB to indicate an amplitude that is (very close to) one half of the amplitude of the 0 dB reference.

Restating (5.32), the equation for dB SPL, as

$$y \text{ dB} = 20 \log_{10} \frac{A'}{A}$$

and simplifying by letting $x = A'/A$, we have y dB $= 20 \log_{10}(x)$. Solving for x, we have

$$x = 10^{y/20}. \tag{B.1}$$

For example, setting $y = -6$ dB, we have $x = 10^{-6/20} = 0.501$. The value of x is the coefficient by which a signal must be multiplied to lower its amplitude by 6 dB. For another example, setting $y = 0$ dB, we have $x = 10^{0/20} = 1$. So multplying by 0 dB does not affect amplitude. Setting $y = -120$ dB, we have $x = 10^{-120/20} = 0.000001$, so multiplying a signal by –120 dB renders it virtually inaudible. Finally, if we wish to amplify a soft sound, scaling it by +6 dB makes it twice as loud. Thus, scaling sounds with decibel coefficients allows us to achieve arbitrary loudness levels for waveforms. So we define

```
Real dB(Real y){
  Return(Pow(10.0, y/20.0));
}
```

For example, `Print(dB(-6))` prints `0.501187`, `Print(dB(0))` prints `1.0`, and `Print(dB(-120))` prints `0.000001`.

Suppose we have the following audio samples for a sound:

```
RealList mySound = {0, 0.16, 0.192, -0.37, -0.45, -0.245, -0.43, 0.09, . . .};
```

We wish to halve the sound's amplitude. Then

```
RealList scaledSound = mySound * dB(-6);
Print(scaledSound);
```

prints `{0.02, 0.08, 0.10, -0.19, -0.23, -0.12, -0.22, 0.05, . . .}`.

See volume 2, chapter 1, for more about sampled signals.

MUSIMAT provides built-in definitions for standard music dynamics levels based on figure 4.7.

```
Real ffff = dB(0), fff = dB(-10), ff = dB(-18), f = dB(-24),
  mf = dB(-32), mp = dB(-40), p = dB(-48), pp = dB(-56),
  ppp = dB(-64);
```

Table B.1
ASCII Character Codes

	0	1	2	3	4	5	6	7	8	9	10	11	12	13	14	15
0	NUL	SOH	STX	ETX	EOT	ENQ	ACK	BEL	BS	HT	LF	VT	FF	CR	SO	SI
1	DLE	DC1	DC2	DC3	DC4	NAK	SYN	ETB	CAN	EM	SUB	ESC	FS	GS	RS	US
2	SP	!	"	#	$	%	&	'	(	)	*	+	,	-	.	/
3	0	1	2	3	4	5	6	7	8	9	:	;	<	=	>	?
4	@	A	B	C	D	E	F	G	H	I	J	K	L	M	N	O
5	P	Q	R	S	T	U	V	W	X	Y	Z	[	\	]	^	_
6	`	a	b	c	d	e	f	g	h	i	j	k	l	m	n	o
7	p	q	r	s	t	u	v	w	x	y	z	{	|	}	~	DEL

Thus `ffff` does not change the amplitude of the signal, but all others attenuate it to varying degrees.

B.3 Unicode (ASCII) Character Codes

The Universal Character Set, or Unicode, encodes virtually all of the world's characters and even leaves room for characters not yet invented. A common subset of Unicode is ASCII (American Standard Code for Information Interchange), which was proposed by ANSI in 1963 and adopted in 1968. Recent standards that refer to ASCII include ISO-14962-1997 and ANSI-X3.4-1986 (R1997). The ASCII code includes many punctuation marks and white space such as blank, tab, and newline (which forces subsequent text onto a new line).

To obtain the integer ASCII number corresponding to a character, first find the row r and column c containing the character in table B.1. The ASCII number of this character is $2^r + c$. For example, the character 'A' corresponds to $2^4 + 1 = 33$.

The characters between 0 and 31 and DEL are reserved for functions that mostly don't concern computer users, except for CR (carriage return) and LF (line feed). SP stands for the space character ' '. This is another one of those tables that you must learn if you expect your geek friends to take you seriously, so place a copy of table B.1 at your bedside or above the mantelpiece, where you can refer to it frequently.

B.4 Operator Associativity and Precedence in MUSIMAT

To keep it simple, the MUSIMAT expressions in this book are formatted to obey simple left-to-right evaluation. In fact, the rules are a little more complex because MUSIMAT is basically C++ in sheep's clothing.

Table B.2
Operator Precedence and Associativity

Operator	Associativity	Description	Examples
( )	left to right	grouping	`a * (x+y) == ax + ay`
–	right to left	negation	`-3 == -1 * 3`
* /	left to right	multiplication and division	`a * b, a / b`
%	left to right	remainder after integer division	`10 % 3 == 1, 12 % 3 == 0`
+ –	left to right	addition and subtraction	`a + b, a - b`
< <= > >=	left to right	less-than, less-than-or-equal, greater-than, greater-than-or-equal	`a < b, a <= b` `a > b, a >= b`
== !=	left to right	equal, not equal	`a == b, a != b`

And	left to right	logical AND	`False And False == False`
			`False And True == False`
			`True And False == False`
			`True And True == True`

Or	left to right	logical OR	`False Or False == False`
			`False Or True == True`
			`True Or False == True`
			`True Or True == True`

=	right to left	assignment	`a = b, a = b + c`

Associativity of operators is generally left to right, except for assignment and negation. For example the expression `a = c = d` assigns the value `d` to `c`, then assigns `c` to `a`, thereby making all three have equal value.

Table B.2 shows MUSIMAT'S simplified operator precedence and associativity in order from highest to lowest. This precedence list is a shortened version derived from C and C++. Since you can't effectively read or write computer programs unless you have memorized these rules of operator precedence and associativity, experts recommend that you study these tables while you brush your teeth every night (Press et al. 1988, 23).

Warning: some expressions that might seem to have self-evident meaning can't be expressed as such in C/C++ and so don't work in MUSIMAT either. Take the expression `c > b > a`, for example. You'd hope it would test whether `b` lies between `a` and `c`. Alas. Consider this example:

```
If (3 > 2 > 1) Print("true") Else Print("false")
```

It first evaluates (3 > 2), which it discovers is `True`, and replaces this expression with the integer 1 (which stands for `True` in C++). It then evaluates the expression (1 > 1) which is `False`. Probably not what we wanted. This example can be rewritten as follows:

```
If (3 > 2 And 2 > 1) Print("true") Else Print("false")
```

which will print `True`.

Glossary

A440 The standard of pitch for Western orchestras, corresponding to 440 Hz.

Acoustics The study of signals and signaling systems where the medium is air.

ADSR Segments of the amplitude envelope named for the initial letters of each segment: attack, decay, sustain, and release.

Amplitude Distance of a wave from its peak height to its point of zero displacement or equilibrium. Also called peak amplitude. Peak-to-peak amplitude is the distance from crest to trough. RMS amplitude is the average energy of a sinusoid, based on its amplitude.

Anechoic chamber A room that is so padded that it produces no echoes, thereby eliminating reverberation; usually also isolated from external noise sources.

Antinode Point where displacement due to vibration is greatest.

Atmosphere Average atmospheric pressure at sea level, with a standardized value of 101,325 Pa.

Band A range of frequencies within a spectrum.

Band center Geometric mean frequency of a band. For a band extending from 707 Hz to 1.414 kHz, the band center frequency is 1000 Hz.

Bandwidth Distance between upper and lower frequency limits of a sound.

Beat Fundamental unit of time measurement, corresponding to the pulse of the music.

Causal System that references only current and past input and past output. Causal systems may not reference future input or current or future output.

Chaotic system A deterministic system that appears to be random such that it is impossible to make long-range predictions about its behavior.

Complex system System that contains elements that are both differentiated (specialized or compartmentalized) and integrated (connected or unified) on all levels of scale.

Compliance The reciprocal of stiffness.

Continuous distribution A distribution where the events in the sample space cannot be individually distinguished. Temperature and frequency are examples of continuous distributions.

Critical bands Channels of frequency-selective psychoacoustic processing that affect our perception of pitch, loudness and masking of components lying within a critical frequency distance (roughly 1/3 of an octave) of one another.

Damping The effect of energy dissipation on a vibrating system.

Decibel Scale used to measure sound level in sound recording and communications, based on the same logarithmic principle as the Richter scale.

Degree Individual element of a scale; also, 1/360 of a circular arc.

Degrees An ordered set of names and positions of the elements of a scale.

Deterministic Characteristic of systems where every cause has a unique effect.

Diatonic scale Seven pitches per octave composed of degrees in the order 2 2 1 2 2 2 1.

Discrete distribution A distribution where the events in the sample space can be individually distinguished. Tossing coins or dice and picking a note on a keyboard are examples of discrete distributions.

Driven harmonic oscillator A vibrating driving force coupled to a driven simple harmonic oscillator, such as a spring/mass combination.

Duration In music, the number of beats a note lasts. Generally, the elapsed time of an event.

Dynamic range Range from the softest to the loudest sound.

Dynamics A field of classical mechanics that studies how force affects motion of material bodies through time.

Efficiency The ratio of useful power output to the total power input.

Elasticity That property of a material that allows it to restore itself to its original shape after being distorted (stretched, compressed, twisted, etc.).

Enharmonic equivalents Chromatic degrees that sound the same pitch despite having different symbols.

Enumeration An itemized list of all possible outcomes; the sum total of such outcomes.

Envelope Characteristic way in which the intensity of a note changes through time.

Equal-tempered interval The semitone, one twelfth of the pitch distance of an octave, the twelfth root of 2.

Equilibrium The state of a system when it has no acceleration; the resultant when the sum of all external forces acting on a body is zero and the sum of the momentum of all parts of the system is zero.

Event The outcome of a random process, such as a roll of the dice.

Expectation A prediction based on current and past experiences. *See also* Surprisal.

Formant Group of frequencies of some particular bandwidth that is emphasized by a resonant system.

Frequency Physical measure of vibrations per second.

Fundamental Lowest pitched partial in a tone.

Gamut Entire range of pitches reachable by an instrument or voice.

Harmonics Frequency components of a complex tone that are positive integer multiples (greater than 0) of a fundamental frequency.

Harmony In general, any simultaneous combination of tones. More narrowly, an agreeable (consonant) combination of tones.

Harmony theory The art of organizing multiple concurrent musical lines to reinforce a feeling of harmonic movement and arrival, suspension and resolution.

Heat capacity ratio The ratio of the specific heat of a gas at constant pressure to the specific heat at a constant volume.

Hertz The unit of one cycle per second, abbreviated Hz.

Histogram A table of event occurrences.

Ideal string String that is perfectly flexible, has constant mass per unit length, and is connected to massive nonyielding supports.

In phase The state of multiple objects that vibrate with the same speed and direction.

Inertial reactance The tendency of a mass to resist change in velocity.

Inharmonic partials Components that are not integer multiples of a fundamental.

Interval Difference in pitch between two tones.

Inversion, of an interval Subtracting an interval from an octave produces its inversion. Intervals of a fifth and fourth are each other's inversions.

JND of loudness Amount by which the intensity of a sound must change for the ear to register a difference in loudness.

JND of pitch Amount by which the frequency of a sound must change for the ear to register a difference in pitch.

Just intervals Intervals made from the ratio of small whole numbers.

Key The degree to which a diatonic scale is transposed.

Key signature Association between the key (the chromatic degree that the scale starts on) and the accidentals required for the corresponding diatonic scale.

Limit of hearing The intensity above which sound is registered as (possibly damaging) pain.

Loudness The subjective experience corresponding most closely to sound intensity.

Mass The quantity of matter contained in an object.

Matter Anything that occupies space and exhibits inertia.

Mean free path The average distance a particle can move in a gas without a collision; in acoustics, the average distance a wave front can travel before being reflected at a wall.

Melody Notes played in sequence.

Metronome mark Indication of which duration symbol gets the beat and how many beats there are per minute.

Microtone Scale degree that is less than a semitone in pitch.

Modes Variations of the diatonic scale that preserve interval order but begin from other than degree 1 of the diatonic scale.

Modulation Changing the effective key signature of a musical work through the introduction of accidentals not in the original key signature.

Monte Carlo method Any technique that uses probability to study complex systems.

***n*-limit** The highest prime factor of any interval in a musical scale; used as a measure of scale complexity.

Node A point where displacement due to vibration is zero.

Normal force A force that is perpendicular to surfaces that are in contact.

Note A tone placed in temporal context by an onset time and duration. *See* Tone.

Octave Ratio of 2/1 between frequencies; the musical quality of equivalence.

Octave equivalence The principle that scale degrees perform the same musical function regardless of the octave in which they are played.

Onset The time when a sound begins; the moment stipulated by the score for a note to begin.

Oscillate To move or swing regularly and continuously from side to side.

Overtones Harmonic components in a tone that are pitched higher than the fundamental.

Partials Individual sinusoids that collectively make up an instrumental tone; also called components.

Period One complete movement through all the phases of a periodic vibration; for a sinusoid, one period corresponds to one complete revolution of a circle.

Permutation The number of possible unique orderings.

Phase The fraction of a complete rotation through which an object has advanced; characteristic points, such as peaks, troughs, and zero-crossings reached periodically each time a wave repeats.

Phon A measure of equal loudness. *See* Sone.

Phon scale A loudness scale that identifies equal loudnesses across all perceivable frequencies and intensities.

Pitch Subjective experience corresponding to the frequency of sounds.

Polyphony The art of sounding more than one musical line concurrently.

Precession time The period required for a higher-frequency vibration to depart from and then return into alignment with a lower-frequency vibration.

Prime number An integer that is not divisible by any other number besides itself and 1.

Probability The relative likelihood of an event, usually expressed as a real number in the range of 0 to 1.

Probability distribution A function, graph, or listing of the probabilities of the sample space that shows how probability is distributed among the possible events.

Programming language A specialized means of describing rule systems and methods.

Psychophysics Psychology of perception, focusing on the boundary between physical and psychological phenomena. Psychoacoustics includes the psychophysics of audition.

Quality factor The ratio of the resonant frequency to the bandwidth 3 dB down from peak amplitude.

Random variable Index of a probability distribution function.

Resonance The tendency of a system to vibrate sympathetically at a particular frequency in response to energy induced at that frequency.

Resonant frequency The frequency that is most effective at enabling a vibrating system to return to its original energy level by dissipation.

Restoring force Internal force that seeks to return an elastic object to its original shape.

Rhythm That which pertains to the temporal quality of musical notes and phrases. Onset and duration largely determine rhythm.

Rubato Gradual perturbations in the tempo.

Sample space The set of possible outcomes.

Scale A named, ordered set of pitches, together with a formula for specifying their frequencies.

Score Combination of notes ordered vertically by pitch and horizontally by time.

Self-similarity Structures that show similarities at all levels of magnification are self-similar.

Series Summation of a repeating pattern of terms. A particular ordering of a set.

Set An unordered collection of any size.

Set class A named group of sets that are equivalent under specific conditions.

Signal A physically detectable quantity such as an acoustical wave that traverses a signaling system.

Signaling system A system that combines time, space, source, medium, and receiver.

Silence Sensory percept of the absence of detectable sound intensity at any frequency.

Simple harmonic motion Vibratory motion in one dimension caused by the interaction of inertia and elastic forces.

Sone A measure of comparative loudness. *See* Phon.

Sonority The sonic character of a musical interval.

Sound pressure level Average pressure variation per unit area.

Spectrum The range of all possible frequencies at all possible intensities.

Staff Five horizontal lines that serve as a grid indicating pitch range (vertically) and relative note onset (horizontally) in common music notation. Attributed to Guido d'Arezzo.

Standard temperature and pressure (STP) One atmosphere of pressure at 0° Celsius (or 273.15 K).

Standing waves Waves constrained by wavelength to match the dimensions of physical boundaries. Waves whose shape remains constant and only their amplitude changes; waves whose height is scaled through time in the direction perpendicular to their length.

Static equilibrium A system in which the sum of applied forces is zero and does not change through time.

Stiffness The ratio of applied force to the resulting displacement.

Surprisal As the probability of an event decreases from 1.0 towards 0, the surprisal goes from zero to infinity. *See* Expectation.

System A combination of interdependent components that can be viewed as a unified whole. Any function that produces one or more outputs based on zero or more inputs.

Tempering The practice of adjusting some of the degrees of the scale to irrational values so as to fit within an overarching order that is still based on simple integer ratios.

Tempo Number of beats per minute.

Threshold of hearing Minimum amount of sound intensity required for a sinusoid to be detected by a listener in a noiseless environment.

Timbre That which allows us to distinguish notes of equal pitch, loudness, and duration; the name of a sound source (such as trumpet, violin) or a quality of a sound source (such as sharp, dull).

Time signature Stipulation of how many beats there are per measure and which note gets the beat. In 3/4 time, there are three beats per measure (indicated by the numerator) and the quarternote gets the beat (indicated by the denominator).

Tonal palette Coloration based on the placement of various-sized intervals in a scale.

Tone Combination of pitch, loudness, and timbre. An ideal tone has constant pitch, loudness, and timbre; conventionally, the term describes any reasonably uniform combination of the three properties. A sound without discernible pitch (such as a drum beat) is not a tone. When placed in a temporal context, a tone becomes a note. *See* Note.

Tone row A series based on a set of pitch classes.

Transpose To start a scale on any chromatic degree but C.

Uniform distribution If all events in a sample space are equally likely, the resulting distribution is said to be uniform.

Unison 1/1 ratio between frequencies. Tones sounding at the same pitch. The musical quality of *identity.*

Wave An organized traveling disturbance in a medium, such as air.

Wave shape Characteristic internal organization of a sound wave, responsible for determining the timbre, or sound quality of a sound.

Well-tempered Characteristic of tuning systems that temper at least some intervals or have reasonably equal-sized semitones.

Wolf fifth Nonharmonic intervals that cause beating between the interval and the overtone series, making it sound unpleasantly like wolves howling.

Work The force applied to move an object times the distance it is moved.

Notes

Preface

1. From a Chinese fortune cookie opened the night the first page was written.

Chapter 2

1. This is a bit of an oversimplification. Our experience of pitch also depends on loudness, among other factors. For the full story, see section 6.5.1.

2. Curiously, the diatonic major scale begins with the letter C, not A. I've never seen a sensible explanation for this fact.

3. The practice of singing aided by solmization syllables was developed by Guido D'Arrezzo, a Franciscan monk of the tenth century. The practice is called solfeggio.

4. For some transpositions, it may be necessary to raise a note that is already sharp, hence the double sharp; similarly, it is sometimes necessary to lower an already flat tone, hence the double flat.

5. Generally, one must study the harmonic semantics of the score to determine whether the major or minor key is indicated by the key signature.

6. 1862–1918. See for example, Debussy's piano prelude *Voiles.*

7. 1917–1982. Monk used whole-tone scales almost as a signature in many of his jazz compositions.

8. The term *overtone* generates confusion in numbering. Note that the first partial is the fundamental, while the second partial is the first overtone. Thus, for example, overtone number 10 is partial number 11. To avoid confusion, I'll generally avoid the term overtone, preferring partial or component. Since the term partial is primarily an adjective, I'll use it only when I think the context is clear.

Chapter 3

1. Helmholtz (1863); second English edition (1885), 250.

2. In 1995 the paleontologist Ivan Turk of the Slovenian Academy of Sciences discovered what appears to be a fragment of a flute made from a cave bear thigh bone in a Neanderthal archaeological site. It was subsequently radio-carbon-dated to be about 43,000 years old. There is an ongoing controversy over whether it is a flute or not, and if so, what scale it would have played. Whether it is proved or not, it suggests we should consider radically revising backward in time what musicologists refer to as early music.

3. The cent scale was developed by Alexander Ellis, who translated into English Helmholtz's treatise *On the Sensations of Tone* (1863), one of the first scientific studies of consonance.

4. The term *diatonic* originally referred to a scale constructed from two (*dia*) tetrachords. The tetrachord was a scale building block in ancient Greek music theory.

5. Robert Fludd, *History of the Macrocosm and Microcosm* (1617). See Debus (1979) and Godwin (1979).

6. Ptolemy, "Harmonics," in Barker, Stevens, and le Huray (1984), 270–360.

7. It is a descending Syntonic comma because the pure major third is smaller than the Pythagorean major third.

8. A function $f(x)$ is said to be monotonic in x if f always changes in the same direction as x.

9. The equation for the fitted curve is $y = 1.9 + 0.12x + 0.18x^2$.

10. Francesco Antonio Vallotti, *Trattato della Scienza Teorica e Pratica della Moderna Musica.* Conceived in 1728, his ideas weren't published until 1779.

11. J. S. Bach, *The Well-Tempered Clavier,* comprising two books (1722 and 1744), each having 24 sets of preludes and fugues in every major and minor key.

12. Simon Stevin's *Van de Spiegheling der Singconst* (*On the Theory of the Art of Singing*), written ca. 1605, was first published in 1884, 264 years after he died. See also Cohen (1987).

13. Partch, from the liner notes of his RCA phonograph record *Castor and Pollux.*

14. I studied sitar in India with S. Dagar and in the United States with Pandit Nikil Banerjee.

15. Kees van Prooijen apparently also discovered the tempered version of this scale in the 1970s.

Chapter 4

1. But there are some interesting cases where this assumption leads into the weeds (see section 9.17.2).

2. For example, consider this ratio of small but nonzero values: a tenth divided by a billionth. Such a ratio is not a small number.

3. It's important to note that the backward velocity is just the velocity between points A and B; it is not about having a negative slope.

4. This is why "speed kills." Reaction time is constant, but the time required to stop is the square of the speed.

5. Actually, log (2) = 0.30103 . . . , but the fractional part beyond the tenths position is often ignored for practical measurements.

Chapter 5

1. If you are uncomfortable with the radian's being a dimensionless number, you probably will seize upon this definition of the radian as proof that its dimension is in units of degrees. However, the degree is also dimensionless. In fact, all angle measures, including trigonometric functions, are dimensionless. Also note that a radian is only approximately 57.3°.

2. The radian was developed by James Thomson in 1873, a professor of mathematics at Queens College, Belfast, Northern Ireland. His brother was the famous physicist William Thomson, Lord Kelvin.

3. It is customary to use t for linear time and T for periodic time.

4. This is the proof that there is no such thing as centrifugal force. If there were, and it applied a force to the object directly away from the axis of rotation, then the object should fly radially away when released, but it does not. Circular motion is the result of a centripetal force applied at right angles to the instantaneous velocity.

5. For example, simple electrical multimeters use this approach when displaying RMS voltage.

Chapter 6

1. Bregman (1990) gives a monumental description of the factors involved in constructing auditory scenes. Handel (1989) and Yost (2000) provide an easier introduction.

2. The majority of cues we use for source identification lie within this frequency band, suggesting that our hearing may have adaptively evolved to be more sensitive to it.

3. The purpose of the ossicle chain in the middle ear as an impedance matching system was first pointed out by Helmholtz (1863).

4. Also called noise-induced temporary threshold shift (NITTS).

5. These symbols are used because when pronounced, Φ (phi) and Ψ (psi) sound like the initial syllables of the words *physical* and *psychological,* respectively.

6. Blind men, touching various parts of an elephant, report conflicting accounts of their experience depending upon the part they touch, then fall into an argument as to whose account is the correct interpretation. The poem by John Godfrey Saxe (1816–1887) describing this event concludes, "So oft in theologic wars, / The disputants, I ween, Rail on in utter ignorance / Of what each other mean, / And prate about an Elephant / Not one of them has seen!"

7. Ernst Weber (1795–1878).

8. A third important attribute is accuracy, not to be confused with precision. Precision has only to do with the fineness of measure. A ruler with very fine gradations may measure precisely, but if it is warped, it will not measure accurately.

9. Imagine a point light source positioned on the y-axis above the spiral in figure 6.5, shining down through the coils onto the floor.

10. The impossible staircase was invented by the Swedish artist Oscar Reutersvard and later independently reinvented by Lionel Penrose and Roger Penrose. It was made famous in M. C. Escher's print *Ascending and Descending.*

11. In fact, the German organist Georg Andreas Sorge published a description of the same phenomenon in 1744, but Tartini's observation is most frequently cited.

12. This is by no means the only possible or the best definition for these terms, but it will serve for this simplified example.

13. I had the privilege of being one of Grey's subjects.

Chapter 7

1. Since the balls represent packets of air rather than individual molecules, we can ignore the random microscopic motion of the individual molecules.

2. Ludwig Boltzmann, Austrian physicist (1844–1906).

3. At a great distance from a sound's origin, a listener experiences the waves to be plane rather than spherical because the circumference of the wave front is by that time very large in comparison to the local experience of it. However, the total wave is still actually spherical. See section 4.24.4.

4. A good modern treatment of the subject is given in Sharp (1996).

5. Since the bars are shorter, they have less mass, but the elasticity of the wire is the same, so the rate of wave propagation increases.

6. Christiaan Huygens, mathematician, physicist, astronomer, lutanist, and music theorist (1629–1695).

7. Named for the British Astronomer Royal Sir George Bidwell Airy (1801–1892).

8. Kids at home: don't try this!

9. For a dramatic telling of the story, see Bliven (1976).

Chapter 8

1. Robert Hooke, physicist, biologist, astronomer, and architect (1635–1703).

2. This expression means "as long as x is much less than 1." This restriction prevents us from having to consider the nonlinear vibratory behavior of pendulums that can swing more widely.

3. Hermann von Helmholtz, a scientist whose contributions spanned physics, biology, and acoustics (1821–1894). His book *On the Sensations of Tone* is still widely referenced.

4. An explosive and racy Châteauneuf, it fairly burst with game, berry, black chocolate, and espresso characteristics. Ripe and sweet-tasting, it had enough opulent fruit to balance the firm tannin structure, like a rose growing up the impenetrable wall of its spectacular finish. (Kids at home: don't try this.)

5. Even in outer space, the internal friction of the spring would eventually dissipate all of the system's energy, but we ignore this effect as well.

6. Pitch is "head over heels" rotation, yaw is spinning side-to-side rotation, and roll is "over your shoulder" rotation. Define three axes through your center of gravity as follows: x is across your body, y is head-to-toe, and z is front to back. Pitch is rotation in x, yaw is rotation in y, and roll is rotation in z.

7. The classical guitarist Andrés Segovia used no amplification during concerts, even though excellent sound reenforcement was available by the end of his career. But there was no need: his sound adequately reached his thousands of listeners, who listened in a hush. True acoustic performance seems like a lost ideal in today's public concerts.

8. This equation has been attributed to Mersenne (from his "laws of stretched strings" in *Harmonie Universelle*) and to Brook Taylor (1685–1731) in 1714.

9. Named after Thomas Young (1773–1829).

10. Published figures vary from about 69 to 79 for aluminum, so 74 is about in the middle.

11. Though a flute may look like it's closed at one end, the fipple of the flute is effectively an opening, so it is open at both ends.

12. The point 3 dB down from the peak energy point is sometimes called the half-power point, a figure used commonly for this purpose by engineers, because 3 dB is equal to the square root of 2.

Chapter 9

1. Augusta Ada Byron King, Countess of Lovelace, note A, 694, in her notes added to the end of her English translation of Luigi F. Menabrea, *Notions sur la Machine Analytique de M. Charles Babbage,* Bibliothèque Universelle de Genève, 41, 352–376. Her translation was published under the pseudonym AAL in Richard Taylor's *Scientific Memoirs,* 3, art. 29, 666–731, under the title "Sketch of the Analytical Engine invented by Charles Babbage, Esq., by L. F. Menabrea of Turin, officer of the Military Engineers," August 1843.

2. The term algorithm derives from the name of ninth-century Persian mathematician, geographer, and astronomer, Abu Jafar Mohammed ibn Musah al-Khorezmi, inventor of modern decimal positional arithmetic and algebra. Al-Khowarizm means citizen of Khowarizm, known today as Khorezm in Uzbekistan. Algorizm, the precursor to the modern term algorithm, is a transliteration of the last part of his name. His treatise on arithmetic was titled *Kitab al jabr w'al-muqabala,* commonly translated as "Rules of restoration and reduction." The word *al-muqabala* is the origin of the term algebra.

3. Barbara Cook Loy, private communication.

4. The precise relation between rate of change and frequency is developed in volume 2.

5. Bailey and Crandall (2001). Though the expansion of π appears to be random, this has not been proven. Expansion of other irrational numbers, such as e and log 2 might also be random but, again, this has not been proven.

6. Notice that Lorenz conjectures that a butterfly might "set off" rather than "cause" a tornado. This is an important distinction, suggesting that the initial conditions serve to select an outcome from many possibilities.

7. An interesting paradox in mathematics concerns the cardinality of the set of points on a line. Georg Cantor established that C, the cardinality of all real numbers (corresponding to the number of points on a line), is greater than $\aleph_0$, the cardinality of all integers. But how much greater is C than $\aleph_0$? In particular, is there a transfinite number between $\aleph_0$ and C? Cantor's continuum hypothesis states that there is no such transfinite number. However, it has been demonstrated that the validity of the continuum hypothesis is undecidable. Using the standard axioms of set theory, Kurt Gödel showed that the continuum hypothesis is impossible to disprove. Later, Paul Cohen showed that it is impossible to prove under the same conditions. Hence, the continuum hypothesis is independent. The independence of the continuum hypothesis has been taken as an exhibit of Gödel's incompleteness theorem, because it is an important question that has been proven to be undecidable, even though the proofs are based on the standard and universally accepted axioms of mathematics.

8. The midpoint of an 88-key keyboard is between E and F above middle C.

9. The analogy between entropy and information has been criticized by some physicists. There are implications in the equation for entropy that are not matched for information. However, this dispute need not concern us here: the analogy between information and entropy has become a fixture in the literature.

10. After R. V. Hartley, who in 1927 proposed using logarithms to measure information.

11. Aristoxenus, "The Harmonics," in Macran (1902), 27–30.

12. A mathematical construction is called pathological if it is created simply to invalidate an otherwise universally valid assertion.

13. Plato, *Laws,* bk 49, in Pangle (1980).

14. J. S. Bach, *389 Choralgesange für Vierstimmigen Gemischten Chor.* Nr. 3765. Breitkopf Edition.

15. Herbert Bielawa, private communication.

16. For example, "up a third" is from a C chord to an E chord; "down a fifth" is from a G chord to a C chord, and so on.

17. Dolson (1989). I am indebted to this article for its intuitive explanation of back propagation.

18. J. S. Bach, *389 Choralgesange für Vierstimmigen Gemischten Chor.* Nr. 3765. Breitkopf Edition.

19. Haus and Sametti (1991, 7). The multiplicity extension is a partial implementation of self-modifying nets, which were introduced by Valk (1978).

20. J. S. Bach, "Canon Perpetuus" from *A Musical Offering.* BWV 1073. London: Boosey and Hawkes, 1952.

21. See, for example, Harel (1987), an important early theoretical paper. For more recent practical developments, see, for example, Samek (2002).

22. Landon (1976, 508–509). Leopold Mozart quoted Haydn's comment in a letter to his daughter. The encounter transpired after Haydn heard Mozart's Bb Maj. Quartet K456, "The Hunt" in 1785. The phrase "knowledge of composition," *kompositionswissenschaft,* means literally "composition science."

23. Flavius Magnus Aurelius Cassiodorus, Senator (ca. 485–ca. 575), Institutiones, II, iii, paragraph 21, in Strunk (1950).

Appendix B

1. A simple MUSIMAT emulator written in C++ is available at http://www.musimathics.com/.

2. Prior to the 1940s, when someone said "computer," they typically referred to a person who performed computations manually or with the aid of a calculating machine. It was not until the 1950s that "robot brains" began to supplant human computers.

3. We can also exit a `Repeat` statement with a `Return` statement.

4. Functions of the same name that vary in the number or type of arguments or type of return value are said to be polymorphic. MUSIMAT manages to keep the various versions separate from each other and to use the correct one in every instance.

5. MUSIMAT Source code is available at http://www.musimathics .com/.

References

Allen, J. B., and S. T. Neely. 1997. "Modeling the Relation Between the Intensity JND and Loudness for Pure Tones and Wide-Band Noise." *Journal of the Acoustical Society of America* 102 (December): 3628–3646.

Ames, Charles. 1983. "Stylistic Automata in Gradient." *CMJ: Computer Music Journal* 7 (4): 45.

———. 1987. "Automated Composition in Retrospect: 1956–1958." *Leonardo* 20 (2): 169–185.

ANSI (American National Standards Institute). 1999. *Acoustical Terminology.* ANSI S1. 1-1994 (R1999).

Antoni, Giovanni degli, and Goffredo Haus. 1982. "Music and Causality." In *Proceedings of the International Computer Music Conference,* 279.

Apel, Willi. 1944. *Harvard Dictionary of Music.* Cambridge, Mass.: Harvard University Press.

Ashmore, J. F. 1987. "A Fast Motile Response in Guinea-Pig Outer Hair Cells: The Cellular Basis of the Cochlear Amplifier." *Journal of Physiology* 388: 323–347.

Atali, Jacques. 1985. *Noise: The Political Economy of Music.* Minneapolis: University of Minnesota Press.

Backus, John. 1961. "Pseudoscience in Music." *Journal of Music Theory* 55: 220–232.

Bailey, D. H., and R. E. Crandall. 2001. "On the Random Character of Fundamental Constants." *Experimental Mathematics* 10 (June): 175–190.

Barbour, J. Murray. 1947. "Bach and the Art of the Temperament." *Musical Quarterly* 33 (January): 64–89.

———. 1953. *Tuning and Temperament: A Historical Survey.* East Lansing: Michigan State College Press.

Barker, Andrew, John Stevens, and Peter le Huray, eds. 1984. *Greek Musical Writings.* Vol. 2: *Harmonic and Acoustic Theory.* Cambridge: Cambridge University Press.

Barnes, John. 1979. "Bach's Keyboard Temperament: Internal Evidence from the Well-Tempered Clavier." *Early Music* 7 (April): 236–249.

Beckman, Petr. 1976. *A History of Pi.* New York: St. Martin's Press.

Békésy, G. von. 1960. *Experiments in Hearing.* New York: McGraw-Hill.

Benade, Arthur H. 1973a. "The Physics of Brasses." *Scientific American* 229 (July): 24–35.

———. 1973b. *Trumpet Acoustics.* Cleveland: Case Western Reserve University.

Benade, Arthur H., and J. S. Murday. 1967. "Measured End Corrections for Woodwind Tone Holes." *Journal of the Acoustical Society of America* 41: 1609.

Beranek, L. L. 1962. *Music, Acoustics, and Architecture.* New York: Wiley.

———. 1986. *Acoustics.* Rev. ed. Melville, N.Y.: American Institute of Physics.

Bismarck, G. von. 1974. "Sharpness as an Attribute of the Timbre of Steady State Sounds." *Acustica* 30: 159.

Bliven, Bruce, Jr. 1976. "Annals of Architecture—A Better Sound." *New Yorker,* November 8, 51–135.

Bobrow, D. G., and D. A. Norman. 1975. "Some Principles of Memory Schemata." In *Representation and Understanding: Studies in Cognitive Science.* Ed. D. G. Bobrow and A. M. Collins. New York: Academic Press.

Bohlen, Heinz. 1978. "13 Tonstufen in der Duodezime." *Acustica* 39. English trans.: "13 Tone Steps in the Twelfth." *Acustica* 87 (2001, no. 5): 617–624.

Boon, Jean Pierre, and Olivier Decroly. 1995. "Dynamical Systems Theory for Music Dynamics." *Chaos* 5: 501–508.

Bosi, Marina, and Richard E. Goldberg. 2003. *Introduction to Digital Audio Coding Standards*. Dordrecht, The Netherlands: Kluwer.

Bregman, Albert S. 1990. *Auditory Scene Analysis: The Perceptual Organization of Sound.* Cambridge, Mass.: MIT Press.

Brün, Herbert. 1970. "From Musical Ideas to Computers and Back." In *The Computer and Music.* Ed. Harry Lincoln. Ithaca, N.Y.: Cornell University Press.

Buchner, Alexander. 1956. *Mechanical Musical Instruments.* Trans. Iris Urwin. London: Batchworth Press.

Burkert, Walter. 1972. *Lore and Science in Ancient Pythagoreanism.* Trans. Edwin L. Minar Jr. Cambridge, Mass.: Harvard University Press.

Cage, John. 1961. *Silence.* Cambridge, Mass.: MIT Press.

Cohen, Alexander, J. Anticaglia, and H. H. Jones. 1970. "Sociocusis: Hearing Loss from Nonoccupational Noise Exposure." *Sound and Vibration* 4 (November): 12–20.

Cohen, H. F. 1987. "Simon Stevin's Equal Division of the Octave." *Annals of Science* 44 (5): 471–488.

Cope, David. 1996. *Experiments in Musical Intelligence.* Middleton, Wisc.: A-R Editions.

———. 1999. *Virtual Mozart.* CRC 2452. Baton Rouge, La.: Centaur Records.

———. 2001. *Virtual Music.* Cambridge, Mass.: MIT Press.

Cowell, Henry. 1930. *New Musical Resources.* New York: Knopf.

Dalmont, J. P., C. J. Nederveen, and N. Joly. 2001. "Radiation Impedance of Tubes with Different Flanges: Numerical and Experimental Investigations." *Journal of Sound and Vibration* 224 (3): 505–534.

Debus, Allen G. 1979. *Robert Fludd and His Philosophical Key.* New York: Neale Watson Academic.

Devlin, Keith. 1994. *Mathematics: The Science of Patterns.* New York: Scientific American Library.

Dolson, Mark. 1989. "Machine Tongues XII: Neural Networks." *CMJ: Computer Music Journal* 13 (3). Also in *Music and Connectionism.* Ed. Peter Todd and Gareth Loy. Cambridge, Mass.: MIT Press, 1991.

Drew, David. 1954/1955. "Messiaen: A Provisional Study. *The Score,* December 1954, 33–49; September 1955, 9–73; December 1955, 41–61.

Ebcioglu, Kemal. 1986. "An Expert System for Harmonizing Chorales in the Style of J. S. Bach." Ph.D. diss., Department of Computer Science, State University of New York, Buffalo, New York. Also in *Understanding Music with AI: Perspectives on Music Cognition,* 294–334. Ed. M. Balaban, K. Ebcioglu, and O. Laske. Menlo Park, Calif.: AAAI Press, 1992.

———. 1988. "An Expert System for Harmonizing Four-Part Chorales." *CMJ: Computer Music Journal* 12 (3): 43–51.

Erickson, R. F. 1975. "The DARMS Project: A Status Report." *Computers and the Humanities* 9 (6): 291–298.

Euler, Leonhard. 1766. "Conjecture sur la raison de quelques dissonances generalement recues dans la musique [Conjecture as to why dissonant tones are generally heard in music]." In *Memoires de l'academie des sciences de Berlin* 20 (1766): 165–173. Also in *Opera Omnia* Ser. 3, Vol. 1, 508–515.

Eves, H., ed. 1972. *Mathematical Circles Squared.* Boston: Prindle, Weber and Schmidt.

Eyring, C. F. 1933. "Methods of Calculating the Average Coefficient of Sound Absorption." *Journal of the Acoustical Society of America* 4: 178–192.

Falconer, K. 1990. *Fractal Geometry.* New York: Wiley.

Fletcher, H. 1940. "Auditory Patterns." *Reviews of Modern Physics* 12: 47–65.

Fletcher, H., and W. J. Munson. 1933. "Loudness, Its Definition, Measurement, and Calculation." *Journal of the Acoustical Society of America* 5: 82–108.

Forte, Allen. 1973. *The Structure of Atonal Music.* New Haven: Yale University Press.

Freyd, Jennifer J. 1987. "Dynamic Mental Representations." *Psychological Review* 94 (4): 427–438.

———. 1993. "Five Hunches about Perceptual Processes and Dynamic Representations." In *Attention and Performance XIV: Synergies in Experimental Psychology, Artificial Intelligence, and Cognitive Neuroscience.* Ed. D. Meyer and S. Kornblum. Cambridge, Mass.: MIT Press.

Fux, Johannes J. 1725. *Gradus ad Parnassum (Steps to Parnassus: The Study of Counterpoint).* Trans. Alfred Mann with John St. Edmunds. New York: W. W. Norton, 1943.

Galilei, Galileo. 1623. *The Assayer.* Rome.

———. 1638. “Two New Sciences Including Centers of Gravity and Force of Percussion.” Trans. Stillman Drake in *Galileo Galilei: Two New Sciences,* 96–108. Madison: University of Wisconsin Press, 1974. Trans. Henry Crew and Alfonso de Salvio in *Dialogues Concerning Two New Sciences.* New York: McGraw-Hill, 1963.

Galilei, Vincenzo. 1581. *Dialogo della Musica Antica e Moderna.* Florence: Marescotti. Trans. Claude V. Palisca as *Dialogue on Ancient and Modern Music.* New Haven: Yale University Press, 2003.

Gardner, Martin. 1978. “Mathematical Games.” *Scientific American* 238 (March).

Genrich, H. J., and K. Lautenbach. 1981. “System Modelling with High-Level Petri Nets.” *Theoretical Computer Science* 13: 109–136.

Gerigk, Herbert. 1934. “Würfelmusik.” *Zeitschrift für Musikwissenschaft* 16 (7/8): 359–363.

Gill, Stephen. 1963. “A Technique for the Composition of Music in a Computer.” *Computer Journal* 6 (2): 129.

Godwin, Joscelyn. 1979. *Robert Fludd, Hermetic Philosopher and Surveyor of Two Worlds.* London: Thames and Hudson.

Goldstein, J. L. 1973. “An Optimum Processor Theory for the Central Formation of the Pitch of Complex Tones.” *Journal of the Acoustical Society of America* 54: 1496–1516.

Greenwood, D. D. 1961a. “Auditory Masking and the Critical Band.” *Journal of the Acoustical Society of America* 33: 484–501.

———. 1961b. “Critical Bandwidth and the Frequency Coordinates of the Basilar Membrane.” *Journal of the Acoustical Society of America* 33 (4): 1344–1356.

Grey, John M. 1975. “An Exploration of Musical Timbre.” Ph.D. diss., Center for Computer Research in Music and Acoustics/Department of Music, Stanford University, Stanford, California. STAN-M-2.

Guinan, J. J., and W. T. Peake. 1967. “Middle Ear Characteristics of Anesthetized Cats.” *Journal of the Acoustical Society of America* 41: 1237–1261.

Haas, H. 1951. “Über den Einfluss eines Einfachechos auf die Horsamkeit von Sprache.” *Acustica* 1: 49–58. Trans. as “The Influence of a Single Echo on the Audibility of Speech.” *Journal of the Audio Engineering Society* 20 (1972): 146–159.

Hall, Donald E. 1980. *Musical Acoustics: An Introduction.* Belmont, Calif.: Wadsworth.

Handel, Stephen. 1989. *Listening.* Cambridge, Mass.: MIT Press.

Harel, D. 1987. “Statecharts: A Visual Formalism for Complex Systems.” *Science of Computer Programming* 8: 231–274.

Haus, Goffredo, and A. Rodriguez. 1993. “Formal Music Representation, a Case Study: The Model of Ravel’s *Bolero* by Petri Nets.” In *Music Processing,* 165–232. Ed. Goffredo Haus. Middleton, Wisc.: A-R Editions.

Haus, Goffredo, and Alberto Sametti. 1991. “ScoreSynth: A System for the Synthesis of Music Scores Based on Petri Nets and a Music Algebra.” *IEEE Computer* (July): 56–59.

Hayburn, Robert F. 1979. *Papal Legislation on Sacred Music.* Collegeville, Minn.: Liturgical Press.

Hayes, William. 1751. “The Art of Composing Music by a Method Entirely New, Suited to the Meanest Capacity.” Cited in R. K. Zaripov, “Cybernetics and Music” *Perspectives of New Music* 7 (1969, no. 2): 115–154, 120.

Hellegouarch, Yves. 2002. “A Mathematical Interpretation of Expressive Intonation.” In *Mathematics and Art: Mathematical Visualization in Art and Education.* Ed. Claude P. Bruter. New York: Springer-Verlag.

Helmholtz, Hermann. 1863. *On the Sensations of Tone.* 2d English ed., 1885. Trans. Alexander J. Ellis based on the 4th German ed., 1877. New York: Dover, 1954.

Hild, Hermann, Johannes Feulner, and Wolfram Menzel. 1991. “HARMONET: A Neural Net for Harmonizing Chorales in the Style of J. S. Bach.” In *Proceedings of Conference on Neural Information Processing Systems.*

Hiller, Lejaren, and Leonard Isaacson. 1959. *Experimental Music.* New York: McGraw-Hill.

Holmes, Bob. 1997. “Requiem for the Soul.” *New Scientist,* August 9.

Hoos, H. H., K. A. Hamel, K. Renz, and J. Kilian. 1998. “The GUIDO Music Notation Format: A Novel Approach for Adequately Representing Score-Level Music.” In *Proceedings of the International Computer Music Conference,* 451–454.

Houtsma, A.J.M. and J. L. Goldstein. 1972. “The Central Origin of the Pitch of Complex Tones: Evidence from Musical Interval Recognition.” *Journal of the Acoustical Society of America* 51: 520–529.

Huron, David. 1991. "Tonal Consonance versus Tonal Fusion in Polyphonic Sonorities." *Music Perception* 9 (2): 135–154.

Huxley, Aldous. 1928. *Point Counter Point.* New York: Harper.

ISO (International Organization for Standardization). 1987. Standard 226. http://www. iso.org/.

———. 1975. Standard 532A. http://www. iso.org/.

James, Jamie. 1993. *The Music of the Spheres.* New York: Grove Press.

Jung, C. G. 1962. *Memories, Dreams, and Reflections.* Ed. Aniela Jaffé. Trans. R. Winston and C. Winston. New York: Random House.

Jungleib, Stanley. 1996. *General MIDI.* Middleton, Wisc.: A-R Editions.

Kachar, B., W. E. Brownell, R. Altschuler, and J. Fex. 1986. "Electrokinetic Changes of Cochlear Outer Hair Cells." *Nature* 322: 365–368.

Kameoka, A., and M. Kuriyagawa. 1969a. "Consonance Theory, Part I: Consonance of Dyads." *Journal of the Acoustical Society of America* 45 (6): 1451–1459.

———. 1969b. "Consonance Theory, Part II: Consonance of Complex Tones and Its Computation Method." *Journal of the Acoustical Society of America* 45 (6): 1460–1469.

Kay, Michael. 1996. "Did Mozart Use the Golden Mean?" *American Scientist,* March/April.

Keislar, Douglas. 1988. "History and Principles of Microtonal Keyboard Design." Department of Music, Stanford University, Stanford, California. STAN-M-45.

Kellner, Herbert A. 1979. "A Mathematical Approach Reconstituting J. S. Bach's Keyboard Temperament." *Bach: The Quarterly Journal of the Riemenschneider Bach Institute* 10 (October): 2–8.

Knuth, Donald E. 1973. *The Art of Computer Programming.* Vol. 1: *Fundamental Algorithms.* Vol. 2: *Seminumerical Algorithms.* Reading, Mass.: Addison-Wesley.

Köchel, Ludwig von. 1862. *Works of Mozart.* 6th ed. Ed. Franz Giegling, Alexander Weinmann, and Gerd Sievers. New York: C. F. Peters Corp., 1964.

Koenig, Gottfried M. 1970. "Project 1." *Electronic Music Reports* 1 (2): 32.

Kohonen, Tuevo. 1989. "A Self-Learning Musical Grammar, or Associative Memory of the Second Kind." In *Proceedings of the International Joint Conference on Neural Networks,* Washington, D.C.

Koza, John R. 1992. *Genetic Programming.* Cambridge, Mass.: MIT Press.

Kruskal, J. B. 1964. "Multidimensional Scaling by Optimizing Goodness of Fit to a Nonmetric Hypothesis." *Psychometrika* 29: 1–27.

Kuttner, Fritz A. 1975. "Prince Chu Tsai-Yu's Life and Work: A Reevaluation of His Contribution to Equal Temperament Theory." *Ethnomusicology* 19 (2): 163–206.

Landon. H. C. Robbins. 1976. *Haydn: Chronicle and Works.* Vol. 2. Bloomington: Indiana University Press.

Langer, S. 1953. *Feeling and Form.* New York: Philosophical Library.

Lazer, A. C., and P. J. McKenna. 1990. "Large Amplitude Periodic Oscillations in Suspension Bridges: Some New Connections with Nonlinear Analysis." *SIAM Review* 32 (4): 537–578.

Lentz, Donald. 1961. *Tones and Intervals of Hindu Classical Music.* Lincoln: University of Nebraska.

Licklider, J.C.R. 1951. "Basic Correlates of the Auditory Stimulus." In *Handbook of Experimental Psychology.* Ed. S. S. Stevens. New York: Wiley.

Ligeti, Gyorgi. 1965. "Metamorphoses of Musical Form." *Die Reihe* 7: 5–19.

Lorenz, Edward. 1972. "Predictability: Does the Flap of a Butterfly's Wings in Brazil Set off a Tornado in Texas?" Paper presented at American Association for the Advancement of Science conference, December.

Lowman, E. L. 1971. "Some Striking Proportions in the Music of Béla Bartók." *Fibonacci Quarterly* 9 (5): 527–528, 536–537.

Loy, Gareth. 1985. "Musicians Make a Standard: The MIDI Phenomenon." *CMJ: Computer Music Journal* 9 (4): 8–26.

Macran, H. S. 1902. *The Harmonics of Aristoxenus.* New York: Oxford University Press.

Majernick, V., and J. Kaluzny. 1979. "On the Auditory Uncertainty Relations." *Acustica* 43: 132.

Mandelbrot, Benoit B. 1977. *The Fractal Geometry of Nature.* New York: W. H. Freeman.

Mathews, Max V., and J. R. Pierce. 1980. "Harmony and Nonharmonic Partials." *Journal of the Acoustical Society of America* 68: 1252–1257.

———. 1989. "The Bohlen-Pierce Scale." In *Current Directions in Computer Music Research.* Ed. M. V. Mathews and J. R. Pierce. Cambridge, Mass.: MIT Press.

Mathews, Max V., L. A. Roberts, and J. R. Pierce. 1984. "Four New Scales Based on Nonsuccessive-Integer-Ratio Chords." *Journal of the Acoustical Society of America* 75 (S10A).

Mathews, Max V., and L. Rossler. 1968. "Graphical Language for the Scores of Computer-Generated Sounds." *Perspectives of New Music* 6 (2): 92.

McClelland, J. L., D. Rumelhart, and G. E. Hinton. 1986. "The Appeal of Parallel Distributed Processing." In *Parallel Distributed Processing: Explorations in the Microstructure of Cognition.* Ed. D. Rumelhart and J. L. McClelland. Cambridge, Mass.: MIT Press.

McKenna, P. J. 1999. "Large Torsional Oscillations in Suspension Bridges Revisited: Fixing an Old Approximation." *American Mathematical Monthly* 106 (January): 1.

Mersenne, Marin. 1635. *Harmonie Universelle, Contenant la Theorie et la Practique de la Musique, ou il est Traite de Consonances, des Dissonances, des Genres, des Modes, de la Composition, de la Voix, des Chants, et de Toutes Sortes d'Instruments Harmoniques.* Paris: Pierre Ballard. Facsimile ed. Francois Lesure. Paris: Editions du Centre National de Recherche Scientifique, 1965. Trans. Roger E. Chapman as *Harmonie Universelle: The Books on Instruments.* The Hague: Nijhoff, 1957.

Messiaen, Olivier. 1942. *Technique de Mon Langage Musical.* Paris: Leduc.

Meyer, Leonard B. 1956. *Emotion and Meaning in Music.* Chicago: University of Chicago Press.

Minsky, Marvin. 1974. "A Framework for Representing Knowledge." MIT Artificial Intelligence Laboratory Memo 306. Also in *The Psychology of Computer Vision.* Ed. P. H. Winston. New York: McGraw-Hill, 1975.

Minsky, Marvin, and Seymour Papert. 1969. *Perceptrons.* Cambridge, Mass.: MIT Press.

Moore, Brian. 1997. *An Introduction to the Psychology of Hearing.* 4th ed. San Diego: Academic Press.

Moore, F. Richard. 1988. "The Dysfunctions of MIDI." *CMJ: Computer Music Journal* 12 (1): 19–28.

———. 1990. *Elements of Computer Music.* Upper Saddle River, N.J.: Prentice-Hall.

Morse, Marston. 1959. "Mathematics and the Arts." *Bulletin of the Atomic Scientist,* February.

Nettheim, Nigel. 1992. "On the Spectral Analysis of Melody." *Interface: Journal of New Music Research* 21: 135–148.

Neumann, John von. 1963. "Various Techniques Used in Connection with Random Digits." In *Collected Works.* Vol. 5, 768–770. Oxford: Pergamon.

Norden, H. 1964. "Proportions in Music." *Fibonacci Quarterly* 2 (3): 219–222.

O'Beirne, Thomas H. 1968. "940,364,969,152 Dice-Music Trios." *Musical Times* 109 (October): 911–913.

Ohm, G. S. 1843. "Über die Definition des Tones, nebst daran geknüfter Theorie der Sirene und ähnlicher tonbildender Vorichtungen." *Annals of Physical Chemistry* 59: 513–565.

Olson, Harry F. 1952. *Music, Physics, and Engineering.* New York: Dover, 1967.

Oppenheim, D. 1996. "DMIX, A Multifaceted Environment for Composing and Performing." *Computers and Mathematics with Applications* 32 (1): 117–135.

Pangle, Thomas L., ed. and trans. 1980. *The Laws of Plato.* New York: Basic Books.

Park, S. K., and K. W. Miller. 1988. "Random Number Generators: Good Ones Are Hard to Find." *Communications of the ACM* 31 (10): 1192–1201.

Partch, Harry. 1947. *Genesis of a Music.* Madison: University of Wisconsin Press. New York: Da Capo Press, 1979.

Perle, George, and Paul Lansky. 1981. *Serial Composition and Atonality.* Los Angeles: University of California Press.

Petri, C. A. 1976. "General Net Theory." In *Proceedings of IBM/University of Newcastle-upon-Tyne Seminar,* 131–169. Also in *Technische Universität Berlin, GMD-Bericht No. 3.* Munich: Oldenbourg Verlag, 1979.

Pierce, John R. 1983. *The Science of Musical Sound.* San Francisco: Scientific American Books.

Pingle, B. A. 1962. *History of Indian Music.* Calcutta: Gupta.

Pirsig, Robert M. 1974. *Zen and the Art of Motorcycle Maintenance.* New York: Morrow.

Plomp, R. 1970. "Timbre as a Multidimensional Attribute of Complex Tones." In *Frequency Analysis and Periodicity Detection in Hearing.* Ed. R. Plomp and G. Smoorenberg. Leiden: Sijthoff.

Plomp, R., and W.J.M. Levelt. 1965. "Tonal Consonance and Critical Bandwidth." *Journal of the Acoustical Society of America* 38: 548–560.

Pope, Stephen. 1986. "Music Notations and the Representation of Musical Structure and Knowledge." *Perspectives of New Music* 24: 156–189.

———, ed. 1991. *The Well-Tempered Object: Musical Applications of Object-Oriented Software Technology.* Cambridge, Mass.: MIT Press.

Potter, Gary M. 1971. "The Role of Chance in Contemporary Music." Ph.D. diss., School of Music, Indiana University, Bloomington.

Press, William H., Brian P. Flannery, Saul A. Teukolsky, and William T. Vetterling. 1988. *Numerical Recipes in C: The Art of Scientific Computing.* Cambridge: Cambridge University Press.

Putz, J. F. 1995. "The Golden Section and the Piano Sonatas of Mozart." *Mathematics Magazine* 68 (4): 275–282.

Rameau, Jean-Philippe. 1722. *Traite de l 'Harmonie.* Trans. Philip Gossett as *Treatise on Harmony.* New York: Dover, 1971.

———. 1737. "Generation Harmonique, ou Traite de Musique Theorique et Pratique—Proposition xii." Trans. D. Hayes in "Rameau's Theory of Harmonic Generation." Ph.D. diss., Department of Music, Stanford University, Stanford, California, 1968.

Ramos, Bartolomé. 1482. "Musica Practica." In *Source Readings in Music History,* 203–204. Ed. Oliver Strunk. New York: W. W. Norton, 1998.

Révész, Geza. 1954. *Introduction to the Psychology of Music.* Norman: University of Oklahoma Press. New York: Dover, 2001.

Roads, Curtis. 1984. "An Overview of Music Representations." In *Musical Grammars and Computer Analysis,* 7–37. Ed. Mario Baroni and Laura Callegari. Firenze: Olschki.

Roberts, L. A., and Max V. Mathews. 1984. "Intonation Sensitivity for Traditional and Nontraditional Chords." *Journal of the Acoustical Society of America* 75: 952–959.

Roederer, Juan. 1973. *Introduction to the Physics and Psychophysics of Music.* London: English Universities Press.

Rossing, Thomas D. 1983. *The Science of Sound.* Reading, Mass.: Addison-Wesley.

Rothstein, Edward. 1995. *Emblems of Mind: The Inner Life of Music and Mathematics.* New York: Times Books.

Ruiz, Pierre M. 1969. "A Technique for Simulating the Vibrations of Strings with a Digital Computer." Master's thesis, Department of Music, University of Illinois, Urbana-Champaign.

Rumelhart, D. E., G. E. Hinton, and R. J. Williams. 1986. "Learning Internal Representations by Error Propagation." In *Parallel Distributed Processing: Explorations in the Microstructure of Cognition.* Ed. D. Rumelhart and J. L. McClelland. Cambridge, Mass.: MIT Press.

Sabine, W. C. 1921. *Collected Papers on Acoustics.* Los Altos, Calif.: Peninsula Publishing, 1993.

Samek, Miro. 2002. *Practical Statecharts in C/C++: Quantum Programming for Embedded Systems.* San Francisco: CMP Books.

Santillana, Giorgio de, and Herta von Dechend. 1969. *Hamlet's Mill.* Boston: David R. Godine.

Scaletti, Carla. 1989. "Composing Sound Objects in Kyma." *Perspectives of New Music* 27: 42–69.

———. 1991. "The Kyma/Platypus Computer Music Workstation." In *The Well-Tempered Object: Musical Applications of Object-Oriented Programming.* Ed. Stephen Pope. Cambridge, Mass.: MIT Press.

Schank, R. C., and R. Abelson. 1976. *Scripts, Plans, Goals, and Understanding.* Hillsdale, N.J.: Erlbaum.

Schenker, Heinrich. 1935. *Der freie Satz: Neue Musikalische Theorien and Phantasien.* Vienna: Universal.

Schillinger, Joseph. 1948. *The Mathematical Basis of the Arts.* New York: DaCapo Press, 1976.

Schouten, J. F., R. J. Ritsma, and B. Lopes Cardozo. 1962. "Pitch of the Residue." *Journal of the Acoustical Society of America* 34: 1418–1424.

Schroeder, Manfred R. 1979. "Binaural Dissimilarity and Optimum Ceilings for Concert Halls: More Lateral Sound Diffusion." *Journal of the Acoustical Society of America* 65: 958.

Seebeck, August. 1841. "Beobachtungen über einige Bedingungen der Entstehung von Tonen [Observations on Some Conditions for the Creation of Tones]." *Annals of Physical Chemistry* 53: 417–436.

Shannon, Claude E. 1948. "A Mathematical Theory of Communication." *Bell System Technical Journal* 27 (July): 379–423; 27 (October): 623–656.

Shannon, Claude E., and Warren Weaver. 1949. *The Mathematical Theory of Communication.* Urbana: University of Illinois Press.

Sharp, D. B. 1996. "Acoustic Pulse Refectometry for the Measurement of Musical Wind Instruments." Ph.D diss., University of Edinburgh.

Shepard, R. N. 1964. "Circularity in Judgments of Relative Pitch." *Journal of the Acoustical Society of America* 36: 2346–2353.

———. 1982. "Structural Representations of Musical Pitch." In *The Psychology of Music,* 344–384. Ed. Diana Deutsch. San Diego: Academic Press.

Slonimsky, Nicholas. 1948. *Slonimsky's Book of Musical Anecdotes.* New York: Allen, Towne, and Heath. New York: Routledge, 2002.

Stevens, S. S. 1956. "Calculation of the Loudness of Complex Noise." *Journal of the Acoustical Society of America* 28: 807–832.

———. 1961. "Procedure for Calculating Loudness: Mark VI." *Journal of the Acoustical Society of America* 33 (11): 1577–1585.

———. 1962. "The Surprising Simplicity of Sensory Metrics." *American Psychologist* 17: 29–39.

Stevens, S. S., and E. B. Newman. 1936. "The Localization of Actual Sources of Sound." *American Journal of Psychology* 48: 297–306.

Stroustrup, Bjame. 1991. *The C++ Programming Language.* Reading, Mass.: Addison-Wesley.

Strunk, Oliver, ed. 1950. *Source Readings in Music History.* 6th rev. ed. New York: W. W. Norton, 1998.

Strutt, John W. 1907. "Our Perception of Sound Direction." *Philosophical Magazine* 13: 214–232.

Stuckenschmidt, H. H. 1969. *Twentieth Century Music.* Trans. Richard Deveson. New York: McGraw-Hill.

Stumpf, Carl. 1883/1890. *Tonspsychologie* [*Tone Psychology*]. Leipzig: Hirzel. 2 vols.

Terhardt, E. 1974. "Pitch, Consonance, and Harmony." *Journal of the Acoustical Society of America* 55: 1061–1069.

———. 1979. "Calculating Virtual Pitch." *Hearing Research* 1: 155–182.

Tiggelen, Philippe J. van. 1987. *Componium: The Mechanical Musical Improvisor.* Louvain-la-Neuve, France: Institut Supérieure d'Archéologie et d'Histoire de l'Art.

Todd, Peter M. 1989. "A Connectionist Approach to Algorithmic Music." *CMJ: Computer Music Journal* 13 (4). Also in *Music and Connectionism.* Ed. Peter Todd and Gareth Loy. Cambridge, Mass.: MIT Press.

Todd, Peter, and Gareth Loy, eds. 1989. *Music and Connectionism.* Cambridge, Mass.: MIT Press.

Todd, Peter M., and Gregory M. Werner. 1998. "Frankensteinian Methods for Evolutionary Music Composition." In *Musical Networks: Parallel Distributed Perception and Performance.* Ed. Niall Griffith and Peter M. Todd. Cambridge, Mass.: MIT Press.

Turing, Alan M. 1950. "Computing Machinery and Intelligence." *Mind* 59: 433–460.

Valk, R. 1978. "Self-Modifying Nets: A Natural Extension of Petri Nets." In *ICALP 1978: Proceedings of the International Conference on Automata, Languages and Programming, Fifth Colloquium,* 464–476.

Vogt, Mauritius. 1719. *Conclave Thesauri Magnae Artis Musicae.* Cited in H. Kirchmeyer. "On the Historical Construction of Rationalistic Music." *Die Reihe* 8 (1962): 11–29, 20.

Voss, Richard F., and J. Clarke. 1975. "1/f Noise in Music and Speech." *Nature* 258: 317–318.

———. 1978. "l/f noise in Music: Music from 1/f Noise." *Journal of the Acoustical Society of America* 63 (1): 258–263.

Wallach, H., E. B. Newman, and M. R. Rosenzweig. 1949. "The Precedence Effect in Sound Localization." *American Journal of Psychology* 52: 315–336.

Ware, J. A., and K. Aki. 1969. "Continuous and Discrete Inverse Scattering Problems in a Stratified Elastic Medium. I: Planes at Normal Incidence." *Journal of the Acoustical Society of America* 45: 911–921.

Warren, R. M. 1970. "Elimination of Biases in Loudness Judgments for Tones." *Journal of the Acoustical Society of America* 48: 1397.

Wiggins, G. A., M. Harris, and A. Smaill. 1989. "Representing Music for Analysis and Composition." In *Proceedings of the 2d International Joint Conference on Artificial Intelligence (IJCAI-89),* Detroit, Workshop on Artificial Intelligence and Music, 63–71.

Wightman, F. L. 1973. "The Pattern Transformation Model of Pitch." *Journal of the Acoustical Society of America* 54: 407–416.

Wilkinson, S. 1988. *Tuning In: Microtonality in Electronic Music.* Milwaukee: Hal Leonard Books.

Xenakis, Iannis. 1955. "The Crisis of Serial Music." *Gravesaner Blätter.*

———. 1971. *Formalized Music.* Bloomington: Indiana University Press.

Yasser, Joseph. 1932. *Theory of Evolving Tonality.* New York: American Library of Musicology. New York: Da Capo Press, 1975.

Yost, William A. 2000. *Fundamentals of Hearing: An Introduction.* 4th ed. New York: Academic Press.

Young, Thomas. 1800. "Of the Temperament of Musical Intervals." *Philosophical Transactions.* Royal Society of London.

Zwicker, E. 1961. "Subdivision of the Audible Frequency Range into Critical Bands (Frequenzgruppen)." *Journal of the Acoustical Society of America* 33: 248.

Zwicker, E., and H. Fastl. 1990. *Psychoacoustics: Facts and Models.* Berlin: Springer-Verlag.

Zwicker, E., and R. Feldtkeller. 1955. "On the Derivation of Critical Bands from the Loudness of Complex Sounds. *Acustica* 5: 40–45.

Zwicker, E., G. Flottorp, and S. S. Stevens. 1957. "Critical Bandwidth in Loudness Summation." *Journal of the Acoustical Society of America* 29: 548–557.

Equation Index

Subject Index